New Advances in Transcendence Theory

T0269328

New Advances in Transcendence Theory

Edited by

ALAN BAKER

Professor of Pure Mathematics
University of Cambridge

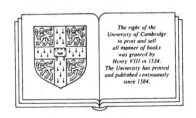

CAMBRIDGE UNIVERSITY PRESS

Cambridge

New York   New Rochelle

Melbourne   Sydney

CAMBRIDGE UNIVERSITY PRESS
Cambridge, New York, Melbourne, Madrid, Cape Town, Singapore, São Paulo, Delhi

Cambridge University Press
The Edinburgh Building, Cambridge CB2 8RU, UK

Published in the United States of America by Cambridge University Press, New York

www.cambridge.org
Information on this title: www.cambridge.org/9780521335454

First published 1988
This digitally printed version 2008

A catalogue record for this publication is available from the British Library

Library of Congress Cataloguing in Publication data
New advances in transcendence theory.
Symposium on Transcendental Number Theory was held
under the auspices of the London Mathematical Society
at the University of Durham in July, 1986.
Bibliography: p.
1. Numbers, Transcendental – Congresses. I. Baker, Alan, 1939–
II. Symposium on Transcendental Number Theory
(1986: University of Durham)
III. London Mathematical Society.
QA247.N49   1988   512'.73   88-11837

ISBN 978-0-521-33545-4 hardback
ISBN 978-0-521-09029-2 paperback

CONTENTS

PREFACE

A highly successful Symposium on Transcendental Number Theory was held under the auspices of the London Mathematical Society at the University of Durham in July, 1986, and the present volume is an account of the proceedings of that meeting. Some fifty mathematicians were present, including most of the leading specialists in the field, and the lectures reflected the remarkable research activity that has taken place in this area in recent years. Indeed, as became apparent, the evolution of transcendence, since the 1960s, into a fertile theory with numerous and widespread applications has been one of the most exciting and important developments of modern mathematics. The conference programme, though comprehensive, was intended to be in no way overcrowded, and it was particularly aimed to create a relaxed atmosphere for the free exchange of ideas. This seems to have worked out well; in fact much valuable material was presented for future study and some original theorems were obtained through informal collaboration during the meeting itself. The invited participants from the USSR were alas unable to come to Durham but they communicated reports subsequently and the editor is grateful to them and indeed to all the distinguished authors for contributing so admirably to this volume.

A conference with a similar theme was held in Cambridge some ten years ago and the proceedings were published under the title *Transcendence Theory: Advances and Applications* (Academic Press, 1977); the present work forms a natural sequel. Again, many papers are concerned with the theory of linear forms in the logarithms of algebraic numbers. In particular, the memoirs of Wüstholz and of Philippon and Waldschmidt both contain definitive results in this context; they eliminate a second order factor from the inequalities that I established at the time of the meeting in Cambridge, and the arguments rest ultimately on the spectacular progress that has been made in recent years, most notably by Wüstholz, concerning multiplicity estimates on group varieties. Studies in the area were initiated by Nesterenko and some new related estimates are given in his paper here. The articles of Bertrand and of Masser illuminate other aspects of proofs in this field, highlighting, for instance, the extensive connections with Kummer theory and elliptic curves. Links with the classic works of Gelfond and Schneider are described in the papers of Feldman and of Waldschmidt, and the current status of the p-adic theory is discussed by Kunrui Yu.

Another major topic is the application of transcendence theory to the study of Diophantine equations. Here the very substantial paper on S-unit equations by Evertse, Györy, Stewart and Tijdeman, and the associated work on decomposable form equations by Evertse and Györy, are particularly welcome. There are, moreover, valuable articles on exponential Diophantine equations by Shorey, on the Thue equation by Schmidt, and on equations over function fields by Mason and by Brindza. The article by Baker and Stewart also relates to Diophantine equations; it is shown that the theory of linear forms in logarithms can be greatly streamlined in certain instances so as to yield surprisingly good numerical bounds.

A subject that has plainly attracted a great deal of research in recent years is the transcendence theory of classical functions, with particular interest focused on hypergeometric functions, on E-functions and on G-functions. The excellent papers by Beukers, Beukers and Wolfart, Galochkin, Shidlovsky and Sprindžuk all cover aspects of this topic. They refer to many recent results, and, taken together, they provide the most complete survey of the field available to date.

Furthermore, this by no means exhausts the range of material that can be found here. Indeed, the paper of Bernik is concerned with the metrical theory of transcendence, an area to which he has made some striking advances; the paper of Brownawell is concerned with the remarkable relation between Hilbert's irreducibility theorem and transcendence; the paper of Erdös is concerned with the questions of irrationality and transcendence; the paper of Loxton is concerned with automata and transcendence and, in particular, with new problems connected with the celebrated Mahler method; the paper of Odoni is concerned with modular forms and transcendence and furnishes the answer to a question of Serre; and the paper of Schinzel continues his fine series of studies on reducibility of polynomials and shows that there is a useful role for transcendence here too. It seems probable that the work as a whole will be of considerable influence in determining the future direction of the theory.

The Symposium was funded by a grant from the Science and Engineering Research Council and this support is acknowledged with gratitude. My colleague, Dr R. C. Mason handled all the domestic and financial arrangements and there is no doubt that the success of the meeting was due in no small measure to his excellent work. The co-operation of Prof. P. Higgins and indeed of all our mathematical colleagues in Durham was invaluable, and particular thanks should be expressed to the Bursar and his staff at Grey College for their helpfulness throughout. Thanks are also due to Dr A. Harris for generously taking on the task of translating articles from Russian into English, to Prof. J. W. S. Cassels for additional advice in this respect, and to Dr D. Tranah of Cambridge University Press for his kind and patient assistance at all stages of production of this volume.

Cambridge, 1987   A. B.

Added in proof. It is with much sadness that the editor records here the death of Prof. V. G. Sprindžuk in July, 1987. His passing is a great loss to mathematics, and we shall remember especially his important contributions to Transcendence Theory.

CONTRIBUTORS

A. BAKER; Department of Pure Mathematics and Mathematical Statistics, University of Cambridge, Cambridge CB2 1SB, England, and Trinity College, Cambridge CB2 1TQ, England.

V. I. BERNIK; Institute of Mathematics, Academy of Sciences, Minsk 220604, USSR.

D. BERTRAND; Université de Paris VI, Mathématiques, T. 46, 4 Place Jussieu, 75252 Paris Cedex 05, France.

F. BEUKERS; Mathematical Institute, University of Utrecht, Budapestlaan 6, PO Box 80.010, 3508 TA Utrecht, The Netherlands.

B. BRINDZA; Mathematical Institute, Kossuth Lajos University, 4010 Debrecen, Hungary.

W. D. BROWNAWELL; Department of Mathematics, College of Science, The Pennsylvania State University, 218 McAllister Building, University Park, Pennsylvania 16802, USA.

P. ERDŐS; Mathematical Institute, Hungarian Academy of Sciences, Reáltanoda u. 13–15, H–1053 Budapest, Hungary.

J.-H. EVERTSE; Department of Pure Mathematics, Centre for Mathematics and Computer Science, Kruislaan 413, 1098 SJ Amsterdam, The Netherlands.

N. I. FELDMAN; Department of Mathematics, University of Moscow, Moscow 119899, USSR.

A. I. GALOCHKIN; Department of Mathematics, University of Moscow, Moscow 119899, USSR.

K. GYŐRY; Mathematical Institute, Kossuth Lajos University, 4010 Debrecen, Hungary.

J. H. LOXTON; School of Mathematics and Physics, Macquarie University, New South Wales 2019, Australia.

xii CONTRIBUTORS

R. C. MASON; Research Department, Credit Suisse First Boston Limited, 22 Bishopsgate, London, England, and Gonville and Caius College, Cambridge, England.

D. W. MASSER; Department of Mathematics, University of Michigan, Ann Arbor, Michigan 48109, USA.

Yu. V. NESTERENKO; Department of Mathematics, University of Moscow, Moscow 119899, USSR.

R. W. K. ODONI; Department of Mathematics, University of Exeter, North Park Road, Exeter EX4 4QE, England.

P. PHILIPPON; Institut Henri Poincaré, 11 rue P. et M. Curie, 75231 Paris Cedex 05, France.

A. SCHINZEL; Instytut Matematyczny, Polskiej Akademii Nauk, skr. pocztowa 137, 00–950 Warszawa, Poland.

W. M. SCHMIDT; Department of Mathematics, University of Colorado, Campus Box 426, Boulder, Colorado 80309, USA.

A. B. SHIDLOVSKY; Department of Mathematics, University of Moscow, Moscow 119899, USSR.

T. N. SHOREY; School of Mathematics, Tata Institute of Fundamental Research, Homi Bhaba Road, Bombay 400 005, India.

V. G. SPRINDŽUK; Institute of Mathematics, Academy of Sciences, Minsk 220604, USSR.

C. L. STEWART; Department of Pure Mathematics, University of Waterloo, Waterloo, Ontario N2L 3G1, Canada.

R. TIJDEMAN, Mathematical Institute, University of Leiden, 2333 CA Leiden, The Netherlands.

R. TIJDEMAN: Mathematical Institute, University of Leiden, 2333 CA Leiden, The Netherlands.

M. WALDSCHMIDT; Institut Henri Poincaré, 11 rue P. et M. Curie, 75231 Paris Cedex 05, France.

J. WOLFART; Fachbereich Mathematik, Goethe Universität, Postfach 11 19 32, D–6000 Frankfurt am Main, Bundesrepublik, Germany.

G. WÜSTHOLZ; Mathematik, ETH-Zentrum, CH-8092, Zürich, Switzerland.

KUNRUI YU; Max-Planck-Institut für Mathematik, Gottfried-Claren-Strasse 26, 5300 Bonn 3, Germany, and Institute of Mathematics, Academia Sinica, Beijing, China.

ON EFFECTIVE APPROXIMATIONS TO
CUBIC IRRATIONALS

A. Baker and C. L. Stewart*

1. Introduction

The problem of obtaining effective measures of irrationality for algebraic irrationals has recently attracted considerable attention. The first result in this field was discovered by Baker [1], [2] in 1964. He used properties of hypergeometric series to obtain effective results for certain fractional powers of rationals. It was shown, in particular, that for all rationals p/q with $q > 0$ we have

$$\left| \alpha - \frac{p}{q} \right| > \frac{c}{q^\kappa}, \tag{1}$$

where $\alpha = \sqrt[3]{2}$, $c = 10^{-6}$ and $\kappa = 2.955$. A similar result was established for instance for $\alpha = \sqrt[3]{19}$ with $c = 10^{-9}$ and $\kappa = 2.56$. This work was recently refined by Chudnovsky [11]; by a careful study of the Padé approximants occurring in the hypergeometric method he obtained more precise values for κ and consequently he was able to deal with a wider range of algebraic numbers. Chudnovsky left the values for c occurring in his results unspecified but these have recently been established in some special cases by Easton [13]. Easton has shown in particular that (1) holds with $\alpha = \sqrt[3]{28}$, $c = 7.5 \times 10^{-7}$ and $\kappa = 2.9$.

The results above improved upon the relatively crude inequality of Liouville established in 1844 to the effect that (1) holds for any algebraic number α, where $\kappa = n$, $n \geq 1$, the degree of α and c is an effectively computable positive number depending only on α. The first general effective improvement on Liouville's theorem was obtained by Baker [3] in 1968 using the theory of linear forms in the logarithms of algebraic

* The research of the second author was supported in part by Grant A3528 from the Natural Sciences and Engineering Research Council of Canada.

numbers. A more precise version of the result was obtained subsequently by Feldman [14] and an explicit formulation of the theorem has recently been given by Győry and Papp [15]. In the present paper we shall sharpen the result of Győry and Papp in the case of cube roots of integers. We shall prove the following result.

Theorem 1. *Let a be a positive integer not a perfect cube, and let $\alpha = \sqrt[3]{a}$. Further let ϵ be the fundamental unit in the field $\mathbf{Q}(\sqrt[3]{a})$. Then (1) holds for all rational numbers p/q, $q > 0$, with $c = 1/(3ac_1)$ and $\kappa = 3 - 1/c_2$, where*

$$c_1 = \epsilon^{(50 \log \log \epsilon)^2}, \qquad c_2 = 10^{12} \log \epsilon. \qquad (2)$$

Here $\mathbf{Q}$ denotes, as usual, the field of rational numbers and by the fundamental unit ϵ in $\mathbf{Q}(\sqrt[3]{a})$ we mean the smallest unit in the field larger than 1. Note that some authors adopt the alternative convention that the fundamental unit lies between 0 and 1. The result of Győry and Papp mentioned above yields a theorem similar to Theorem 1 but with

$$c_2 = 300^{60} \log \epsilon (\log \log \epsilon)^2 \qquad (3)$$

and with a value for c_1 slightly greater than $(40a)^6 \epsilon$. In both (2) and (3) we have made use of the fact, established in §2 below, that $\log \epsilon > 1$ for all fields $\mathbf{Q}(\sqrt[3]{a})$. Although our value for c_2 improves substantially on (3), the value for κ that it furnishes is far from the exponent $2 + \delta$, $\delta > 0$, occurring in the Thue-Siegel-Roth theorem. As is well known the latter theorem is ineffective, that is, it does not provide an explicit value for the constant c in (1). But Bombieri [8] and Bombieri and Mueller [9] have recently shown that in certain special cases effective results can in fact be derived from the Thue-Siegel method. Nevertheless the restrictions attaching to α in their work are very stringent at present.

The inequality established in Theorem 1 is essentially equivalent to an upper bound for the solutions of the Diophantine equation

$$x^3 - ay^3 = n. \qquad (4)$$

We have the following result.

Theorem 2. *Let a and n be positive integers with a not a perfect cube. Then all solutions in integers x and y of (4) satisfy*

$$\max(|x|, |y|) < (c_1 n)^{c_2},$$

where c_1 and c_2 are given by (2).

In order to derive Theorem 1 from Theorem 2 we denote by p/q, $q > 0$, any rational number and we suppose that $|\alpha - p/q| \leq c$; then $|p/q| \leq \alpha + c$, whence

$$\left|\alpha^2 + \alpha(p/q) + (p/q)^2\right| \leq 3\alpha^2 + 3\alpha c + c^2 \leq 3a.$$

This gives

$$\left|a - (p/q)^3\right| \leq 3a|\alpha - p/q|. \tag{5}$$

We now apply Theorem 2 with $n = |p^3 - aq^3|$ and conclude that $q < (c_1 n)^{c_2}$ whence $n > (1/c_1)q^{1/c_2}$. By (5) we have $|\alpha - p/q| \geq n/(3aq^3)$ and our result follows.

The proof of Theorem 2 is based essentially on the methods of [3] and [4]. In particular we reduce the problem to the study of a linear form in three logarithms and we ultimately establish the bound $2 \cdot 10^{12} \log(c_1 n)$ for the size of the integer coefficients in that form. Our exposition will follow the general pattern of the earlier papers but we shall use a simplified auxiliary function, and also a more efficient extrapolation procedure to which Kummer theory can be applied directly. The work here together with the technique of Baker and Davenport [6] would enable the complete list of solutions of (4) to be computed for any moderately sized a and n. Indeed we have $\log \epsilon < (0.37)d^{1/2}(\log d)^2$ where d is the absolute value of the discriminant of $\mathbf{Q}(\sqrt[3]{a})$ (see [18]); thus, since $d \leq 27a^2$ we obtain, for $a > 3$,

$$\log c_1 \leq (50 \log d)^2 \log \epsilon \leq (37 \log a)^4 a.$$

Hence if, for example, $a \leq 10^3$ and $\log n \leq 10^{10}$ then the coefficients of the logarithms in the linear form will have sizes at most 10^{25}.

As a particular instance of Theorem 1 we take $\alpha = \sqrt[3]{5}$; this is the smallest cube root not covered by the papers employing the hypergeometric method. Then $\epsilon = 41 + 24\alpha + 14\alpha^2$ (see [10], Table 2, p. 270) and $\log \epsilon < 5$. Hence we conclude that (1) holds with $c = 10^{-12900}$ and

$$\kappa = 2.9999999999998.$$

We should like to express our thanks to Professor D. Djokovic for his generous assistance in the computational work referred to in §3. The latter was carried out while the first author was visiting the University of Waterloo and he is grateful for their hospitality.

2. Preliminary lemmas

We shall require modified forms of two classical lemmas in transcendence theory. First we obtain the following sharpening of Lemma 4 of Baker and Stark [7].

Lemma 1. *Suppose that α, β are elements of an algebraic number field and that for some positive integer p we have $\alpha = \beta^p$. If a, b are the leading coefficients in the field polynomials defining α, β respectively then $b \leq a^{1/p}$.*

Here the field polynomials are, as usual, powers of the minimal polynomials with degree D, where D denotes the degree of the field. Lemma 4 of [7] gives the weaker inequality $b \leq a^{D/p}$, where a denotes any non-zero integer such that $a\alpha$ is an algebraic integer.

Proof. Let $\alpha^{(1)}, \ldots, \alpha^{(D)}$ and $\beta^{(1)}, \ldots, \beta^{(D)}$ be the field conjugates of α and β respectively. Then b is the least positive integer such that

$$f(x) = b(x - \beta^{(1)}) \ldots (x - \beta^{(D)})$$

has rational integer coefficients. We write

$$g(x) = a(x^p - \alpha^{(1)}) \ldots (x^p - \alpha^{(D)}), \qquad h(x) = \prod_{j=1}^{p} f(x e^{2\pi i j/p}).$$

Since, by hypothesis, $\alpha = \beta^p$ we have

$$b^p g(x) = (-1)^{D(p+1)} a h(x).$$

Arguing as in [7] we deduce from the algebraic generalization of Gauss' lemma that $h(x)$ has relatively prime rational integer coefficients. But $g(x)$ also has rational integer coefficients and so b^p divides a, whence $b \leq a^{1/p}$ as required.

Secondly, we shall establish a version of Siegel's lemma appropriate to our work here. We shall adapt the result of Dobrowolski [12] so as to deal with linear forms with arbitrary algebraic coefficients, not merely algebraic integers. Obviously it would suffice to multiply through each equation by a suitable common denominator but this would be too crude for our purpose. In order to state the lemma, we define K to be an algebraic number field with degree n over $\mathbf{Q}$ and we let $\sigma_1, \ldots, \sigma_n$ be the embeddings of K in the complex numbers. Further we signify by b_{ij}, $1 \leq i \leq N$, $1 \leq j \leq M$, elements of K such that for each j not all b_{ij},

$1 \leq i \leq N$, are zero. We now define c_j, $1 \leq j \leq M$, to be a positive integer such that

$$c_j \sigma_1(b_{i_1,j}) \ldots \sigma_n(b_{i_n,j})$$

is an algebraic integer for all choices of $i_1, \ldots, i_n$.

Lemma 2. *If $N > nM$ then the system of equations*

$$\sum_{i=1}^{N} b_{ij} x_i = 0, \qquad 1 \leq j \leq M,$$

has a solution in rational integers $x_1, \ldots, x_N$, not all 0, with absolute values at most

$$Y = \left(2\sqrt{2}(N+1)Z^{1/(nM)}\right)^{nM/(N-nM)},$$

where

$$Z = \prod_{j=1}^{M} \left(c_j \prod_{k=1}^{n} \max_i |\sigma_k(b_{ij})|\right).$$

Proof. The proof follows almost verbatim that of Dobrowolski [12]. the main idea is to select rational integers $x_1, \ldots, x_N$ by the box principle such that

$$\left| c_j N_{K/\mathbf{Q}}\left(\sum_i b_{ij} x_i\right)\right| < 1, \qquad 1 \leq j \leq M.$$

This differs from [12] by virtue of the presence of c_j; our definition of c_j ensures that the expression on the left of the above inequality is a rational integer. The only significant modification in the proof concerns the quantity

$$C_j = \left(c_j \prod_{k=1}^{n} \max_i |\sigma_k(b_{ij})|\right)^{1/n}$$

which now includes c_j. This leads to the definition

$$\ell_j = (Y^N/Z)^{1/(nM)} C_j,$$

which gives

$$2\sqrt{2}(N+1)Y C_j - \ell_j = 0$$

as in [12]. Further, as there, we note that $C_j \geq 1$ and hence also $Y \geq 1$; this follows from our definition of c_j and the assumption that, for each j, not all b_{ij} are zero.†

We now record three lemmas that will be needed later. Lemma 3 is classical Kummer theory; for a proof see Baker and Stark [7]. Lemma 4 is a famous result of Delaunay and Nagell; for a proof see Nagell [17]. Lemma 5 is due to Ljunggren [16].

Lemma 3. *Let* $\alpha_1, \ldots, \alpha_n$ *be non-zero elements of an algebraic number field* K *and let* $\alpha_1^{1/p}, \ldots, \alpha_{n-1}^{1/p}$ *denote fixed pth roots for some prime* p. *Further, let* $K' = K(\alpha_1^{1/p}, \ldots, \alpha_{n-1}^{1/p})$. *Then either* $K'(\alpha_n^{1/p})$ *is an extension of* K' *of degree* p *or we have*

$$\alpha_n = \alpha_1^{j_1} \ldots \alpha_{n-1}^{j_{n-1}} \gamma^p$$

for some γ *in* K *and some integers* $j_1, \ldots, j_{n-1}$ *with* $0 \leq j_\ell < p$.

Lemma 4. *Let* a *be a positive integer, not a perfect cube. The equation*

$$x^3 - ay^3 = 1$$

has at most one solution in integers x, y *with* $y \neq 0$ *and, for this,* $x - y\sqrt[3]{a}$ *is given by either* $1/\epsilon$ *or* $1/\epsilon^2$, *where* ϵ *is the fundamental unit of* $\mathbf{Q}(\sqrt[3]{a})$ *as in* §1.

Lemma 5. *Let* A, B, C *be positive integers with* $C = 1$ *or* $C = 3$ *and suppose that* A *and* B *are* > 1 *when* $C = 1$. *Suppose further that* AB *is not divisible by 3 when* $C = 3$. *Then the equation*

$$Ax^3 + By^3 = C$$

has at most one solution in integers x, y *and for this,* $C^{-1}(x\sqrt[3]{A} + y\sqrt[3]{B})^3$ *is either* $1/\eta$ *or* $1/\eta^2$ *where* η *is the fundamental unit in* $\mathbf{Q}(\sqrt[3]{(AB^2)})$. *The only exception is the equation* $2x^3 + y^3 = 3$ *which has two solutions, namely* $x = y = 1$ *and* $x = 4$, $y = -5$.

Note that if the condition in Lemma 5 that AB be not divisible by 3 when $C = 3$ is violated then the equation reduces to an equation with

---

† Professor Vaaler has pointed out to us that the result can also be obtained from Theorem 9 of Bombieri and Vaaler, "On Siegel's Lemma", *Invent. Math.* **73** (1983), 11–32, and in fact with $\sqrt{N}$ in place of $2\sqrt{2}(N+1)$.

$C = 1$. Note further that if the condition that A and B be > 1 when $C = 1$ is violated then the equation reduces to one of the kind considered in Lemma 4. Hence, taking into account the possible replacement of x or y by $-x$ or $-y$, we see that the lemmas incorporate all equations $Ax^3 + By^3 = C$, where A, B are any integers and $C = 1$ or $C = 3$.

Now these results of Delaunay, Nagell and Ljunggren can be viewed as providing the complete solution to (4) when n divides the discriminant $-27a^2$ of $x^3 - a$; it is precisely this condition that will arise in our discussion later. In particular, we see that Theorem 2 certainly holds in this case. To verify the assertion, note that if n^2 divides $27a^2$ then n divides $3a$ and so also n divides $3x^3$. We write $3x^3/n = Az^3$ where A, z are integers and A divides $3n^2$. Further we put $B = -3a/n$ and $C = 3$. Then $Az^3 + By^3 = C$ and $AB^2 = (3x/nz)^3 a^2$. Hence Lemmas 4 and 5 give the possible values of y, z and thus also x, in terms of the fundamental unit ϵ in $\mathbf{Q}(\sqrt[3]{a})$.

3. On units in purely cubic fields

Let a be a positive integer, not a perfect cube and let $\alpha = \sqrt[3]{a}$ as in §1. Let ω be a primitive cube root of unity and put $\alpha' = \omega\alpha$, $\alpha'' = \omega^2\alpha$. Further let ϵ be the fundamental unit in $\mathbf{Q}(\alpha)$ with $\epsilon > 1$ as in §1. We define ϵ', ϵ'' to be the conjugates of ϵ corresponding to α', α'' and we put $\rho = \epsilon''/\epsilon'$. Throughout the paper, logarithms will have their principal values.

Lemma 6. *We have $\log\epsilon > 1$. Further, if $\log\epsilon \le 3$ then $\mathbf{Q}(\alpha)$ is $\mathbf{Q}(\sqrt[3]{m})$ where m is one of* 2, 3, 7, 19, 28. *Furthermore, if $\mathbf{Q}(\alpha)$ is not $\mathbf{Q}(\sqrt[3]{28})$ then we have $|\log\rho| < \frac{\pi}{3}\log\epsilon$.*

Proof. Since $\epsilon\epsilon'\epsilon'' = \epsilon|\epsilon'|^2 = 1$, the minimal polynomial defining ϵ has the form

$$x^3 + bx^2 + cx - 1.$$

Here b, c are integers and

$$b = -(\epsilon + \epsilon' + \epsilon''), \qquad c = \epsilon\epsilon' + \epsilon\epsilon'' + \epsilon'\epsilon''.$$

We have

$$-(\epsilon + 2/\epsilon^{1/2}) \le b \le 0, \qquad |c| \le 2\epsilon^{1/2} + 1/\epsilon.$$

If $\log\epsilon \le 3$ these give $-21 \le b \le 0$, $|c| \le 9$. The discriminant of the polynomial is

$$d = b^2c^2 + 4b^3 - 4c^3 - 27 - 18bc.$$

Now the discriminant of $\mathbf{Q}(\sqrt[3]{a})$ divides the discriminant $-27a^2$ of $x^3 - a$ and so it has the form $-3k^2$ for some divisor k of $3a$. Hence $d = -3\ell^2$ for some multiple ℓ of k. A computer search shows that the only possible b, c in the above ranges for which $d = -3\ell^2$ for some integer ℓ are given by $(-3, -3), (-12, -6), (-12, 6), (-14, 2), (-5, -1), (-15, 3), (0, 0)$ and $(-2, 2)$. We find that the corresponding equations have precisely one real root; in the case of the first five pairs on the list the root is given respectively by

$$1 + \sqrt[3]{2} + (\sqrt[3]{2})^2 = 3.84\ldots$$

$$4 + 3\sqrt[3]{3} + 2(\sqrt[3]{3})^2 = 12.48\ldots$$

$$4 + 2\sqrt[3]{7} + (\sqrt[3]{7})^2 = 11.48\ldots$$

$$\tfrac{1}{3}(14 + 5\sqrt[3]{19} + 2(\sqrt[3]{19})^2) = 13.86\ldots$$

$$\tfrac{1}{6}(10 + 4\sqrt[3]{28} + (\sqrt[3]{28})^2) = 5.22\ldots$$

The sixth pair in the first list, that is $(-15, 3)$, corresponds to an equation with real root $1/(\sqrt[3]{2} - 1)^2$; this is the square of the first root above. The last two pairs of admissible values of (b, c), namely $(0, 0)$ and $(-2, 2)$, correspond to reducible equations with real root 1. This establishes the first two assertions of the lemma.

For the last assertion we note that $|\rho| = 1$ and so $|\log \rho| \leq \pi$. Hence the required inequality certainly holds if $\log \epsilon > 3$. If $\log \epsilon \leq 3$ we have the five possibilities for m above, and it is readily checked that the corresponding values of $|\log \rho|/(\pi \log \epsilon)$ are $0.27, 0.13, 0.13, 0.31, 0.50$ to two decimal places. This establishes the result.

Let now $K = \mathbf{Q}(\alpha, \omega)$. We define σ as either $-\omega$ or $-\omega^2$ so that the real numbers $i \log \rho$ and $i \log \sigma$ have opposite signs.

Lemma 7. $K(\rho^{1/2}, \sigma^{1/2})$ is an extension of K of degree 4.

Proof. First we show that $[K(\rho^{1/2}) : K] = 2$. We have $\epsilon\epsilon'\epsilon'' = 1$ and so $\rho = \epsilon''/\epsilon' = 1/\epsilon(\epsilon')^2$. Hence if $K(\rho^{1/2})$ were not an extension of K with degree 2 then we would have $\epsilon = \eta^2$ for some $\eta \in K$. Thus $\epsilon^{1/2}$ is in K. But ϵ is the fundamental unit in $\mathbf{Q}(\alpha)$ whence $\epsilon^{1/2}$ is not in $\mathbf{Q}(\alpha)$ and thus $\mathbf{Q}(\alpha, \epsilon^{1/2})$ is a field with degree 6 over $\mathbf{Q}$. On the other hand, K has degree 6 and is not a real field whence K is not $\mathbf{Q}(\alpha, \epsilon^{1/2})$. Thus $\epsilon^{1/2}$ is not in K, a contradiction.

Secondly, we show that $[K(\rho^{1/2}, \sigma^{1/2}) : K(\rho^{1/2})] = 2$. If this does not hold then $\sigma^{1/2}$ is in $K(\rho^{1/2})$. But $K(\rho^{1/2}) = K(\epsilon^{1/2})$ and so $i =$

$\lambda + \mu\omega$ for some λ, μ in $\mathbf{Q}(\alpha, \epsilon^{1/2})$. This gives $2i = \mu(\omega - \omega^2)$, that is $2 = \pm\sqrt{3}\,\mu$. Hence $\sqrt{3} = \gamma + \delta\epsilon^{1/2}$, where γ, δ are in $\mathbf{Q}(\alpha)$. Thus $3 = \gamma^2 + \epsilon\delta^2 + 2\gamma\delta\epsilon^{1/2}$. If $\gamma\delta \neq 0$ then this implies that $\epsilon^{1/2}$ is in $\mathbf{Q}(\alpha)$, a contradiction. We cannot have $\delta = 0$ for this would give $\sqrt{3} = \gamma$, contrary to the fact that $\mathbf{Q}(\alpha)$ does not have a quadratic subfield. Hence $\gamma = 0$ and thus $\sqrt{3} = \delta\epsilon^{1/2}$. This gives $3 = \delta^2\epsilon$ and consequently, taking norms from $\mathbf{Q}(\alpha)$ to $\mathbf{Q}$, we get $27 = (N\delta)^2$, a contradiction since $N\delta$ is rational. This proves the result.

4. Reduction to a linear form in logarithms

Let n be a positive integer and let x, y be integers satisfying (4). With the notation of §3 we have

$$(x - \alpha y)(x - \alpha'y)(x - \alpha''y) = n.$$

We shall prove that if $n > c_1$ then

$$\max(|x|, |y|) < n^{c_2}. \tag{6}$$

This will suffice to establish Theorem 2. For if $n \leq c_1$ then we put

$$x_1 = Cx, \qquad y_1 = Cy, \qquad n_1 = C^3 n,$$

where $C = [(c_1/n)^{1/3}] + 1$; this gives $x_1^3 - ay_1^3 = n_1$ with $n_1 > c_1$, whence, by (6), we have

$$\max(|x_1|, |y_1|) < (c_1^{1/3} + n^{1/3})^{3c_2}$$

and Theorem 2 follows.

We now show that we can assume that the quotient

$$\nu = (x - \alpha''y)/(x - \alpha'y)$$

is not a unit in K. Put

$$x_2 = x/(x, y), \qquad y_2 = y/(x, y), \qquad n_2 = n/(x, y)^3.$$

Then x_2, y_2 are relatively prime and we have $x_2^3 - ay_2^3 = n_2$. Further we have

$$(\alpha'' - \alpha')x_2 = (x_2 - \alpha'y_2)(\alpha'' - \alpha'\nu),$$
$$(\alpha'' - \alpha')y_2 = (x_2 - \alpha'y_2)(1 - \nu).$$

Hence, if ν is a unit, then, taking norms from K to $\mathbf{Q}$, we find that $N(x_2 - \alpha' y_2)$ divides $N(\alpha'' - \alpha')N(x_2)$ and $N(\alpha'' - \alpha')N(y_2)$. But $N(x_2)$ and $N(y_2)$ are relatively prime whence $N(x_2 - \alpha' y_2)$ divides $N(\alpha'' - \alpha')$, that is n_2^2 divides $27a^2$. We have $(x, y) \leq n$ and hence $|x| \leq n|x_2|$, $|y| \leq n|y_2|$. Now Lemmas 4 and 5 give bounds for x_2, y_2 in terms of the fundamental unit in $\mathbf{Q}(\alpha)$ as in §2, and Theorem 2 follows in this case.

We define $\beta = (x - \alpha y)\epsilon^j$, where j is the integer such that

$$1 \leq n^{-1/3}|\beta| < \epsilon.$$

We put

$$\beta' = (x - \alpha' y)\epsilon'^j, \qquad \beta'' = (x - \alpha'' y)\epsilon''^j.$$

Then $\beta\beta'\beta'' = n$ and since $|\beta'| = |\beta''|$ we obtain

$$\epsilon^{-1/2} < n^{-1/3}|\beta'| \leq 1.$$

We shall assume in the sequel that

$$j > 2(10^{12} - 1)\log n \tag{7}$$

and we shall ultimately derive a contradiction. This will suffice to prove Theorem 2; for we have $|\beta'| \leq n^{1/3}$, whence

$$|x - \alpha' y| \leq n^{1/3}|\epsilon'|^{-j} = n^{1/3}\epsilon^{j/2}.$$

Thus if (7) does not hold then $|x - \alpha' y| \leq n^{c_2 - 2/3}$. But since the imaginary part of ω is $\pm\sqrt{3}/2$ this gives $|\alpha y| < (2/\sqrt{3})n^{c_2 - 2/3}$. We have $n > c_1$ and $|x| \leq |\alpha y| + |x - \alpha' y|$, and (6) follows.

We now consider the number $\lambda = -\omega\beta''/\beta'$. Ideally we would like $\lambda^{1/2}$ to generate an extension of $K(\rho^{1/2}, \sigma^{1/2})$ of degree 2; but this is not necessarily so. We overcome the problem by substituting τ for λ as described below. Our argument is apparently novel and more efficient than those applied previously in this context. Let $v \geq 0$ be an integer such that

$$\lambda = \rho^{t'}\sigma^{t''}\tau^t, \tag{8}$$

where $t = 2^v$ and t', t'' are integers with $0 \leq t' < t$, $0 \leq t'' < t$ and τ is in K. Plainly at least one such v exists since we can take $t' = t'' = 0$ and $t = 1$. We proceed to prove that $t < 3\log n$. Now λ is an element of K and the leading coefficient in the field polynomial of λ divides n^2. Since ρ and σ are units, the same holds for the field polynomial of τ^t. It follows from Lemma 1 that the leading coefficient, say q, in the field polynomial of τ satisfies $q \leq n^{2/t}$. Suppose now that $t \geq 3 \log n$. Then, since q is assumed to be positive, we have $q = 1$. Hence τ is an algebraic integer and thus also λ is an algebraic integer. But we have $\nu = -\omega^2 \rho^{-j} \lambda$ and it follows that ν is an algebraic integer. On the other hand, it is an immediate consequence of the definition of ν that its norm is 1. Thus ν is a unit contrary to our assumption above. We shall suppose henceforth that v is the largest integer such that (8) holds. Then, by Lemma 3, $\tau^{1/2}$ generates an extension of $K(\rho^{1/2}, \sigma^{1/2})$ of degree 2. Further, by Lemma 7, we see that $K(\rho^{1/2}, \sigma^{1/2}, \tau^{1/2})$ is an extension of K with degree 8.

We require estimates for the conjugates of τ. For this purpose we observe that the field conjugates of ρ are ϵ''/ϵ', ϵ'/ϵ'', ϵ/ϵ', ϵ/ϵ'', ϵ'/ϵ, ϵ''/ϵ and these have absolute values $1, 1, \epsilon^{3/2}, \epsilon^{3/2}, \epsilon^{-3/2}, \epsilon^{-3/2}$ respectively. Further, from our estimates for $|\beta|$, $|\beta'|$ above we see that four of the conjugates of λ have absolute values at most 1 and the other two have absolute value at most $\epsilon^{3/2}$. Hence from (8) we see that two of the conjugates of τ^t have absolute values at most $\epsilon^{(3/2)(t'+1)}$ and the remainder have absolute value at most 1. Since $t' + 1 \leq t$, it follows that two of the conjugates of τ have absolute value at most $\epsilon^{3/2}$ and the remainder have absolute value at most 1.

We now derive the basic inequality involving a linear form in logarithms. We have the identity

$$\beta \epsilon^{-j}(\alpha' - \alpha'') + \beta' \epsilon'^{-j}(\alpha'' - \alpha) + \beta'' \epsilon''^{-j}(\alpha - \alpha') = 0.$$

Hence

$$\frac{\beta''}{\beta'}\left(\frac{\epsilon'}{\epsilon''}\right)^j \frac{\alpha - \alpha'}{\alpha'' - \alpha} + 1 = \frac{\beta}{\beta'}\left(\frac{\epsilon'}{\epsilon}\right)^j \frac{\alpha'' - \alpha'}{\alpha'' - \alpha}$$

and thus

$$\lambda \rho^{-j} - 1 = \left(\beta/(\omega\beta')\right)(\epsilon'/\epsilon)^j.$$

As above we have $|\beta/\beta'| < \epsilon^{3/2}$ and $|\epsilon'/\epsilon| = \epsilon^{-3/2}$. This gives

$$|\lambda\rho^{-j} - 1| < \epsilon^{-(3/2)(j-1)}.$$

We substitute for λ from (8) and obtain

$$|\rho^{t'-j}\sigma^{t''}\tau^t - 1| < \epsilon^{-(3/2)(j-1)}.$$

Since, for any complex number z, the inequality $|e^z - 1| < 1/4$ implies that $|z - ik\pi| \leq 4|e^z - 1|$ for some rational integer k, we deduce that

$$|r \log \rho + s \log \sigma - t \log \tau| < 4\epsilon^{-(3/2)(j-1)}, \tag{9}$$

where $r = j - t'$ and s is a rational integer. We recall here that $t = 2^v < 3\log n$ and that the logarithms have their principal values. Since $0 \le t' < t$, we see from (7) that $0 < r \le j$. Further we observe that

$$|s\log\sigma| \le \pi(r+t) + 1$$

and thus, since $\log\sigma = \pm\frac{\pi}{3}i$, we have

$$|s| \le 3(r+t) + 1 \le 3j + 10\log n.$$

5. The auxiliary function

We shall now assume that (7) and (9) hold and that $n > c_1$, and we shall eventually deduce a contradiction. By virtue of the results referred to in §1 we can suppose that $Q(\alpha)$ is not $Q(\sqrt[3]{2})$ or $Q(\sqrt[3]{28})$ (see [2], [13]).

We put $u = \max(1, v)$ and $h = 500u$. Further we put $L = \frac{2}{5}(j/h)\log\epsilon$ and we write

$$L_1 = [10^{-2}L/\log\epsilon], \qquad L_2 = [10^{-2}L], \qquad L_3 = [2\cdot 5^7 Lh^2/j].$$

Then for any non-negative integers m_1, m_2 we define the function

$$f(z;m_1,m_2) = \sum_{\lambda_1=0}^{L_1}\sum_{\lambda_2=0}^{L_2}\sum_{\lambda_3=0}^{L_3} p(\lambda)\Delta(t\gamma_1;m_1)\Delta(t\gamma_2;m_2)\rho^{\gamma_1 z}\sigma^{\gamma_2 z},$$

where

$$\gamma_1 = \lambda_1 + (r/t)\lambda_3, \qquad \gamma_2 = \lambda_2 + (s/t)\lambda_3,$$

and the $p(\lambda) = p(\lambda_1, \lambda_2, \lambda_3)$ are integers to be determined later. The Δ-polynomials are defined, as usual, by

$$\Delta(x;0) = 1, \quad \Delta(x;k) = (x+1)\ldots(x+k)/k!, \qquad k \ge 1,$$

and z^w means $e^{z\log w}$ where the logarithm has its principal value. We also introduce the function

$$g(z;m_1,m_2) = \sum_{\lambda_1=0}^{L_1}\sum_{\lambda_2=0}^{L_2}\sum_{\lambda_3=0}^{L_3} p(\lambda)\Delta(t\gamma_1;m_1)\Delta(t\gamma_2;m_2)\rho^{\lambda_1 z}\sigma^{\lambda_2 z}\tau^{\lambda_3 z}.$$

The coefficients $p(\lambda)$ are chosen so that $g(\ell; m_1, m_2) = 0$ for all odd integers ℓ with $1 \leq \ell \leq 2h$ and all non-negative integers m_1, m_2 with $m_1 + m_2 \leq L$. The number of such m_1, m_2 is $H = (1/2)(L + 1)(L + 2)$ and thus we have to solve $M = Hh$ linear equations in the $N = (L_1 + 1)(L_2 + 1)(L_3 + 1)$ unknowns $p(\lambda)$. By the definition of L we have $N > (25/4)L^2 h$ and, from (7), it follows that $N > 12M$. We shall apply Lemma 2 with $K = \mathbf{Q}(\alpha, \omega)$ so that $n = 6$; we conclude that there exist rational integers $p(\lambda)$, not all 0, such that

$$|p(\lambda)| \leq 2\sqrt{2}(N + 1)Z^{1/(6M)} \tag{10}$$

and our purpose now is to determine a bound for the quantity Z referred to in the lemma.

First we shall establish estimates for the Δ-polynomials. We shall write, for brevity,

$$U(m_1, m_2) = \max_{\lambda_1, \lambda_2, \lambda_3} \left| \Delta(t\gamma_1; m_1) \Delta(t\gamma_2; m_2) \right|.$$

Lemma 8. *We have*

$$U(m_1, m_2) \leq (2 \cdot 10^{11} t)^{m_1 + m_2} 2^{2L}.$$

Proof. We begin by noting that

$$L_3/L \leq 2 \cdot 5^7 h^2/j = 2^{-7} 5 \cdot 10^{12} u^2/j. \tag{11}$$

Since $u^2 \leq t$ except for $u = 3$, $t = 8$ we have $u^2/t \leq 9/8$ and this gives

$$L_3/L \leq 2^{-10} \cdot 45 \cdot 10^{12}(t/j) \leq 45 \cdot 10^9(t/j). \tag{12}$$

Further, from (7) and (11), we obtain

$$L_3/L \leq u^2/(50 \log n) \leq 9t/(400 \log n). \tag{13}$$

We can now estimate γ_1, γ_2. We have

$$|\gamma_1| \leq L_1 + (r/t)L_3 \leq L_1 + (j/t)L_3$$

and so from (12) we see that

$$|\gamma_1| \leq L_1 + 45 \cdot 10^9 L \leq 5 \cdot 10^{10} L.$$

Similarly we have

$$|\gamma_2| \leq L_2 + (|s|/t)L_3 \leq L_2 + ((3j + 10\log n)/t)L_3$$

and so from (12) and (13) we see that $|\gamma_2| \leq 15 \cdot 10^{10} L$. Thus we obtain

$$\left|\Delta(t\gamma_1; m_1)\right| \leq t^{m_1} 10^{11m_1} \Delta(L/2; m_1) = t^{m_1} 10^{11m_1} 2^{L/2+m_1}.$$

Similarly we obtain

$$\left|\Delta(t\gamma_2; m_2)\right| \leq t^{m_2} 10^{11m_2} 2^{3L/2+m_2},$$

and the lemma follows.

As a corollary we see that if $m_1 + m_2 \leq L/2$ then $U(m_1, m_2) \leq \Delta$, where

$$\Delta = (2^5 \cdot 10^{11}t)^{L/2}.$$

Now we have $t = 2^v \leq 2^u \leq e^{(0.7)u}$ and $(2^5 \cdot 10^{11})^{1/2} < e^{14.4}$. Hence we obtain $\Delta \leq e^{14.75Lu}$ and since $h = 500u$, this gives $\Delta \leq e^{(0.03)Lh}$.

We also wish to estimate

$$U = \prod_{m_1, m_2} U(m_1, m_2),$$

where the product is taken over all non-negative integers m_1, m_2 with $m_1 + m_2 \leq L$. For this purpose we observe that

$$\sum_{m_1=0}^{L} \sum_{m_2=0}^{L-m_1} (m_1 + m_2) = \tfrac{2}{3}LH$$

where $H = \tfrac{1}{2}(L+1)(L+2)$ as above; indeed the left-hand side is

$$\tfrac{1}{2}\sum_{m_1=0}^{L} (L+m_1)(L-m_1+1) = (1/2)L(L+1)\{(L+1)+1/2-(1/6)(2L+1)\}.$$

Thus from Lemma 8 we obtain

$$U \leq (2^4 \cdot 10^{11}t)^{(2/3)LH}.$$

Now $t \leq e^{(0.7)u}$ and $(2^4 \cdot 10^{11})^{2/3} < e^{18.75}$. Hence we have

$$U \leq e^{19.22LHu} \leq e^{(0.0385)LHh}. \tag{14}$$

Lemma 9. *For all* λ_1, λ_2, λ_3, *we have*

$$|p(\lambda)| \leq N^{-1} e^{(0.052)Lh}.$$

Proof. We apply Lemma 2 with

$$Z = \prod_{\ell, m_1, m_2} \{q^{L_3 \ell}(U(m_1, m_2))^6 P\},$$

where

$$P = \prod_{k=1}^{6} \max_{\lambda_1, \lambda_2, \lambda_3} |\sigma_k(\rho^{\lambda_1 \ell} \sigma^{\lambda_2 \ell} \tau^{\lambda_3 \ell})|.$$

Here we recall that q is the leading coefficient in the field polynomial for τ. Hence

$$q^{L_3 \ell} \sigma_1(\tau^{\lambda_{3,1} \ell}) \ldots \sigma_6(\tau^{\lambda_{3,6} \ell})$$

is an algebraic integer for all integer choices of $\lambda_{3,k}$ $(1 \leq k \leq 6)$ with $0 \leq \lambda_{3,k} \leq L_3$. Since clearly $U(m_1, m_2)$ is a rational integer and ρ and σ are units we see that the numbers $q^{L_3 \ell}$ have the property required of the c_j in Lemma 2.

In the expression for Z above, the product is over all odd integers ℓ with $1 \leq \ell < 2h$ and all non-negative integers m_1, m_2 with $m_1 + m_2 \leq L$. Note that the sum of the integers ℓ is

$$\sum_{j=0}^{h-1}(2j + 1) = 2((1/2)h(h-1)) + h = h^2.$$

To estimate P we recall that two of the conjugates of τ have absolute values at most $\epsilon^{3/2}$ and that the remainder have absolute values at most 1; moreover the same holds for the conjugates of ρ. Since also σ is a root of unity it follows that

$$P \leq \epsilon^{3(L_1 + L_3)\ell}.$$

Now by the definition of U we obtain

$$Z \leq q^{L_3 H h^2} U^{6h} \epsilon^{3(L_1 + L_3)H h^2}.$$

Hence, since $M = Hh$, we deduce from (10) that

$$|p(\lambda)| \leq 2\sqrt{2}(N + 1)q^{L_3 h/6} U^{1/H} \epsilon^{(1/2)(L_1 + L_3)h}.$$

We have $q \leq n^{2/t}$ and thus, by (13), $q^{L_3 h/6} \leq e^{(3/400)Lh}$. Further, by (14), $U^{1/H} \leq e^{(0.0385)Lh}$. Furthermore, we have $\epsilon^{(1/2)L_1 h} \leq e^{(1/200)Lh}$. We shall verify in a moment that, since $n > c_1$,

$$u^2 / \log n < 1/(10.4 \log \epsilon). \tag{15}$$

This together with (13) gives $\epsilon^{(1/2)L_3 h} \leq e^{(1/1040)Lh}$. Hence, on combining our estimates, we get

$$|p(\lambda)| \leq 2\sqrt{2}(N+1)e^{(0.05197)Lh}.$$

Then Lemma 9 follows since clearly $2\sqrt{2}(N+1)N < e^{(0.00001)Lh}$.

It remains to verify (15). Since $2^u < 3 \log n$ we have $u < \psi(n)$, where

$$\psi(n) = (1/\log 2)(\log 3 + \log \log n).$$

Now $(\psi(n))^2 / \log n$ is a decreasing function of n for $n > c_1$, and thus it suffices to prove that

$$(\psi(c_1))^2 / \log c_1 < 1/(10.4 \log \epsilon).$$

We have

$$\log c_1 = (50 \log \log \epsilon)^2 \log \epsilon$$

and thus we require that

$$50 \log \log \epsilon > (\sqrt{10.4}/\log 2)(\log 3 + \log \log c_1).$$

The expression on the right is

$$42 + 5 \log \log \epsilon + 10 \log \log \log \epsilon,$$

with constants rounded up slightly, and if $\log \epsilon \geq 3$ then the desired inequality is obvious. If $\log \epsilon < 3$ we have the five possibilities for ϵ listed in the proof of Lemma 6. We have already remarked that we can exclude the fields $\mathbf{Q}(\sqrt[3]{2})$ and $\mathbf{Q}(\sqrt[3]{28})$; and the desired inequality is readily checked for the three remaining values of ϵ.

6. Basic estimates

Our purpose here is to establish the main estimates needed for the extrapolation algorithm described in the next section. The object is to prove that $g(\ell/2; m_1, m_2) = 0$ for all odd integers ℓ with $1 \leq \ell < 4h$ and all non-negative integers m_1, m_2 with $m_1 + m_2 \leq L/2$. Accordingly we shall suppose that

$$g = g(\ell/2; m_1, m_2) \neq 0$$

for some such ℓ, m_1, m_2 and we shall ultimately obtain a contradiction.

First we note that g is an algebraic number in the field $K(\rho^{1/2}, \sigma^{1/2}, \tau^{1/2})$, and consequently g has degree at most 48. We proceed to estimate the field norm $N(g)$ of g. By Lemma 9 we have $|p(\lambda)| \leq N^{-1} X$, where $X = e^{(0.052)Lh}$. Further, as in the proof of the lemma, we see that

$$\prod_{k=1}^{6} \max_{\lambda_1, \lambda_2, \lambda_3} \left| \sigma_k(\rho^{\lambda_1 \ell/2} \sigma^{\lambda_2 \ell/2} \tau^{\lambda_3 \ell/2}) \right| \leq \epsilon^{(3/2)\ell(L_1 + L_3)}.$$

Furthermore it is clear that one of the conjugates of g is in fact the complex conjugate $g(-\ell/2; m_1, m_2)$. Hence we obtain

$$|N(g)| \leq |g|^2 (X\Delta)^{46} \epsilon^{12\ell(L_1 + L_3)}.$$

Now we have $X\Delta \leq e^{(0.082)Lh}$ and so $(X\Delta)^{46} \leq e^{(3.772)Lh}$. Also, as in Lemma 9, we see that if $\ell < 4h$, then $\epsilon^{12\ell L_1} \leq e^{(0.48)Lh}$ and $\epsilon^{12\ell L_3} \leq e^{(0.093)Lh}$. This gives

$$|N(g)| \leq |g|^2 e^{(4.345)Lh}$$

To obtain a lower bound for $|N(g)|$, we observe that $\tau^{\lambda_3 \ell/2}$ can be expressed as τ^{λ} or $\tau^{\lambda+1/2}$, where λ is an integer with $0 \leq \lambda \leq L_3 \ell/2$. Now, since $\ell < 4h$, it follows that $q^{16 L_3 h} N(g)$ is an algebraic integer. By supposition $g \neq 0$, and hence

$$|N(g)| \geq q^{-16 L_3 h} \geq e^{-(0.72)Lh}.$$

On comparing estimates we obtain $|g|^2 \geq e^{-(5.065)Lh}$ and so $|g| \geq e^{-(2.533)Lh}$.

This gives a similar estimate for

$$f = f(\ell/2; m_1, m_2).$$

Indeed, for any complex number z we have $|e^z - 1| \leq |z| e^{|z|}$ and thus, by (9), we obtain

$$\left| (\rho^{r/t} \sigma^{s/t})^{\lambda_3 \ell/2} - \tau^{\lambda_3 \ell/2} \right| \leq (9 L_3 h) \epsilon^{-(3/2)(j-1)}.$$

Now $9L_3h \leq e^{(0.001)Lh}$ and, by the definition of L, we have $\epsilon^{(3/2)j} = e^{(15/4)Lh}$. Hence the number on the right is at most $e^{-(3.7)Lh}$. This gives

$$|f - g| \leq X\Delta e^{-(3.7)Lh} \leq e^{-(3.6)Lh} \tag{16}$$

and so certainly $|f - g| \leq |g|/2$. It follows that $|f| \geq |g|/2$, whence

$$|f| > e^{-(2.54)Lh}. \tag{17}$$

We shall also require an upper bound for $|f(z; m_1, m_2)|$ with $m_1 + m_2 \leq L/2$. By the definition of σ, the numbers $i \log \rho$ and $i \log \sigma$ take opposite signs. Hence

$$\left|\rho^{\lambda_1 z}\sigma^{\lambda_2 z}\right| \leq \max\left(e^{L_1|z\log\rho|}, e^{L_2|z\log\sigma|}\right).$$

We have $|\log \sigma| = \pi/3$ and, by Lemma 6, if $\mathbf{Q}(\alpha)$ is not $\mathbf{Q}(\sqrt[3]{28})$, as we can assume, then $|\log \rho| < (\pi/3)\log\epsilon$. thus we obtain

$$\left|\rho^{\lambda_1 z}\sigma^{\lambda_2 z}\right| \leq e^{(\pi/3)L_2|z|}.$$

Further we have, by (9),

$$\left|\log(\rho^{r/t}\sigma^{s/t}) - \log\tau\right| < 4\epsilon^{-(3/2)(j-1)}$$

for some value of the first logarithm. This gives

$$\left|(\rho^{r/t}\sigma^{s/t})^{\lambda_3 z}\right| \leq e^{(|\log\tau|+0.001)L_3|z|}$$

and since $|\log\tau| \leq \pi$, the number on the right is at most $e^{(3.15)L_3|z|}$. It follows that

$$|f(z; m_1, m_2)| \leq X\Delta e^{((\pi/3)L_2 + (3.15)L_3)|z|}.$$

Now $X\Delta \leq e^{(0.082)Lh}$ and $L_2 \leq 10^{-2}L$, whence $e^{(\pi/3)L_2} \leq e^{(0.0105)L}$. Further, by (13) and (15), we have $L_3/L \leq 1/(520\log\epsilon)$. If we exclude the fields $\mathbf{Q}(\sqrt[3]{2})$ and $\mathbf{Q}(\sqrt[3]{28})$ which, as we noted in §5, we may, then $\log\epsilon \geq \log(11.48) > 2.44$. Hence $L_3/L \leq 1/(1268)$ and so $e^{(3.15)L_3} \leq e^{(0.0025)L}$. We conclude that

$$|f(z; m_1, m_2)| \leq e^{(0.082)Lh+(0.013)L|z|}. \tag{18}$$

7. Extrapolation

Let ℓ be any odd integer with $1 \leq \ell \leq 4h$. Suppose that m_1, m_2 are non-negative integers with $m_1 + m_2 \leq L/2$ and let $f(z) = f(z; m_1, m_2)$. Our purpose here is to obtain an upper bound for $f = f(\ell/2)$ which is stronger than the lower bound given by (17). Thus we shall conclude that $g = 0$ as required.

We shall denote the mth derivative of $f(z)$ by $f_m(z)$. Our first objective is to estimate $f_m(\ell')/m!$, where ℓ' is any odd integer with $1 \leq |\ell'| < 2h$ and m is any integer with $0 \leq m \leq L/2$. We have

$$f_m(\ell')/m! = \sum (\mu_1! \mu_2!)^{-1} (\log \rho)^{\mu_1} (\log \sigma)^{\mu_2} f'(\ell'; m_1', m_2'),$$

where the sum is over all non-negative integers μ_1, μ_2 with $\mu_1 + \mu_2 = m$ and $m_1' = m_1 + \mu_1$, $m_2' = m_2 + \mu_2$. Here $f'(\ell'; m_1', m_2')$ is defined like $f(\ell'; m_1', m_2')$ but with $\Delta(t\gamma_j; m_j + \mu_j)$ replaced by $\gamma_j^{\mu_j} \Delta(t\gamma_j; m_j)$. Now the auxiliary function was constructed so that $g(\ell'; m_1', m_2') = 0$ for positive ℓ', and in fact this holds also for negative ℓ', since $g(-\ell'; m_1', m_2')$ is a conjugate of $g(\ell'; m_1', m_2')$. Further arguing inductively with respect to $\mu_1 + \mu_2$ and observing that $\Delta(t\gamma_j; m_j)$ is a polynomial in γ_j with coefficients independent of the λ's we deduce that $g'(\ell'; m_1', m_2') = 0$, where g' is the analogue of f'. Hence we obtain

$$\left| f_m(\ell')/m! \right| \leq A \left| f(\ell'; m_1, m_2) - g(\ell'; m_1, m_2) \right|^*,$$

where

$$A = \max_{\lambda_1, \lambda_2, \lambda_3} \sum \left| (\mu_1! \mu_2!)^{-1} (\gamma_1 \log \rho)^{\mu_1} (\gamma_2 \log \sigma)^{\mu_2} \right|$$

and the $*$ signifies that each term in the sum over λ_1, λ_2, λ_3 representing $f - g$ is to be replaced by its absolute value. We have $|\gamma_1| \leq 5 \cdot 10^{10} L$ and $|\log \rho| \leq \pi$, and thus

$$\left| (\gamma_1 \log \rho)^{\mu_1} / \mu_1! \right| \leq (5 \cdot 10^{10} \pi L)^{\mu_1} / \mu_1! \leq 10^{10\mu_1} e^{5\pi L}.$$

Similarly since $|\log \sigma| \leq \pi/3$ we have

$$\left| (\gamma_2 \log \sigma)^{\mu_2} / \mu_2! \right| \leq 10^{10\mu_2} e^{5\pi L}.$$

Hence, since $m \leq L/2$ and $h \geq 500$, we obtain

$$A \leq L\, 10^{10m} e^{10\pi L} < e^{(0.1)Lh},$$

and it follows from the estimates of §6 (cf. (16)) that

$$\left| f_m(\ell')/m! \right| \leq e^{-(3.5)Lh}. \tag{19}$$

Now let $S = [L/2]$ and let

$$F(z) = ((z^2 - 1^2)(z^2 - 3^2)\ldots(z^2 - (2h-1)^2))^{S+1}.$$

Further let Γ and $\Gamma_{\ell'}$ be the circles $|z| = 76h$ and $|z-\ell'| = 1/4$, described in the positive sense. By Cauchy's theorem we have

$$\frac{1}{2\pi i}\int_\Gamma \frac{f(z)dz}{(z-\ell/2)F(z)} = \frac{f(\ell/2)}{F(\ell/2)} + \frac{1}{2\pi i}\sum{}' \sum_{m=0}^{S} \frac{f_m(\ell')}{m!}\int_{\Gamma_{\ell'}} \frac{(z-\ell')^m dz}{(z-\ell/2)F(z)},$$

where $\sum'$ signifies summation over all odd integers ℓ' with $1 \le |\ell'| < 2h$. We require an estimate for the last integral, and for this purpose we note that

$$(F(z))^{1/(S+1)} = 2^{2h}\prod_{k=-h}^{h-1}(z' + k)$$

where $z' = \frac{1}{2}(z+1)$. Further, for z on $\Gamma_{\ell'}$, we have $|z - \ell'| = 1/4$ and hence $|z' - \ell''| = 1/8$, where $\ell'' = \frac{1}{2}(l'+1)$. Thus we obtain

$$|z' + k| = |(z' - \ell'') + (k + \ell'')| \ge |k + \ell''| - 1/8.$$

It follows that if $k > -\ell'' + 1$ then we have $|z' + k| > k + \ell'' - 1$, and if $k < -\ell'' - 1$, then $|z' + k| > -k - \ell'' - 1$. Since also $|z' + k| \ge 7/8$ for $k = -\ell'' \pm 1$ and $|z' + k| = 1/8$ for $k = -\ell''$, we obtain

$$|F(z)|^{1/(S+1)} \ge 2^{2h}(1/8)(7/8)^2(h + \ell'' - 2)!(h - \ell'' - 1)!,$$

where, for brevity, we have adopted the convention that $(-1)! = 1$. The number on the right is at least $(2h-3)!(7/8)^2$ and since $(2h)^3(7/8)^{-2} < e^{(0.05)h}$ this gives

$$|F(z)|^{1/(S+1)} \ge (2h)!e^{-(0.05)h}.$$

Now clearly, for z on $\Gamma_{\ell'}$, we have

$$|(z-\ell')^m/(z-\ell/2)| \le 4.$$

Further, the number of terms in the double sum above is $2h(S+1) < e^{(0.02)Lh}$. Hence, from (19), the absolute value of the sum is at most

$$e^{-(3.4)Lh}((2h)!)^{-(S+1)}. \tag{20}$$

It is readily verified that, for $z = \ell/2$, we have

$$\left|F(z)\right|^{1/(S+1)} \leq 2^{2h}\left(h + [z']\right)!\left(h - [z']\right)! \leq 2^{2h}(2h)!.$$

This gives

$$\left|F(\ell/2)\right| \leq 2^{Lh+2h}\left((2h)!\right)^{S+1},$$

and hence (20) is at most

$$e^{-(2.7)Lh}\left|F(\ell/2)\right|^{-1}.$$

Let now θ and Θ denote respectively the upper bound of $\left|f(z)\right|$ and the lower bound of $\left|F(z)\right|$ with z on Γ. Since $2|z - \ell/2|$ with z on Γ exceeds the radius of Γ, we obtain

$$\left|f(\ell/2)\right| \leq (2\theta/\Theta)\left|F(\ell/2)\right| + e^{-2.7Lh}.$$

On noting that

$$\left|z^2 - k^2\right| \geq |z|^2\left(1 - k/|z|\right)$$

for each odd integer k with $1 \leq k < 2h$, and recalling that Γ has radius $76h$, we deduce that

$$\Theta \geq \left((37/38)^{1/2}76h\right)^{2h(S+1)}.$$

Thus, from the trivial estimate

$$\left|F(\ell/2)\right| \leq (2h)^{2h(S+1)},$$

it follows that

$$\Theta/\left|F(\ell/2)\right| \geq \left((37/38)^{1/2}38\right)^{2h(S+1)} \geq e^{3.62Lh}.$$

But from (18) we have $\theta \leq e^{1.07Lh}$ and hence

$$\left|f(\ell/2)\right| \leq 2e^{-2.55Lh} + e^{-2.7Lh}.$$

This contradicts (17), and the contradiction implies that $g = 0$, as required.

8. Kummer theory

The equation $g(\ell/2; m_1, m_2) = 0$, where ℓ is any odd integer with $1 \leq \ell < 4h$ and m_1, m_2 are non-negative integers with $m_1 + m_2 \leq L/2$, can be replaced by eight equations formed by restricting λ_1, λ_2, λ_3 to run through residue classes (mod 2). This is a consequence of the fact, established in §4, that $K(\rho^{1/2}, \sigma^{1/2}, \tau^{1/2})$ is an extension of K with degree 8. Hence for any λ'_1, λ'_2, λ'_3 given by 0 or 1 we have

$$\sum_{\mu_1=0}^{L_1} \sum_{\mu_2=0}^{L_2} \sum_{\mu_3=0}^{L_3} p(\mu)\Delta(\gamma'_1; m_1)\Delta(\gamma'_2; m_2)\rho^{\mu_1 \ell/2}\sigma^{\mu_2 \ell/2}\tau^{\mu_3 \ell/2} = 0,$$

where $\mu_j = \lambda'_j + 2\lambda_j$, $1 \leq j \leq 3$, $p(\mu) = p(\mu_1, \mu_2, \mu_3)$ and

$$\gamma'_1 = \mu_1 + (r/t)\mu_3, \qquad \gamma'_2 = \mu_2 + (s/t)\mu_3;$$

it is understood that λ_1, λ_2, λ_3, are allowed to run through all integers compatible with the ranges of μ_1, μ_2, μ_3. The above equation gives

$$\sum_{\lambda_1=0}^{L'_1} \sum_{\lambda_2=0}^{L'_2} \sum_{\lambda_3=0}^{L'_3} p'(\lambda)\Delta(\gamma'_1; m_1)\Delta(\gamma'_2; m_2)\rho^{\lambda_1 \ell}\sigma^{\lambda_2 \ell}\tau^{\lambda_3 \ell} = 0$$

where $L'_j = [(L_j - \lambda'_j)/2]$, $1 \leq j \leq 3$. The coefficients $p'(\lambda) = p'(\lambda_1, \lambda_2, \lambda_3)$ are a subset of the original $p(\lambda)$ and we can suppose that λ'_1, λ'_2, λ'_3 are chosen such that the $p'(\lambda)$ are not all 0. Furthermore it is clear that $\Delta(\gamma'_1; m_1)$ and $\Delta(\gamma'_2; m_2)$ are polynomials in γ_1 and γ_2 with degrees m_1 and m_2 and with coefficients independent of the λ's. Hence, arguing by induction with respect to $m_1 + m_2$, we see that they can be replaced by $\Delta(\gamma_1; m_1)$ and $\Delta(\gamma_2; m_2)$. Thus we have shown that there is a function

$$g^{(1)}(z) = \sum_{\lambda_1=0}^{L'_1} \sum_{\lambda_2=0}^{L'_2} \sum_{\lambda_3=0}^{L'_3} p'(\lambda)\Delta(\gamma_1; m_1)\Delta(\gamma_2; m_2)\rho^{\lambda_1 z}\sigma^{\lambda_2 z}\tau^{\lambda_3 z}$$

such that $g^{(1)}(\ell) = 0$ for all odd integers ℓ with $1 \leq \ell < 4h$ and all non-negative integers m_1, m_2 with $m_1 + m_2 \leq L/2$; and here we have $L'_j \leq L_j/2$, $1 \leq j \leq 3$.

The argument can now be repeated by induction and we deduce that for each integer $J = 0, 1, \ldots$ there exist integers $p^{(J)}(\lambda)$, not all 0, given by a subset of the original $p(\lambda)$, such that the function

$$g^{(J)}(z) = \sum_{\lambda_1=0}^{L_1^{(J)}} \sum_{\lambda_2=0}^{L_2^{(J)}} \sum_{\lambda_3=0}^{L_3^{(J)}} p^{(J)}(\lambda)\Delta(\gamma_1; m_1)\Delta(\gamma_2; m_2)\rho^{\lambda_1 z}\sigma^{\lambda_2 z}\tau^{\lambda_3 z}$$

satisfies $g^{(J)}(\ell) = 0$ for all odd integers ℓ with $1 \leq \ell < 2^{J+1}h$ and all non-negative integers m_1, m_2 with $m_1 + m_2 \leq (1/2)^J L$; and we have

$$L_j^{(J)} \leq (1/2)^J L_j, \qquad 1 \leq j \leq 3.$$

But when J is large enough it follows that $L_j = 0$, $1 \leq j \leq 3$, and since then $p^{(J)}(0) \neq 0$, we plainly have a contradiction. This proves the theorems.

References

[1] A. Baker, Rational approximations to certain algebraic numbers, *Proc. London Math. Soc.* **4** (1964), 385–398.

[2] A. Baker, Rational approximations to $\sqrt[3]{2}$ and other algebraic numbers, *Quart J. Math. Oxford*, **15** (1964), 375–383.

[3] A. Baker, Contributions to the theory of Diophantine equations I: On the representation of integers by binary forms, *Phil. Trans. Royal Soc.* **A 263** (1968), 173–191.

[4] A. Baker, A sharpening of the bounds for linear forms in logarithms I, II, III *Acta. Arith.* **21** (1972), 117–129; **24** (1973), 33–36; **27** (1975), 247–252.

[5] A. Baker, The theory of linear forms in logarithms. *Transcendence theory: advances and applications*, edited by A. Baker and D. W. Masser, Academic Press, 1977, pp. 1–27.

[6] A. Baker and H. Davenport, The equations $3x^2 - 2 = y^2$ and $8x^2 - 7 = z^2$, *Quart. J. Math. Oxford* **90** (1969), 129–137.

[7] A. Baker and H. M. Stark, On a fundamental inequality in number theory, *Ann. Math.* **94** (1971), 190–199.

[8] E. Bombieri, On the Thue-Siegel-Dyson theorem, *Acta. Math.* **148** (1982), 255–296.

[9] E. Bombieri and J. Mueller, On effective measures of irrationality for $\sqrt[r]{(a/b)}$ and related numbers, *J. reine angew. Math.* **342** (1983), 173–196.

[10] J. W. S. Cassels, The rational solutions of the Diophantine equation $Y^2 = X^3 - D$, *Acta Math.* **82** (1950), 243–273.

[11] G. V. Chudnovsky, On the method of Thue-Siegel, *Ann. Math.* **117** (1983), 325–382.

[12] E. Dobrowolski, On a question of Lehmer and the number of irreducible factors of a polynomial, *Acta. Arith.* **34** (1979), 391–401.

[13] D. Easton, Effective irrationality measures for certain algebraic numbers, *Math. Comp.*, **46** (1986), 613–622.

[14] N. I. Feldman, An effective refinement of the exponent in Liouville's theorem (in Russian), *Izv. Akad. Nauk. SSSR*, **35** (1971), 973–990.

[15] K. Győry and Z. Z. Papp, Norm form equations and explicit lower bounds for linear forms with algebraic coefficients, *Studies in Pure Mathematics* (to the memory of Paul Turán) (Budapest, 1983), pp. 245–257.

[16] W. Ljunggren, On an improvement of a theorem of T. Nagell concerning the Diophantine equation $Ax^3 + By^3 = C$, *Math. Scand.* **1** (1953), 297–309.

[17] T. Nagell, Solution complète de quelques équations cubiques à deux indéterminées, *J. de Math.*, **4** (1925), 209–270.

[18] C. L. Siegel, Abschätzung von Einheiten, *Nachrichten Akad. Wiss. Göttingen*, Math-phys. Klasse 1969, No. 9, 71–86.

2

APPLICATIONS OF MEASURE THEORY
AND HAUSDORFF DIMENSION TO THE THEORY OF
DIOPHANTINE APPROXIMATION

V. I. Bernik

In this survey we touch upon results obtained mainly in recent years in questions of the application of the concepts of measure and Hausdorff dimensionality in the theory of Diophantine approximation on manifolds. The monographs of Sprindzhuk [25, 27] and his survey [28] summarize the research completed up to 1979.

A decisive impact on the formation of this theory was made by research done in the 1960s by Sprindzhuk [24–26], Schmidt [39], and Baker and Schmidt [31].

Let $\psi(x)$ be a monotone decreasing function, and $f_1(x)$, ..., $f_n(x)$ be real-valued functions of x. Define $\mathcal{L}_n(\psi, f_1, \ldots, f_n)$ as the set of $x \in \mathbf{R}$ for which the inequality

$$|a_n f_n(x) + \ldots + a_1 f_1(x) + a_0| < \psi(H) \tag{1}$$

has an infinite number of solutions in integral vectors $\bar{a} = (a_n, a_{n-1}, \ldots, a_1, a_0)$, $H = \max\limits_{1 \leq i \leq n} |a_i|$. Denote the Lebesgue measure and Hausdorff dimension of the set A by μA and $\dim A$.

Sprindzhuk's Theorem. For $w > n$ we have

$$\mu \mathcal{L}_n(H^{-w}, x, \ldots, x^n) = 0. \tag{2}$$

Schmidt's Theorem. If $w > 2$, $f_1(x)$, $f_2(x)$ are three times continuously differentiable functions and the curvature of the curve

$$\Gamma = \big(f_1(x), f_2(x)\big) \tag{3}$$

is non-zero almost everywhere, then $\mu \mathcal{L}_2\big(H^{-w}, f_1(x), f_2(x)\big) = 0$.

The Baker-Schmidt Theorem. For $w > n$

$$\frac{n+1}{w+1} \leq \dim \mathcal{L}_n(H^{-w}, x, \ldots, x^n) \leq 2\frac{n+1}{w+1} \tag{4}$$

§1 The method of essential and inessential domains

This method was introduced by Sprindzhuk in the proof of Mahler's conjecture (2), posed in 1932 on the classification of real and complex numbers. The application and development of the method facilitated the proof of several quite deep results.

Theorem 1. [5]. Let $P(x) \in \mathbf{Z}[x]$, and let $w_n(\overline{\omega})$ be the least upper bound of those $w > 0$ for which the inequality

$$\prod_{i=1}^{k} |P(\omega_i)| < H(P)^{-w} \tag{5}$$

has an infinite number of solutions. Then for almost all $\overline{\omega} = (\omega_1, \ldots, \omega_k)$

$$w_n(\overline{\omega}) = n - k + 1.$$

Theorem 1 proves in a more general formulation conjecture C of Sprindzhuk [24]. The fundamental idea of the proof consists in the introduction of a classification of polynomials, connected with the mutual arrangement of the roots of the polynomials $P(x)$ in the complex plane and with the order of the approximations ω_i corresponding to a root of $P(x)$.

In [12] Zheludevich significantly generalised Theorem 1 and proved a further conjecture of Sprindzhuk [28]. Let k, ℓ, m, n, $n_1, n_2, \ldots, n_m$, $M = k + \ell + n_1 + \ldots + n_m$ be integral non-negative numbers such that

$$\max(k + 2\ell, n_1, \ldots, n_m) \leq n,$$

and let $p_1, \ldots, p_m$ be distinct primes. Consider the direct product $\Omega = \prod_{v \in V} K_v$ of fields K_v with valuations $|\cdot|_v$, $v \in V = \{1, 2, \ldots, M\}$, as follows:

$$K_v = \mathbf{R}, \qquad |\cdot|_v = |\cdot| \qquad (1 \leq v \leq k)$$

$$K_v = \mathbf{C}, \qquad |\cdot|_v = |\cdot|^2 \qquad (k+1 \leq v \leq k+\ell)$$

$$K_v = \mathbf{Q}_{p_1}, \qquad |\cdot|_v = |\cdot|_{p_1} \qquad (k+\ell+1 \leq v \leq k+\ell+n_1)$$

$$\ldots \ldots$$

$$K_v = \mathbf{Q}_{p_m}, \qquad |\cdot|_v = |\cdot|_{p_m} \qquad (M - n_m + 1 \leq v \leq M)$$

In the space Ω, considered as a product of spaces, we introduce in the natural way the measure $\mu = \prod_{v \in V} \mu_v$, where μ_v is the Haar measure on K_v.

Theorem 2. If $w_n(\overline{\omega})$ is the least upper bound for those $w > 0$ for which the inequality

$$\prod_{v \in V} |P(\omega_v)|_v < H(P)^{-w}, \qquad H(P) = \max_{1 \leq i \leq n} |a_i|$$

has an infinite number of solutions in integral polynomials

$$P(x) = a_n x^n + \ldots + a_1 x + a_0, \qquad \deg P \leq n,$$

then $w_n(\overline{\omega}) = n + 1 - k - 2\ell$ for almost all $\overline{\omega} \in \Omega$ (inthe sense of μ).

Making use of Theorem 2, Zheludevich [11] obtained results about the approximation by coordinates of almost all points Ω by means of algebraic elements of bounded degree.

Sprindzhuk's Theorem was made more precise by Baker [30], who proved that if $\sum_{q=1}^{\infty} \psi(q) < \infty$, then

$$\mu \mathcal{L}_n \big(\psi^n(H), x, \ldots, x^n \big) = 0. \qquad (6)$$

His conjecture [30] consisted in the possible replacement in (6) of $\psi^n(H)$ by $H^{-n+1} \psi(H)$

Theorem 3. [8]. $\mu \mathcal{L}_n \big(H^{-n+1} \psi(H), x, \ldots, x^n \big) = 0.$

The essential point in the proof of Theorem 3 is the avoidance of mathematical induction in the method of essential and inessential domains used in [24] and [30]. In each class of polynomials, the degree of two polynomials taking small values on a fairly long interval is reduced maximally, so far as is possible. The following Lemma also plays an important rôle.

Lemma 1 [7]. Let $\delta > 0$ be a real number, s be a natural number, and $H = H(\delta, s)$ be a sufficiently large natural number. Further, let $P(x)$, $Q(x) \in \mathbb{Z}[x]$ be two coprime polynomials, and

$$\max\big(H(P), H(Q) \big) = H^\mu, \qquad \deg P(x) \leq s, \qquad \deg Q(x) \leq s.$$

If for all ω from some interval $I = [-s, s]$, $\mu I = H^{-\eta}$, $\eta > 0$, the following inequality is satisfied

$$\max(|P(\omega)|, |Q(\omega)|) < H^{-\tau}, \qquad \tau > 0,$$

then
$$\tau + \mu + 2\max(\tau + \mu - \eta, 0) < 2\mu s + \delta.$$

Lemma 1 has by now been generalized in various directions in connection with the study of inequalities (1) in the field of complex numbers, in the field of p-adic numbers, etc. One of its possible strengthenings is contained in [6].

The following conjecture of Sprindzhuk [24, 25] is very interesting and up to now has not been proved: to generalize Mahler's conjecture to polynomials in several variables.

Up to the present, the conjecture has in practice been proved for polynomials of degree at most four [23] and for polynomials of an arbitrary degree but of a special type [2], [40]. If we look at (7) for $(x, y) \in Q_p^2$, then it is possible to prove a stronger result. Thus Yanchenko [29], making use of a technique of differential algebra developed by Nesterenko [29], showed that the inequality

$$\left|P(x, y)\right|_p < H^{-cn^2}$$

has only a finite number of solutions for sufficiently large c for almost all $(x, y) \in Q_p^2$.

The question about the study of (1) for integral algebraic numbers $(a_0, a_1, \ldots, a_n)$ of a certain finite extension K of the field of rational numbers was also discussed in [25]. Recent results of Golubevaya (K imaginary quadratic) and Markovich [14, 15] (second degree polynomials) allow us to hope that Sprindzhuk's Theorem may be generalized in this direction.

§2. An application of bounds of trigonometric sums and integrals

A systematic application of the method of trigonometric sums in problems of the metric theory of numbers was begun by Sprindzhuk [25, 26], although earlier Kubilius applied it to specific problems. We mention some new results not given in [27].

Theorem 4. Let $\sum_{q=1}^{\infty} \psi^2(q) < \infty$. Then for almost all α the inequality

$$\max(\|\alpha q\|, \|\alpha^2 q\|) < \psi(q) \tag{8}$$

has only a finite number of solutions.

With the help of the method explained in [1], Theorem 4 reduces to the estimation of a trigonometric integral. The individual estimates of the integrals do not turn out good enough for the proof of the theorem. Therefore, with the help of Abel's transform, an averaging of these estimates is constructed.

Yu Kunrui [36] considered a more general inequality than (8), namely

$$\|\alpha q\| \|\alpha^2 q\| < \varphi(q), \tag{9}$$

making use of a transference theorem. He solved the dual problem to (9) with the help of the summation of the discriminants of quadratic polynomials. In this paper he also proved several complete metric results on second degree polynomials. Mashanov [16] introduced a variation into [3] and proved the following theorem.

Theorem 6. Let $\varphi(q)$ be monotone decreasing and $\sum_{q=1}^{\infty} \varphi(q) \ln q < \infty$. Then for almost all α the inequality (9) has only a finite number of solutions.

Kovalevskaya [13] considered a more general inequality, analogous to [9]. Let the surface S be given by the inequality $\tau = f(x, y)$, where $f(x, y)$ is a three times continuously differentiable function in $\mathbf{R}^2$.

Theorem 7. For any $\epsilon > 0$, for almost all $(x, y) \in \mathbf{R}^2$, the inequality

$$\|xq\| \|yq\| \|f(x, y)q\| < q^{-1-\epsilon}$$

has only finitely many solutions in the following cases:

1) $f''_{xx}(x, y) \neq 0$ (or $f''_{yy}(x, y) \neq 0$) almost everywhere in $\mathbf{R}^2$, $f''_{xy}(x, y) \neq 0$ almost everywhere in $\mathbf{R}^2$.

2) One of the conditions in 1) is not fulfilled, but the Gaussian curvature is almost everywhere different from zero.

3) $f(x, y) = h(x)$ (or $f(x, y) = g(y)$), $h''(x) \neq 0$, $(g''(y) \neq 0)$ almost everywhere in $\mathbf{R}$.

§3. **Rational points near smooth curves.** The method of essential and inessential domains allows us to reduce the study of inequality (1) to the study of the system of inequalities

$$\left| a_n f_n(x) + \ldots + a_1 f_1(x) + a_0 \right| < \psi(H)$$
$$\left| a_n f'_n(x) + \ldots + a_1 f'_1(x) \right| < H^{-\lambda} \tag{10}$$

where $\lambda > 0$. In spite of the fact that an additional condition has appeared, to analyze the system (10) is practically as hard as inequality (1). The case $f_1(x) = x, \ldots, f_n(x) = x^n$ of (10) is considered in [24, 25] in the solution of Mahler's problem (classes of the second kind). In [27] a new method was applied to (10) which allows us to prove Schmidt's Theorem and transfer it to the field of p-adic numbers [17]. Schmidt based the proof on the following arithmetic lemma [38].

Lemma 2. Let $0 < \lambda_1 < \lambda_2$ and for a natural number h let G_h be the domain in $\mathbf{R}^2$ defined by the inequalities $1 \leq x \leq h$, $\lambda_1 x \leq y \leq \lambda_2 x$. Let further $g(x, y)$ be a real-closed function defined for $x > 0$, $\lambda_1 x \leq y \leq \lambda_2 x$, homogeneous of degree 1 such that

$$\frac{\partial}{\partial y} g(x, y) = \ell(y/x),$$

where the function $\ell(t)$ is defined in the interval $\lambda_1 \leq t \leq \lambda_2$, twice continuously differentiable, and with derivative $\ell'(t)$ not vanishing in this interval.

For ν in the interval $1 < \nu < 3/2$ denote by $N_\nu(h)$ be the number of integers $(x, y) \in G_h$, for which

$$\|g(x, y)\| < x^{-\nu}. \tag{11}$$

Then we have

$$N_\nu(h) \ll h^{3/2 + (3/2 - \nu)/7}.$$

R. Baker [35] made Lemma 2 more precise and widened the range of ν in (11). This allowed a significant advance in the study of metric properties of plane curves.

Theorem 8. Let $\sum_{n=1}^\infty \psi(n) < \infty$. Then

$$\mu\big(\psi^2(H), f_1(x), f_2(x)\big) = 0.$$

Theorem 9. For $w > 2$

$$\dim \mathcal{L}_2\big(H^{-w}, f_1(x), f_2(x)\big) = \frac{3}{w+1}.$$

§4. Hausdorff dimension.

Obtaining upper bounds for Hausdorff dimension is connected with the economical covering of the sets under investigation by a system of intervals. Therefore all the methods §1–3 allow us to give nontrivial upper bounds for Hausdorff dimension. In [31] a useful method in the theory of Diophantine approximation is given to obtain lower bounds. For this it is necessary to prove the regularity of the set of zeros of the function $F(x) = a_n f_n(x) + \ldots + a_1 f_1(x) + a_0$. By this means, for example, a lower bound is obtained for the dimension in Theorem 9. In the papers [34] and [37], the concept of a regular system is used.

Theorem 10 [11]. Let M_ν be a set, $0 \leq x \leq 1$, for which the system of inequalities

$$\max\left(\|xq\|, \|\sqrt{1-x^2}\,q\|\right) < q^{-v}$$

has an infinite number of solutions in natural numbers. Then for $v > 1$

$$\frac{1}{2(v+1)} \leq \dim M_v \leq \frac{1}{v+1}. \tag{12}$$

The lower bounds in (12) are obtained by constructing a regular system, and the upper bounds by the method of trigonometric sums with the use of the distribution of rational points on a circle.

Theorem 11. ([4, 7], the Baker-Schmidt conjecture. For $w > n$

$$\dim \mathcal{L}_n(H^{-w}, x, \ldots, x^n) = \frac{n+1}{w+1}. \tag{13}$$

The lower bound in (13) is taken from the theorem of Baker-Schmidt. In estimating from above, for the first time nontrivial upper bounds depending on λ in (10) were obtained for the number of irreducible polynomials of bounded degree and height which have values of the derivative at a root not exceeding a given value.

A p-adic generalization of the Theorem of Jarník-Besicovitch (Melnichuk) and Baker-Schmidt was obtained.

Theorem 12. [9]. Let A_w be the set of $w \in \mathbb{Q}_p$ for which the inequality

$$|P(w)|_p < H(P)^{-w}$$

has an infinite number of solutions in $P(x) \in \mathbf{Z}[x]$. Then for $w > n + 1$

$$\dim A_w \geq \frac{n+2}{w+1}.$$

For the construction of the regular system in Theorem 12, the following was used.

Theorem 13 [10]. Let $\delta > 0$ and B be the set of $w \in \mathbf{Q}_p$ for which the system of inequalities

$$\left|P(w)\right|_p < H(P)^{-n-1}$$

$$\left|P'(w)\right|_p < H(P)^{-\delta}$$

has an infinite number of solutions in $P(x) \in \mathbf{Z}[x]$. Then the Haar measure of the set B is zero.

§5. Conjectures.

It seems to us that the conjecture of Sprindzhuk [24, 25, 27, 28] and Baker [30] in the metric theory of Diophantine approximation on manifolds are quite complicated and require new deep methods for their solution. Solutions of the following conjectures may possibly result from a synthesis of known methods.

Conjecture 1. To prove that for some $0 < \gamma < 1$ the set $\mathcal{L}_n(H^{-n} \ln^{-\gamma} H, x, \ldots, x^n)$ contains almost all points x.

Along the lines of Conjecture 1 there exist the results of Khinchine and R. Baker [33].

Conjecture 2. Under sufficiently general analytic assumptions about the functions $f_1(x), \ldots, f_n(x)$, to prove that for some $w < n^\lambda$, $\lambda < 2$, we have $\mu \mathcal{L}_n(H^{-w}, f_1(x), \ldots, f_n(x)) = 0$.

At present only results with an exponent w of order n^2 are known.

Conjecture 3. Is the condition about the monotonicity of the function $\psi(x)$ in Theorem 4 essential, or can it be dropped?

Conjecture 4. If $\sum_{q=1}^{\infty} \psi^2(q) = \infty$, then is inequality (8) satisfied for almost all α infinitely often?

Problems analogous to Conjecture 3 have still been solved only for linear approximations [27], and Conjecture 4 for linear approximations and specially constructed surfaces [27].

In conclusion we note that the results of the metric theory of Diophantine approximation on manifolds have applications in a number of problems of mathematical physics [20], [21, 22].

References

[1] V. I. Bernik, E. I. Kovalevskaya, Extremal problems of certain surfaces in n-dimensional Euclidean space, *Mat. Zametki*, 1974, **15**, No. 2, pp. 247–254.

[2] V. I. Bernik, Induced extremal surfaces, *Mat. Sbornik*, 1977, vol. 103/145, No. 4(8), pp. 480–489.

[3] V. I. Bernik, On the exact order of the approximation of almost all points of a parabola, *Mat. Zametki*, 1979, **26**, Issue 5, pp. 657–665.

[4] V. I. Bernik, On the conjecture of Baker-Schmidt, *Dokl. Akad. Nauk BSSR*, 1979, **23**, No. 5, pp. 392–395.

[5] V. I. Bernik, A metric theorem on the simultaneous approximation of zero by the values of integral polynomials, *Izv. Akad. Nauk SSSR*, Ser. Mat., 1980, **44**, No. 1, pp. 24–45.

[6] V. I. Bernik, F. F. Zheludevich, On integral polynomials taking small values in a certain interval, *Izv. Akad. Nauk BSSR*, Ser. Fiz-Mat, 1981, No. 3, pp. 27–33.

[7] V. I. Bernik, An application of Hausdorff dimension in the theory of Diophantine approximation, *Acta Arithmetica*, 1983, **42**, pp. 219–253.

[8] V. I. Bernik, A proof of Baker's conjecture in the theory of transcendental numbers, *Dokl. Akad. Nauk SSSR*, **277**, No 5 1984, pp. 1036–1039.

[9] V. I. Bernik, I. L. Morotskaya, Diophantine approximation in Q_p and Hausdorff dimension, *Izv. Akad. Nauk. BSSR*, Ser. Fiz-Mat, 1986, No. 3, pp. 3–9.

[10] V. I. Bernik, Integral polynomials realizing Minkowski's theorem about linear forms, *Dokl. Akad. Nauk BSSR*, 1986, **30**, No. 5, pp. 403–405.

[11] F. F. Zheludevich, On the simultaneous approximation of numbers by means of the roots of integral polynomials, *Izv. Akad. BSSR*, 1984, No. 2, pp. 14–20.

[12] F. F. Želudevič, Simultane diophantische Approximationen abhängiger Größen in mehreren Metriken, *Acta Arithmetica*, 1986, No. 3 pp. 87–98.

[13] E. I. Kovalevskaya, Strong extremality of surfaces consisting of regular points, *Dokl. Akad. Nauk BSSR*, 1986, **30**, No. 8, pp. 692–695.

[14] N. I. Markovich, Extremality of one manifold in $\mathbf{R}^3$, *Izv. Akad. Nauk BSSR*, Ser. Fiz-Mat, 1986, No. 3, pp. 18–21.

[15] N. I. Markovich, On the approximation of zero by the values of quadratic polynomials, *Izv. Akad. Nauk BSSR*, Ser. Fiz-Mat, 1986, No. 4, pp. 18–22.

[16] V. I. Mashanov, Baker's problem in the metric theory of Diophantine approximation, *Izv. Akad. Nauk BSSR*, Ser. Fiz-Mat 1987, No. 1, 34–38.

[17] Yu. V. Melnichuk, Diophantine approximations on curves and Hausdorff dimension, *Dokl. Akad. Nauk Ukr. SSR*, Ser. A, 1978, No. 9, pp. 793–796.

[18] Yu. V. Melnichuk, Diophantine approximations in a neighbourhood and Hausdorff dimension, *Mat. Zametki*, 1979, **26**, No. 3, pp. 347–354.

[19] Yu. V. Nesterenko, The measure of algebraic independence of almost all pairs of p-adic numbers, *Mat. Zametki*, 1984, **36**, No. 3, pp. 295–304.

[20] B. I. Ptashnik, Improper boundary problems for differential equations with partial derivatives, Kiev: *Naukova Dumka*, 1984, 264pp.

[21] M. M. Skriganov, Finiteness of the number of gaps in the spectrum of multidimensional polyharmonic operators with periodic potential, *Mat. Sbornik*, 1980, Vyp. 1, pp. 133–145.

[22] M. M. Skriganov, General properties of the spectrum of differential and pseudodifferential operators with periodic coefficients and some problems of geometric numbers, *Dokl. Akad. Nauk SSSR*, 1981, **256**, No. 1, pp. 47–51.

[23] R. Y. Slesoraitene, An analogue of the theorem of Mahler-Sprindzhuk for polynomials of the fourth degree in two variables,

theses of papers of the all-union conference 'The theory of numbers and its applications', Tbilisi, 1985, pp. 232–233.

[24] V. G. Sprindzhuk, Proof of Mahler's conjecture on the size of the set of S-numbers, *Izv. Akad. Nauk SSSR*, Ser. Mat., 1965, **29**, No. 2, pp. 379–436.

[25] V. G. Sprindzhuk, Mahler's problem in the metric theory of numbers, Minsk: *Nauka i Technika*, 1967, 194.

[26] V. G. Sprindzhuk, The method of trigonometic sums in the metric theory of Diophantine approximtion of dependent values, *Proceedings of the Mathematical Steklov Institute*, 1972, **128**, No. 2, pp. 212–228.

[27] V. G. Sprindzhuk, The metric theory of Diophantine approximation, Moscow: *Nauka*, 144pp.

[28] V. G. Sprindzhuk, Achievements and problems in the theory of Diophantine approximation, *Uspekhi Mat. Nauk*, 1980, **35**, No. 4, pp. 3–68.

[29] A. Ya. Yanchenko, On the measure of mutual transcendence of almost all pairs of p-adic numbers, *Vestn. Mosk. Gos. Univ.* Ser. Mat., Mechanika, 1982, No. 3, pp. 62–65.

[30] A. Baker, On a theorem of Sprindzhuk, *Proc. Royal Soc.*, 1966, **A292**, No. 1428, pp. 92–104.

[31] A. Baker, W. M. Schmidt, Diophantine approximation and Hausdorff dimension, *Proc. London Math. Soc.* (3), 1970, **21**, pp. 1–11.

[32] R. C. Baker, Sprindzhuk's theorem and Hausdorff dimension, *Mathematica, 1976*, **23**, pp. 184–197.

[33] R. C. Baker, Metric diophantine approximation on manifolds, *J. London Math. Soc.*, 1976, **14**, pp. 43–48.

[34] R. C. Baker, Singular n-tuples and Hausdorff dimension, *Math. Proc. Camb. Phil. Soc.*, **1977, 81**, pp. 377–385.

[35] R. C. Baker, Dirichlet's theorem on diophantine approximation, *Math. Proc. Cambridge Phil. Soc.*, 1978, **83**, pp. 37–59.

[36] Yu Kunrui [= Kunrui Yu], A note on a problem of Baker in metrical number theory, *Math. Proc. Camb. Phil. Soc.*, 1981, **90**, 215–227.

[37] A. D. Pollington, The Hausdorff dimension of a set of non-normal well approximable numbers, *Lect. Notes Math.*, 1979, **751**, pp. 256–264.

[38] W. M. Schmidt, Über Gitterpunkte auf gewissen Flächen, *Monatsh. Math.*, 1964, Bd. 68, No. 1, pp. 59–74.

[39] W. M. Schmidt, Metrische Sätze über simultane Approximation abhänger Grossen, *Monats. Math.*, 1964, Bd. 68, No. 2, pp. 154–166.

[40] G. V. Chudnovsky, Contributions to the theory of transcendental numbers, *Math. surveys and monographs*, **19**, American Math. Soc., Providence, 1980, pp. 450.

3

GALOIS REPRESENTATIONS AND
TRANSCENDENTAL NUMBERS

D. Bertrand

§1. Introduction

Let G be a g-dimensional commutative algebraic group defined over a number field k, and embedded in a projective space $\mathbf{P}^N$. The concluding step of several transcendence proofs can be summarized as follows: given a hypersurface Z of degree D in $\mathbf{P}^N$, defined over k and not containing G, and a k-rational point P in $G \cap Z$, find a point P' in G not lying on Z, but still sufficiently close to Z; in particular, some interpolation process must link P' to P.

Two methods have been developed to deal with this problem. The first consists in searching for P' in the orbit of P under the action of the ring $\operatorname{End} G$ of endomorphisms of G. Suppose for instance that G is a (split) torus L, i.e. a power $\mathbf{G}_m^g$ of the multiplicative group $\mathbf{G}_m$, so that the point P is represented by a g-tuple $(\alpha_1, \ldots, \alpha_g)$ of non-zero elements of k, and that these numbers are multiplicatively independent — a situation which may serve as a guideline throughout this report. A Van der Monde determinant as in Chapter III of [6] then shows that in the natural embedding of G in $\mathbf{P}^g$, a multiple $P' = nP$ of P of order $n < \binom{D+g}{g}$ does not belong to Z. The recent zero estimates of transcendence theory generalize this result in an essentially optimal way: for example, the theorem of Masser and Wüstholz [51] provides a bound of the type $n \leq cD^g$, with a constant c depending only on G, if the smallest algebraic subgroup G_P of G containing P is G itself (this is the natural extension to an arbitrary group G of the hypothesis on $\alpha_1, \ldots, \alpha_g$ made above), and Philippon's theorem [53] sharpens this estimate when multiplicities in given directions are involved.

The second method amounts to considering a division point $P' = \frac{1}{n}P$ of P. In the study of abelian integrals, where it has proved specially useful, the corresponding interpolation process was introduced by Baker in [4]. The non-vanishing argument itself is due to Coates [24], and runs as follows. Suppose that g coordinates of P' generate linearly disjoint extensions of k of degrees $> c'D$ for some large constant c'. Then, the value at P' of a suitable defining equation of Z is an algebraic number of high degree, which cannot consequently be 0. In order to check the hypothesis made on P' (which, in our typical situation, is satisfied as soon as n is larger than a constant multiple of D), one is led to extend to G the classical Kummer theory on $\mathbf{G}_m$ and its powers. This has been achieved in a series of works, due mainly to Ribet, which we describe in the second part of this report.

Kummer theory led to important sharpenings in the study of linear forms in logarithms, both in the classical case and in the abelian one (see, in particular, [5], [7]; resp. [26], [44]). However, in view of the recent work of Wüstholz ([74], [75]) and of Philippon and Waldschmidt ([54], [55]) on Baker's method, the technique of zero estimates mentioned above seems to supersede most of these advances, and is certainly more efficient in the case of general groups G, so that it is now difficult to guess what influence Kummer theoretical methods will retain in transcendence theory. We therefore content ourselves in this article with a short list of references (end of §2) for such applications.

Relations between Kummer theory and transcendence, on the other hand, are not restricted to applications of the former to the latter. Reversing points of view, Lang suggested in [36] to prove Kummer theoretical results using transcendence methods (see also Masser [43]). In fact, Kummer theory can be viewed as a special version of Hilbert's irreducibility theorem, according to which generic Galois groups remain unchanged when specialized at certain points, and in this perspective, Siegel's work on G-functions had already shown the adequacy of transcendence methods to such problems. This range of ideas is illustrated in the third part of our report by two results, due respectively to Masser and D. and G. Chudnovsky, which concern the torsion points of abelian varieties and which have not (yet) been fully reached by algebraic techniques.

Even if no logical link is to remain between the two theories under review here, the similarity of form that their respective statements have now acquired is worth pointing out. In particular, the group G_P introduced above plays a fundamental rôle in both (compare, for instance, Theorem 2 and 9 below). In the last part of this report, we give a conjectural description of this group in terms of its degree as a subvariety of $\mathbf{P}^N$. As a matter of fact, G_P is a natural extension of the linear dependence relation group studied by Masser in [48]. The first description of this relation group was obtained by Baker [3] by means of a suitable modification of his transcendence method, and, although of a purely algebraic nature, our conjecture may well follow again from from a transcendence technique, as given, for instance, in [54]. Finally, we close the report with a further conjecture on the group G_P and some of its variations, this time in relation to questions of algebraic independence in logarithms.

The author wishes to thank M. Laurent and D. Masser for their help during the preparation of this paper.

§2. Kummer theory on algebraic groups

Let k be a number field, with absolute Galois group $\mathcal{G} = \mathrm{Gal}(\overline{k}/k)$, and let G be a connected commutative algebraic group, defined over k. For any positive integer n, the morphism of multiplication by n on $G(\overline{k})$ is surjective. Its kernel G_n is isomorphic to $(\mathbf{Z}/n\mathbf{Z})^b$, where b denotes the first Betti number of G: if G is a torus $\mathbf{G}_m^g$ (resp. an abelian variety), b (resp. $b/2$) is equal to the dimension of G. The group $\mathcal{G}$ acts on G_n through a representation

$$\rho_n : \mathcal{G} \to \mathrm{Aut}(V_n) = GL_b(\mathbf{Z}/n\mathbf{Z}),$$

whose image $\mathcal{G}_n = \mathcal{G}_n(G)$ is isomorphic to the Galois group over k of the field $k(G_n)$ generated by the points of order dividing n in $G(\overline{k})$, and whose kernel is $\mathrm{Gal}(\overline{k}/k(G_n))$.

Let now P be a point in $G(k)$. For any n-th division point $P_n = \frac{1}{n}P$ of P in $G(\overline{k})$, the map

$$\xi(P) : \sigma \to \sigma(P_n) - P_n$$

defines a homomorphism of $\mathrm{Ker}\,\rho_n$ in G_n, which does not depend on the choice of P_n, and whose image $\chi_n(P) = \chi_n(G, P)$ is isomorphic to the Galois group over $k(G_n)$ of the field $k(G_n, P_n)$ generated by the n-th division points in P. Kummer theory consists of describing this image, and is based on the following remarks. On the one hand, since G_n is an abelian group, $\mathcal{G}_n$ has a natural action on $\mathrm{Gal}(k(G_n, P_n)/k(G_n))$ (resp. on $\chi_n(P)$) by conjugation (resp. via ρ_n), and a straightforward computation shows that the map $\xi_n(P)$ is $\mathcal{G}_n$-equivalent. In particular, $\chi_n(P)$ is a $\mathcal{G}_n$-submodule of G_n. On the other hand, *let G_P be the smallest algebraic subgroup of G containing P* (i.e. the Zariski closure of the group $\mathbf{Z}P$ in G), let G_P^o be its connected component through the origin and denote by $\nu = \nu(P)$ the index of G_P^o in G_P. Since G_P and G_P^o are defined over k, the group $\chi_n(P)$ will lie in the n-th torsion subgroup $(G_P^o)_n$ of G_P^o as soon as n is prime to ν.

Suppose until further notice that G is the product of an abelian variety A by a torus L, so that after making a finite extension of k if necessary, it satisfies Poincaré's complete reducibilty theorem. Building upon previous work of Bashmakov [10], Tate [24] and himself ([58], [11]; see also [13]), Ribet proved in [60] that for all sufficiently large prime numbers ℓ, $\chi_\ell(P)$ is indeed as large as allowed by the constraints above –at least as soon as the representation ρ_ℓ satisfy certain certain axioms, which the work of Faltings [30] and Serre [66] eventually turned into theorems. His method naturally extends to the following theorem.

Theorem 1 (see Ribet [60]). *Let $G = A \times L$, and assume that the algebraic subgroup G_P associated to P is connected. There exists a positive number $c = c(G, P)$ such that for all positive integers n, the group $\mathrm{Gal}\big(k(G_n, P_n)/k(G_n)\big)$ is isomorphic to a subgroup $\chi_n(P)$ of $(G_P)_n$ of index bounded by c.*

In particular, on denoting by $b(P)$ the first Betti number of G_P^o, the degree of the extension $k(P_n)$ of k grows at least as a constant multiple of $n^{b(P)}$, and this remains true even if the assumption $\nu(P) = 1$ is dropped.

By standard topological arguments, Theorem 1, in a slightly sharper form, can be deduced from the case where n runs through the powers ℓ^s of a given prime number ℓ. Denote by $T_\ell(G)$ the Tate-module formed by the projective limit of the group G_{ℓ^s}. The maps ρ_{ℓ^s}, $\xi_{\ell^s}(P)$ give rise to continuous homomorphisms

$$\rho_{(\ell)} : \mathcal{G} \to \rho_{(\ell)}(\mathcal{G}_{(\ell)}) \subset \underset{\mathbf{Z}_\ell}{\mathrm{Aut}}\big(T_\ell(G)\big)$$

$$\xi_{(\ell)}(P); \mathrm{Ker}\, \rho_{(\ell)} \to \xi_{(\ell)}(P)(\mathrm{Ker}\, \rho_{(\ell)}) = \chi_{(\ell)}(P) \subset T_\ell(G).$$

The latter image, which is related to the affine Tate-module introduced by Bogomolov in [21], is the Galois group of the field of ℓ^∞-division points of P over the field generated by the ℓ-primary part $G_{(\ell)} = \bigcup_s G_{\ell^s}$ of $G(\overline{k})$. It is again a $\mathcal{G}_{(\ell)}$-submodule of $T_\ell(G)$, admitting an open subgroup contained in the Tate-module of the connected algebraic subgroup G_P^o, and in fact we have:

Theorem 2. *Let $G = A \times L$, and let P be a point in $G(k)$. For all prime numbers ℓ, the group $\chi_{(\ell)}(P)$ is commensurable with $T_\ell(G_P^o)$, and coincides with $T_\ell(G_P^o)$ when ℓ is sufficiently large.*

The proof of Theorem 2 follows Ribet's method [60] (see also [37], [38]): since the representations $\rho_{(\ell)}$, ρ_ℓ are semi-simple and satisfy Tate's conjecture [30], and in view of the finiteness or vanishing of the cohomology groups $H^1\big(\mathcal{G}_{(\ell)}, T_\ell(G)\big)$ ([62], [21], [66]), one is reduced to showing that an endomorphism α of G_P sends P to a highly divisible point in $G_P(k)$ if and only if α itself is highly divisible in $\text{End}\, G_P$; in practice, this will mean that $\alpha = 0$. An effective version of this statement requires a precise description of the groups G_P, and we return to this point in Theorem 8 below.

In contrast to these results, the Galois group $\mathcal{G}_{(\ell)}$ of the fields of ℓ^∞-torsion points are not yet known in full generality. By cyclotomic theory, $\mathcal{G}_{(\ell)}$ is an open subgroup of $\mathbf{Z}_\ell^*$ (of index 1 for large ℓ) when G is a split torus. When G is an abelian variety A, every algebraic subvariety in its different powers provides a constraint on $\mathcal{G}_{(\ell)}$. For instance, fix a k-rational ample divisor Y on A, and denote by φ_Y the associated isogeny from A to its dual $\hat{A}$ (the restrictions imposed by other polarizations would follow from those given by the graphs in A^2 of the endomorphisms of A). Then φ_Y endows $T_\ell(A)$ with an alternating form, and $\mathcal{G}_{(\ell)}$ is contained in the corresponding group $GSp(\mathbf{Z}_l)$ of symplectic similitudes. More generally, the set of all Hodge cycles on the powers of A imposes constraints on $\mathcal{G}_{(\ell)}$, and an open subgroup of $\mathcal{G}_{(\ell)}$ is contained in the group of $\mathbf{Q}_\ell$-points of the corresponding Mumford-Tate group $\mathcal{G}^H(A)$ of A (see [29], [64]). For later comparisons, we record the following properties of $\mathcal{G}_{(\ell)}$, extending Serre's classical results [63] on elliptic curves.

Theorem 3 (Serre [66]). *Let $G = A$, and denote prime numbers by ℓ. Then*

   i) *$\mathcal{G}_{(\ell)}$ contains an open subgroup of the group $\mathbf{Z}_\ell^*$ of homotheties on $T_\ell(A)$ (Bogomolov [21]), of index $c(\ell)$ bounded independently of ℓ.*

   ii) *there exists a finite extension k' of k depending only on A and k such that the extensions $k'(\mathcal{G}_{(\ell)})$ are linearly disjoint over k'.*

   iii) *assume that the ring $\text{End}\, A$ of $\bar{k}$-endomorphisms of A is reduced to $\mathbf{Z}$, and that A is an abelian surface, or has odd dimension. Then, $\mathcal{G}_{(\ell)}$ is open in $GSp(\mathbf{Z}_\ell)$, and coincides with $GSp(\mathbf{Z}_\ell)$ when ℓ is sufficiently large.*

We now turn to the case of a general connected commutative algebraic group G. Since extensions by additive groups do not affect the Galois groups studied here, we can suppose that G is an extension of an abelian variety A by a torus $L = \mathbf{G}_m^r$. Assume for simplicity that A admits a principal polarization φ_Y, so that the class of G in $\text{Ext}(A, L)$ is parameterized by a point Q in A^r. Then, the field $k(G_n)$ generated by the n-torsion points of G coincides with the field $k(A_n, \frac{1}{n}Q)$ (see [60]), and $\mathcal{G}_n(G)$ is an extension of $\mathcal{G}_n(A)$ by the group $\chi_n(A^r, Q)$ studied in

Theorem 1. Its properties, therefore, easily follow from those of $\mathcal{G}_n(A)$, as given by Theorem 3 (see [33] in the case of elliptic curves). As for the division points of an element P in $G(k)$, Ribet's Theorem 4.3 in [60] implies that Theorem 1 still holds on G when the projection $\pi(P)$ of P on A and the coordinates $Q_1, \ldots, Q_r$ of Q in A^r are linearly independent over End A. Obvious discrepancies between $\chi_n(G, P)$ and $(G_P^0)_n$, however, follow from the description of $k(G_n)$ above when this independence condition is not satisfied (see [33]). We refer to [34], [61] for a subtler type of degeneracy, related to the theory of 1-motives, when the linear relation linking $\pi(P)$, $Q_1, \ldots, Q_r$ involve endomorphisms of A which are antisymmetric with respect to φ_Y: for instance, if $r = 1$ and A is an elliptic curve with complex multiplications by $\mathbf{Q}(\sqrt{-D})$, the Galois groups $\chi_n(G, P)$ are trivial for some choices of P in the fiber above each multiple of the point $\sqrt{-D} \cdot Q$. By analogy with the groups G_P and $\mathcal{G}^H(A)$, it would be interesting to analyse which algebraic group $\mathcal{G}_P(G)$, if any, governs such degeneracies.

We close this section with a list of applications of Theorems 2 and 3 (or, rather, their previous versions) to transcendence theory. In the case of a torus, Theorem 2 is of course classical, and was applied extensively in the theory of linear forms in logarithms: see in particular [68], [8], [9] for division points of prime order, [5], [69], [7], [71] for an ℓ-adic type argument. Applications of Theorem 2 to the theory of abelian integrals can be found in [24], [26], [11], [1], [18], [4]. Theorem 3, in the case of elliptic curves, and its analogue for abelian varieties of CM type, was used in [25], [45] for torsion points of prime order (see [12] for an ℓ-adic type argument). One of the main drawbacks of these theorems for such applications is their ineffectiveness with respect to the abelian variety A. We return to this point in the next section of this report, and refer to [13] for a detailed discussion in the case of elliptic curves.

§3. Torsion points and transcendence methods

We now restrict ourselves to the case when G is a g-dimensional abelian variety A, defined over the number field k, and describe two applications of transcendence methods to the study of the Galois group $\mathcal{G}_n = \mathrm{Gal}(k(A_n)/k)$ over k of the field $k(A_n)$ of n-torsion points on A.

The first is due to Masser [49], and deals with the cardinality of the orbits of $\mathcal{G}_n$. It relies in fact, as in [46], on an estimate for the points of small heights on A, from which we extract the following corollary, extending P. Cohen's bound [27] in the case of elliptic curves, and sharpening those of [46] and [14] in the general case.

Theorem 4 (Masser [49]). *There exists a positive number $c_{A,k}$ such that for all non-zero torsion points e in $A(\bar{k})$ of order $n = n(e)$, the degree $d = d(e)$ over k of the field $k(e)$ of definition of e satisfies*

$$d > c_{A,k} n^{1/g} / \log n.$$

Moreover, $c_{A,k}$ is effectively computable in terms of the degree of k over $\mathbf{Q}$ and the height of a set of equations defining A in a suitable projective embedding.

Sharper bounds, of the type $d > c_{A,k,\epsilon} n^{1-\epsilon}$ for any positive number ϵ, of course follow from the first and second parts of Theorem 3 –and the exponent $1 - \epsilon$ can be strengthened to $2 - \epsilon$ when A has no factor of CM type [66]. The argument recalled in [14], p. 25 (and its analogue for abelian varieties with many real multiplications) shows that, in any dimension, each of these results is essentially best possible. In contrast to Theorem 4, however, the constants they involve are in general not effective, although an explicit expression for $c_{A,k,\epsilon}$, now independent of A, has been derived by A. Silverberg [67] when A is of CM type. We refer to [47], [28] (and [27] when $g = 1$) for a more intrinsic description of the constant $c_{A,k}$ in Theorem 4.

Masser's method of proof, which we now sketch in the situation of Theorem 4, relies on the zero estimates alluded to in the introduction. Reduce to the case where A is a simple abelian variety, and fix an embedding of $\bar{k}$ in $\mathbf{C}$, an invariant metric on $A(\mathbf{C})$, a basis $\mathcal{D}$ of of the k-space of invariant derivations on A, and g algebraically independent k-rational functions $x_1, \ldots, x_g$ on A, whose polar divisors avoid the finite subgroup A_e of $A(\bar{k})$ generated by e. Denote by $c_1, c_2, \ldots$ sufficiently large constants, and for $d > c_1$, put

$$L = c_2 d, \quad N = c_3 d^{1/2}, \quad T_0 = c_4 d^2, \quad T = c_5 d / \log d.$$

Using Waldschmidt's technique of [72], §3, Masser constructs a non-zero polynomial R in $2g$ variables, of partial degrees $\leq L$, with coefficients in $k(e)$ of height $\leq c_6 d$, such that the coefficients of order $\leq T_0$ of the Taylor expansion at 0 (relatively to $\mathcal{D}$) of the rational function

$$F = R(x_1, \ldots, x_g, x_1 \circ N, x_g \circ N)$$

have absolute values at most $\exp(-d/c_6)$. An interpolation process then implies that F vanishes with multiplicity at least T at all points of A_e whose distance to O is $< 1/c_7$, and F gives rise to a non-zero polynomial $\tilde{R}(x_1, \ldots, x_g)$ of degree $< c_9 L N^2$ vanishing on all A_e with multiplicity at least T. Since A is simple, the required upper bound for the cardinality n of A_e now follows from Philippon's multiplicity estimates [53], or alternatively, from the result of [73], combined with the following corollary of Masser and Wüstholz' zero estimates, where we denote by $\deg V$ the degree of a subvariety V in a given projective space.

Theorem 5 (see [51], [53], [39]). *Fix a projectively normal embedding of the simple abelian variety A, associated to a symmetric divisor H on A, and let D be a positive integer. Let further Z be an effective divisor on A, linearly equivalent to $D.(H)$. Then, any finite subgroup of A lying on Z has order at most $(\deg A).(2D)^g$.*

Question: Up to multiplication by $g!$, the last expression is the square root of the degree of the isogeny φ_Y associated to the divisor $Y = 2Z$ on A. Is there a more direct connection between the subgroups studied in Theorem 5 and the kernel of φ_Y, and in particular its maximal isotropic subgroups (see [52], pp. 88, 150, 233)? A proof of Theorem 5 based on the theory of theta functions and their pfaffians would also be of interest.

Our second illustration of transcendence methods concerns a special type of abelian variety A. It relies on a solution of the isogeny conjecture, established by D. and G. Chudnovsky [23] in connection with the theory of G-functions. For an elliptic curve E defined over $\mathbf{Q}$, we denote by $H(E)$ the inverse of the area of a fundamental domain of the period lattice of E with respect to a Néron differential form on a minimal model of E over $\mathbf{Z}$.

Theorem 6 (D. and G. Chudnovsky [23]; see also [41]): *let E_1 and E_2 be two elliptic curves defined over $\mathbf{Q}$, and let ℓ be a prime number. Assume that the representations $\rho^1_{(\ell)}$, $\rho^2_{(\ell)}$ of $\mathrm{Gal}(\overline{\mathbf{Q}}/\mathbf{Q})$ on the Tate modules $T_\ell(E_1)$, $T_\ell(E_2)$ have identical characters. Then, E_1 and E_2 are isogenous over $\mathbf{Q}$. Moreover, $\rho^1_{(\ell)}$, $\rho^2_{(\ell)}$ satisfy the hypothesis above as soon as E_1 and E_2 have the same number of points modulo each prime $p \leq c_\epsilon(H(E_1)H(E_2))^{1+\epsilon}$; here, ϵ denotes any positive number and c_ϵ is an effective constant depending only on ϵ.*

The first conclusion of Theorem 6 is of course an immediate consequence of Faltings' solution of Tate's conjecture on the abelian variety $E_1 \times E_2$ [30], and the main value of Theorem 6 lies here in the simplicity of its proof, once the following property of the formal groups of E_1 and E_2 is recalled (see [41]): under the assumption of Theorem 6, there exist non-constant rational functions x_1, x_2 on E_1, E_2 and a power series h with integral coefficients such that on denoting by ϵ_i the exponential map on E_i, $x_1 \circ \epsilon_1$ can be expressed as $h \circ x_2 \circ \epsilon_2$. One now completes the argument with the help of the following variant of the Schneider-Lang transcendence criterion.

Theorem 7 ([23], [41]). *Let f_1, f_2 be power series with rational integral coefficients. Assume the existence of a local analytic isomorphism u such that $f_1 \circ u$ and $f_2 \circ u$ extend to meromorphic functions of finite order of growth on $\mathbf{C}$. Then, f_1 and f_2 are algebraically dependent over $\mathbf{Q}$.*

Replacing the hypothesis on f_1, f_2 by G-function type conditions (see [23], Theorem 1.1) leads to the effective refinement given by the second assertion of Theorem 6. Note that a bound of a similar type had in fact already been obtained by Serre ([65], pp. 196/636 and 715), but none is yet known to cover the other one in full generality. We refer to M. Laurent's exposition in [41] for a more complete comparison of these results.

Returning to the degree of torsion points, we record in relation with Theorem 3 the following consequence of Theorem 6. Let A be the abelian variety formed by the product of g elliptic curves defined over $\mathbf{Q}$. Assume that the ring End A of $\overline{\mathbf{Q}}$-endomorphisms of A is reduced to $\mathbf{Z}^g$ (i.e. that over $\overline{\mathbf{Q}}$, the curves are non-isogenous and do not admit complex multiplications –see the appendix to [59] for the case of complex multiplications). Then, the Galois group $\mathcal{G}_{(\ell)}$ of ℓ^∞-torsion points on A is commensurable with $\mathbf{Z}_\ell^* \times \left(SL_2(\mathbf{Z}_\ell) \right)^g$, and when n runs through all positive integers, the order of $\mathcal{G}_n$ grows at least as a constant multiple of $n^{3g+1}/\log n$: for $g = 2$, this follows from Serre's results (see [63], Theorem 6, and in particular, assertion (iv)) together with Theorem 6, while the general case reduces to $g = 2$ thanks to a well-known argument on powers of simple groups, known as Kolchin's theorem [35] by transcendence specialists and as Ribet's lemma [57] by algebraists (see [64], 4.5). Again, one could alternatively appeal to Faltings' theorem.

To close this section on a less competitive tone, we note that the techniques of §2 and of §3 can sometimes be combined. For instance, on replacing by isogenies the translations occurring in the proof of Theorem 5, M. Hindry [32] has recently derived from Theorem 3 a quantitative version of the following theorem of Raynaud [56]: the Zariski closure of the intersection of the torsion subgroup of $A(\overline{k})$ with a subvariety defined over k is composed of a finite number of translations of abelian subvarieties of A. In the situation, say, of Theorem 5, and under the assumption that Z is defined over k, Hindry's result provides a bound for the order of the torsion points on A lying on Z (as opposed to the finite subgroups of Theorem 5), which can be computed effectively in terms of g, D, $\deg A$ and the constants $c(\ell)$, $[k' : k]$ of Theorem 3. Hindry's method completes a program of Lang ([38], p. 221) extending Bogomolov's ℓ-adic argument in [21], and in view of Theorems 2 and 5 above, leads to a similar sharpening of Theorem 5 of [21] on the Manin-Mumford-Lang conjecture (see also [56]). Moreover, it is capable of generalizations to arbitrary algebraic groups G (in the case of tori, a complete solution of the conjecture, based on both classical Kummer theory and the results of Evertse, Sclickewei and van der Poorten on equations in units, was established by M. Laurent in [40]).

§4 Linear and algebraic dependence on algebraic groups

We now come back to the general situation of §2. Thus, a k-rational point P is given on the g-dimensional connected commutative algebraic group G. Because of its rôle in Kummer theory, it is important to have an explicit description of the algebraic group G_P generated by P in G. With this aim in mind, we fix an embedding φ of G in a projective space $\mathbf{P}^N$, and we denote by $\deg_\varphi G_P$ the degree of the Zariski closure of $\varphi(G_P)$ in $\mathbf{P}^N$. We let $h_\varphi(P)$ represent the (absolute and logarithmic) Weil height of $\varphi(P)$, while the Weil height on $\overline{k}$ is simply denoted by h.

When G is a torus $\mathbf{G}_m^g$, the group G_P coincides with the intersection of the kernels of the characters of G vanishing on $P = (\alpha_1, \ldots, \alpha_g)$. In other words, G is defined by the equations $x_1^{\lambda_1} \ldots x_g^{\lambda_g} = 1$, where $(\lambda_1, \ldots, \lambda_g)$ runs through the subgroup of $\mathbf{Z}^g$ of elements satisfying $\alpha_1^{\lambda_1} \ldots \alpha_g^{\lambda_g} = 1$. The reader will here recognize the "relation group" studied by Masser in these Proceedings [48]. The results of his paper lead to the following conjecture.

Conjecture 1. *There exists an effective constant c, depending only on G, φ and k, such that for all k-rational points P with positive heights, the degree of the algebraic subgroup G_P satisfies*

$$\deg_\varphi G_P \leq c\big(h_\varphi(P)\big)^\gamma,$$

where $\gamma = \gamma(P)$ denotes the dimension of G_P.

Here is a list of remarks in support of this conjecture:

i) the conjecture is trivially satisfied by vector groups $G = \mathbf{G}_a^g$.

ii) when G is a torus, it is equivalent to the existence, established in [48], of a basis of the relation group enjoying the following property:

the product of the maximum of the absolute values of the coordinates of its different elements does not exceed a constant multiple of $\left(h_\varphi(P)\right)^\gamma$.

iii) the conjecture also holds on abelian varieties. This follows again, although in a less direct way, from the main proposition of [48]. Conversely, Conjecture 1 implies an estimate of the type discussed in the previous remark for the generalized relation group that one naturally associates to points on powers of a simple abelian variety (see [50], [19], [17]).

iv) when P is a torsion point e, the degree of G_e coincides with the order of e, for which an effective upper bound in terms of G and k can indeed be deduced from Theorem 4. More generally, it can be proved that the index $\nu(P)$ in G_P of its connected component G_P^o is bounded uniformly on $G(k)$, so that it suffices to establish Conjecture 1 in the special situation where G_P is connected.

v) assume the conjecture has been verified in the given embedding φ. It then holds with the same constant c in the projective embedding obtained by composing φ with any Veronese map of large degree D. Indeed, both terms of the inequality will be multiplied by D^γ under this transformation.

A finer description of the group G_P, involving multiprojective embeddings, can be obtained when G is a power of a simple algebraic group. Suppose for instance that G is a torus G_m^g in its natural embedding in $(\mathbf{P}^1)^g$. Then, for any element σ of the set $S_{\gamma,g}$ of sequences of γ integers in $\{1,\ldots,g\}$, with complement σ', the multidegree $\deg_{\sigma'} G_P$ is well-defined (see [51], [19], [55]), and satisfies:

Theorem 8. *There exists a positive number c, effectively computable in terms of g and of the degree of k over $\mathbf{Q}$, such that*

$$\deg_{\sigma'} G_P \le c \prod_{i \in \sigma} h(\alpha_i)$$

for all points $P = (\alpha, \ldots, \alpha_g)$ in $(k^*)^g$ with $\gamma = \gamma(P) > 0$, and all elements σ in $S_{\gamma,g}$.

In terms of the "relation group", Theorem 8 gives an upper bound for the minors of order $g - \gamma$ of the matrices attached to its generators. In particular, it generalizes Theorem 3A of [42], which holds when G_P is a hypersurface. Assume now that $\alpha_1, \ldots, \alpha_g$ are multiplicatively independent, and denote by Γ the group they generate in k^*. Theorem 8, applied to G_m^{g+1}, yields an upper bound for the index of Γ in its division group in k^*, often referred to as Cassels' remark (see, however, [22]). This, and the corresponding result on abelian varieties, is the statement which was alluded to after the enunciation of Theorem 2. From the point of view of Kummer theory, it leads to an effective determination of the dependence in P of the constant $c(G, P)$ occurring in Theorem 1.

Theorem 8 can be proved along the lines of [48]. Another proof, based on the dual idea of parametrizing the k-rational points of the connected component G_P^o by those of the torus G_m^γ, is given in [17]. This point of view, which in a sense was already taken in [16] and [19], also provides a proof of Conjecture 1 in the case of abelian varieties. We merely mention its interpretation of both sides of the inequalities as volume elements, by-passing any reference to the geometry of numbers.

As is well known, the study of the "relation group" has accompanied the progress of Baker's method (see for instance [3], [31], [20], [37], [18]). Not surprisingly, the group G_P plays a fundamental rôle in the recent advances on linear forms in generalized logarithms (and its occurrence in the zero estimates that they rely upon may also justify Conjecture 1). The analogy with Kummer theory is particularly striking in the following example. Fix an ultrametric place v of k, and assume that the point P lies in the domain of definition of the logarithm map $\log_{G,v}$ from the Lie group $G(k_v)$ into the k_v-points of the tangent space $t(G)$ of G at the origin. Then, the smallest k-linear subspace $t_P = t_{P,v}$ of $t(G)$ whose k_v-points contain $\log_{G,v}(P)$ is contained in $t(G_P) = t(G_P^o)$, and in fact, we have:

Theorem 9 (Wüstholz [74]; see also [15]). *For a general group G and a point P in $G(k)$, the k-vector spaces t_P and $t(G_P^o)$ coincide.*

An analogous statement holds in the complex case. However, $t_P = t_P(u)$ now depends on the choice of a logarithm u of P. Such a choice determines an action of $\mathbf{Q}$ on P (i.e. a way of dividing P in $G(\overline{k})$), and G_P^o must be replaced by the Zariski closure of the orbit of P under this action.

In a parallel direction, the Mumford-Tate group $\mathcal{G}^H(A)$ of an abelian variety A defined over k, which conjecturally governs the Galois groups of its torsion points, should play a fundamental rôle in questions of *algebraic independence*. Indeed, a full set of periods and quasi-periods of A (with respect to a fixed embedding of $\overline{k}$ in $\mathbf{C}$) correspond to a complex point η_A on a k-variety $\mathcal{P}$ isomorphic to $\mathcal{G}^H(A)$ over $\mathbf{C}$ ([29], Corollary 6.4), and a conjecture of Grothendieck asserts that η_A is a generic point of a component of $\mathcal{P}$. In other words, the transcendence degree of the field

$k(\eta_A)$ over k does not exceed, and should be equal to, the dimension of $\mathcal{G}^H(A)$. As we saw in §2, a similar assertion holds for the dimensions of the Galois groups $\mathcal{G}_{(\ell)}(A)$.

It is therefore tempting to ask for a direct link between the transcendence degree of $k(\eta_A)$ over k and the dimension of $\mathcal{G}_{(\ell)}(A)$. In fact, the techniques mentioned in the last paragraph of §2 (and those of §3 in the opposite direction) do point towards such a relation. We conclude this report by specifying the transcendental side of this question, and by extending it to the Kummer-theoretic groups $\chi_{(\ell)}(G, P)$ –not to mention the conjectural group $\mathcal{G}_P(G)$ –studied in §2. For a k-point P on a general group G, and a fixed embedding of $\overline{k}$ in $\mathbf{C}$, write $k(\omega_G)$ for the field generated by the periods of G, and $k(\omega_G, \log_G P)$ for its compositum with the field of definition of any (or every) choice u of a complex logarithm of P. Restricting to extensions G of A by a torus L, we denote by $\tilde{A}$ the universal vectorial extension of A (i.e. the extension of A by G_a^g parameterized by a basis of $H^1(A, \mathcal{O}_A)$; its field of periods $k(\omega_{\tilde{A}})$ coincides with $k(\eta_A)$), and by $\tilde{G}$ the pull-back of G to $\tilde{A}$. Let finally $\tilde{P}$ be a point of $\tilde{G}(k)$ above P. As briefly mentioned in §2, the Galois group $\mathcal{G}_n(\tilde{G})$ (resp. $\chi_n(\tilde{G}, \tilde{P})$) coincides with $\mathcal{G}_n(G)$ (resp. $\chi_n(G, P)$). In a similar vein, the field $k(\omega_{\tilde{G}}, \log_{\tilde{G}} \tilde{P})$ depends only on P, so that the following statements make sense.

Conjecture 2. *Let G be an extension of A by L, and let P be a k-rational point on G. Assume the existence of real numbers $c > 0$, $\kappa = \kappa(G)$, $\beta = \beta(G, P)$ such that for all positive integers n:*

(a) *the degree of $k(G_n)$ over k is $\geq cn^\kappa$; then, the transcendence degree of $k(\omega_{\tilde{G}})$ over k is $\geq \kappa$.*

(b) *the degree of $k(G_n, \frac{1}{n}P)$ over $k(G_n)$ is $\geq cn^\beta$; then, the transcendence degree of $k(\omega_{\tilde{G}}, \log_{\tilde{G}} \tilde{P})$ over $k(\omega_{\tilde{G}})$ is $\geq \beta$.*

We close with a list of remarks in support of Conjecture 2

(i) Part (b) of the conjecture is trivial when P is a torsion point.

(ii) Suppose that $G = A$, i.e. that $L = \{1\}$. Part (a) then follows from the conjectures on abelian varieties above, and, by a classical result of Chudnovsky (see [70]), is known to hold when A is an elliptic curve with complex multiplications (see [2]– and [57]– for a functional analogue on one-parameter families of abelian varieties with real multiplications).

(iii) In view of the description, given in §2, of $\mathcal{G}_n(G)$, and of a standard computation of the periods of G, Part (a) for a general torus L is equivalent to the full conjecture for $L = \{1\}$.

(iv) Suppose that the abelian variety A is 0. Then, Part (a) becomes Lindemann's theorem, while Part (b) follows from Schanuel's conjecture [70], according to which arbitrary logarithms of multiplicatively independent algebraic numbers are algebraically independent over $\mathbf{Q}(\pi)$.

(v) Suppose that $G = A \times L$, so that Part (a) reduces to the case of abelian varieties, while by Theorem 1, β can be taken as the first Betti number $b(P)$ of the group G_P. Part (b) then implies that the ideal of definition of any logarithm $\tilde{u}$ of $\tilde{P}$ over the field $k(\omega_{\tilde{G}})$ is generated by equations of degree ≤ 1 (and in fact, as in Schanuel's conjecture, by the k-linear equations defining $t(\tilde{G}_{\tilde{P}})$ in $t(\tilde{G})$, for a suitable choice of $\tilde{P}$ and $\tilde{u}$). For instance, if $L = \{1\}$ and A is the g-th power of an elliptic curve with complex multiplications by $\mathbf{Q}(\sqrt{-D})$, Part (b) asserts that in the usual notations of transcendence theory, the transcendence degree over $\mathbf{Q}(\omega_1, \eta_1)$ of the field $\mathbf{Q}(\omega_1, \eta_1, u_1, \ldots, u_g, \zeta(u_1), \ldots, \zeta(u_g))$ is equal to the dimension over $\mathbf{Q}$ of the $\mathbf{Q}(\sqrt{-D})$-vector space generated by the classes of $u_1, \ldots, u_g$ in $\mathbf{C}/\mathbf{Q}(\sqrt{-D})\omega_1$.

(vi) In contrast, consider the degenerate example on an extension G of an elliptic curve with complex multiplications given at the end of §2, where the Galois groups $\chi_n(G, P)$ — but not the group G_P itself — are trivial for all n. It can then be checked that $\log_{\tilde{G}} \tilde{P}$ is defined over $k(\omega_{\tilde{G}})$, so that the choice $\beta = 0$ cannot be improved in Part (b) of the conjecture. This shows that Riemann-type relations occur even in the presence of non-torsion points, and that Conjecture 2 is in general best possible.

A natural candidate for the conjectural group $\mathcal{G}_P(G)$ of §2 is the Mumford-Tate group of the 1-motive associated to the homomorphism from $\mathbf{Z}$ to G which P defines. See J.-L. Brylinski: 1-motifs et formes automorphes, in *Journées automorphes*, Publ. Math. Univ. Paris VII, No. 15, 1983, 43–106, for the basic properties of this group, extending the main result of [29]. A precise computation in the situation of Theorem 2 above, and an application to a geometric analogue of Conjecture 2, will be found in Y. André: Mumford-Tate groups and the theorem of the fixed part, *Preprint Max-Planck-Inst*, Bonn, 1987.

References

[1] M. Anderson, Inhomogeneous linear forms in algebraic points of an elliptic function, in *Transcendence Theory; Advances and Applica-

tions, A. Baker, D. Masser eds., Academic Press, 1977 (Chapter 7).

[2] Y. André, Sur certaines algèbres de Lie associées aux schémas abéliens, *C.R.A.S. Paris*, **299**, 1984, 137–140.

[3] A. Baker, Linear forms in the logarithms of algebraic numbers, IV, *Mathematika*, **15**, 1968, 204–216.

[4] A. Baker, On the periods of the Weierstrass $\wp$-function, *Symposia Math.* IV, INDAM Rome, 1968, 155–174 (see also: On the quasi-periods of the Weierstrass ζ-function, *Göttinger Nachr.*, 1969, 145–157).

[5] A. Baker, A sharpening of the bound for linear forms in logarithms, III, *Acta Arith.* **27**, 1975, 247–252.

[6] A. Baker, *Transcendental Number Theory* (2nd edition), Cambridge University Press, 1979.

[7] A. Baker, The theory of linear forms in logarithms, in *Transcendence Theory: Advances and Applications*, A. Baker and D. Masser eds., Academic Press, 1977 (Chapter 1).

[8] A. Baker and H. Stark, On a fundamental inequality in number theory, *Ann. Math.*, **94**, 1971, 190–199.

[9] A. Baker and C. L. Stewart, On effective approximations to cubic irrationals, *New Advances in Transcendence Theory*, (A. Baker ed.), Cambridge Univ. Press, 1988, Chapter 1.

[10] M. Bashmakov, The cohomology of abelian varieties over a number field, *Russian Math. Surveys*, **27**, 1972, 25–70.

[11] D. Bertrand, Sous-groupes à un paramètre p-adique de variétés de groupe, *Ivent. math.*, **40**, 1977, 171–193.

[12] D. Bertrand, Une mesure de transcendence liée à la torsion des courbes elliptiques, *Gr. Et. Anal. ultramétrique* (Secr. math., Paris), 1977–78, No. 14.

[13] D. Bertrand, Kummer theory on the product of an elliptic curve by the multiplicative group, *Glasgow Math. J.*, **22**, 1981, 83–88.

[14] D. Bertrand, Galois orbits on abelian varieties and zero estimates, in *Diophantine Analysis*, J. Loxton–A. van der Poorten eds., *London Math. Soc. Lecture Notes* No. **109**, Cambridge Univ. Press, 1986, 21–35.

[15] D. Bertrand, Lemmes de zéros et nombres transcendants; in *Sém. Bourbaki*, 1985–86, No. 652 (= Astérisque, No. 145–146, 1987, 21–44).

[16] D. Bertrand, La théorie de Baker revisitée, in *Problèmes Diophantiens* 84–85, No. 2, Publ. Math. Univ. Paris VI, No. 73, 1985.

[17] D. Bertrand, Minimal heights and polarizations on abelian varieties, Preprint MSRI, Berkeley, June 1987.

[18] D. Bertrand and Y. Flicker, Linear forms on abelian varieties over local fields, *Acta Arith.*, 38, 1980, 47–61.

[19] D. Bertrand and P. Philippon, Sous-groupes algébriques de groupes algébriques comutatifs, manuscript, 1986.

[20] A. Bijlsma and P. Cijsouw, Degree-free bounds for dependence relations, *J. Austral. M. S.*, 31, 1981, 496–507.

[21] F. Bogomolov, Points of finite order on an abelian variety, *Math. USSR Izvestija* 17, 1981, 55–72.

[22] J. W. S. Cassels, Review of [38], *Bulletin of L.M.S.*, 1984, 637–640.

[23] D. and G. Chudnovsky, Padé approximations and Diophantine geometry, *Proc. Natl. Ac. Sc. USA*, 82, 1985, 2212–2216.

[24] J. Coates, An application of the division theory of elliptic curves to diophantine approximation, *Invent. math.*, 11, 1970, 167–182.

[25] J. Coates, Linear forms in the periods of the exponential and elliptic functions, *Invent. math.*, 12, 1971, 290–299.

[26] J. Coates and S. Lang, Diophantine approximation on abelian varieties with complex multiplication, *Invent. math.* 34, 1976, 129–133.

[27] P. Cohen, Explicit calculation of some effective constants in transcendence proofs, Ph.D. Dissertation, Nottingham, 1985 (Chapter 3).

[28] S. David, Fonctions theta et points de torsion sur les variétés abéliennes, *C.R.A.S. Paris*, 1987, to appear.

[29] P. Deligne, Hodge cycles on abelian varieties (Notes by J. S. Milne), Springer Lecture Notes No. 900, 9–100 (1982).

[30] G. Faltings, Endlichkeitssätze für abelsche Varietäten über Zahlkörpern, *Invent. math.* 73, 1983, 349–366.

[31] N. I. Fel'dman, Estimate for a linear form of logarithms of algebraic numbers, *Math. USSR Sb.*, 5, 1968, 291–307.

[32] M. Hindry, Points de torsion sur les variétés, *C.R.A.S. Paris*, **304**, 1987, 311–314.

[33] O. Jacquinot, Théorie de Kummer sur une extension d'une courbe elliptique par G_m, *Progress in Mathematics*, **31**, 1983, 125–132, Birkhäuser.

[34] O. Jacquinot and K. Ribet, Deficient points on extensions of abelian varieties by G_m, *J. Number Th.*, **25**, 1987, 327–352.

[35] E. Kolchin, Algebraic groups and algebraic dependence, *Amer. J. Math.*, **90**, 1968, 1151–1164.

[36] S. Lang, Division points of elliptic curves and abelian functions over number fields, *Amer. J. Math.* **97**, 1975, 124–132.

[37] S. Lang, *Elliptic curves, Diophantine Analysis*, Springer, 1978.

[38] S. Lang, *Fundamentals of Diophantine Geometry*, Springer, 1983.

[39] H. Lange and W. Ruppert, Complete systems of addition laws on abelian varieties, *Invent. math.*, **79**, 1985, 603–610.

[40] M. Laurent, Equations diophantiennes exponentielles, *Invent. math.*, **78**, 1984, 299–327.

[41] M. Laurent, Une nouvelle démonstration du théorème d'isogénie, d'après D. et G. Chudnovsky, in *Sém. Th. Nombres, Paris*, 1985–86, to appear.

[42] J. Loxton and A. van der Poorten, Multiplicative dependence in number fields, *Acta. Arith.*, **62**, 1983, 291–302.

[43] D. Masser, Division fields of elliptic functions, *Bulletin L.M.S.* **9**, 1977, 49–53.

[44] D. Masser, Diophantine approximation and lattices with complex multiplication, *Invent. math.*, **45**, 1978, 61–82.

[45] D. Masser, On quasi-periods of abelian functions with complex multiplication, *Mém. SMF*, No. **2**, 1980, 55–68.

[46] D. Masser, Small values of the quadratic part of the Néron-Tate height on an abelian variety, *Compo. Math.*, **53**, 1984, 153–170.

[47] D. Masser, Small values of heights on families of abelian varieties, manuscript, 1986.

[48] D. Masser, Linear relations on algebraic groups, *New Advances in Transcendence Theory* (A. Baker, ed.), Cambridge Univ. Press, 1988, Chapter 15.

[49] D. Masser, Letter to the author, dated Nov. 17, 1986.

[50] D. Masser and G. Wüstholz, Fields of large transcendence degree generated by values of elliptic functions, *Invent. math.*, **72**, 1983, 407–464.

[51] D. Masser and G. Wüstholz, Zero estimates on group varieties, II, *Invent. math.*, *80*, 1985, 233–267.

[52] D. Mumford, *Abelian Varieties*, Oxford University Press, 1970.

[53] P. Philippon, Lemmes de zéros dans les groupes algébriques commutatifs, *Bull. SMF*, **114**, 1986, 355–383.

[54] P. Philippon and M. Waldschmidt, Formes linéaires de logarithmes sur les groupes algébriques commutatifs, manuscript, 1986.

[55] P. Philippon and M. Waldschmidt, Lower bound for linear forms in logarithms, *New Advances in Transcendence Theory*, (A. Baker, ed.) Cambridge Univ. Prss, 1988, Chapter 18.

[56] M. Raynaud, Sous-variétés d'une variété abélienne et points de torsion, *Progress in Mathematics*, **35**, 1983, 327–352, Birkhäuser (see also: Around the Mordell conjecture for function fields and a conjecture of Serge Lang, *Springer Lecture Notes*, No. **1016**, 1982, 1–19).

[57] K. Ribet, Galois action on division points of abelian varieties with many real multiplications, *Amer. J. Math.*, **98**, 1976, 751–804.

[58] K. Ribet, Dividing rational points on abelian varieties of CM-type, *Compo. Math.* **33**, 1976, 64–74.

[59] K. Ribet, Division fields of abelian varieties with complex multiplication, *Mém. SMF*, No. **2**, 1980, 75–94 (see also, in appendix: Deux lettres de Serre, ibid., 95–102).

[60] K. Ribet, Kummer theory on extensions of abelian varieties by tori, *Duke Math. J.* 46, 1979, 745–761.

[61] K. Ribet, Cohomological realization of a family of 1-motives, *J. Number Th.* **25**, 1987, 152–161.

[62] J-P. Serre, Sur les groupes de congruence des variétés abéliennes II, *Izv. AN SSSR*, **35**, 1971, 731–735.

[63] J-P. Serre, Propriétés galoisiennes des points d'ordre fini des courbes elliptiques, *Invent. math.*, **15**, 1972, 259–331.

[64] J-P. Serre, Représentations ℓ-adiques, in *Kyoto Int. Symp. Algebraic Number Th.*, 1977, 177–193.

[65] J-P. Serre, Quelques applications du théorème de Cheboratev, *Publ. Math. IHES*, **54**, 1981, 123–201 (= *Oeuvres*, III, 563–641 and 715).

[66] J-P. Serre, Résumé des cours de 1985–86, Ann. Collège de France, 1986.

[67] A. Silverberg, Points de torsion des variétés abéliennes de type CM, in *Problèmes Diophantiens* 85–86, No. 7, Publ. math. Univ. Paris VI, No. **79**, 1986.

[68] H. Stark, A transcendence theorem for class number problems I, *Ann. Math.*, **94**, 1971, 153–173.

[69] R. Tijdeman, On the equation of Catalan, *Acta. Arith.*, **29**, 1976, 197–209.

[70] M. Waldschmidt, Les travaux de G. V. Chudnovsky sur les nombres transcendants; in Sém. Bourbaki, 1975–76, No. 488 (= *Springer Lecture Notes* No. **567**, 1977, 274–292).

[71] M. Waldschmidt, A lower bound for linear forms in logarithms, *Acta Arith.*, **37**, 1980, 257–283.

[72] M. Waldschmidt, Transcendence et exponentielles en plusieurs variables, *Invent. math.*, **63**, 1981, 97–127.

[73] G. Wüstholz, Nullstellenabschätzungen auf Varietäten, *Progress in Mathematics*, **22**, 1982, 359–362, Birkhäuser.

[74] G. Wüstholz, Recent progress in transcendence theory, *Springer Lecture Notes* No. **1068**, 1984, 280–296.

[75] G. Wüstholz, A new approach to Baker's theorem on linear forms in logarithms, III, *New Advances in Transcendence Theory* (A. Baker, ed.), Cambridge Univ. Press, 1988, Chapter 25.

4

SOME NEW RESULTS
ON ALGEBRAIC INDEPENDENCE OF E-FUNCTIONS

F. Beukers

Introduction

A famous theorem of Lindemann and Weierstrass (1885) states that
for any set of Q-linearly independent algebraic numbers $\alpha_1, \ldots, \alpha_n$ the
numbers $\exp(\alpha_1), \ldots, \exp(\alpha_n)$ are algebraically independent over Q. In
a classical paper in 1929, C. L. Siegel [14] developed a method for demon-
strating the algebraic independence of values of E-functions satisfying
certain differential equations over $C(z)$. Here, an E-function is an analytic function given by a power series of the form

$$\sum_{n=0}^{\infty} \frac{a_n}{n!} z^n, \qquad a_n \in K, [K : Q] < \infty$$

such that there exists a constant $c > 0$ satisfying:

  i) $\overline{|a_n|} \ll c^n$ for all n,

  ii) $\underset{i=0,\ldots,n}{\text{l.c.m.}} (\text{denominator } a_i) \ll c^n$ for all n.

Examples are of course $\exp(z)$ and $J_0(z)$, the Bessel-function of order
zero given by

$$J_0(z) = \sum_{n=0}^{\infty} \left(\frac{1}{n!}\right)^2 \left(-\frac{z^2}{4}\right)^n = \sum_{n=0}^{\infty} \binom{2n}{n} \left(-\frac{1}{4}\right)^n \frac{z^{2n}}{(2n)!}.$$

Siegel was able to show the algebraic independence of $\exp(\alpha_1)$, $\ldots$,
$\exp(\alpha_n)$, $J_0(\xi)$, $J_0'(\xi)$ over Q if $\alpha_1, \ldots, \alpha_n$ are as above and ξ is a non-
zero algebraic number. Another result is that $J_0(\xi_1)$, $\ldots$, $J_0(\xi_n)$, $J_0'(\xi)$,
$\ldots, J_0'(\xi_n)$ are algebraically independent if $\xi_1, \ldots, \xi_n$ are non-zero algebraic numbers whose squares are pairwise distinct.

In his book *Transcendental Numbers* [15] in 1949, Siegel formalized
his approach and used the following setting:

Let $f_1, \ldots, f_n$ be a set of E-functions satisfying a system of linear first order differential equations of the form

$$(*) \quad \frac{d}{dz} \begin{pmatrix} f_1 \\ \vdots \\ f_n \end{pmatrix} = \begin{pmatrix} a_{11} & \cdots & a_{1n} \\ \vdots & \ddots & \vdots \\ a_{n1} & \cdots & a_{nn} \end{pmatrix} \begin{pmatrix} f_1 \\ \vdots \\ f_n \end{pmatrix} \qquad a_{ij} \in \overline{\mathbf{Q}}(z) \text{ for all } i, j$$

Let $\alpha \in \overline{\mathbf{Q}}$ be such that $\alpha \neq 0$ and $a_{ij}(\alpha) \neq \infty$ for all i, j. Then we have the following Theorem.

Theorem (Siegel, 1949). *Let the system* $(*)$ *be* **Siegel normal**. *Then for every homogeneous* $P \in \mathbf{Z}[X_1, \ldots, X_n]$, $P \not\equiv 0$ *we have* $P(f_1(\alpha), \ldots, f_n(\alpha)) \neq 0$. *Moreover, there exist completely effective lower bounds for* $|P(f_1(\alpha), \ldots, f_n(\alpha))|$.

Of course the concept 'Siegel normal' is ours and not Siegel's. It stands for the phrase 'the power products $f_1^{\nu_1} \ldots f_m^{\nu_m}$ $(\nu_1 + \ldots + \nu_m = \nu)$ form a normal system for all $\nu = 1, 2, 3, \ldots$.' formulated in [15] Ch. 2.8. We refer to Siegel's work for a further discussion. In section 2 of this paper we give a precise reformulation of the concept 'Siegel normality'. Also note that here we have restricted ourselves to homogeneous polynomials P, hence we like to speak of homogeneous algebraic independence. This is no restriction since algebraic independence of $a_1, \ldots, a_n$ is equivalent to homogeneous algebraic independence of $1, a_1, \ldots, a_n$.

Although the concept of Siegel normality was introduced to guarantee the non-vanishing of a certain determinant in the proof, it eluded verification for any further functions beyond those already considered in 1929. However, in a fundamental advance Shidlovsky proved

Theorem (Shidlovsky, 1954). *If* $f_1(z), \ldots, f_n(z)$ *are homogeneously algebraically independent over* $\mathbf{C}(z)$, *then* $f_1(\alpha), \ldots, f_n(\alpha)$ *are homogeneously algebraically independent over* $\mathbf{Q}$.

After this simplification many authors have developed techniques to obtain the algebraic independence of functions. For example, V. A. Oleinikov [9] and V. Ch. Salikov [10] define criteria implying the algebraic independence of components of solutions of $(*)$. For an overview of early developments see the review paper of Shidlovsky and Fel'dman [13]. We also refer to Mahler's book on transcendence [7] Ch. 7 for some very interesting examples.

It should be mentioned that Shidlovsky's method does not yield completely effective lower bounds for $|P(f_1(\alpha), \ldots, f_n(\alpha))|$. In order to obtain such effective measures one can use zero-estimates such as [8], where the ineffectivity is localized to one constant. See [3] for a detailed exposition. Another approach is to use the zero-estimates in [2] which can be expressed as functions of the parameters of the system (∗). Recently, Yebbou [16] gave an effective procedure to compute these parameters.

Despite all efforts to establish algebraic independence of functions, a uniform approach seems yet missing. As indicated in [5] such an approach can be effected through the use of the well-known classification and representation theory which exists for algebraic groups over C. These algebraic groups arise from the Picard-Vessiot extension corresponding to (∗) and its differential Galois group, two important concepts in the theory of linear differential equations. They play the same rôle as the splitting field of a polynomial and the corresponding ordinary Galois group. In Section 1 we introduce their precise definitions and state some properties.

In a paper written by W. D. Brownawell, G. Heckman and the present author [1] we used the theory of algebraic groups to establish practical criteria for Siegel normality of (∗). The results are mentioned in Section 2. An application of Theorems 2.1 and 2.2 to generalized hypergeometric functions is stated below. For its proof which is based on ideas of N. M. Katz and O. Gabber, we refer to [1] and to joint work of Katz, Pink and Gabber (to appear).

Application. A parameter set S is a set of real numbers of the form $\{\mu_1 \ldots, \mu_p : \nu_1, \ldots, \nu_q\}$ where $q > p \geq 0$, $q \geq 2$ and $\nu_q = 1$. The set S is called *admissible* if it satisfies the following conditions.

(A) $\nu_j - \mu_i \notin \mathbf{Z}$ for $1 \leq i \leq p$, $1 \leq j \leq q$.

(B) There does not exist $d \in \mathbf{N}$, $d > 1$ dividing both p and q such that the sets $\{\nu_1, \ldots, \nu_q\}$ and $\{\mu_1, \ldots, \mu_p\}$ are modulo $\mathbf{Z}$ unions of sets of the form $\{\beta, \beta + \frac{1}{d}, \ldots, \beta + \frac{d-1}{d}\}$.

(C) Either

    (i) all sums $\nu_i + \nu_j$, $1 \leq i \leq j \leq q$ are distinct modulo $\mathbf{Z}$,

  or

    (ii) $p = 0$, q is odd or 2,

  or

    (iii) (Katz, Gabber) $q - p$ is odd,

  or

(iv) (Katz, Gabber) S is not selfsimilar (see below) and $(q, p) \neq (9, 3)$.

Two parameter sets $S = \{\mu_1, \ldots, \mu_p : \nu_1, \ldots, \nu_q\}$, $S' = \{\mu_1', \ldots, \mu_r' : \nu_1' \ldots, \nu_s'\}$ are called *similar* if

(a) $p = r$, $q = s$.

(b) there exist $\mu, \nu \in \mathbf{R}$ and a choice of $\pm$ sign such that, on renumbering if necessary,

$$\mu_i' \equiv \mu \pm \mu_i \pmod{\mathbf{Z}}, \qquad i = 1, \ldots, p,$$

$$\nu_i' \equiv \nu \pm \nu_i \pmod{\mathbf{Z}}, \qquad i = 1, \ldots, q.$$

For a given parameter set $S = \{\mu_1, \ldots, \mu_p : \nu_1, \ldots, \nu_q\}$ consider the function

$$f_S(z) = \sum_{n=0}^{\infty} \frac{(\mu_1)_n \ldots (\mu_p)_n}{(\nu_1)_n \ldots (\nu_{q-1})_n} \frac{(-z)^{(q-p)n}}{n!}$$

where $(\alpha)_n = \alpha(\alpha + 1) \ldots (\alpha + n - 1)$. Hence $f_S(z)$ is a generalized hypergeometric function of type $_pF_{q-1}$ with z replaced by $(-z)^{(q-p)}$. Let $S_1, \ldots, S_r$ be (possibly equal) admissible parameter sets with rational μ's and ν's and $a_1, \ldots, a_r \in \overline{\mathbf{Q}}^*$ (non-zero algebraic numbers). If the parameter sets S_i and S_j, $i \neq j$, are similar of length p, q then we assume that $\pm a_i/a_j$ are not $(q - p)$th roots of unity. Let $b_1, \ldots, b_s$ be $\mathbf{Q}$-linear independent algebraic numbers. Then the numbers

$$e^{b_1}, \ldots, e^{b_s}$$

$$f_{S_1}(a_1), f_{S_1}'(a_1), \ldots, f_{S_1}^{(q(1)-1)}(a_1)$$

$$\ldots \ldots \ldots \ldots \ldots$$

$$f_{S_r}(a_r), f_{S_r}'(a_r), \ldots, f_{S_r}^{(q(r)-1)}(a_r)$$

where $q(i)$ is the q-parameter of S_i, are all algebraically independent over $\mathbf{Q}$.

Remark. In our original notion of admissibility only possibilities C(i), C(ii) occurred. However, N. M. Katz kindly informed us that he and O. Gabber have managed to give a complete description of the Galois groups of hypergeometric differential equations with $q > p$. As an offspring of their theory we have added the very general possibilities C(iii) and C(iv) in our notion of admissibility. In fact, it turns out that in all admissible cases the differential Galois group contains $SL(q, \mathbf{C})$. This has important consequences for the algebraic independence of the solutions of a differential equation and their derivatives. See the remark following Theorem 1.1.

In order to indicate the limits of applicability of our criteria, we note that according to Theorem 2 of Section 2, Siegel normality of a system (∗) implies that the corresponding Galois group is reductive. But there are many applications of Shidlovsky's theorem in which the Galois group is not reductive, hence the system not Siegel normal. A simple example is the system

$$
z\frac{d}{dz}\begin{pmatrix} y_1 \\ \vdots \\ y_n \end{pmatrix} = \begin{pmatrix} 0 & & & & \\ z & z & & 0 & \\ & 1 & & & \\ & & \ddots & & \\ 0 & & 1 & 0 & \\ & & & 1 & 0 \end{pmatrix}\begin{pmatrix} y_1 \\ \vdots \\ y_n \end{pmatrix}
$$

of which $\left(1, \sum_1^\infty z^r/r!, \sum_1^\infty z^r/r.r!, \ldots, \sum_1^\infty z^r/r^n.r!\right)^t$ is a solution. This system can be solved by iterated integration and its Galois group is solvable. However, the components of the solution just given are homogeneously algebraically independent E-functions (see [13]). It would be interesting to have some better understanding of the non-reductive case as well.

1. Linear Systems of First Order Differential Equations and Differential Galois Groups

Consider the system (∗) from the introduction where we now assume $a_{ij} \in \mathbf{C}(z)$. It is clear that the components of the solutions are usually not contained in $\mathbf{C}(z)$, but in some extension of it. We formalize this as follows. Let F be a differential field, that is, a field equipped with a derivation $\partial : F \to F$ satisfying $\partial(x+y) = \partial x + \partial y$, $\partial(xy) = x\partial y + y\partial x$. Suppose also that F contains $\mathbf{C}(z)$ and that ∂ restricted to $\mathbf{C}(z)$ is exactly d/dz. Finally suppose that $\partial x = 0$, $x \in F \Leftrightarrow x \in \mathbf{C}$. Examples are the field of functions meromorphic on some open disc, or $\mathbf{C}(z)$, or $\mathbf{C}(z, e^z)$ all equipped with $\partial = d/dz$. Finally suppose that F contains the components of n $\mathbf{C}$-linearly independent solutions of (∗). Such a field can always be found in our case. Take for example the field of Laurent expansions around a point which is not a pole of any a_{ij}, or take for F the field generated by elements of the form $(z-a)^\lambda \times$ Taylor series in $(z - a)$ if a is a regular singularity of $(*)$ (see [12], Ch XV). Now let F_0 be the subfield of F obtained by adjoining to $C(z)$ all components of n linearly independent solutions of $(*)$. This differential field F_0 is called the Picard-Vessiot extension of the system $(*)$. Up to differential isomorphism F_0 is independent of the choice of F (see [6]. Ch VI, Prop. 13).

An $n \times n$-matrix Y formed by writing down n linearly independent solutions is called a *fundamental solution matrix* of $(*)$. The determinant of Y is called the *Wronskian determinant* $W = \det Y$. It is determined up to a constant factor. One easily verifies $W'/W = \operatorname{trace} A$. The *differential Galois group* of $F_0/C(z)$ is defined by

$$\operatorname{Gal}(F_0/C(z)) = \left\{\sigma : F_0 \to F_0 \text{ automorphism} : \sigma.\frac{d}{dz} = \frac{d}{dz}\sigma,\ \sigma z = z\right\}.$$

Let $\underline{y}$ be any solution of the system $(*)$ abbreviated as $\underline{y}' = A\underline{y}$, and let $\sigma \in \operatorname{Gal}(F_0/C(z))$. Since σ and d/dz commute, we see that $(\sigma y)' = \sigma(y') = \sigma(Ay) = A\sigma y$, hence σy is again a solution of $(*)$. Thus we see that $\mathcal{G} = \operatorname{Gal}(F_0/C(z))$ acts as a linear group on $V(\simeq C^n)$, the C-vector space of solutions of $(*)$. We call $G \hookrightarrow GL(V)$ the natural representation of the differential Galois group G. To be even more explicit, let Y be a fundamental solution matrix $(*)$, and $\sigma \in \operatorname{Gal}(F_0/C(z))$. Then there exists an $n \times n$ matrix Σ with entries in C such that $\sigma(Y) = Y.\Sigma$, i.e. the Galois group acts on the right of Y. The following basic result of Kolchin [4] has its roots in the work of Picard and Vessiot.

Theorem 1.1

i) *The group* $\operatorname{Gal}(F_0/C(z))$ *is a linear algebraic subgroup of* $GL(V)$.

ii) *The transcendence degree of* $F_0/C(z)$ *equals the dimension of* $\operatorname{Gal}(F_0/C(z))$.

iii) $C(z)$ *is the maximal fixed field of* $\operatorname{Gal}(F_0/C(z))$.

Remark. Notice that if $SL(n, C) \subset G$, then $\deg \operatorname{tr}(F_0/C(z)) \geq n^2 - 1$, i.e. at most one algebraic relation can exist between the components of a fundamental solution matrix. It is not hard to see that this relation occurs if and only if the Wronskian determinant is algebraic over $C(z)$.

The rest of this section consists of some definitions in order to understand the statements of section 2. First of all, we call $(*)$ *linearly irreducible* if G acts irreducibly on V. It is a well-known fact that a linear differential equation $\mathcal{L}y = 0$, $\mathcal{L} \in C(z)[d/dz]$ of order n can be rewritten as an $n \times n$-system of first order equations whose solutions are given by $(y, y', \ldots, y^{(n-1)})^t$ as y runs through the solutions of $\mathcal{L}y = 0$. In this particular case linear irreducibility is equivalent to the fact that $\mathcal{L}$ has no factorization in $\mathbf{C}(z)[d/dz]$ into factors of degree less than n in d/dz.

In algebraic group theory one often speaks of *reductive* groups. It is an elementary fact that a group G is reductive if and only if there exists a faithful representation of G which is completely reducible. One may use the latter property as a working definition.

By a permutation of $(*)$ we mean a system obtained from $(*)$ by subjecting rows and columns of the coefficient matrix of $(*)$ to the same permutation.

Consider two systems (A) and (B) of size $n \times n$ and $m \times m$. Let G, H be their corresponding Galois groups and V, W their solution spaces. A direct sum of these systems is simply a permutation of the $(n+m) \times (n+m)$ system of first order equations whose coefficient matrix is obtained by putting the coefficient matrix of each equation on the diagonal of an $(n+m) \times (n+m)$ matrix and choosing all other coefficients zero. Its space of solutions is clearly isomorphic to $V \oplus W$. Its Galois group is a subgroup of $G \times H$ and it acts on the solutions via the direct sum representation $V \oplus W$ of $G \times H$. Similarly we can define a direct sum of any number of systems.

Let $(x_1, \ldots, x_n)^t$ be a solution of (A) and $(y_1, \ldots, y_n)^t$ a solution of (B). Form the product columns consisting of all nm products $x_i y_j$, $i = 1, \ldots, n; j = 1, \ldots, m$). As we let $(x_1, \ldots, x_n)^t$ and $(y_1, \ldots, y_m)^t$ run through the solution spaces V of (A) and W of (B), their product columns span an nm dimensional $\mathbf{C}$-vector space isomorphic to $V \otimes W$. The $nm \times nm$ system of first order linear equations having all these product columns as solutions is called the tensor product of (A) and (B). Its Galois group is a subgroup of $G \times H$ which acts on $V \otimes W$ by the tensor product representation of the natural representations of G and H. Similarly we define tensor products of any number of systems.

To every solution $y = (y_1, \ldots, y_n)^t$ of (A) we associate the product column of length $N = \binom{m+n-1}{n-1}$ consisting of all monomials of degree m in $y_1, \ldots, y_n$. Let $V_{(m)}$ be the $\mathbf{C}$-linear span of such product columns as we let y run through V, the space of solutions of (A). It is not hard to show that $V_{(m)}$ is isomorphic to the mth symmetric tensor product $S^m V$ of V. We identify $V_{(m)} \simeq S^m V$. The Galois group of (A) acts on $S^m V$ via the mth symmetric power of the natural representation of G on V. Moreover, the elements of $S^m V$ satisfy $N \times N$ system of linear first order equations, which we call the mth symmetric power of (A).

Let Y be a fundamental solution matrix of (A). Then (A) and (B)

are said to be *cogredient* if there exists an $n \times n$ matrix M, $\det M \neq 0$ with entries in $\mathbf{C}(z)$ and a function $Q(z)$ whose logarithmic derivative is in $\mathbf{C}(z)$ such that $Q.MY$ is a solution matrix of (B). We call two cogredient systems *equivalent* if we have $Q \in \mathbf{C}(z)$. Let $(A)^d$ be the dual system obtained by taking minus the transpose of the coefficient matrix of (A). Notice that

$$0 = \frac{d}{dz}Y.Y^{-1} = Y\frac{d}{dz}Y^{-1} + \left(\frac{d}{dz}Y\right)Y^{-1} = Y\frac{d}{dz}Y^{-1} + AY.Y^{-1}.$$

Hence

$$\frac{d}{dz}Y^{-1} = -Y^{-1}A$$

and on taking the transpose,

$$\frac{d}{dz}(Y^{-1})^t = -A^t(Y^{-1})^t.$$

So the dual system $(A)^d$ has $(Y^{-1})^t$ as solution matrix. We call (A) and (B) *contragredient*, if $(A)^d$ is cogredient with (B).

2. Characterizations of Siegel Normal Systems.

In modern terms, Siegel's original notion of normality can be formulated as follows:

For every $m \in \mathbf{N}$ the direct sum of the 1st, 2nd, $\ldots$, m-th symmetric power of $(*)$ is a direct sum of irreducible non-equivalent systems.

However, here we put forward the homogeneous notion which, as we remarked before, is more general and more natural.

Definition. *The system $(*)$ of first order differential equations is called Siegel normal if each symmetric power of $(*)$ is a direct sum of irreducible non-equivalent systems.*

A direct consequence of Siegel normality is that the non-zero components of any solution do not satisfy any homogeneous polynomial relation over $\mathbf{C}(z)$. The converse is not true, as is shown by the example

$$\frac{d}{dz}\begin{pmatrix} y_1 \\ y_2 \end{pmatrix} = \begin{pmatrix} 0 & 1/z \\ 0 & 0 \end{pmatrix}\begin{pmatrix} y_1 \\ y_2 \end{pmatrix}$$

having the general solution $(\alpha \log z + \beta, \alpha)^t$, α, $\beta \in \mathbf{C}$. Clearly, the non-zero components of each solution are homogeneously algebraically independent over $\mathbf{C}(z)$, but the system is not Siegel normal. Another such example is given by the system at the end of our introduction. In Theorem 2.2 we will see, however, that we do have a converse if we require the Galois group of $(*)$ to be reductive. Our first theorem characterizes irreducible Siegel normal systems.

Theorem 2.1. *Let* (A) *be an* $n \times n$ *system,* V *its space of solutions and* G *the Galois group. Suppose* $\dim V = n \geq 2$. *Then the following properties are equivalent:*

(i) (A) *is irreducible and Siegel normal.*

(ii) *Every non-trivial solution of* (A) *has components which do not satisfy any non-trivial homogeneous polynomial relation over* $\mathbf{C}(z)$.

(iii) G *divided out by its centre is infinite, and* G *acts irreducibly on* S^2V, *the second symmetric tensor power of* V.

(iv) G *contains* $SL(n, \mathbf{C})$ *or* $\mathrm{Sp}(n, \mathbf{C})$.

Remark. $\mathrm{Sp}(n, \mathbf{C})$ is the symplectic group which exists only if $n = 2m$ is even. It can be defined as $\{T \in SL(n, \mathbf{C}) : T^t E T = E\}$ where

$$E = \begin{pmatrix} 0 & I_m \\ -I_m & 0 \end{pmatrix}$$

and I_m is the $m \times m$ identity matrix.

The next theorem gives two characterizations for general Siegel normal systems.

Theorem 2.2 *Let* (A) *be a system of first order differential equations. Then the following statements are equivalent:*

(i) (A) *is Siegel normal*

(ii) (A) *has a reductive Galois group and the non-zero components of each solution do not satisfy any non-trivial homogeneous polynomial relation over* $\mathbf{C}(z)$.

(iii) (A) *is a direct sum of* r *non-cogredient and non-contragredient irreducible Siegel normal systems* (A_i) *of size* $n_i \times n_i$, $n_i \geq 2$, $(i = 1, \ldots, r)$ *and* s *one by one systems* $y' = B_j y$, $(j = 1, \ldots, s)$, *such that the coefficients* $B_j \in \mathbf{C}(z)$ *do not satisfy any relation of the form*

$$\sum_{j=1}^{s} a_j B_j = \frac{1}{f} \frac{d}{dz} f$$

with $a_j \in \mathbf{Z}$ not all zero, $\sum_{j=1}^{s} a_j = 0$ and $f \in \mathbf{C}(z)^*$.

As an illustration of how to apply Theorem 2.2 we indicate the proof of the following statement.

Let $\xi_1, \ldots, \xi_n, \alpha_1, \ldots, \alpha_m \in \overline{\mathbf{Q}}$ such that $\xi_i^2 \neq \xi_j^2$ for all $i \neq j$ and $\alpha_1, \ldots, \alpha_m$ are $\mathbf{Q}$-linearly independent. Then $J_0(\xi_1), J_0'(\xi_1), \ldots, J_0(\xi_n), J_0'(\xi_n), \exp(\alpha_1), \ldots, \exp(\alpha_m)$ are algebraically independent over $\mathbf{Q}$.

Note that this is also a consequence of the result mentioned in the introduction. For its proof we start with the system of first order differential equations having the vector solution

$$\left(J_0(\xi_1 z), J_0'(\xi_1 z), \ldots, J_0(\xi_n z), J_0'(\xi_n z), \exp(\alpha_1 z), \ldots, \exp(\alpha_m z), 1 \right)^t.$$

Note that this system is direct sum of n 2×2 systems derived from the Bessel equation in the obvious way and $m + 1$ 1×1 equations $y' = \alpha_i y, i = 0, \ldots, m$, where we have taken $\alpha_0 = 0$. According to Siegel's theorem we have to prove Siegel normality for this system to obtain our statement. In order to do this we shall use Theorem 2.2(iii).

One easily verifies that the conditions on the B_j are equivalent to $\mathbf{Q}$-linear independence of $\alpha_1, \ldots, \alpha_m$. A necessary and sufficient condition for the non-cogredience and non-contragredience of the 2×2 systems is that $\xi_i^2 \neq \xi_j^2$ for all $i \neq j$. This can be verified in a surprisingly easy way by looking at the asymptotic expansions of the solutions of the Bessel equation at ∞. Finally we must verify that a 2×2 system coming from a Bessel equation is Siegel normal. This was already shown by Siegel and Kolchin [5]. But now we can use Theorem 2.1(iii) or (iv). We shall do both.

First note that a second solution of the Bessel equation is $Y_0(z) = J_0(z) \log z + f(z)$ where $f(z)$ is holomorphic around $z = 0$. Secondly, the Bessel equation is irreducible over $\mathbf{C}(z)$ which implies that its differential Galois group acts irreducibly on the two-dimensional space V of solutions. In particular G is reductive.

Note that $S^2 V$ is isomorphic to the $\mathbf{C}$-span of the products $Y_0^2, Y_0 J_0, J_0^2$. Suppose G acts reducibly on $S^2 V$. Since G is reductive the representation is completely reducible and so there must be a one-dimensional G-stable subspace of $S^2 V$. In other words, there exist $a, b, c, \in \mathbf{C}$, not all zero, such that G acts on $aY_0^2 + bY_0 J_0 + cJ_0^2$ by scalar multiplication. In particular this expression must be stable under monodromy and since Y_0 contains $\log z$ we see that $a = b = 0$. So G acts on J_0^2 by scalars. Hence $(J_0^2)'/J_0^2 = 2J_0'/J_0$ is fixed under G, and thus belongs to $\mathbf{C}(z)$, contradicting the irreducibility of the Bessel equation. Hence G acts irreducibly on S^2V and according to Theorem 2.1(iii) we have Siegel normality.

We can also argue as follows. Since one of the solutions contains a logarithm, the Galois group contains a unipotent element, that is, equivalent to $\begin{pmatrix} 1 & 1 \\ 0 & 1 \end{pmatrix}$. It is easy to verify that any algebraic subgroup of $GL(2,\mathbb{C})$ acting irreducibly on $\mathbb{C}^2$ and containing a unipotent element must contain $SL(2,\mathbb{C})$. Hence, according to Theorem 2.1(iv) we again have Siegel normality.

References

[1] Beukers, F., Brownawell, W. D., Heckman, G., *Siegel normality*, to appear in *Annals of Mathematics*.

[2] Bertrand, D., Beukers, F., Équations différentielles linéaires et ma-jorations de multiplicités, *Ann. Scient. École Norm. Sup.* **18** (1985), 181–192.

[3] Brownawell, W. D., Effectivity in independence measures for values of E-functions, *J. Austral. Math. Soc.* **39** (1985), 227–240.

[4] Kolchin, E. R., Algebraic matrix groups and the Picard-Vessiot the-ory of homogeneous linear ordinary differential equations, *Ann. of Math.* **49** (1948), 1–42.

[5] Kolchin, E. R., Algebraic groups and algebraic independence, *Amer. J. Math.* **90**, (1968), 1151–1164.

[6] Kolchin, E. R., Differential algebra and algebraic groups, *Academic Press* (1973), New York, London.

[7] Mahler, K., Lectures on transcendental numbers, *Lecture Notes in Mathematics* **546**, Springer, Berlin.

[8] Nesterenko, Yu. V., Bounds on the order of zeros of a class of func-tions and their application to the theory of transcendental numbers, *Math. USSR Izv.* **11** (1977), 253–284.

[9] Oleinikov, V. A., Transcendence and algebraic independence of val-ues of E-functions representing a solution of a third order linear differential equation. *Dokl. Akad. Nauk SSSR* **166** (1966), 540–543 = *Soviet Math. Dokl.* **7** (1966), 118–121.

[10] Salichov, V. Ch., On algebraic irreducibility of a collection of differ-ential equations, *Math. USSR Izv.* **26** (1986), 185–200.

[11] Shidlovsky, A. B., A criterion for algebraic independence of the values of a class of entire functions, *Izv. Akad. Nauk SSSR Ser. Math.* **23** (1959), 35–66 = *Amer. Math. Soc. Transl.* (2) **22** (1962), 339–370.

[12] Ince, E. L., *Ordinary differential equations*, Dover, New York (1956).

[13] Shidlovsky, A. B., Fel'dman, N. I., The development and present state of the theory of transcendental numbers, *Russian Math. Surveys* **22** (1967), 1–79. Translated from *Uspeki Mat. Nauk SSSR* **22** (1967) 3–81.

[14] Siegel, C. L., Über einige anwendungen diophantischer Approximationen, *Abh. Preuss. Akad. Wis. Phys. Math. Kl. Berlin* (1929), reprinted in C. L. Siegel Gesammelte Abhandlungen I, Springer (1966), 209–266.

[15] Siegel, C. L., Transcendental numbers, *Ann. of Math. Studies* **16**, Princeton University Press, Princeton, 1949.

[16] Yebbou, J. Calcul de facteurs déterminants, submitted to *J. Diff. Equations*.

5

ALGEBRAIC VALUES OF HYPERGEOMETRIC FUNCTIONS

F. Beukers and J. Wolfart

A natural question one can can ask in transcendence theory is about values of Taylor series at algebraic points. To be more precise, for the rest of this introduction let

$$f(z) = \sum_{n=0}^{\infty} f_n z^n$$

be a transcendental function with Taylor coefficients $f_n \in K$, $[K : \mathbf{Q}] < \infty$ and radius of convergence $\rho > 0$.

Question: Does it always follow from $\xi \in \overline{\mathbf{Q}}$, $0 < |\xi| < \rho$ that $f(\xi) \notin \overline{\mathbf{Q}}$? ($\overline{\mathbf{Q}}$ denotes the set of numbers algebraic over $\mathbf{Q}$.)

In general the answer is negative, as illustrated by the following counterexample.

Theorem (Stäckel, 1895). *There exists an entire analytic transcendental function $f(z)$ with Taylor coefficients in $\mathbf{Q}$ such that $f(\xi) \in \overline{\mathbf{Q}}$ for all $\xi \in \overline{\mathbf{Q}}$.*

For a history of this theorem and subsequent ones, see [11] Ch. 3. In order to make the above question into a sensible one, there should be some structure hypotheses on $f(z)$. The most common ones are that $f(z)$ is either a solution of a linear differential equation or a linear functional equation. For an account of the latter possibility we refer to [10]. Here we adopt the first possibility and from now on we assume the

D.E.-hypothesis: There exist $p_i(z) \in \overline{\mathbf{Q}}[z]$, $i = 0, \ldots, n$ such that

$$p_n(z)f^{(n)} + p_{n-1}(z)f^{(n-1)} + \ldots + p_1(z)f' + p_0(z)f = 0.$$

In addition we need some hypotheses on the arithmetic nature of the coefficients f_n. The most common ones that are studied, read

E-function hypothesis: $f_n = a_n/n!$ and there exists $c > 1$ such that

(i) $\lceil \overline{a_n} \rceil = O(c^n)$ (ii) $\underset{i=1,\ldots,n}{\text{l.c.m.}}$ (denominator a_i) $= O(c^n)$

where $\lceil \overline{x} \rceil$ denotes the maximal absolute value of the conjugates of $x \in \overline{\mathbf{Q}}$ and 'l.c.m.' stands for 'lowest common multiple'.

G-function hypothesis: there exists $c > 1$ such that

(i) $\lceil \overline{f_n} \rceil = O(c^n)$ (ii) $\underset{i=1,\ldots,n}{\text{l.c.m.}}$ (denominator f_i) $= O(c^n)$

Examples of *E*-functions are of course e^z and the Bessel function

$$J_0(z) = \sum_{n=0}^{\infty} \frac{(-z^2/4)^n}{(n!)^2} = \sum_{n=0}^{\infty} \binom{2n}{n} \left(-\frac{1}{4}\right)^n \frac{z^{2n}}{(2n)!}.$$

A standard example of *G*-functions is given by Gauss' hypergeometric function

$$F(\alpha,\beta,\gamma|z) = 1 + \frac{\alpha\beta}{\gamma \cdot 1}z + \frac{\alpha(\alpha+1)\beta(\beta+1)}{\gamma(\gamma+1)1 \cdot 2}z^2 + \ldots \tag{1}$$

where $\alpha.\beta, \gamma \in \mathbf{Q}$, $\gamma \neq 0, -1, -2, \ldots$.

The arithmetic nature of values at algebraic points of both types of function, can be studied using a method initiated by C. L. Siegel. For *E*-functions the method was completed successfully by Shidlovsky and for *G*-functions the method was developed notably by Galoschkin, Bombieri and Chudnovsky.

Theorem (Shidlovsky, 1954). *Let $f(z)$ be an E-function which is not a polynomial and which satisfies the D.E. hypothesis. Then, for any $\xi \in \overline{\mathbf{Q}}$ with $\xi p_n(\xi) \neq 0$ we have $f(\xi) \notin \overline{\mathbf{Q}}$.*

For a proof see [11] Ch. 6, Theorem 10.

Theorem (Chudnovsky, 1983). *Let $f(z)$ be a transcendental G-function satisfying the D.E. hypothesis. For any $t > 1$ there exists $c = c(f,t)$ such that for any $\xi \in \overline{\mathbf{Q}} \setminus \{0\}$ of degree $\leq t$, it follows from $|\xi| < \exp(-c(\log\lceil\overline{\xi}\rceil)^{4n/4n+1})$, that $f(\xi)$ is not algebraic of degree $\leq t$.*

For a proof see [3] Theorem 1 and for some more explicit bounds [1], §12.

Note that the situation for G-functions is much less satisfactory than for E-functions, although the general expectation was that a similar theorem should hold for G-functions as well. However, very recently the second author [19] has provided a class of counterexamples to this expectation. It consists of certain hypergeometric functions as defined in (1).

Theorem (Wolfart, 1985). *Let* $F(z) = F(\alpha, \beta, \gamma | z)$, α, β, γ *be as above. We distinguish the following possibilities:*

(A) $F(z)$ *is algebraic over* $\mathbf{C}(z)$. *Then* $\xi \in \overline{\mathbf{Q}} \Rightarrow F(\xi) \in \overline{\mathbf{Q}}$.

(B) *The monodromy group of* $F(z)$ *is an arithmetic hyperbolic triangle group and* $\gamma < 1, 0 < \alpha < \gamma\, 0 < \beta < \gamma, 1 - \gamma + |\alpha - \beta| + |\gamma - \alpha - \beta| < 1$. *Then there is a set* $E \subset \overline{\mathbf{Q}}$, *dense in* $\mathbf{C}$, *such that* $F(\xi) \in \overline{\mathbf{Q}}$ *whenever* $\xi \in E$.

(C) *None of the above. Then there are only finitely many* $\xi \in \overline{\mathbf{Q}}$ *such that* $F(\xi) \in \overline{\mathbf{Q}}$.

A complete list of algebraic hypergeometric functions can be inferred from Schwarz' list as given in [7] §57 or [19] §6. A theorem of Takeuchi [16] says that there are 85 arithmetic triangle groups. From this list one can infer the possibilities corresponding to case (B).

The proof of the transcendence of values of $F(\alpha, \beta, \gamma | z)$, $z \in \overline{\mathbf{Q}}$ has only recently become possible through the use of a deep theorem of Wüstholz [20] quoted in §1. The other surprising result of this theorem is the existence of the infinite set E in case (B), that is, certain hypergeometric G-functions assume algebraic values at algebraic points infinitely often.

In this paper we explain the basic ideas that enter the proof of Wolfart's theorem by means of the example $F(\frac{1}{12}, \frac{5}{12}, \frac{1}{2} | z)$. Let $J(\tau)$ be the modular invariant normalised by $J(i) = 1$. Then the result proved in this paper will be

Theorem 1 *Let* $z \in \mathbf{C}$, $|z| < 1$. *Then*

$$z \in \overline{\mathbf{Q}}, \quad F\left(\frac{1}{12}, \frac{5}{12}, \frac{1}{2} \middle| z\right) \in \overline{\mathbf{Q}}$$

$$\Longleftrightarrow \exists \tau, \tau \in \mathbf{Q}(i), \operatorname{Im}\tau > 0 \text{ such that } z = 1 - J(\tau)^{-1}.$$

In §1 we prove the '⟹' part where we closely follow the method from [18]. It is here that Wüstholz's transcendence theorem is crucial. Then, in §2, we prove the '⟸' part by an *ad hoc* method which uses the fact that $F\left(\frac{1}{12},\frac{5}{12},\frac{1}{2}|z\right)$ is almost uniformised by $z = 1 - J(\tau)^{-1}$. The ideas contained in this method are classical and were known at the beginning of this century. Yet, to our knowledge, the '$\Leftarrow$' implication of Theorem 1 was not noticed.

By entirely analogous methods one can prove:

Theorem 2. *Let* $z \in \mathbf{C}$, $|z| < 1$. *Then*

$$z \in \overline{\mathbf{Q}}, \quad F\left(\frac{1}{12},\frac{5}{12},\frac{1}{2}\,\middle|\,z\right) \in \overline{\mathbf{Q}}$$

$$\Longleftrightarrow \exists\, \tau,\ \tau \in \mathbf{Q}(\sqrt{-3}),\ \operatorname{Im}\tau > 0\ \textit{such that}\ z = \frac{J(\tau)}{J(\tau)-1}.$$

As amusing examples we verify in §2 the following special values.

Theorem 3.

$$F\left(\frac{1}{12},\frac{5}{12},\frac{1}{2}\,\middle|\,\frac{1323}{1331}\right) = \frac{3}{4}\sqrt[4]{11}, \qquad F\left(\frac{1}{12},\frac{7}{12},\frac{2}{3}\,\middle|\,\frac{64000}{64009}\right) = \frac{2}{3}\sqrt[6]{253}.$$

Theorems 1, 2, 3 are examples of case (B) of Wolfart's theorem. They arise essentially through the theory of complex multiplication. In case (C) there may also be hypergeometric functions having algebraic values at algebraic points. As explained in [18] the finitely many exceptional algebraic ξ are mapped by the triangle function associated to $F(z)$ on some fixed point of the maximal triangle group extending the monodromy group of F, at least if α, β, γ, $\alpha - \gamma$, $\beta - \gamma \notin \mathbf{Z}$. Generically this monodromy group is maximal itself, and then the only possible exceptional arguments are $\xi = 0\ (F(0) = 1)$, $\xi = 1$ or $\xi = \infty$. In fact, sometimes (not always!) $F(1)$ is algebraic as in the example

$$F(2a, 2b, a + b + \tfrac{1}{2}|1) = \cos \pi(a - b)/\cos \pi(a + b).$$

However, if the monodromy group is not maximal, additional fixed points are to be considered, such as those corresponding to $\xi = 1/2$, $-1/3$, $1/9$. Then F can be expressed by other hypergeometric functions having this extended monodromy group, using quadratic or higher transformations (compare [5]). With Gauss' formula for the values of these hypergeometric functions at the argument 1 and the classical Γ-function identities one obtains sometimes algebraic values such as the following ones:

Theorem 4. $F(1-3a, 3a, a|\tfrac{1}{2}) = 2^{2-3a}\cos\pi a$, $F(2a, 1-4a, 1-a|\tfrac{1}{2}) = 4^a\cos\pi a$,

$$F\left(\frac{7}{48}, \frac{31}{48}, \frac{29}{24}\,\bigg|-\frac{1}{3}\right) = 2^{5/24}\cdot 3^{-11/12}\cdot 5\sqrt{\frac{\sin\pi/24}{\sin 5\pi/24}},$$

$$F\left(-\frac{7}{6}, -\frac{2}{3}, \frac{1}{18}\,\bigg|\frac{1}{9}\right) = 2^{29/2}\cdot 3^{-7/6}\sin 4\pi/9.$$

Using the formulas in [5] Vol I, §2.8 the reader can prove these identities as an exercise.

For the exceptional arguments ξ considered in Theorems 1 and 2 it is easy to prove that with $\xi \to 0$ the degree $D = \big[\mathbf{Q}(\xi) : \mathbf{Q}\big]$ goes to infinity as $|\xi|^{-1}$ and $|\xi|^{-2/3}$ respectively. Therefore we are far from the bounds given in Chudnovsky's or Bombieri's theorems which show the non-existence of such exceptional ξ in much smaller discs with radius c^{-D}. Thus one may be tempted to believe that the theorems on G-functions can be improved considerably.

Finally, we would like to point out that it may be possible to find other G-function solutions of second order equations assuming algebraic values at infinitely many algebraic arguments. This might be done through the use of one-dimensional families varieties characterized by their endomorphism algebras. In a future paper we hope to elaborate on these ideas.

1. TRANSCENDENTAL VALUES.

In this and the next section we will mainly consider the following example. Let $F(z)$ denote the hypergeometric function as defined in (1) with parameters $\alpha = 1/12$, $\beta = 5/12$, $\gamma = 1/2$. Felix Klein's lectures [7] from 1893/4 show that it is a classical and well-known fact that up to normalising factors such hypergeometric functions can be written as period integrals. In our example this period is

$$\oint x^{\beta-1}(1-x)^{\gamma-\beta-1}(1-zx)^{-\alpha}dx = \oint x^{-7/12}(1-x)^{-11/12}(1-zx)^{-1/12}\,dx$$

$$= \oint \frac{dx}{y}$$

on the curve $X(12, z)$ whose affine part is given by the equation

$$y^{12} = x^7(1-x)^{11}(1-zx).$$

The cycle of integration can be taken as a 'Pochhammer cycle' on $X(12, z)$ whose image under the natural projection $X(12, z) \to \overline{\mathbf{C}}$: $(x, y) \to x$ (a covering ramified in 0, 1, $1/z$ and ∞) looks like:

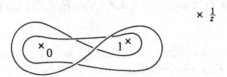

For $z = 0$ we obtain, up to an algebraic factor, Euler's integral representation of the beta-function

$$\oint \eta_2 := B(\beta, \gamma - \beta) = B(5/12, 1/12).$$

Therefore, with $\eta_1 := dx/y$ on $X(12, z)$ and up to an algebraic factor, we have

$$F(z) = \oint \eta_1 \Big/ \oint \eta_2.$$

Of course, we are interested only in algebraic arguments z, and we can suppose $z \neq 0$, 1, ∞. Then $X(12, z)$ and η_1 are defined over $\overline{\mathbf{Q}}$, as η_2, too. To prove the transcendence of $F(z)$ one has only to check that the periods $\oint \eta_1$ and $\oint \eta_2$ are linearly independent over $\overline{\mathbf{Q}}$. To this aim we apply a consequence of a much more general theorem of Wüstholz ([20] §1) which extends previous results of Th. Schneider [13], Masser [12] and Laurent [9]. If we define abelian varieties only up to isogeny, we can formulate this consequence as follows:

Proposition. *Let $\oint \eta_i$ be periods $\neq 0$ of first or second kind on abelian varieties A_i (all defined over $\overline{\mathbf{Q}}$), A_2 simple. Then the quotient $\oint \eta_1 / \oint \eta_2$ can only be algebraic in the obvious case when A_2 is isogenous to a factor of A_1.*

To control this necessary condition, one has to construct abelian varieties A_i on which our $\oint \eta_i$ live as periods; in our present example they are in fact both of first kind.

By classical Γ-function identities, $B(5/12, 1/12)$ equals $B(1/4, 1/2)$ up to an algebraic factor. The latter number is a period of the elliptic curve $y^2 = x^4 - 1$. Hence we can take A_2 isogenous to $\mathbf{C}/\mathbf{Z}[i]$.

The construction of A_1 is more complicated. Of course we start with the Jacobian of the curve $X(12, z)$. Neglecting the algebraic structure we consider $\operatorname{Jac} X(12, z)$ as complex torus $\mathbf{C}^g/\Lambda^*$, where g is the genus of $X(12, z)$ and Λ^* the following period lattice: we choose a basis B^* of differentials of the first kind on the curve and integrate them to obtain

$$\Lambda^* = \{(\oint_\gamma \omega)_{\omega \in B^*} \in \mathbf{C}^g : \gamma \in H_1(X(12, z), \mathbf{Z})\}.$$

The obvious automorphism of $X(12, z)$, $\kappa : (x, y) \to (x, \zeta_{12}^{-1} y)$, $\zeta_{12} = e^{\pi i/6}$, facilitates this construction considerably. By F. Klein [7], a basis of $H_1(X(12, z), \mathbf{Z})$ can be chosen among the Pochhammer cycles around 0 and 1 and those around $1/z$ and ∞; there are 12 of each type, and they are permuted by κ. Of course, as on the homology of the curve, κ acts also in a natural way on the differentials of the first kind. Their basis can be chosen among the eigen-differentials

$$\omega_n = \frac{g(x) dx}{y^n} \quad \text{with } g \in \overline{\mathbf{Q}}(x), \ n \in \mathbf{Z}/12\mathbf{Z}$$

of κ (this is a classical and very general procedure used already by Chevalley and Weil [4]). To every $\gamma \in H_1(X(12, z), \mathbf{Z})$ therefore corresponds a unique pair $(u, v) \in \mathbf{Z}[\zeta_{12}]$ such that for every n prime to 12

$$\oint_\gamma \omega_n = \sigma_n(u) \int_0^1 \omega_n + \sigma_n(v) \int_{1/z}^\infty \omega_n;$$

here, $\int_0^1$ and $\int_{1/z}^\infty$ stand for integration over two fixed Pochhammer cycles around 0, 1 and $1/z$, ∞ respectively. The symbol σ_n stands for the Galois-automorphism given by $\sigma_n : \zeta_{12} \to \zeta_{12}^n$. The n not prime to 12 are less interesting. They are pullbacks of differentials on curves $X(D, z) : y^D = x^7(1 - x)^{11}(1 - zx)$ covered by $X(12, z)$ for all proper divisors $D|12$. All Jacobians of these curves can be considered as factors of $\operatorname{Jac} X(12, z)$. For our purposes only their common complement counts, which we will call A_1 from now on:

$$\operatorname{Jac}(12, z) = A_1 \oplus \sum_{D|12, D \neq 12} \operatorname{Jac} X(D, z).$$

By the general theory of abelian varieties and by our hypothesis $z \in \overline{\mathbf{Q}}$, all these Jacobians are defined over $\overline{\mathbf{Q}}$, and their complement as well.

Now we have to calculate those $\omega_n \in B^*$ with $n \in (\mathbf{Z}/12\mathbf{Z})^*$ since they form a basis B of the differentials of the first kind on A_1. Using appropriate uniformizing variables in $x = 0, 1, 1/z$ and ∞ and calculating orders, one finds the following four basis differentials:

$$\omega_1 = \eta_1 = \frac{dx}{y} = x^{-7/12}(1-x)^{-11/12}(1-zx)^{-1/12}\,dx$$

$$\omega_5 = \frac{x^2(1-x)^4\,dx}{y^5} = x^{-11/12}(1-x)^{-7/12}(1-zx)^{-5/12}\,dx$$

$$\omega_{-5} = \frac{y^5\,dx}{x^3(1-x)^5(1-zx)} = x^{-1/12}(1-x)^{-5/12}(1-zx)^{-7/12}\,dx$$

$$\omega_{-1} = \frac{y\,dx}{x(1-x)(1-zx)} = x^{-5/12}(1-x)^{-1/12}(1-zx)^{-11/12}\,dx.$$

Remarks.

1) As predicted by the general theory, the κ-eigenspaces in the space of differentials of our A_1 are one-dimensional (compare [18] §4 or [4]).

2) In the same way as $\omega_1 = \eta_1$ the ω_n correspond to certain hypergeometric functions, only for different parameters α, β, γ. Since later the parameter-differences $\lambda := 1 - \gamma$, $\mu := \alpha - \beta$, $\nu := \gamma - \alpha - \beta$ will become important, we give their values only:

n	λ	μ	ν
1	1/2	1/3	0
5	1/2	-1/3	0
-5	-1/2	1/3	0
-1	-1/2	-1/3	0

$$(2)$$

As explained above, we can now describe A_1 as a four-dimensional complex torus $\mathbf{C}^4/\Lambda_1^*$ by its period lattice

$$\Lambda_1^* = \left\{ \left(\sigma_n(u) \int_0^1 \omega_n + \sigma_n(v) \int_{1/z}^\infty \omega_n \right)_{n=1,5,-5,-1} : u,v \in \mathbf{Z}[\zeta_{12}] \right\}.$$

Dividing each component by $\int_0^1 \omega_n$ and introducing

$$D_n(z) := \int_{1/z}^\infty \omega_n \Big/ \int_0^1 \omega_n,$$

we can replace Λ_1^* by the isomorphic lattice

$$\Lambda_1 = \{ (u + vD_1(z),\ \sigma_5(u) + \sigma_5(v)D_5(z),\ \overline{\sigma_5(u)} + \overline{\sigma_5(v)}D_{-5}(z),$$
$$\overline{u} + \overline{v}D_{-1}(z)) : u,v \in \mathbf{Z}[\zeta_{12}] \}. \quad (3)$$

Since $\int_{1/z}^{\infty} \omega_n$ and $\int_0^1 \omega_n$ are, as functions of z, linearly independent solutions for the same hypergeometric differential equation, their quotient $D_n(z)$ is a Schwarz' triangle function (H. A. Schwarz 1873 [14]; see also [7], [2] or the short resumé in [19] §4). $D_n(z)$ is a biholomorphic map of the upper half plane $\mathcal{H}$ onto a curvilinear triangle in $\mathbf{C}$ with vertices $D_n(0)$, $D_n(\infty)$, $D_n(1)$ and angles $|\lambda|\pi$, $|\mu|\pi$, $|\nu|\pi$ respectively (compare the definition of λ, μ, ν and their list (2)). Kummer's classical identities and Gauss' formula for $F(\alpha, \beta, \gamma|1)$ allow us to compute the shape of these triangles explicitly. In our case we obtain

$$D_1(z) = D_5(z) = \frac{1}{D_{-5}(z)} = \frac{1}{D_{-1}(z)}$$

for all z and in particular

$$D_1(0) = 0, \qquad D_1(\infty) = -i(2 - \sqrt{3}), \qquad D_1(1) = -1.$$

In fact, $D_1(\mathcal{H})$ is a hyperbolic triangle in the unit disc. On the other hand, $D_1(\mathcal{H})$ is mapped to the left half F^+ of the well-known fundamental domain of the elliptic modular group by the fractional linear transformation μ,

$$w \to \tau := -i\,\frac{w-1}{w+1}.$$

Since the elliptic modular invariant $J : \mathcal{H} \to \mathbf{C}$, normalised by $J(i) = 1$, $J(e^{2\pi i/3}) = 0$ and $J(i\infty) = \infty$, maps F^+ biholomorphically onto $\mathcal{H}$, we obtain by the uniqueness part of the Riemann mapping theorem: $\mu \circ D_1$ is the inverse function of $1 - J^{-1}$. So $z = 1 - J(\tau)^{-1}$.

Proof of the '⇒' part of Theorem 1. Assume that both z and $F(1/12, 5/12, 1/2|z)$ are algebraic. Then, according to Wüstholz' Proposition, A_1 contains a factor isogenous to A_2. The lattice of A_1 as given by (3), reads:

$$\Lambda_1 = \left\{ u + v\omega, \sigma_5(u) + \sigma_5(v)\omega, \overline{\sigma_5(u)} + \overline{\sigma_5(v)}\frac{1}{\omega}, \overline{u} + \overline{v}\frac{1}{\omega} : u, v \in \mathbf{Z}[\zeta_{12}] \right\}$$

where $\omega = D_1(z)$. The lattice of A_2 equals $\mathbf{Z}[i]$. Isogeny implies that we have a C-linear map $\phi : \mathbf{C} \to \mathbf{C}^4$ such that $\phi : \mathbf{Z}[i] \to \Lambda_1$. Suppose

$$\phi : 1 \to (u + v\omega,\ \sigma_5(u) + \sigma_5(v)\omega, \ldots) \quad \text{for some } u, v \in \mathbf{Z}[\zeta_{12}]$$

$$\phi : i \to (u' + v'\omega,\ \sigma_5(u') + \sigma_5(v')\omega, \ldots) \tag{4}$$

$$\text{for some } u', v' \in \mathbf{Z}[\zeta_{12}].$$

By C-linearity we have:

$$\phi(i) = i\phi(1) = \left(i(u + v\omega),\ i(\sigma_5(u) + \sigma_5(v)\omega), \ldots \right).$$

It is a straightforward matter to verify that this can only be compatible with (4) if $\omega \in \mathbf{Q}(i)$. Hence $\tau \in \mathbf{Q}(i)$, which proves our assertion.

Remarks. (1). There is an obvious converse to the Proposition involving only some extra conditions on the differentials η_i and the cycles of integration. Along the same lines of proof, this leads to the '$\Leftarrow$' part of Theorem 1. In fact, every $\tau \in \mathbf{Q}(i) \cap \mathcal{H}$ gives an algebraic argument $z = 1 - J(\tau)^{-1}$ such that the value $F(z)$ is also algebraic. But this method doesn't give any information on the algebraic nature of $F(z)$, so we will treat this question in Section 2 by a different idea.

(2). For the comparison with the relations considered by Bombieri one should note that the second solution

$$z^\lambda F(1 + \alpha - \gamma, 1 + \beta - \gamma, 2 - \gamma | z) = zF\left(\frac{11}{12}, \frac{7}{12}, \frac{3}{2} \bigg| z \right)$$

of the differential equation for $F(z)$ has, of course, the same monodromy group. However, this second solution takes transcendental values for every algebraic $z \neq 0$ because it is the quotient of a period of the first kind by a period of the second kind, and such quotients are always transcendental by [20] §1. In the terminology of the introduction, this $F(\alpha', \beta', \gamma' | z) = F(11/12, 7/12, 3/2 | z)$ doesn't satisfy $\gamma' < 1$.

2. ALGEBRAIC VALUES

In this section we shall prove the '$\Leftarrow$' part of Theorem 1.

Proposition 2.1. *Let $J(\tau)$ be the modular J-invariant on $\mathcal{H} = \{\tau\,|\,\mathrm{Im}\,\tau > 0\}$ normalised such that $J(i) = 1$. Let $\eta(\tau) = q^{1/24} \prod_{n=1}^{\infty}(1 - q^n)$, $q = e^{2\pi i \tau}$ be the Dedekind η-function. Then we have in the neighbourhood of $\tau = i$ given by $\left| 1 - J(\tau)^{-1} \right| < 1$ the relation*

$$F\left(\frac{1}{12}, \frac{5}{12}, \frac{1}{2} \bigg| 1 - \frac{1}{J(\tau)} \right) = A(\tau + i)\eta(\tau)^2 \left(J(\tau) \right)^{1/12}$$

where $A \in \mathbf{C}$.

Proof. Consider the standard family of elliptic curves

$$y^2 = 4x^3 - 27\frac{J}{J-1}(x+1) \tag{5}$$

parametrised by $J = J(\tau)$. According to Klein-Fricke [6] p. 34 the periods Ω of dx/y are multivalued functions of J satisfying the differential equation

$$\frac{d^2\Omega}{dJ^2} + \frac{1}{J}\frac{d\Omega}{dJ} + \frac{(31J-4)}{144J^2(J-1)^2}\Omega = 0. \tag{6}$$

It is not hard to verify that $(1 - J^{-1})^{1/4} F(1/12, 5/12, 1/2|1 - J^{-1})$ is a solution of it.

Now consider the lattice $\mathbf{Z} + \mathbf{Z}\tau$. Th elliptic curve having this lattice as a period lattice is

$$y^2 = 4x^3 - g_2(\tau)x - g_3(\tau) \tag{7}$$

where g_2, g_3 are the standard Eisenstein series of weight 4 and 6 respectively. For each τ the curve (5) is isomorphic to (7) and the period lattice of (5) is obtained from $\mathbf{Z} + \mathbf{Z}\tau$ by multiplication with $((J-1)g_2/27J)^{1/4} = 2\pi \cdot 3^{-3/4}(J-1)^{1/4}J^{-1/6}\eta^2$ according to Tate's formulas [17] §2. Hence, $(J-1)^{1/4}J^{-1/6}\eta^2$ and $\tau(J-1)^{1/4}J^{-1/6}\eta^2$ considered as multivalued functions of J, satisfy (6) and thus there exist A, $B \in \mathbf{C}$ such that

$$\left(\frac{1-J}{J}\right)^{1/4} F\left(\frac{1}{12}, \frac{5}{12}, \frac{1}{2}\bigg|1 - \frac{1}{J}\right) = (J-1)^{1/4}J^{-1/6}\eta(\tau)^2(A\tau + B)$$

which implies

$$F\left(\frac{1}{12}, \frac{5}{12}, \frac{1}{2}\bigg|1 - \frac{1}{J}\right) = J^{1/12}\eta(\tau)^2(A\tau + B). \tag{8}$$

Notice that the left-hand-side of (8) considered as a function of τ around i is invariant with respect to $\tau \to -1/\tau$. Using $\eta(-1/\tau)^2 = \tau\eta(\tau)^2/i$ we see that the right-hand-side is invariant if and only if $B = Ai$. Hence our Proposition follows.

Remark. In a similar way as above one shows the existence of $A \in \mathbf{C}$ such that

$$F\left(\frac{1}{12}, \frac{7}{12}, \frac{2}{3}\bigg|\frac{J}{J-1}\right) = A\left(\tau + \frac{1}{2} + \frac{1}{2}\sqrt{-3}\right)(1 - J)^{1/12}\eta(\tau)^2 \tag{9}$$

in the neighbourhood of $\tau = \omega = -\frac{1}{2} + \frac{1}{2}\sqrt{-3}$ given by $\left| J/(J-1) \right| < 1$.

Proposition 2.2. Let A, B, C, $D \in \mathbf{Z}$ and $d = \det \begin{pmatrix} A & B \\ C & D \end{pmatrix} > 0$.
Denote $T\tau = (A\tau + B)/(C\tau + D)$. Then $\sqrt{d}\,\eta(T\eta)/\eta(\tau)$ satisfies a monic polynomial equation with coefficients in $\mathbf{Z}[12^3 J]$.

Proof. This is a well-known fact from the theory of modular forms. A proof can be found in [8] p. 163.

Proof of the '$\Leftarrow$' implication of Theorem 1. Suppose $z = (Ai + B)/D$, A, B, $D \in \mathbf{Z}$, $(A, B, D) = 1$. Writing $F(z) = F(1/12, 5/12, 1/2|z)$ it follows from Proposition 2.1 that

$$\frac{F\left(1 - J\left(\frac{A\tau + B}{D}\right)^{-1}\right)}{F\left(1 - J(\tau)^{-1}\right)} = \frac{A\tau + B + Di}{D(\tau + i)} \left(\frac{\eta\left(\frac{A\tau + B}{D}\right)}{\eta(\tau)}\right)^2 \left(\frac{J\left(\frac{A\tau + B}{D}\right)}{J(\tau)}\right)^{\frac{1}{12}}.$$

Substitute $\tau = i$. By the theory of complex multiplication it is known that $J((Ai + B)/D) \in \overline{\mathbf{Q}}$. Proposition 2.2 implies that $\eta((Ai + B)/D)/\eta(i) \in \overline{\mathbf{Q}}$ since $J(i) = 1$. Thus our implication follows. ∎

Proof of Theorem 3. Apply the above method and take $A = 2$, $B = 0$, $D = 1$. It is not hard to verify that

$$J(\tau) = \frac{4}{27} \frac{(\phi + 1)^3}{\phi} \quad \text{where} \quad \phi(\tau) = 256 \left(\frac{\eta(2\tau)}{\eta(\tau)}\right)^{24}.$$

Hence $J(i) = 1$ implies $\eta(i) = 1/2$, which implies $\left(\eta(2i)/\eta(i)\right)^2 = 2^{-3/4}$. Furthermore, $J(2i) = 2^3 \cdot 3^3 \cdot 11^3/1728 = (11/2)^3$. Hence

$$F\left(\frac{1}{12}, \frac{5}{12}, \frac{1}{2} \middle| 1 - \left(\frac{2}{11}\right)^3\right) = \frac{3i}{2i}\left(\frac{1}{2}\right)^{3/4}\left(\frac{11}{2}\right)^{1/4} = \frac{3}{4}\sqrt[4]{11}.$$

For the second identity we use exactly the same argument starting from identity (9). One takes $A = 3$, $B = 1$, $D = 1$ and substitutes $\tau = \omega$ afterwards. Given the following formulas the reader should be able to verify the second statement.

$$J(\tau) = \frac{1}{64} \frac{(\phi + 1)(1 + 9\phi)^3}{\phi}, \qquad \phi(\tau) = -27\left(\frac{\eta(3\tau + 1)}{\eta(\tau)}\right)^{12},$$

$$J(\omega) = 0 \Rightarrow \left(\frac{\eta(3\omega + 1)}{\eta(\omega)}\right)^2 = 3^{-5/6},$$

$$1 - J(3\omega + 1) = (11 \cdot 23/3)^2, \qquad J(3\omega + 1) = -2^9 \cdot 5^3 \cdot 3^{-2}. \qquad ∎$$

80 HYPERGEOMETRIC FUNCTIONS

References

[1] E. Bombieri, On G-functions, pp. 1–67 from *Recent progress in analytic number theory*, vol 2, Durham 1979, Academic Press, London 1981.

[2] C. Carathéodory, *Funktiontheorie* II, Birkhäuser 1961.

[3] D. V. Chudnovsky, G. V. Chudnovsky, Applications of Padé-approximations to Diophantine inequalities in values of G-functions. *Lecture Notes in Mathematics* **1135**, 9–51, Springer, 1985.

[4] Cl. Chevalley, A. Weil, Über das Verhalten der Integrale 1. Gattung bei Automorphismen des Funktionenkörpers, *Abh. Math. Sem. Hamburg* **10** (1934), 358–361.

[5] A Erdélyi, M. Magnus, F. Oberhettinger, F. G. Tricomi, *Higher transcendental functions*, Bateman manuscript project, McGraw-Hill, 1953.

[6] F. Klein, *Vorlesungen über die Theorie der elliptischen Modulfunktionen*, ausgearbeitet und vervollständigt von R. Fricke, Leipzig, 1890.

[7] F. Klein, *Vorlesungen über die hypergeometrische Funktion*, Springer, 1933.

[8] S. Lang, *Elliptic functions*, Addison-Wesley, 1973.

[9] M. Laurent, Sur la transcendence du rapport de deux intégrales eulériennes, pp. 133–141 in: *Approximations diophantiennes et nombres transcendants*, Luminy 1982, *Progress in Math.* **31** Birkhäuser 1983.

[10] J. H. Loxton, A. J. van der Poorten, Arithmetic properties of the solution of a class of functional equations, *J. reine angew. Math.* **330** (1982), 159–172.

[11] K. Mahler, Lectures on transcendental numbers, *Lecture Notes in Math.* **546**, Springer, 1976.

[12] D. Masser, The transcendence of certain quasi-periods associated with abelian functions in two variables, *Compos. Math.* **35** (1977), 239–278.

[13] Th. Schneider, Zur Theorie der abelschen Funktionen und Integrale, *J. reine angew. Math.* **183** (1941), 110–128.

[14] H. A. Schwarz, Über die jenige Fälle, in denen die Gaussische hypergeometrische Reihe eine algebraische Funktion ihres vierten Elements darstellt, *J. reine angew. Math.* **75** (1873), 292–335.

[15] G. Shimura, On analytic families of polarized abelian varieties and automorphic functions, *Ann. of Math.* **78** (1963), 149–192.

[16] K. Takeuchi, Arithmetic triangle groups, *J. Math. Soc. Japan* **29** (1977), 91–106.

[17] J. Tate, The arithmetic of elliptic curves, *Inv. Math.* **23** (1974), 179–206.

[18] J. Wolfart, Werte hypergeometrischer Funktionen, preprint.

[19] J. Wolfart, Fonctions hypergéométriques, arguments exceptionnels et groupes de monodromie, in *Publ. Math. Univ. P. et M. Curie, Problèmes diophantiens*, 1985–1986, IX.1–IX.24.

[20] J. Wolfart, G. Wüstholz, Der Überlagerungsradius gewisser algebraischer Kurven und die Werte der Betafunktion an rationalen Stellen, *Math. Ann.* **273** (1985), 1–15.

6

SOME NEW APPLICATIONS
OF AN INEQUALITY OF MASON

B. Brindza

1. Introduction

Several Diophantine problems can be reduced to the so-called superelliptic (or hyperelliptic, if $m = 2$) equation

$$f(x) = y^m \tag{1}$$

where $f(x)$ is a given polynomial with rational (or algebraic) coefficients, $m > 1$ is a given positive integer and the unknowns x, y are rational (or algebraic) integers. In the study of this equation the distribution of the multiplicities of the zeros of $f(x)$ plays an important role. First we present some lower bounds for the numbers of distinct and simple zeros of polynomials, obtained by using some inequalities of Mason and Brownawell-Masser concerning unit equations over function fields. Then we combine these results with known results on equation [1] proved by the Gelfond-Baker method, to obtain various generalizations and applications.† We note that several other applications of Mason's inequality have been earlier obtained by Mason to Thue, superelliptic, norm form and decomposable form equations (cf. Mason's paper in this volume). Moreover, we apply the above-mentioned inequality of Mason to derive an effective upper bound for the degrees of the polynomial solutions x, y, z and the exponent n of the equation

$$F(x, y) = z^n$$

over function fields where $F(x, y)$ is a binary form.

The proofs of the results presented in the paper will be given elsewhere (see [5] and [6]).

---

† This part of the research has been done during a stay at the University of Leiden, (cf. [6]).

2. The superelliptic equation

We assume throughout that the polynomial $f \in \mathbf{Z}[X]$ has the representation

$$f(X) = a \prod_{i=1}^{n} (X - \alpha_i)^{r_i}$$

where a is different from zero and $\alpha_1, \ldots, \alpha_n$ are distinct algebraic numbers. Improving a well-known result of Siegel [20], in 1963 LeVeque [14] showed in a more general form that equation (1) has only finitely many rational integer solutions x and y, unless

$$\left\{ \frac{m}{(m_1, r_1)}, \ldots, \frac{m}{(m_1, r_n)} \right\}$$

is a permutation of one of the n-tuples

$$\{t, 1, \ldots, 1\}, \quad t \geq 1, \quad \text{or} \quad \{2, 2, 1, \ldots, 1\}$$

when (1) may have infinitely many solutions.

LeVeque's theorem is ineffective. A. Baker [1] proved the first effective result for [1] in the most important special case when f has at least two (or three if $m = 2$) simple zeros. Later, Sprindžuk [22] and Trelina [24] extended Baker's result to the more general case when the ground ring is the ring of integers or S-integers in an arbitrary but fixed algebraic number field. In 1983 I proved the effective version of LeVeque's result (see [2] and [21]). The proof involved the Baker method.

In the following paragraphs I shall formulate some common applications of this theorem and Mason's inequality.

3. The equation $\mathbf{F}^n(x) + \mathbf{G}^m(x) = \mathbf{y}^z$

In 1946 Inkeri [11] showed that, for a given prime $p \geq 3$ there exists at most a finite number of positive integer triplets $\{x, y, z\}$ which satisfy the equation

$$x^p + y^p = z^p, \qquad (x, y, z) = 1$$

for which one of the differences $|x - y|$, $z - x$, $z - y$ is less than a given positive number. His proof is very long; later Everett [9] gave a new proof for this theorem using Roth's famous theorem on approximation of algebraic numbers. Some years ago Stewart [23] and Inkeri-van der Poorten [13] independently showed that, for any positive number M all positive integer solutions $x, y, z > 1$, $n > 2$ of the equation

$$x^n + y^n = z^n \quad \text{with} \quad |x - y| \leq M$$

are bounded by an effectively computable constant which depends only on M. Later Inkeri [12] studied the more general equation

$$f^n(x) + g^n(x) = z^n \qquad \text{in } x, z \in \mathbf{Z} \qquad (2)$$

where f and g are non-constant and relatively prime polynomials from $\mathbf{Z}[X]$. Under some assumptions made on f and g, he gave an effective upper bound for the solutions depending only on n, f, g. In 1984 I proved that the conditions in Inkeri's theorem can be omitted. Later Győry, Tijdeman and I [7] obtained some similar and more general results.

Now, consider the more general equation

$$aF^n(x) + bG^m(x) = y^z \qquad \text{in } x, y, z \in \mathbf{Z} \text{ with } y, z > 1 \qquad (3)$$

where F and G are non-constant relatively prime polynomials in $\mathbf{Z}[X]$ and $n, m \geq 2$ and a, b are given non-zero integers. The equations (2) and (3) can be considered as superelliptic equations and as was mentioned above, the applicability of the Baker method depends on the factorization of the left-hand sides. Mason's inequality made it possible to obtain a sharp lower bound for the number of simple zeros of the polynomial $P = aF^n + bG^m$.

Theorem 1. (Brindza [6]). *If $\frac{2}{n} + \frac{2}{m} \leq 1$ and $n \deg(F) \geq m \deg(G)$ then the polynomial P has at least*

$$2 + n\left(1 - \frac{2}{n} - \frac{2}{m}\right) \deg F$$

simple zeros.

$n \deg F$ is a trivial upper bound for the number of simple zeros of P, hence our theorem shows that "almost" every zero of P is simple and

$$\deg P \geq 2 + n\left(1 - \frac{2}{n} - \frac{2}{m}\right) \deg(F).$$

The condition $\frac{2}{n} + \frac{2}{m} \leq 1$ is necessary because it can be written in the form $4 \leq (n-2)(m-2)$ and if $(n-2)(m-2) = 3, 2$ or 1 then P may have no simple zero at all. Combining this result with my above-mentioned result on superelliptic equations we have the following theorem.

Theorem 2. (Brindza [6]). *If $(n-2)(m-2) \geq 4$ then all rational integer solutions x, y, z of the equation (3) with $z > 2$ and $|y| > 1$ satisfy*

$$\max\{|x|, |y|, z\} < C_1$$

where C_1 is an effectively computable number depending only on the polynomial P.

We note that using an improvement of Brownawell and Masser concerning S-unit equations over function fields, the above-quoted theorems can be extended to more general equations of the type

$$f_1^{k_1}(x) + \ldots + f_n^{k_n}(x) = u_1^{w_1} \ldots u_s^{w_s} y^z \quad \text{in } x, y, z, w_1, \ldots, w_s \in \mathbb{Z}$$

where $u_1, \ldots, u_s$, $s \geq 0$, are distinct primes (cf. Brindza [6]).

4. Power values of the sum $1^k + 2^k + \ldots + x^k$

The arithmetical properties of the sum

$$S_k(x) = 1^k + 2^k + \ldots + x^k$$

have been studied by several authors. It is well-known that $S_k(x)$ can be expressed by Bernoulli polynomials, namely

$$S_k(x) = \frac{1}{k+1}\big(B_{k+1}(x+1) - B_{k+1}(0)\big)$$

holds for every positive integer x where B_i denotes the i-th Bernoulli polynomial. Hence the equation

$$S_k(x) = y^z, \tag{4}$$

where k is a given positive integer, is a superelliptic equation again. Schäffer [17] determined all the pairs $\{k, z\}$ for which (4) has infinitely many positive integer solutions x and y, and in other cases he gave an upper bound for the number of solutions. Later Győry, Tijdeman and Voorhoeve [10] proved the following nice effective generalization.

Theorem A. *Let s be a given positive square-free integer and r an arbitrary but fixed integer. If $k \notin \{1, 3, 5\}$ then all integer solutions x, y, $z > 1$ of the equation*

$$s \cdot S_k(x) + r = y^z \tag{5}$$

satisfy

$$\max\{x, y, z\} < C_2$$

where C_2 is an effectively computable constant depending only on k.

In the proof they used Baker's effective result on the superelliptic equation. Using the effective version of LeVeque's theorem I showed in [3] that a similar result can be proved on the more general equation

$$F\big(S_k(x)\big) = y^z$$

where F is a given polynomial with rational integer coefficients. The existence of at least one simple zero of F was enough to give an effective bound for x, y and z. For instance, in Theorem A, s can be chosen to be an arbitrary non-zero integer. Similarly, Győry, Tijdeman and Voorhoeve [25] proved the surprising result that the additive term r in (5) can be replaced by an arbitrary but fixed polynomial $R(X) \in \mathbf{Z}[X]$, and under a necessary condition on k, the more general equation so obtained has also finitely many integer solutions x, y, $z > 1$. Their theorem is ineffective for x and y because they used LeVeque's ineffective result. Its effective version formulated above renders the theorem of Győry, Tijdeman and Voorhoeve effective (cf. Brindza [3]).

5. Mason's inequality and the equation $F(x, y) = z^n$

Let k be an algebraically closed field of characteristic zero and $k(t)$ be the rational function field k. Further, let $\mathbf{K}$ be a finite algebraic extension of $k(t)$ of genus g. The height of a non-zero element α of $\mathbf{K}$ is defined by

$$H_{\mathbf{K}}(\alpha) = \sum_v \max\big\{0, v(\alpha)\big\}$$

where v runs through the (additive) valuations of $\mathbf{K}/k$ with value group $\mathbf{Z}$. The following inequality is due to Mason.

Theorem B. *Let S be a finite set of valuations on $\mathbf{K}$, and suppose that γ_1, γ_2 and γ_3 are non-zero elements of $\mathbf{K}$ such that $v(\gamma_1) = v(\gamma_2) = v(\gamma_3) = 0$ for all valuations $v \in S$ and*

$$\gamma_1 + \gamma_2 + \gamma_3 = 0.$$

Then either γ_1/γ_2 is an element of k, or

$$H_{\mathbf{K}}(\gamma_1/\gamma_2) \leq \#S + 2g - 2$$

where $\#S$ denotes the cardinality of S.

Mason [16] extended his result to the more general unit equation

$$\gamma_1 + \gamma_2 + \dots + \gamma_n = 0, \qquad n \geq 3$$

and under a natural condition he gave an upper bound for

$$\max_{1 \leq i < j \leq n} H_{\kappa}\left(\gamma_i/\gamma_j\right).$$

Later, his theorem was improved by Brownawell and Masser [8].

Let R denote the polynomial ring $C[t]$ and let $F(X, Y)$ be a binary form with coefficients in R. Consider the equation

$$F(x, y) = z^n \tag{6}$$

where x, y, z are relatively prime polynomials from R and $n > 1$ is a given integer. This equation can be considered as a common generalization of the superelliptic equation and Thue equation. In the special case when $n = \deg(F)$ we have the Thue equation

$$F\left(x/z, y/z\right) = 1$$

where the unknowns x/z and y/z are rational functions in $C[t]$. In 1978 W. M. Schmidt [19] proved the following nice result.

Theorem C. *If F is irreducible (over R) then all relatively prime solutions x, y, $z \in R$ of (6) satisfy*

$$\max\{\deg(x), \deg(y), \deg(z)\} < D \cdot C_3$$

where D is the maximum of degrees of the coefficients of F and C_3 is an effectively computable constant depending only on n and $\deg(F)$, provided

$$n = 2, \quad \deg(F) \geq 17;$$
$$n = 3, \quad \deg(F) \geq 4;$$

or

$$n > 3, \quad \deg(F) \geq 3.$$

Using Theorem B of Mason, I have proved the following result.

Theorem 3. (Brindza [5]). *If $\deg(F) \geq 3$ and $z \notin C$ then*

$$n \leq 30D \cdot \deg(F) + 6g$$

and

$$\max\{\deg(x), \deg(y), \deg(z)\} \leq (D + g)30 \deg F$$

where g denotes the genus of the splitting field of F.

We remark that this theorem is proved in [5] in a more general from when x and y are S-integers in an algebraic function field L.

It is clear that if $z \in C$ then n may be arbitrarily large. In this case (6) is a Thue equation. Schmidt [18] and Mason [15], independently, gave upper bounds for the heights of its solutions in the more general situation when x and y are algebraic functions in an arbitrary but fixed algebraic function field of one variable over the constant field k.

References

[1] Baker A., Bounds for the solutions of the hyperelliptic equation, *Math. Proc. Camb. Phil. Soc.* **65**, (1969), 439–444.

[2] Brindza B., On S-integral solutions of the equation $f(x) = y^m$, *Acta Math. Hung.* **44** (1984), 133–139.

[3] Brindza B., On some generalizations of the equation $1^k + 2^k + \ldots + x^k = y^z$, *Acta. Arith.* **39** (1984), 99–107.

[4] Brindza B., On a Diophantine equation connected with the Fermat equation, *Acta Arith.* **39** (1984), 357–363.

[5] Brindza B., On the equation $F(x,y) = z^m$ over function fields, *Acta Math. Hung.* to appear.

[6] Brindza B., Zeros of polynomials and exponential Diophantine equations, *Compositio Math.* to appear.

[7] Brindza B., Győry K.. Tijdeman R., The Fermat equation with polynomial values as base variables, *Invent. Math.* **80** (1985), 139–151.

[8] Brownawell W. D. and Masser D. W., Vanishing sums in function fields, *Math. Proc. Camb. Phil. Soc.* to appear.

[9] Everett C. J., Fermat's conjecture, Roth's theorem, Pythagorean triangles and Pell's equation. *Duke Math. J.* **40** (1973), 801–804.

[10] Győry K., Tijdeman R. and Voorhoeve M., On the equation $1^k + 2^k \ldots + x^k + r = y^z$, *Acta Arith.* **37** (1980), 233–240.

[11] Inkeri K., Untersuchungen über die Fermatsche Vermutung, *Ann. Acad. Soc. Fenn. Ser. AI No* **33** (1976), 251–256.

[12] Inkeri K., A note on Fermat's conjecture, *Acta Arith.* **39** (1976), 251–256.

[13] Inkeri K. and van der Poorten A. J., Some remarks on Fermat's conjecture, *Acta Arith.* **36** (1980), 107–111.

[14] LeVeque W. J., On the equation $f(x) = y^m$, *Acta Arith.* **9** (1964), 209–219.

[15] Mason R.C., *Diophantine equations over function fields*, LMS Lecture Notes 96, Cambridge University Press, 1984.

[16] Mason R.C., Norm form equations I, *J. Number Theory* **22** (1986), 190–207.

[17] Schäffer J., The equation $1^p + 2^p + \ldots + n^p = m^q$, *Acta Math.* **95** (1956), 155–159.

[18] Schmidt W. M., Thue's equation over function fields, *J. Austral. Math. Soc. Ser A* **25** (1978), 385–422.

[19] Schmidt W. M., Polynomial solutions of $F(x,y) = z^n$, *Queen's Papers in Pure and Appl. Math.*, **54** (1980), 33–65.

[20] Siegel C. L., The integer solutions of the equation $y^2 = ax^n + bx^{n-1} + \ldots + k$, *J. London Math. Soc.* **1** (1926), 66–68.

[21] Shorey T. N. and Tijdeman R., *Exponential Diophantine equations*, Cambridge University Press, 1986.

[22] Sprindžuk V. G., A hyperelliptic Diophantine equation and class numbers, *Acta Arith.* **30** (1976), 95–108.

[23] Stewart C.L., A note on the Fermat equation, *Mathematika*, **24** (1977), 130–132.

[24] Trelina L. A., On S-integral solutions of the hyperelliptic equation, *Dokl. Akad. Nauk. BSSR.* **22** (1978), 881–884.

[25] Voorhoeve M., Győry K. and Tijdeman R., On the Diophantine equation $1^k + 2^k + \ldots + x^k + R(x) = y^z$, *Acta Math.* **143** (1979), 1–8.

7

ASPECTS OF THE HILBERT NULLSTELLENSATZ

W. Dale Brownawell*

I. Historical Remarks

In 1893 D. Hilbert proved his celebrated theorem showing how the zeros (Nullstellen) of a polynomial determine whether it has a power in a given ideal. We state the theorem [12] in its affine form over C.

Hilbert Nullstellensatz. *Let the polynomial $P(\mathbf{x})$ vanish at all the common zeros in $\mathbf{C}^n$ of the polynomials $P_1(\mathbf{x})$, ..., $P_m(\mathbf{x})$ from $\mathbf{C}[\mathbf{x}] = \mathbf{C}[x_1, \ldots, x_n]$. Then for some $e \in \mathbf{N}$, $P^e \in (P_1, \ldots, P_m)$.*

This theorem formed one of the three pillars upon which Hilbert based his theory of invariants and polynomial ideals. Like the other two members of this triad, the Hilbert polynomial and the finite generation of polynomial ideals over fields, the Nullstellensatz was ineffective; no procedure was given to determine either the exponent e or the coefficients used to express P^e in terms of the generators P_1, ..., P_m. These innovations of Hilbert met initially with resistance because constructions of that era in algebra were expected to involve algorithms.

In 1929 J. L. Rabinowitsch published a remarkable half page paper [22], showing that the full Hilbert Nullstellensatz follows from the special case (in one more variable) where $P = 1$. Consequently the following result is also now often referred to as the Nullstellensatz, with no evident trace of irony even though it deals precisely with the case that there are no zeros.

Nullstellensatz. *If the ideal $I = (P_1, \ldots, P_m)$ in $\mathbf{C}[\mathbf{x}]$ has no zero in $\mathbf{C}^n$, then $I = \mathbf{C}[\mathbf{x}]$, i.e. there are $A_1, \ldots, A_m \in \mathbf{C}[\mathbf{x}]$ with*

$$A_1 P_1 + \ldots + A_m P_m = 1. \tag{1}$$

Nullstellensatz $\Longrightarrow$ Hilbert Nullstellensatz (Rabinowitsch Trick): Set $P_0 = 1 - x_{n+1} P$ with a new variable x_{n+1}, and let B_0, ..., B_m be

---

* Research supported in part by NSF.

polynomials such that $1 = B_0 P_0 + \ldots + B_m P_m$. In this formal identity set $x_{n+1} = 1/P$ and clear out the denominator of the resulting rational function by multiplying both sides with the appropriate power of P.

K. Henzelt, E. Noether and G. Hermann put many of the basic procedures of the theory of polynomial ideals over a field k on a constructive footing in [9], [10]. (This is especially striking in the light of Noether's much reported disdain for any algebraic deliberations not aimed at revealing "structure", in particular for explicit calculation.) The first, fundamental step in Hermann's considerations was to generate all solutions of a system of t linear equations over a polynomial ring over a field, i.e. to develop linear algebra over a polynomial ring. In particular, Hermann gave a procedure, whose origins go back to Hilbert's treatment ([11], p. 493) of the homogeneous case, for generating all polynomial solutions of a system of linear equations

$$P_{i1} X_1 + \ldots + P_{im} X_m = C_i,$$

where C_i, $P_{ij} \in k[\mathbf{x}]$ have degree $\leq D$. The case $t = 1$, $C_1 = 1$ is the situation of the Nullstellensatz.

If the rank of the corresponding homogeneous system is m, then one simply solves by Cramer's rule. Otherwise Hermann's procedure uses a suitable linear change of variables to guarantee that $1 \cdot x_n^{\text{maximal power}}$ is a term in the expansion of a maximal non-singular subdeterminant of the P_{ij}. Then applications of Cramer's rule show that any solution is the sum of a particular solution with degree in x_n less than (rank of system) $\cdot D$ plus an "obvious" solution to the corresponding homogeneous equation. Having strictly bounded the degree in x_n for a particular solution by (rank) $\cdot D$, one can rewrite the system as a system of at most $(1 + \text{rank}) \cdot Dt \leq t(t+1)D$ equations involving coefficients of degree $\leq D$ in $n-1$ variables. One repeats until a system of equations is reached with constant coefficients. Thus letting $T = t + 1$, one obtains a solution to the original system of equations satisfying

$$\deg X_j \leq (TD) + (TD)^2 + \ldots + (TD)^{2^{n-1}}.$$

M. Reufel [24] was apparently the first to emphasize the algorithmic character of Hermann's considerations. See [23] for a detailed description of the development of Hermann's method in general and in particular of the reliability of bounds appearing in the literature dating back to [11].

Hermann's process looks plausibly wasteful as an approach to (1). D. Lazard took up the question of obtaining better bounds in [13], where he obtained significant strengthenings in certain cases. D. W. Masser and G. Wüstholz employed Hermann's approach to polynomials over algebraic number fields in [14] to carry out effective elimination as part of their method for algebraic independence. We state a special case of their Theorem IV.

Theorem (Masser-Wüstholz). *Let* $P_1, \ldots, P_k \in \mathbf{Z}[x_1, \ldots, x_n]$ *of degree* $\leq D$ *and* $\log \text{height} \leq h$ *have no common zeros in* $\mathbf{C}^n$. *Then there are* $a \in \mathbf{N}$ *and* $A_1, \ldots, A_k \in \mathbf{Z}[x_1, \ldots, x_n]$ *of degree* $\leq (8D)^{2N+1}$, *where* $N = 2^{n-1}$, *such that*

$$A_1 P_1 + \ldots + A_k P_k = a,$$

$$\log a, \ \log ht A_i \leq (8D)^{4N-1}(h + 8D \log 8D).$$

As a consequence of the obvious upper bound for $|A_i(\omega)|$, they deduced an inequality, which in our case becomes

$$\log \max\{|P_i(\omega)|\} \geq -(8D)^{4N-1}(h + 8D \log 8D)$$
$$-(8D)^{2N+1}\log(n+1)|\omega| - \log k. \quad (2)$$

where $|\omega| = \max\{1, |\omega_i|\}$ for any $\omega = (\omega_1, \ldots, \omega_n) \in \mathbf{C}^n$.

Since the exponential nature of N resulted in a lower bound for the transcendence degree which was approximately the logarithm of the result hoped for, there was a strong incentive to improve the bound on the degrees in the Nullstellensatz. That desire was made less urgent by P. Philippon's generalization [19] to several variables of Gelfond's criterion for algebraicity. Thus Philippon [19], [20] and M. Waldschmidt [26] established all the independence results one had hoped to attain from a sharpening of the Nullstellensatz. In the meantime we have obtained a local version [6] of (2) essentially with N replaced by the optimal n. This also provides the foundations for a revival [7], [8] of the approach of [14] to obtain the applications of [19], [26] for exponential functions and for elliptic functions with algebraic invariants.

Until very recently, the only hope for proving inequalities of type (2) seemed to be through the Nullstellensatz. Thus at the time, Philippon's criterion seemed especially fortunate for the theory of algebraic independence in light of the example of E. Mayr and A. Meyer [15]. That example had established the daunting fact that the order of growth for $\deg A_i$ obtained by Hermann's method is essentially unavoidable even to express linear elements of I in terms of the generators P_i. Still, some held that the situation of the Nullstellensatz is so special that A_i might exist for (1) with $\deg A_i \leq p(n)D^{r(n)}$, where $r(n)$ is a polynomial in n and possibly m. The example of Masser and Philippon with P_i given by

$$x_1^D,\, x_2^D - x_1,\, \ldots,\, x_{n-1}^D - x_{n-2},\, 1 - x_{n-1}x_n^{D-1}$$

shows that in general a solution of (1) must satisfy $\max \deg A_i \geq D^n - D^{n-1}$.

II. Effective Nullstellensatz

In [5] it was shown that $r(n)$ can be taken $\leq \min(m, n)$:

Theorem 1. *Let the polynomials P_i in the Nullstellensatz have $\deg P_i \leq D$. Let $\mu = \min\{m, n\}$. Then A_i can be selected in (1) with*

$$\deg A_i \leq \mu n D^\mu + \mu D.$$

It is one of the purposes of this note to provide an outline of the proof of Theorem 1 (at the end of Section IV below). As mentioned above, through the Rabinowitsch technique, Theorem 1 bounds the exponent e occurring in the Hilbert Nullstellensatz by an expression of order $D^{\mu+1}$, where $\max\{\deg P, \deg P_i\} \leq D$. Thus unfortunately the degree of P enters into the bound to the power $\mu + 1$. However, it is not hard to see the existence of a bound independent of P, say the maximal exponent of all primary components in an irredundant decomposition. Although we do not know how to obtain such a value for e effectively, we show here that e can be bounded by an expression involving a factor of $\deg P$ to the first power only.

Theorem 2. *Let the polynomials P_i in the Hilbert Nullstellensatz have $\deg P_i \leq D$ and $\deg P \leq D_0$. Then $P^e \in (P_1, \ldots, P_m)$ for $e \in \mathbf{N}$ with*

$$e \leq n(q+1)(D_0+1)D^\mu + q \cdot \max\{D-1, D_0\},$$

where $q = \min\{n+1, m\}$ and $\mu = \min\{m, n\}$.

The proof here of Theorem 2 differs from that of [5] chiefly in the specification of the initial element of a certain auxiliary regular sequence. Thus we shall be able to relegate some technical points to references from [5]. Nevertheless we shall have to recall in the next section the basic definitions of Chow forms, of resultants of Chow forms and ordinary forms, and of norms of Chow forms. The properties of resultants and norms will appear as analogues of familiar results for classical resultants. One principal difference is that now the resultant does not eliminate specific variables unless the original Chow form is of dimension 0. Rather, dimensions of common zeros decrease. The major benefit is that degrees of resultants are not squared at each step of decreasing dimension, as was the case in classical approaches to algebraic independence using ordinary resultants (or even semi-resultants).

III. Basic Properties of Chow Forms

Let $\mathcal{P}$ be a homogeneous prime ideal of $k[x_0, \ldots, x_n]$ of rank $r \leq n$, where k is a field of characteristic zero. Then the condition that the $n + 1 - r$ linear forms

$$L_j(\mathbf{x}) = u_{j0}x_0 + \ldots + u_{jn}x_n, \qquad j = 1, \ldots, n + 1 - r,$$

share a common non-trivial zero with $\mathcal{P}$ in the algebraic closure of k is given by a single polynomial equation $F(\mathbf{u}_1, \ldots, \mathbf{u}_{n+1-r}) = 0$, where F is unique up to a non-zero constant factor. The polynomial F, which we shall call the *Chow form* of $\mathcal{P}$, is irreducible, homogeneous in each set of variables $\mathbf{u}_j = (u_{j0}, \ldots, u_{jn})$, and invariant up to a constant factor under any permutation of $\mathbf{u}_1, \ldots, \mathbf{u}_{n+1-r}$. If $x_i \notin \mathcal{P}$, then F can be written as

$$F = a \prod_\gamma (u_{n+1-r,0}\alpha_0^\gamma + \ldots + u_{n+1-r,n}\alpha_n^\gamma),$$

where $\alpha_i = 1$, a is a polynomial in $k(\mathbf{u}_1, \ldots, \mathbf{u}_{n-r}) = k_r$, $K = k_r(\alpha_0, \ldots, \alpha_n)$ has degree $[K : k_r] = \deg \mathcal{P}$, and γ runs through all the k_r-embeddings of K into an algebraic closure of k_r. [17] is a good reference, although this was all known classically.

Our demonstration hinges critically on a construction introduced by Yu. V. Nesterenko [17], [18] to carry out elimination for proofs of algebraic independence. If $Q \in k[x_0, \ldots, x_n]$ is homogeneous, then we use the above factorization for F to define the *resultant* $\mathrm{Res}(F, Q)$ of F and Q as

$$\mathrm{Res}(F, Q) = a^{\deg Q} \prod Q(\alpha_0^\gamma, \ldots, \alpha_n^\gamma).$$

Clearly if $r = n$, then $\mathrm{Res}(F, Q) \in \mathbb{C}$ and if $Q \in \mathcal{P}$, then $\mathrm{Res}(F, Q) = 0$. Nesterenko showed that if $Q \notin \mathcal{P}$, and $r < n$, then $\mathrm{Res}(F, Q)$ is a product involving precisely the Chow forms of the minimal prime ideals of $(Q, \mathcal{P})$.

This leads us to extend the notion of Chow form multiplicatively. We say that F is a Chow form of rank r based on the homogeneous prime ideals $\mathcal{P}_1, \ldots, \mathcal{P}_s$ if the ideals are of rank r, and F is a power product of their Chow forms. Then when $r \leq n$, $\mathrm{Res}(F, Q)$ of the (general) Chow form F and an ordinary form Q can be defined by multiplicativity, and if $r < n$ and $Q \notin \bigcup \mathcal{P}_i$, then $\mathrm{Res}(F, Q)$ is a Chow form of rank $r + 1$, based on the isolated components of the various ideals $(\mathcal{P}_i, Q)$.

It was long known that if F is the Chow form of a homogeneous prime ideal $\mathcal{P}$ of rank r, then a canonical basis for $\mathcal{P}$ (or at worst $\mathcal{P}$ intersected with some embedded components) can be obtained in the following way: Let $S^{(j)} = (s_{kl}^{(j)})$ be skew symmetric matrices in the new variables $s_{kl}^{(j)}$, $0 \leq k < l \leq n$, $j = 1, \ldots, n + 1 - r$. Then the polynomials in $\mathbf{x}$ obtained as coefficients in $F(S^{(1)}\mathbf{x}, \ldots, S^{(n+1-r)}\mathbf{x})$ of distinct monomials in the $s_{kl}^{(j)}$ generate $\mathcal{P}$, or its intersection with embedded ideals (e.g. Lemma 11, [16]). Thus we have a method for choosing a canonical "basis" for $\mathcal{P}$. Nesterenko recognized that this gives a way of defining a normalized absolute value of an unmixed ideal, when $k \subset \mathbf{C}$, which we assume from now on. In this spirit we define for any Chow form F:

$$\|F\|_\omega = \frac{H(F(S^{(1)}\omega, \ldots, S^{(n+1-r)}\omega))}{H(F)\|\omega\|^{\mathrm{Deg}\, F}},$$

where $\|\omega\| = \max |\omega_i|$ for $\omega = (\omega_0, \ldots, \omega_{n+1}) \in \mathbf{C}^{n+1} \setminus 0$, $\mathrm{Deg}\, F = $ total $\deg F$, and H denotes ordinary height. The argument for Lemma 6 of [5] shows the next result, where we use the notation

$$\|Q\|_\omega = \frac{|Q(\omega)|}{H(Q)\|\omega\|^{\deg Q}}.$$

The result corresponds to classical inequalities for ordinary resultants and ordinary absolute values.

Lemma. (i) *Let $\mathrm{Res}(F, Q)$ be non-constant. Then each*

$$\deg_{\mathbf{u}_i} \mathrm{Res}(F, Q) = \deg Q \cdot \deg_{\mathbf{u}_i} F.$$

(ii) *Let $r \leq n$ and let $R^*(F, Q)$ denote a Chow form obtained from $\mathrm{Res}(F, Q)$ by omitting an arbitrary number of factors, of total degree Γ, whose corresponding prime ideals $\mathcal{P}$ contain x_0. Then if $\omega \in \mathbf{C}^{n+1}$ with $\omega_0 \neq 0$,*

$$\|R^*(F, Q)\|_\omega \leq c(\|\omega\|/|\omega_0|)^\Gamma \max\{\|F\|_\omega, \|Q\|_\omega\},$$

where $c > 0$ depends only on F and Q.

This crucial proposition was essentially established by Nesterenko when F is the Chow form of a prime ideal and no factors are removed. Using Gelfond's classical inequality for the height of polynomial factors, the general form was deduced in [5], although the statement there made the unnecessary assumption that for $R^*(F,Q)$ all factors of $\text{Res}(F,Q)$ whose underlying prime ideals contain x_0 be omitted. Successive application of the lemma gives the following key result, where we start the induction with Proposition 1 of [18] whose proof implies that
$$\left\|(^hQ_1)\right\|_\omega = \left\|^hQ_1\right\|_\omega (n+1)^{2nD_1} \text{ and } \deg_{u_i}(^hQ_1) = D_1.$$

Proposition. *Let* $Q_1, \ldots, Q_k \in \mathbf{C}[\mathbf{x}]$, *of degrees* $D_1, \ldots, D_k$, *respectively, be a regular sequence without common zeros in* $\mathbf{C}^n$. *Then there exists a constant* $C > 0$, *depending only on* $Q_1, \ldots Q_k$ *such that for all* $\omega = (1, \omega_1, \ldots, \omega_n) \in \mathbf{C}^{n+1}$, *we have*

$$\max \left\|^hQ_i\right\|_\omega \geq C\|\omega\|^{-(n-1)D_1\ldots D_\kappa},$$

where $\kappa = \min\{k, n\}$.

IV. Brief Proofs

Since the proof of Theorem 2 is an elaboration of that found in [5] for Theorem 1, we have included only as much detail for the common parts as is needed to make the outline of the argument clear here.

Set $Q_1' = 1 - x_{n+1}P$ and complete to a maximal regular sequence $Q_1', \ldots, Q_i'$ from $\mathcal{M} = \mathbf{Z}(1 - x_{n+1}P) + \mathbf{Z}P_1 + \ldots + \mathbf{Z}P_m$. Then the coefficients of $(1 - x_{n+1}P)$ in $Q_2', \ldots, Q_i'$ can be set equal to zero to obtain a new regular sequence $Q_1 (= Q_1'), Q_2, \ldots, Q_i$ from $\mathcal{M}$. Since $1 - x_{n+1}P, P_1, \ldots, P_m$ have no common zeros, they lie in no (proper) prime ideal of $\mathbf{C}[x_1, \ldots, x_{n+1}]$. Thus by, say, Lemma 1, p. 438 of [14], there is an element of $\mathcal{M}$ outside any finite list of (proper) prime ideals. Consequently by the maximality of i, $Q_1, \ldots, Q_i$ have no common zeros in $\mathbf{C}^{n+1}$. Moreover $i \leq n + 1$. Thus the Proposition gives a lower bound

$$C'\|\omega'\|^{1-n(D_0+1)D^\mu} \leq \max|Q_i(\omega')|$$
$$\leq C'' \max\{|P_k(\omega')|, |1 - \omega_{n+1}P(\omega')|\},$$

where $\|\omega'\| = \max\{|\omega_j|, 1\}$ for $\omega' = (\omega_1, \ldots, \omega_{n+1})$ and C', C'' are independent of ω'.

To complete the proof we appeal to a result of H. Skoda [26] which implies that if $K \geq 0$ is large enough so that for $\epsilon > 0$,

$$\int_{\mathbf{C}^{n+1}} |P|^{-2(1+\epsilon)q-2} |\mathbf{x}|^{-2K} d\lambda = 1 < \infty,$$

where $|P|^2 = |1 - x_{n+1}P|^2 + |P_1|^2 + \ldots + |P_m|^2$, $|\mathbf{x}|^2 = 1 + |x_1|^2 + \ldots + |x_{n+1}|^2$, and $q = \min\{n+1, m\}$, then there are polynomials B_0, $\ldots$, $B_m \in \mathbf{C}[x_1, \ldots, x_{n+1}]$ with

$$\deg B_k \leq q(1+\epsilon) \max\{D_0 + 1, D\} + K - n - 1$$

such that
$$1 = B_0(1 - x_{n+1}P) + B_1 P_1 + \ldots + B_m P_m.$$

The lower bound obtained in the preceding paragraph shows that convergence of the integral is guaranteed when, say, $K = \big(q(1+\epsilon)+1\big)\big(n(D_0 + 1)D^\mu - 1\big) + n + 1 + \epsilon$. Taking $\epsilon < 1/\{2qn(D_0 + 1)D^\mu\}$, we obtain B_k with

$$\deg B_k \leq n(q+1)(D_0 + 1)D^\mu + q \cdot \max\{D_0, D - 1\}.$$

Setting $x_{n+1} = 1/P$ and clearing out the denominator gives the desired bound on e. ∎

The proof of Theorem 1 (see [5] for details) is quite similar, but it begins with a non-extendable regular sequence from $\mathbf{Z}P_1 + \ldots + \mathbf{Z}P_m$. From our Proposition above, we deduce that when $D \geq 1$,

$$\|\omega'\|^{1-(n-1)D^\mu} \leq C^* \max|P_k(\omega')|,$$

where $C^* > 0$ and $\|\omega'\| = \max\{1, |\omega_j|\}$ for $\omega' = (\omega_1, \ldots, \omega_n) \in \mathbf{C}^n$. Let $q = \min\{n, m - 1\}$. Skoda's result gives a solution of (1) with

$$\deg A_k \leq q(1+\epsilon)D + K - n - 1.$$

The bound claimed for $\deg A_k$ follows when ϵ is small enough if we take

$$K = \big(q(1+\epsilon)+1\big)\big((n-1)D^\mu - 1\big) + n + \epsilon. \qquad \blacksquare$$

It is clear that the original sequence can be required to begin with a specified non-zero polynomial P_i. Thus the degree of P_i will enter into the bound obtained for the degrees of a particular solution A_k of (1) to the first power only. Of course if P_1, $\ldots$, P_m happen to form a regular sequence already, each degree will enter to the first power only.

V. Concluding Comments

A. Remarks

1. I am very much indebted to C. Berenstein and A. Yger for pointing out that lower bounds from our original version of the Proposition give upper bounds on degrees for A_k satisfying (1). In fact the result of [2] based on the work of Berndtsson and Andersson [4] gives polynomial solutions of (1) of slightly higher degree, but whose coefficients are explicit integrals.

2. The Proposition can also be approached through Philippon's development [19], [20], [21] of the Chow form. The main differences are the introduction of higher degree forms in the place of the L_j above (so such a Chow form has a multidegree) and the use of the more precise Mahler measure throughout rather than the height. Since any improvement for the present questions occurs in the constants which we cannot in general control anyway, the final results for questions we have considered here about the Nullstellensatz are equivalent.

B. Open Questions

1. It is an open problem to investigate the algebraic or arithmetic nature of the integrals in [2] representing the coefficients of the A_k. It is highly non-trivial to show even that they lie in the coefficient field of the polynomials [3].

2. The proofs were given above and in [5] for coefficient field C so that we could take advantage of the techniques of analysis. By the Lefschetz principle, these results hold over any algebraically closed field of characteristic 0. In fact appropriate formulations then hold for any coefficient field of characteristic 0. The situation for finite characteristic is not clear. Standard techniques show that for fixed n and for given bounds for the degrees, there are only finitely many primes p for which the bounds given above do not hold. As far as I know, it is an open question whether there are ever in fact any exceptional p.

3. Our results amount to an assertion of the solvability of certain systems of linear equations over the coefficient field. When that coefficient field is an algebraic number field, say **Q**, then Cramer's rule allows us to bound the logarithmic height of a solution to, say, (1) by $cD^{m^2}h$, where $\max \log D \cdot \operatorname{ht} P_i \leq h$, and c depends only on n.

That result is too weak to obtain even a global version of the in-equality of [6]. Can one obtain a bound of the "right" form $cD^m h$ for the height of a solution?

4. Can one obtain a purely algebraic proof of the bounds of our theorems? Such a proof may help resolve the preceding two open questions.

5. Can one obtain any effective upper bound at all for e in the Hilbert Nullstellensatz which is independent of $\deg P$? Must one bound the exponents of all components of I, including embedded ones? (See e.g. [1]). *Added in proof*: Quite recently Question 5 has been resolved without bounding the exponents of embedded components. Combining the idea behind Theorem 2 above with a refinement of Rabinowitsch's technique shows, among other things, that e can be chosen as $e = n(\mu+1)D^\mu + \mu$. See W. D. Brownawell, Borne effective pour l'exposant dans le théorème des zéros, *Comptes Rendus de l'Acad. Sci. Paris*, 1987.

References

[1] D. Bayer and M. Stillman, On the complexity of computing syzygies, submitted to *J. Symb. Comp.*

[2] C. Berenstein and D. C. Struppa, On explicit solutions to the Bezout equation, *Systems & Control Letters* 4 (1984), 33–39.

[3] C. Berenstein and A. Yger, Le problème de la déconvolution, *J. Funct. Anal.* 54, (1983), 113–160.

[4] B. Berndtsson and M. Andersson, Henkin-Ramirez formulas with weight factors, *Ann. Inst. Fourier* 32 (1982), 91–110.

[5] W. D. Brownawell, Bounds for the degrees in the Nullstellensatz, *Annals of Math.*, to appear.

[6] W. D. Brownawell, A local Diophantine Nullstellen inequality, *J. Am. Math. Soc.*, to appear.

[7] W. D. Brownawell, Large transcendence degree revisited I, Exponential and non-CM cases, Bonn Transcendence Conference 1985.

[8] W. D. Brownawell and R. Tubbs, Large transcendence degree revisited II, CM case, Bonn Transcendence Conference 1985.

[9] K. Henzelt and E. Noether, Zur Theorie der Polynomideale und Resultanten, *Math. Ann.* 88 (1922), 53–79.

[10] G. Hermann, Die Frage der endlich vielen Schritte in der Theorie der Polynomideale, *Math. Ann.* **95** (1926), 736–788.

[11] D. Hilbert, Über die Theorie der algebraischen Formen, *Math. Ann.* **36** (1890), 473–534.

[12] D. Hilbert, Über die vollen Invariantensysteme, *Math. Ann.* **42** (1893), 313–373.

[13] D. Lazard, Algèbre linéaire sur $K[X_1, \ldots, X_n]$ et élimination, *Bull. Soc. Math. France* **105** (1977), 165–190.

[14] D. W. Masser and G. Wüstholz, Fields of large transcendence degree generated by values of elliptic functions, *Invent. Math.* **72** (1983), 407–464.

[15] E. Mayr and A. Meyer, The complexity of the word problem for commutative semigroups and polynomial ideals, *Advances in Math.* **46** (1982), 305–329.

[16] Yu. V. Nesterenko, Estimates for the orders of zeros of functions of a certain class and applications in the theory of transcendental numbers, *Izv. Akad. Nauk SSSR Ser. Mat.* **41** (1977), 253–284 = *Math. USSR Izv.* **11** (1977), 239–270.

[17] Yu. V. Nesterenko, Bounds for the characteristic function of a prime ideal, *Mat. Sbornik* **123**, No. 1, (1984), 11–34 = *Math. USSR Sbornik* **51** (1985), 9–32.

[18] Yu. V. Nesterenko, On algebraic independence of algebraic powers of algebraic numbers, *Mat. Sbornik* **123**, No. 4, (1984), 435–459 = *Math. USSR Sbornik* **51** (1985), 429–454.

[19] P. Philippon, Critères pour l'indépendance algébrique, *Inst. Hautes Etudes Sci. Publ. Math.* No. **64**, 1986, 5–52.

[20] P. Philippon, Sur les mesures d'indépendance algébrique, in *Séminaire de Théorie des Nombres, Paris 1983-1984*, C. Goldstein, ed., Birkhäuser Verlag, Boston, Basel, Stuttgart, 1985.

[21] P. Philippon, A propos du texte de W. D. Brownawell: Bounds for the degrees in the Nullstellensatz.

[22] J. L. Rabinowitsch, Zum Hilbertschen Nullstellensatz, *Math. Ann.* **10** (1929), 520.

[23] B. Renschuch, Beiträge zur konstruktiven Theorie der Polynomideale, XVII/1,2, *Wissenschaftliche Zeitschrift, Pädagogische*

Hochschule "Karl Liebknecht", Potsdam, Jahrgang 24/1980, Heft 1, 87–99, Jahrgang 25/1981, Heft 1, 125–136.

[24] M. Reufel, Konstruktionsverfahren bei Moduln über Polynomringen, *Math. Z.* **90** (1965), 231–250.

[25] H. Skoda, Applications des techniques L^2 à la théorie des idéaux algèbre de fonctions holomorphes avec poids, *Ann. Sci. Ecole Norm. Sup.* (4e sér.), **5** (1972), 545–579.

[26] M. Waldschmidt, Groupes algébriques et grands degrés de transcendance, *Acta. Math.* **156** (1986), 253–302.

8

ON THE IRRATIONALITY OF CERTAIN SERIES: PROBLEMS AND RESULTS

P. Erdős

During my long life I have spent lots of time thinking about irrationality of series. The reader with a little maliciousness may say "spent and wasted" since I have never discovered any new general methods nor had any spectacular success like Apéry. Nevertheless I hope to convince the reader that not all of it was completely wasted. First of all, I state some of my previous results several of which were obtained with E. Straus. I state many unsolved problems and also prove some new theorems.

I proved more than 30 years ago [1] that for every integer $t > 1$,

$$\sum_{n=1}^{\infty} \frac{d(n)}{t^n} = \sum_{n=1}^{\infty} \frac{1}{t^n - 1}$$

($d(n)$ is the number of divisors of n) is irrational. Chowla conjectured that the same holds for every rational $t > 1$. This is almost certainly true but is unattackable by my methods. It is very annoying that I cannot prove that $\sum_{n=1}^{\infty} \frac{1}{2^n-3}$ and $\sum_{n=2}^{\infty} \frac{1}{n!-1}$ are both irrational (one of course expects that $\sum_{n=1}^{\infty} \frac{1}{2^n+t}$ and $\sum_{n=1}^{\infty} \frac{1}{n!+t}$ are irrational and in fact transcendental for every integer t.) Peter Borwein just informed me (June 1987) that he proved that $\sum \frac{1}{2^n+r}$ is irrational for every rational r. Denote by $\nu(n)$ the number of distinct prime factors of n; $\varphi(n)$ is Euler's φ function and $\delta_k(n)$ the sum of the k-th powers of divisors of n. It is very annoying that I cannot prove that $\sum_{n=1}^{\infty} \frac{\nu(n)}{2^n}$ is irrational; perhaps here I am overlooking a simple argument. $\sum_{n=1}^{\infty} \frac{\varphi(n)}{2^n}$ and $\sum_{n=1}^{\infty} \frac{\delta(n)}{2^n}$, $\delta(n) = \delta_1(n)$, are no doubt also irrational but this is probably unattackable by my methods.

Kac and I [2] proved that $\sum_{n=1}^{\infty} \frac{\delta_k(n)}{n!}$ is irrational for $k = 1$ and $k = 2$. Our proof does not seem to work for $k > 2$, but perhaps we again are overlooking a simple argument.

Straus and I [3] proved that if $1 < a_1 < a_2 < \ldots$ is a sequence of integers then

$$\sum_{n=1}^{\infty} \frac{d(n)}{a_1 a_2 \ldots a_n}$$

is irrational; we conjectured that it suffices to assume that $a_n \to \infty$ but could not prove it. We also conjectured that if $a_{n+1} \geq a_n$ then

$$\sum_{n=1}^{\infty} \frac{\varphi(n)}{a_1 \ldots a_n} \quad \text{and} \quad \sum_{n=1}^{\infty} \frac{\delta(n)}{a_1 \ldots a_n} \qquad (1)$$

are irrational, but we could only prove (1) if we made some further assumptions on the growth properties of the a's. Observe that $a_n = \varphi(n) + 1$, $a_n = \delta(n) + 1$ shows that $a_n \to \infty$ does not suffice for the irrationality of (1). I further proved that if $p_1 < p_2 < \ldots$ is the sequence of primes then $\sum_{n=1}^{\infty} \frac{p_n^k}{n!}$ is irrational for every k [4]. I could not prove that $\sum \frac{p_n^k}{2^n}$ is irrational for every k. This is probably very difficult already for $k = 1$. It seems reasonable to expect that if $g_n \geq 2$, $g_n/p_n \to 0$ then

$$\sum_{n=1}^{\infty} p_n / g_1 \ldots g_n \qquad (2)$$

is irrational, but I can prove the irrationality of (2) only under much more restrictive conditions; $g_n = p_n + 1$ shows that some growth condition is needed for the irrationality of (2).

A few years ago I proved [5] that if $n_{k+1} - n_k \to \infty$ then

$$\sum_{k=1}^{\infty} \frac{n_k}{2^{n_k}} \qquad (3)$$

is irrational. The proof is not entirely trivial. I think (3) remains true if we assume only that $n_k/k \to \infty$, but unless I am overlooking a simple argument my proof breaks down. In this connection there are two questions which I could not settle and which particularly annoy me. The first question states: Does there exist a sequence $n_1 < n_2 < \ldots$ for which $\limsup n_{k+1} - n_k = \infty$ but $\sum_{k=1}^{\infty} \frac{n_k}{2^{n_k}}$ is rational? The answer is almost certainly affirmative. The second question states: Let $q_1 < q_2 \ldots$ be the sequence of square free numbers. Then $\sum q_n/2^{q_n}$ surely must be irrational, and in fact this should hold if the q's are an arbitrary subsequence of the square free numbers.

Several related problems are stated in my book with Graham [6]. Does the equation

$$\frac{n}{2^n} = \sum_{k=1}^{t} a_k/2^{a_k}, \qquad t > 1,$$

have a solution for infinitely many n,† or perhaps for all n? Is there a rational x for which

$$x = \sum_{k=1}^{\infty} a_k/2^{a_k}$$

has two solutions?

It is a simple exercise to prove that if $n_k^{1/2^k} \to \infty$ then $\sum \frac{1}{n_k}$ is irrational, and it is easy to see that the condition $n_k^{1/2^k} \to \infty$ cannot be weakened. On the other hand I proved [7] that if

$$\limsup n_k^{1/2^k} = \infty \quad \text{and} \quad n_k > k^{1+\epsilon} \qquad (4)$$

for all k then $\sum \frac{1}{n_k}$ is irrational. My proof is not entirely trivial. It is probable that many other theorems of this kind can be proved. In (4) the condition $n_k > k^{1+\epsilon}$ is essentially best possible.

Once I asked: Assume that $\sum \frac{1}{n_k}$ and $\sum \frac{1}{n_k-1}$ are both rational. How fast can n_k tend to infinity? I was (and am) sure that $n_k^{1/k} \to \infty$ is possible but $n_k^{1/2^k}$ must tend to 1. Unfortunately almost nothing is known. David Cantor observed that

$$\sum_{k=3}^{\infty} \frac{1}{\binom{k}{2}} \quad \text{and} \quad \sum_{k=3}^{\infty} \frac{1}{\binom{k}{2}+1}$$

are both rational and we do not know any sequence with this property which tends to infinity faster than polynomially. Stolarsky asked the following pretty question: is it true that if $\sum \frac{1}{n_k} < \infty$ then there is always an integer t, $t \neq n_k$, for which $\sum_{k=1}^{\infty} \frac{1}{n_k-t}$ is irrational?

Straus and I proved that the set of points in the plane of the form

$$x = \sum \frac{1}{n_k}, \qquad y = \sum \frac{1}{n_k+1}$$

---

† A simple proof of this statement was communicated to me by Cusick (June 1987). The question for all n remains open.

contains open sets. Probably the analogous result holds for r dimensions. We never published our proof since we did not work out the r-dimensional case.

Straus and I proved [8] that if $\limsup n_k^2/n_{k+1} \leq 1$ and $N_k = [n_1, \ldots, n_k]$ and

$$\limsup \frac{N_k}{n_{k+1}} \left(\frac{n_k^2}{n_{k+1}} - 1 \right) \leq 0 \qquad (5)$$

then $\sum 1/n_k$ is irrational except if $n_{k+1} = n_k^2 - n_k + 1$ for all $k > k_0$. Perhaps our result remains true without the assumption (5).

We defined a sequence $n_1 < n_2 < \ldots$ to be an *irrationality sequence* if, for every sequence of integers t_k, $\sum_{k=1}^{\infty} \frac{1}{t_k n_k}$ is irrational. Observe that $n!$ is not an irrationality sequence since $\sum \frac{1}{(n+2)n!} = 1$. We conjectured that and I later proved, [7], that $n_k = 2^{2^k}$ is an irrationality sequence.

It is not clear if the irrationality sequence must increase very rapidly. I have not been able to find an irrationality sequence for which $n_k^{1/2^k} \to 1$. Graham and I observed that if n_k is an irrationality sequence then $n_k^{1/k} \to \infty$. We do not know if there is an irrationality sequence $n_1 < n_2 < \ldots$ for which $(n_i, n_j) = 1$ and $\limsup n_k^{1/2^k} < \infty$.

Graham and I modified the definition of an irrationality sequence. Let us try to call a sequence $a_1 < a_2 \ldots$ an irrationality sequence if, for every $b_n/a_n \to 1$, $\sum_{n=1}^{\infty} \frac{1}{b_n}$ is irrational. The trouble with this definition is that we do not know a simple non-trivial irrationality sequence, for example, we cannot prove that 2^{2^n} is an irrationality sequence of this kind. On the other hand it is not difficult to show that if $\liminf n_k^{\frac{1}{2^k}} > 1$ and $\lim n_k^{\frac{1}{2^k}}$ does not exist then $\sum_{k=1}^{\infty} \frac{1}{n_k}$ is irrational and hence $\{n_k\}$ is an irrationality sequence of this kind; but perhaps it is not an irrationality sequence with our original definition.

Another possibility would be to call $\{a_n\}$ an irrationality sequence if for every $|b_n| < C$, $\sum_{n=1}^{\infty} \frac{1}{a_n + b_n}$ is irrational. In this case we proved that 2^{2^n} is an irrationality sequence but we cannot decide if 2^n or $n!$ is an irrationality sequence. Is there an irrationality sequence a_n of this type which increases exponentially? It is not hard to show that it cannot increase slower than exponentially. As stated previously, Borwein showed that 2^n is an irrationality sequence of this kind.

The following further problems stated in [6] are perhaps interesting: let $n_1 < n_2 < \ldots$. Is it then true that $\sum \frac{1}{2^{n_k} - 1}$ is irrational, or perhaps $\sum \frac{1}{2^{n_k} - t_k}$ is irrational for every $|t_k| < C$?

Let $n_k \to \infty$ rapidly; then $\sum_{k=1}^{\infty} \frac{1}{n_k n_{k+1}}$ is irrational. Probably $\liminf n_k^{1/2^k} > 1$ should suffice.

It is not hard to prove that $\sum \frac{1}{2^{n_k}}$ is transcendental if $n_k/k^{\ell} \to \infty$ for every ℓ. Perhaps the weaker condition $\frac{1}{k} n_k \to \infty$ suffices. On the other hand we do not know of any algebraic number for which $\limsup(n_{k+1} - n_k) = \infty$, but in fact one would expect that every algebraic number which is irrational has this property. Many of these problems seem hopeless at present, but perhaps one can prove that if $n_k > ck^2$ then $\sum_{k=1}^{\infty} \frac{1}{2^{n_k}}$ is not the root of any quadratic polynomial.

Let $p_1 < p_2 < \ldots$ be an infinite sequence of primes. It is a simple exercise to prove that if $a_1 < a_2 < \ldots$ is the sequence of integers composed of the p's then

$$\sum_{n=1}^{\infty} \frac{1}{[a_1, \ldots, a_n]}$$

is irrational, where $[a_1, \ldots, a_n]$ is the least common multiple of $a_1, \ldots, a_n$. This result probably remains true if the number of primes p_i is finite but of course greater than 1.

We are going to prove the following Theorem. Let $a_1 < a_2 < \ldots$ be an infinite sequence of integers. Assume that for every $x > x_0$ and some $\varepsilon > 0$

$$A(x) = \sum_{a_\ell < x} 1 > (1 - \log 2 + \varepsilon)x. \tag{6}$$

Then

$$\sum_{n=1}^{\infty} \frac{1}{c(n)}$$

is irrational, where $c(n)$ is the least common multiple of the integers $a_i < n$.

We present the proof of Halberstam who simplified and clarified my somewhat carelessly presented proof.

We need the following simple lemma.

Lemma. Denote by $P(n)$ the greatest prime factor of n; then if $\eta = \eta(\varepsilon) > 0$ is sufficiently small we have

$$t = t(x) = \sum_{\substack{a_\ell < x \\ P(a_\ell) > x^{\frac{1}{2}+\eta}}} 1 > \tfrac{1}{2}\varepsilon x.$$

The proof follows easily from

$$\sum_{x^{1/2} < p < x} \frac{1}{p} = \log 2 + o(1).$$

The details are left to the reader.

Let $a_{n_1} < \ldots < a_{n_t} \leq x$, $t > \frac{1}{2}\epsilon x$, be the integers for which $P(a_{n_i}) > x^{\frac{1}{2}+\eta}$. By the Lemma these a's exist. Let now p be a prime greater than $x^{\frac{1}{2}+\eta}$. There clearly are at most $\frac{x}{p} \leq x^{\frac{1}{2}-\eta}$ of the a_{n_i} which are multiples of p and since an integer not exceeding x is divisible by at most one of these primes there are at least $\frac{1}{2}\epsilon x^{\frac{1}{2}(1+\eta)}$ distinct primes p_i for which there is an integer a_{n_i} satisfying $P(a_{n_i}) = p_i > x^{\frac{1}{2}+\eta}$ and we can assume that a_{n_i} is chosen minimally.

Denote now by I_r the interval

$$\left(rx^{\frac{1}{2}}, (r+1)x^{\frac{1}{2}}\right), \qquad 1 \leq r \leq x^{\frac{1}{2}}.$$

There clearly is at least one r, say r_0, for which there are at least $u > \frac{1}{2}\epsilon x^{\frac{1}{2}}$ integers a_{n_i} in I_{r_0} each of which have a prime factor $p_i > x^{\frac{1}{2}+\eta}$ and for distinct i's the p_i's are distinct. Denote these a's by $r_0 x^{\frac{1}{2}} < b_1 < \ldots < b_u < (r_0+1)x^{\frac{1}{2}}$.

Now we are ready to prove our Theorem. Assume that

$$\sum_{n=1}^{\infty} \frac{1}{c(n)} = \ell_1/\ell_2, \qquad (\ell_1, \ell_2) = 1. \tag{7}$$

Multiply both sides by $\ell_2 c(b_1 - 1)$. We immediately obtain from (7)

$$\ell_2 c(b_1 - 1) \sum_{n \geq b_1} \frac{1}{c(n)} \geq 1. \tag{8}$$

Now by the definition of b_1, $b_1 \equiv 0 \pmod{p}$ for some $p > x^{\frac{1}{2}+\eta}$ and b_1 is the smallest $a \equiv 0 \pmod{p}$. Thus

$$\frac{c(b_1 - 1)}{c(b_1)} \leq \frac{1}{p} \leq \frac{1}{x^{\frac{1}{2}+\eta}}. \tag{9}$$

Write now (8) in the form

$$\ell_2 c(b_1 - 1)(\Sigma_1 + \Sigma_2) \geq 1, \tag{10}$$

where we place in Σ_1 the integers n in I_{r_0}; each such n satisfies $n \geq b_1$ and there are at most $x^{\frac{1}{2}} + 1$ such n's. Thus

$$c(b_1 - 1)\Sigma_1 < \frac{x^{\frac{1}{2}} + 1}{x^{\frac{1}{2}+\eta}} < 2x^{-\eta}. \tag{11}$$

Now we have to estimate $c(b_1 - 1)\Sigma_2$. If n is in Σ_2 we can of course assume $n > (r_0 + 1)x^{1/2}$, i.e. n lies beyond I_{r_0}. But since $b_1, b_2, \ldots, b_u$ are in I_{r_0} and each is divisible by a distinct prime $> x^{\frac{1}{2}+\eta}$, we have (for large x)

$$\frac{c(b_1 - 1)}{c(n)} < \left(\frac{1}{x^{1/2}}\right)^u < \left(\frac{1}{x^{1/2}}\right)^{\frac{1}{2}\epsilon x^{\eta/2}} < x^{-10}.$$

Thus we evidently have

$$c(b_1 - 1) \sum_{(r_0+1)x^{1/2}<n\leq x^2} \frac{1}{c(n)} < \frac{x^2}{x^{10}} = x^{-8}. \tag{12}$$

Finally, suppose $n > x^2$. Write

$$\sum_{n>x^2} \frac{1}{c(n)} = \sum_{r=1}^{\infty} \sum_{x^{2^r}<n\leq x^{2^{r+1}}} \frac{1}{c(n)}. \tag{13}$$

We obtain from our Lemma and by the argument we just used that

$$c([y^{1/2}]) \sum_{y<n\leq y^2} \frac{1}{c(n)} < y^{-8}. \tag{14}$$

Thus from (13) and (14) we obtain

$$\sum_{n\geq x^2} \frac{c(b_1 - 1)}{c(n)} < \sum_{r=1}^{\infty} (x^{2^r})^{-8} < x^{-8}. \tag{15}$$

Then (11), (12) and (15) clearly contradict (10) which completes the proof of our Theorem. It can be shown without much difficulty that the Theorem does not remain true if $A(x) < x(1 - \log 2 + \varepsilon)$.

References

[1] P. Erdős, On arithmetical properties of Lambert series, *J. Indian Math. Soc.* **12** (1948), 63–66.

[2] P. Erdős and M. Kac, *Amer. Math. Monthly*, **61** (1954) Problem 4518, p. 264.

[3] P. Erdős and E. Straus, Some number theoretic results, *Pacific J. of Math.* **36** (1971), 635–646 and On the irrationality of certain series *ibid* **55** (1974), 85–92.

[4] P. Erdős, Sur certaines series a valeur irrationnelle, *Enseignement Math.* **4** (1958), 93–100.

[5] P. Erdős, Sur l'irrationalité d'une certaine série, *C. R. Acad. Sci. Paris, Sér I* **292** (1981), 765–768.

[6] P. Erdős and R. L. Graham, Old and new problems and results in combinatorial number theory, *Monographie No. 38 de L'Enseignement Mathématique Genéve* 1980 (imp Kundig).

[7] P. Erdős, Some problems and results on the irrationality of the sum of infinite series, *Journal of Math. Sciences* **10** (1975), 1–7.

[8] P. Erdős and E. Straus, On the irrationality of certain Ahmes series, *J. Indian Math. Soc.* **27** (1963), 129–133.

For some further results on irrationality see P. Erdős, On the irrationality of certain series, *Indagationes Math.* **19** (1957), 212–219 and *Math. Student* **36** (1968), 222–226.

9

S-UNIT EQUATIONS AND THEIR APPLICATIONS

J.-H. Evertse[1], K. Győry[2], C. L. Stewart[3], R. Tijdeman

Contents

§0. Introduction.

This paper gives a survey of the remarkable results on S-unit equations and their applications which have been obtained, mainly in the

[1] Research supported by the Netherlands Organization for the Advancement of Pure Research (Z.W.O.)

[2] Research supported in part by the Hungarian National Foundation for Scientific Research Grant 273

[3] Research supported in part by Grant $A3528$ from the Natural Sciences and Engineering Research Council of Canada

last five years. It is impossible to cover all applications within the scope of this paper, but the wide range of applications illustrates how fundamental these developments are. The developments are still going on and several of the mentioned results are new.

In §1 we introduce notation which will be used throughout the paper. In §§2–5 five theorems on S-unit equations are treated. The Main Theorem on S-Unit Equations (Theorem 1), which deals with general S-unit equations

$$x_0 + x_1 + \ldots + x_n = 0 \quad \text{in } S\text{-units } x_0, x_1, \ldots, x_n, \qquad (0.1)$$

is stated in §2 and its deduction from the Subspace Theorem is sketched in §4. Theorem 1' is a version of Theorem 1 dealing with arbitrary finitely generated multiplicative subgroups of $C \setminus \{0\}$. Theorems 2–5 stated in §3 deal with S-unit equations in two variables,

$$\alpha_1 x_1 + \alpha_2 x_2 = 1 \quad \text{in } S\text{-units } x_1, x_2 \qquad (0.2)$$

where α_1 and α_2 are constants. Theorem 2 gives an upper bound for the number of solutions of (0.2). A proof of it, this time not derived from the theory of hypergeometric functions, but from a variant of Roth's theorem, is given in §4. Theorem 3 was proved during the conference in Durham. It says that apart from finitely many equivalence classes of equations only, equation (0.2) has at most two solutions. Its proof is sketched in §5. Theorems 1–3 are ineffective and hence the methods do not yield upper bounds for the sizes of the solutions. In contrast, Theorems 4 and 5 are effective. They are based on Baker's method concerning linear forms in logarithms of algebraic numbers. Theorem 4 gives an upper bound for the sizes of the solutions of (0.2). Theorem 5, which is new, is an effective, but weaker version of Theorem 3. The proofs of Theorems 4 and 5 in the rational case are given in §5. The formulations of Theorems 1–5 in the case of rational integers are given as Corollaries 1.3 and 2–5. Both §5 and §6 deal with rational integers and can be read independently of the rest of the paper. They are meant for those readers who want to understand and apply the results on S-unit equations for rational integers only.

In §§6–9 applications of Theorems 1–5 are given which are more or less straightforward. Theorems 6 and 7 in §6 are new. Theorem 7 resolves a conjecture of D. Newman on the number of representations of an integer in the form $2^\alpha 3^\beta + 2^\gamma + 3^\delta$ where α, β, γ and δ are non-negative integers. Theorem 8 in §7 is also new. It gives a result on groups which has been applied in the study of ellipticity problems in group theory.

Theorem 9 in §8 is an extension of a result of Evertse [20]. It implies several known results on recurrence sequences as is shown in §§8, 9.

The more classical applications of S-unit equations are mentioned in §§10–12. There is a strong connection between the theory of S-unit equations and the theory of decomposable form equations (which covers the Thue-Mahler equations). The two theories are in fact equivalent (cf. §11). Siegel proved the finiteness of the number of solutions of unit equations in two variables via Thue equations. The opposite approach has also proved applicable, even for decomposable form equations in more than two unknowns. There are several consequences of unit equations which can be deduced via complicated systems of unit equations. All these results can be proved by using the same "intermediate" results, Theorems 10 and 11, which are applications of Theorems 4 and 2 and are presented in §10. The versions of Theorems 10 and 11 presented here had only appeared in Hungarian [42] before. These results are improvements of results in Győry [35]. Theorem 12 in §10 is an application of Theorem 11 to irreducibility of polynomials. Theorems 13–15 in §11 provide general finiteness results for decomposable form equations which imply several known results on Thue equations, Thue-Mahler equations, norm form equations, discriminant form equations and index form equations. Theorems 16–18 in §12 give finiteness results for algebraic integers and polynomials with a given non-zero discriminant. They have many applications in algebraic number theory.

Finally, in §13, a remarkable application of the Main Theorem on S-unit Equations to algebraic independence of function values due to Nishioka [60, 61] is mentioned. Nishioka solved in this way a conjecture of D. W. Masser and a more general problem which had been open for several years.

For more information on S-unit equations and their applications, see [25], [36], [51] and [77].

The authors thank the organisers of the conference in Durham, A. Baker and R. C. Mason, for the excellent opportunity offered to the authors to discuss mathematics and to work together. They further thank F. Beukers and P. Erdős for valuable discussions and Lianxiang Wang for remarks on an early draft of the paper.

§1. Notation and simple observations

The notation introduced in this paragraph will be used throughout the paper without further mention. Let K be an algebraic number field with ring of integers $\mathcal{O}_K$. Let d, h_K, r_K and R_K denote the degree, class number, unit rank and regulator of K, respectively. Let M_K be the set of *places* on K (i.e. equivalence classes of multiplicative valuations on K). A place v is called *finite* if v contains only non-archimedean valuations and *infinite* otherwise. K has only finitely many infinite places. The rational number field $\mathbf{Q}$ has only one infinite place ∞, containing the ordinary absolute value, and a finite place for each prime number p. In ∞ we choose a representative $|\,.\,|_\infty$ which is equal to the ordinary absolute value. In the place corresponding to p (which is also denoted by p) we choose the valuation $|\,.\,|_p$ such that $|p|_p = p^{-1}$ as representative. In each place v of M_K we choose a valuation $|\,.\,|_v$ as follows. Let $p \in M_{\mathbf{Q}}$ be such that $v|p$ (i.e. the restrictions to $\mathbf{Q}$ of the valuations in v belong to p; in particular v is infinite if and only if $v|\infty$). We put $d_v = [K_v : \mathbf{Q}_p]$, where K_v and $\mathbf{Q}_p$ denote the completions of K at v and $\mathbf{Q}$ at p, respectively. In v we choose the valuation $|\,.\,|_v$ satisfying

$$|\alpha|_v = |\alpha|_p^{d_v/d} \quad \text{for each } \alpha \text{ in } \mathbf{Q}. \tag{1.1}$$

By these choices for the valuations we have the *Product Formula*

$$\prod_{v \in M_K} |\alpha|_v = 1 \quad \text{for } \alpha \in K^*. \tag{1.2}$$

Here and elsewhere we put $V^* = V \setminus \{0\}$ for any set V. Put

$$s(v) = \begin{cases} 0 & \text{if } v \text{ is a finite place,} \\ 1/d & \text{if } K_v = \mathbf{R}, \\ 2/d & \text{if } K_v = \mathbf{C}. \end{cases}$$

Then $\sum_{v \in M_K} s(v) = 1$ and

$$|\alpha_1 + \ldots + \alpha_r|_v \leq r^{s(v)} \max(|\alpha_1|_v, \ldots, |\alpha_r|_v) \tag{1.3}$$

$$\text{for } \alpha_1, \ldots, \alpha_r \in K \text{ and } v \in M_K.$$

The *height function* $h(\,.\,)$ on K is defined by

$$h(\alpha) = \prod_{v \in M_K} \max(1, |\alpha|_v) \quad \text{for } \alpha \in K.$$

This height depends only on α, and not on the choice of the algebraic number field K. The following elementary properties of h can be proved.

$$h(\alpha^{-1}) = h(\alpha) \quad \text{for } \alpha \in K^*,$$

$$h(\alpha_1 \ldots \alpha_r) \leq h(\alpha_1) \ldots h(\alpha_r) \quad \text{for } \alpha_1, \ldots, \alpha_r \in K, \tag{1.4}$$

$$h(\alpha_1 + \ldots + \alpha_r) \leq r h(\alpha_1) \ldots h(\alpha_r) \quad \text{for } \alpha_1, \ldots, \alpha_r \in K,$$

$$h(\alpha) = 1 \quad \text{if and only if } \alpha = 0 \text{ or a root of unity.}$$

Other heights of algebraic numbers α which are often used in diophantine approximation, are $\overline{|\alpha|}$ (the maximum of the absolute values of the conjugates of α over $\mathbf{Q}$) and $H(\alpha)$ (the maximum of the absolute values of the coefficients of the minimal polynomial of α over $\mathbf{Z}$). If α is an algebraic number of degree m, then

$$\overline{|\alpha|} \leq \left(h(\alpha)\right)^m \leq \overline{|\alpha|}^{\,m}, \quad \text{if } \alpha \text{ is an algebraic integer,}$$

$$2^{1-m} H(\alpha) \leq \left(h(\alpha)\right)^m \leq \sqrt{m+1}\, H(\alpha), \quad \text{if } \alpha \text{ is an arbitrary} \quad (1.5)$$

$$\text{algebraic number.}$$

The first inequality is obvious, while the second follows from Lang [51] Ch. 3, Theorem 2.8. Consequently, for each positive number C there are only finitely many α in K with $h(\alpha) \leq C$ and these belong to an effectively determinable finite subset of K.

Let S_∞ be the set of all infinite places on K, and let S be a finite subset of M_K containing S_∞. Let s denote the cardinality of S. An element α of K is called an S-unit if $|\alpha|_v = 1$ for each $v \notin S$ (i.e. $v \in M_K \setminus S$). The S-units form a finitely generated multiplicative group of rank $s-1$ which is denoted by U_S. If S contains no finite places, then U_S is just the group of units, U_K, of $\mathcal{O}_K$. Note that if $\alpha \in U_S$, then, by (1.2) and (1.4),

$$\prod_{v \in S} |\alpha|_v = 1, \qquad h(\alpha) = \prod_{v \in S} \max(1, |\alpha|_v). \qquad (1.6)$$

Suppose that the finite places in S correspond to the prime ideals $\wp_1, \ldots, \wp_t$ and that these prime ideals lie above rational primes not exceeding $P(\geq 2)$. An element α of K is called an S-integer if $|\alpha|_v \leq 1$ for all $v \notin S$. The S-integers form a ring which is denoted by $\mathcal{O}_S$. If $\alpha \in K$ then the principal ideal (α) can be written uniquely as a product of two ideals $\mathcal{A}_1$, $\mathcal{A}_2$ where $\mathcal{A}_1$ is composed of $\wp_1, \ldots, \wp_t$ and $\mathcal{A}_2$ is composed solely of prime ideals different from $\wp_1, \ldots, \wp_t$. We define $N_S(\alpha)$, which is sometimes called the S-norm of α, by $N_S(\alpha) = N_{K/\mathbf{Q}}(\mathcal{A}_2)$. This function N_S has several useful properties. We have

$$N_S(\alpha) = \left(\prod_{v \in S} |\alpha|_v\right)^d \quad \text{for all } \alpha \text{ in } K.$$

Further N_S is multiplicative, $N_S(\alpha) \geq 1$ if $\alpha \in \mathcal{O}_S$, and $N_S(\alpha) = 1$ if $\alpha \in U_S$. Finally we note that if $S = S_\infty$, then $N_S(\alpha) = \left|N_{K/\mathbf{Q}}(\alpha)\right|$.

We shall deal with the (*general homogeneous*) *S-unit equation*

$$\alpha_0 x_0 + \ldots + \alpha_n x_n = 0 \quad \text{in } x_0, x_1, \ldots, x_n \in U_S \qquad (1.7)$$

where $\alpha_0, \alpha_1, \ldots, \alpha_n \in K^*$. In the study of this equation we can identify pairwise linearly dependent non-zero points in K^{n+1}, that is, consider solutions in the n-dimensional projective space $\mathbf{P}^n(K)$. Points in $\mathbf{P}^n(K)$, so-called *projective points*, are denoted by $X = (x_0 : x_1 : \ldots : x_n)$, where the homogeneous coordinates are in K, and are determined up to a multiplicative factor in K. Alternatively we can divide all coefficients α_i by α_0 and all variables x_i by $-x_0$ and study the *inhomogeneous S-unit equation*

$$\alpha_1 x_1 + \ldots + \alpha_n x_n = 1 \quad \text{in } x_1, x_2, \ldots, x_n \in U_S.$$

Since U_S is finitely generated, S-unit equations are in fact exponential diophantine equations. Most of our attention will be focussed on the (*inhomogeneous*) *S-unit equation in two variables*,

$$\alpha_1 x + \alpha_2 y = 1 \quad \text{in } x, y \in U_S. \qquad (1.8)$$

It is implicit in the work of Mahler [56] and explicitly stated by Lang [50] that (1.8) has only finitely many solutions. Denote the number of solutions of (1.8) by $\nu(\alpha_1, \alpha_2)$.

In §3 we shall give upper bounds for $\max(h(x), h(y))$ and for $\nu(\alpha_1, \alpha_2)$ when x, y satisfy (1.8). In view of the symmetry in (1.7) we can distinguish equivalence classes of equations such that the sets of solutions of two equations from the same class are isomorphic: two tuples $(\alpha_0, \alpha_1, \ldots, \alpha_n)$ and $(\beta_0, \beta_1, \ldots, \beta_n)$ in $(K^*)^{n+1}$ (resp. the corresponding homogeneous S-unit equations) are called *S-equivalent* if there is a permutation σ of $\{0, 1, \ldots, n\}$, a $\lambda \in K^*$ and S-units $\epsilon_0, \epsilon_1, \ldots, \epsilon_n$ such that

$$\beta_i = \lambda \epsilon_i \alpha_{\sigma(i)} \quad \text{for } i = 0, , \ldots, n.$$

Observe that the solution $(\epsilon_{\sigma^{-1}(0)} \tilde{x}_0 : \epsilon_{\sigma^{-1}(1)} \tilde{x}_1 : \ldots : \epsilon_{\sigma^{-1}(n)} \tilde{x}_n)$ of $\alpha_0 x_0 + \alpha_1 x_1 + \ldots + \alpha_n x_n = 0$ corresponds to the solution $(\tilde{x}_{\sigma(0)} : \tilde{x}_{\sigma(1)} : \ldots : \tilde{x}_{\sigma(n)})$ of $\beta_0 x_0 + \beta_1 x_1 + \ldots + \beta_n x_n = 0$ so that there is indeed a simple bijection between the solutions of both equations. Transferring the concept of S-equivalence to the inhomogeneous case, we find that the S-equivalence class of equation (1.8) consists of the following six classes of inhomogeneous S-unit equations:

$$\alpha_1 \epsilon_1 x + \alpha_2 \epsilon_2 y = 1,$$

$$\alpha_1^{-1}\alpha_2\epsilon_1 x + \alpha_1^{-1}\epsilon_2 y = 1,$$

$$\alpha_2^{-1}\epsilon_1 x + \alpha_1\alpha_2^{-1}\epsilon_2 y = 1,$$

$$\alpha_2\epsilon_1 x + \alpha_1\epsilon_2 y = 1, \qquad (1.9)$$

$$\alpha_1^{-1}\epsilon_1 x + \alpha_1^{-1}\alpha_2\epsilon_2 y = 1,$$

$$\alpha_1\alpha_2^{-1}\epsilon_1 x + \alpha_2^{-1}\epsilon_2 y = 1,$$

where ϵ_1 and ϵ_2 are arbitrary S-units.

We now show that if U_S is infinite (which is the case if $s > 1$), then there are infinitely many S-equivalence classes of S-unit equations with at least two distinct solutions. Let $\xi \in U_S$, $\xi \neq 1$. For each η in U_S with $\eta \neq \xi$, $\eta \neq 1$ we define α_1, α_2 by

$$\alpha_1 = \frac{\eta - 1}{\eta - \xi}, \qquad \alpha_2 = \frac{\xi - 1}{\xi - \eta}.$$

Then (1.1) and (ξ, η) are distinct solutions of $\alpha_1 x + \alpha_2 y = 1$ in $x, y \in U_S$. The equations $\alpha_1 x + \alpha_2 y = 1$ constructed in this way must belong to infinitely many S-equivalence classes, since the number of equations constructed in this way is infinite, but each S-equivalence class contains only finitely many equations with solution $(1,1)$. This last fact follows from applying Lang's result to (1.9) with $x = y = 1$, α_1 and α_2 fixed and ϵ_1, $\epsilon_2 \in U_S$ variables.

§2. The General Case: The Main Theorem on S-Unit Equations

In this paragraph we deal with equations (1.7). The results in this paragraph are all based on p-adic versions of the Thue-Siegel-Roth-Schmidt method. Both Schlickewei [68], [69], [70] and Dubois and Rhin [14] gave such a p-adic version and used it to prove that, for any given set of prime numbers $T = \{p_1, \ldots, p_t\}$, the equation

$$x_0 + x_1 + \ldots + x_n = 0 \qquad \text{in } x_0, x_1, \ldots, x_n \in \mathbf{Z} \qquad (2.1)$$

has only finitely many solutions $x_0, x_1, \ldots, x_n$ each composed of primes from T such that

$$\gcd(x_i, x_j) = 1 \qquad \text{for } i \neq j. \qquad (2.2)$$

Actually they proved the following more general result. Let Δ, δ be real constants with $\Delta > 0$, $0 \leq \delta < 1$. Then the number of solutions of (2.1) satisfying (2.2) and

$$\prod_{k=0}^{n}\left(|x_k| \prod_{p \in T} |x_k|_p\right) \leq \Delta\left(\max(|x_0|, |x_1|, \ldots, |x_n|)\right)^{\delta} \qquad (2.3)$$

is finite. The restriction of pairwise coprimality may be too severe, but some restriction is needed in view of the equation $x_0 + x_1 + \ldots + x_5 = 0$ with $T = \{2, 3\}$ which has the solution $x_0 = 2^{k+1}$, $x_1 = 2^k$, $x_2 = -3 \cdot 2^k$, $x_3 = 2^3 3^\ell$, $x_4 = 3^\ell$, $x_5 = -3^{\ell+2}$ for all positive integers k, ℓ.

Van der Poorten and Schlickewei [67] proved that (2.1) has only finitely many solutions $x_0, x_1, \ldots, x_n$ each composed of primes from T such that

$\gcd(x_0, \ldots, x_n) = 1$ and no proper non-empty subsum

$$x_{i_1} + \ldots + x_{i_k} \text{ of } x_0 + x_1 + \ldots + x_n \text{ vanishes.} \quad (2.4)$$

Condition (2.4) is necessary and sufficient. Their result holds for algebraic number fields (cf. Corollary 1.1) and even for finitely generated subgroups of $\mathbf{C}^*$ (cf. Theorem 1′), but they have not yet published the complete proofs of their claim. Independently of van der Poorten and Schlickewei, Evertse [20] proved that (2.1) has only finitely many solutions satisfying (2.3) and (2.4) and extended this result to algebraic number fields. By using these results of van der Poorten and Schlickewei and Evertse, a further extension for subgroups of $\mathbf{C}^*$ of finite rank was given by Laurent [52].

To state Evertse's result in full generality we need some more notation. For any projective point $\mathbf{x} = (x_0 : x_1 : \ldots : x_n)$ in $\mathbf{P}^n(K)$ and for any $v \in M_K$ we put $|\mathbf{x}|_v = \max(|x_0|_v, \ldots, |x_n|_v)$. We define the *projective height*[1] of $\mathbf{x}$ as

$$\mathcal{H}(\mathbf{x}) = \prod_{v \in M_K} |\mathbf{x}|_v. \quad (2.5)$$

This height is well-defined, since it is independent of the multiplicative factor by the Product Formula. There is a simple relation between the height h and the projective height $\mathcal{H}$, namely

$$h(\alpha) = \mathcal{H}(1 : \alpha) \qquad \text{for } \alpha \in K. \quad (2.6)$$

Let, as always, S be a finite subset of M_K containing all infinite places. Let Δ, δ be real constants with $\Delta > 0$, $\delta \geq 0$. A projective point $\mathbf{x} \in \mathbf{P}^n(K)$ is called (Δ, δ, S)-*admissible*[1] if its homogeneous coordinates can be chosen such that

---

[1] The valuation $\|\,.\,\|_v$ in [20] is not the same as the valuation $|\,.\,|_v$. The relation between them is given by $\|\alpha\|_v = |\alpha|_v^d$ for $\alpha \in K$. Hence the notation of (Δ, δ, S)-admissibility here corresponds with (Δ^d, δ, S)-admissibility in Evertse's paper.

(i) all x_k are S-integers

and

(ii) $\displaystyle\prod_{v\in S}\prod_{k=0}^{n}|x_k|_v \leq \Delta\big(\mathcal{H}(\mathbf{x})\big)^{\delta}.$

Clearly the homogeneous coordinates of $(1,0,S)$-admissible projective points can all be chosen to be S-units.

Theorem 1. (The Main Theorem on S-Unit Equations for Algebraic Number Fields) (Evertse [20]).

Let $\Delta > 0$, $0 \leq \delta < 1$. *There are only finitely many* (Δ, δ, S)-*admissible projective points* $\mathbf{x} = (x_0 : x_1 : \ldots : x_n) \in \mathbf{P}^n(K)$ *satisfying*

$$x_0 + x_1 + \ldots + x_n = 0 \qquad (2.7)$$

but

$$x_{i_1} + \ldots + x_{i_k} \neq 0 \text{ for each proper, non-empty subset}$$

$$\{i_1, \ldots, i_k\} \text{ of } \{0, 1, \ldots, n\} \quad (2.8)$$

We express (2.8) succinctly by saying that no subsum of $x_0 + \ldots + x_n$ vanishes. When we use the word 'subsum' we exclude the full and empty sum.

For general homogeneous S-unit equations (1.7) we derive the following consequence of Theorem 1.

Corollary 1.1. *Let* $\alpha_0, \alpha_1, \ldots, \alpha_n \in K^*$. *There are only finitely many projective points* $\mathbf{x} = (x_0 : x_1 \ldots : x_n) \in \mathbf{P}^n(K)$ *with* $x_0, x_1, \ldots, x_n \in U_S$ *such that* $\alpha_0 x_0 + \alpha_1 x_1 + \ldots + \alpha_n x_n = 0$, *but no subsum of* $\alpha_0 x_0 + \alpha_1 x_1 + \ldots + \alpha_n x_n$ *vanishes.*

This implies for inhomogeneous S-unit equations:

Corollary 1.2. *Let* $\alpha_1, \alpha_2, \ldots, \alpha_n \in K^*$. *There are only finitely many tuples* $(x_1, \ldots, x_n) \in U_S^n$ *such that* $\alpha_1 x_1 + \ldots + \alpha_n x_n = 1$, *but no subsum of* $\alpha_1 x_1 + \ldots + \alpha_n x_n$ *vanishes.*

By a specialisation argument, Theorem 1 can be extended as follows.

Theorem 1′ (The Main Theorem on S-Unit Equations for Groups) (Van der Poorten and Schlickewei [67]).

Let G be a finitely generated multiplicative subgroup of $\mathbf{C}^*$. There are only finitely many projective points $\mathbf{x} = (x_0 : x_1 : \ldots : x_n) \in \mathbf{P}^n(G)$ satisfying (2.7) and (2.8).

Laurent [52] proved Theorem 1' in the more general case of multiplicative subgroups of $\mathbf{C}^*$ of finite rank. He used it to prove a special case of a conjecture of S. Lang which is an assertion on commutative algebraic groups.

§3. Upper bounds in the two variables case

In this section we deal with the S-unit equation in two variables

$$\alpha_1 x + \alpha_2 y = 1 \qquad \text{in } x, y \in U_S, \qquad (1.8)$$

where $\alpha_1, \alpha_2 \in K^*$. It is implicit in the work of Siegel [78, 79] that equations of the form (1.8) have only finitely many solutions in units x, y, and implicit in the work of Mahler [56] that (1.8) has only finitely many solutions (in S-units x, y). As remarked before, Lang [50] proved this result explicitly. Siegel developed the so-called Thue-Siegel method involving hypergeometric functions. By combining his method with ideas of Mahler about p-adic approximation of algebraic numbers, Evertse proved the following result on the number of solutions $\nu(\alpha_1, \alpha_2)$ of (1.8).

Theorem 2. (Evertse [19]).

$$\nu(\alpha_1, \alpha_2) \leq 3 \times 7^{d+2s}.$$

This bound has the remarkable feature of being dependent only on the degree of K and the cardinality of S. Theorem 2 is a considerable improvement and generalisation of a result of Lewis and Mahler [54] who derived an upper bound for $\nu(1,1)$ in the rational case which depends on the primes involved in S and not only on their number. Independently of Evertse, and by a different method, Silverman [81] showed $\nu(1,1) \leq C \times 2^{20s}$. Here and elsewhere C is a constant, the value of which may be different at each occurrence. Later, Evertse and Győry [22] derived an upper bound for the number of solutions of (1.8) independent of α_1 and α_2 in the general case that the variables x, y belong to a finitely generated multiplicative subgroup of $\mathbf{C}^*$.

The dependence of Evertse's bound on the degree of K and the cardinality of S is necessary. Nagell [59] proved that for $d \geq 5$ there exists a number field K of degree d such that $x + y = 1$ has at least

$3(2d - 3)$ solutions in units x, y of K. Erdős, Stewart and Tijdeman [17] proved that in the case $K = \mathbf{Q}$ the equation $x + y = 1$ can have more than $\exp(Cs^{1/2}/\log s)$ solutions x, $y \in U_S$. This implies that the best improvement of Theorem 2 one can hope for is $\nu(\alpha_1, \alpha_2) \leq \exp(s^{1/2})$. According to a conjecture which Stewart presented during the conference, the exponent $\frac{1}{2}$ should be replaced by $\frac{2}{3}$. In great contrast to this result is the observation made during the conference that for most pairs α_1, α_2 we have $\nu(\alpha_1, \alpha_2) \leq 2$.

Theorem 3 (Evertse, Győry, Stewart, Tijdeman [26]).

There are only finitely many S-equivalence classes of equations (1.8) *with more than two solutions.*

As observed at the end of §1 there are infinitely many S-equivalence classes of equations (1.8) with two solutions. The proof of Theorem 3 is based on Corollary 1.1. Its principle will be explained in §5. Theorem 3 can be extended to finitely generated multiplicative subgroups of $\mathbf{C}^*$.

Up to now all the upper bounds we have mentioned were proved by ineffective methods. This has the important disadvantage that it is impossible to derive upper bounds for the solutions themselves or to decide from the proof that for given α_1, α_2 (1.8) has no more than two solutions. Skolem [83], using Skolem's method, and Cassels [9], using Gelfond's results, showed how certain classes of S-unit equations in rationals can be solved effectively, at least in principle. The important breakthrough was Baker's method for estimating linear forms in logarithms and its p-adic analogue by Coates. Implicitly in Coates' work on the Thue-Mahler equation [11] there are S-unit equations in two variables and upper bounds for their solutions. The first explicit mention of such an application is in Sprindžhuk [84]. Győry [34] worked out an explicit upper bound for the heights of the solutions of (1.8). We state his result in a slightly different and less precise form. To state his result we transform (1.8) into an equivalent equation. By multiplying α_1 and α_2 by the product of their denominators, (1.8) transforms into an equivalent equation of the form

$$\alpha_1 x + \alpha_2 y = \alpha_0 \qquad \text{in } x, y \in U_S \qquad (3.1)$$

where $\alpha_1, \alpha_2, \alpha_0 \in \mathcal{O}_K \setminus \{0\}$.

Theorem 4 (Győry [34]). *Let $\epsilon > 0$. Every solution (x, y) of* (3.1) *satisfies*

$$\max(h(x), h(y)) < \exp(s^{C(K,\epsilon)s} P^{d+\epsilon} \log A) \qquad (3.2)$$

where $A = \max(h(\alpha_1), h(\alpha_2), h(\alpha_0), 3)$ *and* $C(K, \epsilon)$ *is an expression, explicitly given in [34], involving the parameters* d, h_K, r_K *and* R_K *of* K *and* ϵ.

It is most likely that the right-hand side of (3.2) cannot be improved on in an essential way when we use the presently available estimates for linear forms in logarithms of algebraic numbers. However, if we assume that (3.1) has at least $s + 2$ solutions, then it is possible, after having replaced $(\alpha_1, \alpha_2, \alpha_0)$ by an appropriate S-equivalent triple, to derive a result similar to (3.2) with a bound independent of A. A first step in this direction was made by Győry. Recall the definition of $N_S(\alpha)$ given in §1. Győry [34] proved the following statement in a more precise form. Here and in the sequel we use $C(K)$ for an effectively computable number depending only on K which may have a different value at each occurrence.

Let $0 < \epsilon < 1$. *For each triple* $(\alpha_1, \alpha_2, \alpha_0)$ *of elements in* $\mathcal{O}_K \setminus \{0\}$ *with*

$$\min(N_S(\alpha_1), N_S(\alpha_2)) \leq N_S(\alpha_0)^{1-\epsilon} \tag{3.3}$$

such that (3.1) has at least $s + 3t + 1$ *solutions, we have*

$$N_S(\alpha_0) \leq \exp\left\{\epsilon^{-1} s^{C(K)s} P^{d+1} \log \frac{2}{\epsilon}\right\}.$$

An upper bound for $N_S(\alpha_0)$ *of the same form can be given if* $\max(\log N_S(\alpha_1), \log N_S(\alpha_2)) \leq (\log N_S(\alpha_0))^{1-\epsilon}$ *and there are at least* $s + t + 1$ *solutions.*

There are infinitely many S-equivalence classes which have a representative satisfying (3.3), but there are also infinitely many S-equivalence classes which do not have such a representative. (If $p_1, \ldots, p_t$ are the rational primes in $\wp_1, \ldots, \wp_t$ and $P = p_1 \ldots p_t$, then for all sufficiently large positive integers a the triples $(Pa + 1, 2Pa - 1, 2Pa + 1)$ will be pairwise S-inequivalent and they do not satisfy (3.3).)

Recently we considerably relaxed Győry's conditions and moreover slightly improved upon the bound for the number of required solutions.

Theorem 5 (Evertse, Győry, Stewart, Tijdeman [26]). *For each* $(\alpha_1, \alpha_2, \alpha_0) \in (\mathcal{O}_K \setminus \{0\})^3$ *such that (3.1) has at least* $s + 2$ *solutions, there exists an* S-equivalent triple $(\beta_1, \beta_2, \beta_0) \in (\mathcal{O}_K \setminus \{0\})^3$ *such that*

$$\max(h(\beta_1), h(\beta_2), h(\beta_0)) \leq \exp\{s^{C(K)s} P^{d+1}\}. \tag{3.4}$$

Since there are only finitely many S-equivalence classes which have a representative satisfying (3.4) (cf. (1.5)), this result implies that (3.1) has at most $s + 1$ solutions for all but the finitely many S-equivalence classes determined by (3.4). It follows from Theorem 4 that the solutions of $\beta_1 x + \beta_2 y = \beta_0$ in $x, y \in U_S$ subject to (3.4) satisfy

$$\max(h(x), h(y)) \le \exp\{s^{C(K)s} P^{2(d+1)}\}. \tag{3.5}$$

§4. On the proofs of Theorems 1 and 2

In this section we shall describe some ideas behind the proofs of Theorems 1 and 2.

Theorem 1 (the Main Theorem on S-Unit Equations) is a consequence of the Subspace Theorem of Schmidt and Schlickewei, stated below. We use the notation introduced in §§1, 2. By a *projective subspace* we shall mean a set of the type

$$\{\mathbf{x} = (x_0 : \ldots : x_n) \in \mathbf{P}^n(K) : \ell_1(\mathbf{x}) = \ldots = \ell_r(\mathbf{x}) = 0\}$$

where $\ell_1, \ldots, \ell_r$ are linear forms in $K[X_0, \ldots, X_n]$.

Subspace Theorem. *Let K be an algebraic number field, S a finite set of places on K with $S_\infty \subseteq S$, and $n \ge 1$ an integer. For each v in S, let $\{\ell_{iv}\}_{i=0}^{n_v}$ be a collection of linear forms in $K[X_0, \ldots, X_n]$ of rank $n_v + 1$; thus $n_v \le n$ for $v \in S$. Then for every $c > 0$ and $\epsilon > 0$, the solutions of the inequality*

$$\prod_{v \in S} \prod_{i=0}^{n_v} \frac{|\ell_{iv}(\mathbf{x})|_v}{|\mathbf{x}|_v} \le c\mathcal{H}(\mathbf{x})^{-n-1-\epsilon} \quad \text{in } \mathbf{x} \in \mathbf{P}^n(K) \tag{4.1}$$

are contained in finitely many proper, projective subspaces of $\mathbf{P}^n(K)$.

This theorem was proved by Schmidt [73, 74] in case that S contains only infinite places, and by Schlickewei [69] in full generality.

We remark that Schlickewei's formulation of the Subspace Theorem is different from ours. Schlickewei considered the inequality

$$\prod_{v \in S} \prod_{i=0}^{n_v} |\ell_{iv}(\mathbf{x})|_v \le c\|x\|^{-\epsilon} \quad \text{in } \mathbf{x} \in \mathcal{O}_K^{n+1} \tag{4.2}$$

where $\|\mathbf{x}\| = \max_{i,j} |\sigma_j(x_i)|$ and $\sigma_1, \ldots, \sigma_d$ are the different $\mathbf{Q}$-isomorphisms of K. Inequality (4.1) is easily reduced to an inequality of type (4.2) and vice versa, by observing that there are positive integers c_1, c_2, c_3 depending on K only, such that each $\mathbf{x} \in \mathbf{P}^n(K)$ can be represented by homogeneous coordinates $(x_0 : x_1 : \ldots : x_n)$ for which

$$x_0, \ldots, x_n \in \mathcal{O}_K,$$

$$N_{K/\mathbf{Q}}((x_0, \ldots, x_n)) \leq c_1,$$

$$c_2 \|\mathbf{x}\| \leq \mathcal{H}(\mathbf{x}) \leq c_3 \|\mathbf{x}\|.$$

Sketch of proof of Theorem 1. We shall proceed by induction on n. For $n = 1$, Theorem 1 is trivial. Suppose that Theorem 1 has been proved for equations $x_0 + \ldots + x_{n'} = 0$ with $n' < n$ (induction hypothesis). Consider the equation

$$x_0 + x_1 + \ldots + x_n = 0 \text{ in } S\text{-integers } x_0, x_1, \ldots, x_n \qquad (4.3)$$

satyisfying

$$\prod_{v \in S} \prod_{i=0}^{n} |x_i|_v \leq \Delta \mathcal{H}(\mathbf{x})^{1-\epsilon}, \qquad (4.4)$$

where S is a finite subset of M_K containing all infinite places, $\Delta > 0$ and $\epsilon = 1 - \delta$. Now (4.4) can be rewritten as

$$\prod_{v \in S} \left(\frac{|x_0|_v \ldots |x_{n-1}|_v |x_0 + \ldots + x_{n-1}|_v}{|\tilde{\mathbf{x}}|^{n+1}} \right) \leq c \mathcal{H}(\tilde{\mathbf{x}})^{-n-\epsilon} \qquad (4.5)$$

where $\tilde{\mathbf{x}} = (x_0 : \ldots : x_{n-1}) \in \mathbf{P}^{n-1}(K)$ and $c = n\Delta$. Consider the solutions of (4.5) with $|\tilde{\mathbf{x}}|_v = |x_{i(v)}|_v$, where $(i(v))_{v \in S}$ is any fixed tuple of subscripts taken from $\{0, 1, \ldots, n-1\}$. For $v \in S$, let $\{\ell_{iv}\}_{i=0}^{n-1}$ be the set of linear forms

$$\{X_0, X_1, \ldots, X_{n-1}, X_0 + X_1 + \ldots + X_{n-1}\} \setminus \{X_{i(v)}\}.$$

Then (4.5) can be written as

$$\prod_{v \in S} \prod_{i=0}^{n-1} \frac{|\ell_{iv}(\tilde{\mathbf{x}})|_v}{|\tilde{\mathbf{x}}|_v} \leq c \mathcal{H}(\tilde{\mathbf{x}})^{-n-\epsilon}. \qquad (4.6)$$

By the subspace Theorem, the solutions of (4.6) in $\tilde{\mathbf{x}} \in \mathbf{P}^{n-1}(K)$ are contained in finitely many proper projective subspaces of $\mathbf{P}^{n-1}(K)$. This implies that the solutions of (4.3) satisfying (4.4) are contained in finitely many linear subspaces of K^{n+1} of the type

$$\{x \in K^{n+1} : \alpha_0 x_0 + \alpha_1 x_1 + \ldots + \alpha_{n-1} x_{n-1} = 0\}$$

where $\alpha_0, \ldots, \alpha_{n-1} \in K$ and $(\alpha_0, \ldots, \alpha_{n-1}) \neq (0, \ldots, 0)$. Fix $\alpha_0, \ldots, \alpha_{n-1}$, fix a non-empty subset J of $\{0, \ldots, n-1\}$, and consider all solutions of (4.3) satisfying

$$\sum_{j \in J} \alpha_j x_j = 0, \quad \text{but no subsum of} \sum_{j \in J} \alpha_j x_j \text{ vanishes.}$$

By the induction hypothesis, there is a finite number of tuples $(\beta_j)_{j \in J}$ such that each x_j $(j \in J)$ can be written as $x_j = \xi \beta_j$ where ξ is some S-integer. By substituting this into (4.3) we obtain

$$\left(\sum_{j \in J} \beta_j \right) \xi + \sum_{j \in \{0, \ldots, n\} \setminus J} x_j = 0.$$

By applying the induction hypothesis again we conclude that there are only finitely many possible projective solutions $(\xi : x_{i_1} : \ldots : x_{i_t})$ where $\{i_1, \ldots, i_t\} = \{0, \ldots, n\} \setminus J$. Combining this with the facts that the numbers of possible tuples $(\alpha_0, \ldots, \alpha_{n-1})$, sets J and tuples β_j $(j \in J)$ are finite, we obtain that the total number of solutions of (4.3) subject to (4.4) is finite. ∎

We shall now sketch the ideas behind the proof of Theorem 2. We use the notation of §§1–3. For $i = 1, 2, \ldots$ let $c_i(\ldots)$ denote effectively computable numbers depending only on the parameters written within the parentheses. Hence c_6 and c_{13} are absolute constants.

Equation (1.8) can be rewritten as

$$y_0 + y_1 + y_2 = 0 \text{ in } (y_0 : y_1 : y_2) \in \mathbf{P}^2(K)$$

$$\text{with } y_i / \nu_i \ S\text{-units for } i = 0, 1, 2, \quad (4.7)$$

where $\nu_0, \nu_1, \nu_2 \in K^*$ are fixed. Put

$$A = \prod_{v \in S} |\nu_0 \nu_1 \nu_2|_v \times \prod_{v \notin S} \left\{ \max(|\nu_0|_v, |\nu_1|_v, |\nu_2|_v) \right\}^3.$$

A straightforward computation shows

$$\prod_{v \in S} \frac{|y_0 y_1 y_2|_v}{\{\max(|y_0|_v, |y_1|_v, |y_2|_v)\}^3} = A\mathcal{H}(\mathbf{y})^{-3},$$

where $\mathbf{y} = (y_0 : y_1 : y_2)$. Let $(i(v))_{v \in S}$ be subscripts such that $|y_{i(v)}|_v = \min(|y_0|_v, |y_1|_v, |y_2|_v)$. If $|y_{i(v)}|_v \leq |y_{j(v)}|_v \leq |y_{k(v)}|_v$ with $\{i(v), j(v), k(v)\} = \{1, 2, 3\}$, then $|y_{k(v)}|_v \leq 2^{s(v)}|y_{j(v)}|_v$ by (1.3). Hence

$$\prod_{v \in S} \frac{|y_{i(v)}|_v}{|\mathbf{y}|_v} \leq 2A\mathcal{H}(\mathbf{y})^{-3}. \tag{4.8}$$

One possible way to deal with inequalities of type (4.8) is to use a quantitative version of Roth's theorem of the type below. If $\ell(X_0, X_1) = \alpha_0 X_0 + \alpha_1 X_1$, we put $|\ell|_v = \max(|\alpha_0|_v, |\alpha_1|_v)$.

Roth's Theorem. *Let K be an algebraic number field of degree d. Let $S \subset M_K$ be a finite set of cardinality s, containing all infinite places. Let $C \geq 1$ be a constant. Let $F(X_0, X_1) \in \mathbf{Z}[X_0, X_1]$ be a binary form of degree m, of which the absolute values of the coefficients are at most M. Finally, let $\{\ell_v\}_{v \in S}$ be a set of linear forms in $K[X_0, X_1]$, all dividing F. Then the number of solutions of the inequality*

$$\prod_{v \in S} \frac{|\ell_v(\mathbf{x})|_v}{|\ell_v|_v |\mathbf{x}|_v} \leq c\mathcal{H}(\mathbf{x})^{-2-\epsilon} \tag{4.9}$$

in $\mathbf{x} \in \mathbf{P}^1(K)$ with $\mathcal{H}(\mathbf{x}) \geq c_1(d, m, \epsilon)(C + M + 1)^{C_2(d, m, \epsilon)}$ is at most $c_3(d, m, \epsilon) \times (c_4(\epsilon))^s$.

As far as we know, no explicit proof of this result has been published. In [51] Ch. 7, Lang proved that (4.9) has only finitely many solutions. It is possible to prove Roth's theorem above by making explicit all arguments in Lang's proof, and combining this with ideas of Davenport and Roth [13]. In [82] Theorem 2.1, Silverman stated and sketched a proof of a result which is equivalent to Roth's Theorem stated above.

Roth's Theorem is a quantitative version of the Subspace Theorem in the special case $n = 1$, $n_v = 0$ for $v \in S \setminus S_\infty$. It would be of great interest to derive a quantitative version of the same type for the general Subspace Theorem.

Using Roth's Theorem with $F(X_0, X_1) = X_0 X_1(-X_0 - X_1)$, it follows that inequality (4.8) has at most $c_5(d) \times c_6^s$ solutions with

$$\mathcal{H}(\mathbf{y}) \geq c_7(d)(2A)^{c_8(d)}.$$

Our purpose is to obtain an upper bound for the number of solutions of (4.8) which is independent of A. Below we state a "gap principle" which enables us to derive such a bound. First we reduce (4.8) to a finite number of systems of inequalities

$$\frac{|y_{i(v)}|_v}{|y|_v} \leq \left(2A\mathcal{H}(\mathbf{y})^{-3}\right)^{\Gamma_v} \qquad \text{for } v \in S, \qquad (4.10)$$

where $\Gamma_v \geq 0$ for $v \in S$, $\sum_{v \in S} \Gamma_v = B$ for some B with $\frac{2}{3} < B < 1$, and the tuple $(\Gamma_v)_{v \in S}$ can be chosen from a set of cardinality at most $\left(c_7(B)\right)^s$. This can be achieved by taking a sufficiently fine grid of Γ_v's (cf. [19] Lemma 4).

Gap Principle. *Let* $\mathbf{y}^{(1)}$, $\mathbf{y}^{(2)}$ *be different solutions of (4.7) satisfying the same system of inequalities (4.10), and suppose that* $\mathcal{H}(\mathbf{y}^{(1)}) \leq \mathcal{H}(\mathbf{y}^{(2)})$. *Then*

$$\mathcal{H}(\mathbf{y}^{(2)}) \geq 2^{-B-1} A^{1-B} \left(\mathcal{H}(\mathbf{y}^{(1)})\right)^{3B-1}. \qquad (4.11)$$

Proof. Let $\mathbf{y}^{(1)} = (y_0^{(1)} : y_1^{(1)} : y_2^{(1)})$, $\mathbf{y}^{(2)} = (y_0^{(2)} : y_1^{(2)} : y_2^{(2)})$. Put

$$\Delta_v = \frac{|y_i^{(1)} y_j^{(2)} - y_j^{(1)} y_i^{(2)}|_v}{|\mathbf{y}^{(1)}|_v \times |\mathbf{y}^{(2)}|_v} \qquad \text{for } v \in M_K,$$

where i, j are distinct elements of $\{0, 1, 2\}$. Obviously, Δ_v is independent of the choice of i, j. For $v \in S$, take $i = i(v)$, $j \neq i(v)$. Then, by (4.10), for $v \in S$,

$$\Delta_v \leq 2^{s(v)} \max \left(\frac{|y_{i(v)}^{(1)}|_v}{|\mathbf{y}^{(1)}|_v}, \frac{|y_{i(v)}^{(2)}|_v}{|\mathbf{y}^{(2)}|_v} \right)$$

$$\leq 2^{s(v)} \left(2A\mathcal{H}(\mathbf{y}^{(1)})^{-3}\right)^{\Gamma_v}.$$

Hence

$$\prod_{v \in S} \Delta_v \leq 2 \left(2A\mathcal{H}(\mathbf{y}^{(1)})^{-3}\right)^{\sum_{v \in S} \Gamma_v} \qquad (4.12)$$

$$= 2 \left(2A\mathcal{H}(\mathbf{y}^{(1)})^{-3}\right)^B.$$

For $v \notin S$ we have, on choosing i, j such that $|\nu_i|_v \leq |\nu_j|_v \leq |\nu_k|_v$ for $k \neq i, j$,

$$\Delta_v \leq \frac{|\nu_i \nu_j|_v}{\{\max(|\nu_0|_v, |\nu_1|_v, |\nu_2|_v)\}^2}$$

$$= \frac{|\nu_0 \nu_1 \nu_2|_v}{\{\max(|\nu_0|_v, |\nu_1|_v, |\nu_2|_v)\}^3}.$$

Together with the Product Formula this shows that

$$\prod_{v \in M_K \setminus S} \Delta_v \leq A^{-1}.$$

By combining this with (4.12) and the Product Formula we obtain

$$\frac{1}{\mathcal{H}(\mathbf{y}^{(1)})\mathcal{H}(\mathbf{y}^{(2)})} = \prod_{v \in M_K} \Delta_v \leq 2^{1+B} A^{B-1} \mathcal{H}(\mathbf{y}^{(1)})^{-3B}.$$

This implies (4.11). ∎

Since in (4.11) the exponent of A is positive and that of $\mathcal{H}(\mathbf{y}^{(1)})$ is greater than one, the number of solutions of (4.10) with

$$\mathcal{H}(\mathbf{y}) < c_7(d)(2A)^{c_8(d)}$$

is bounded above by $c_9(B,d)$. We infer that the total number of solutions of (4.8) is at most $c_{10}(B,d)(c_{11}(B))^s$. By choosing B appropriately and observing that equation (4.7) can be reduced to at most 3^s different inequalities (4.8), we conclude that the number of solutions of (4.7) is at most $c_{12}(d) \times c_{13}^s$.

We remark that it is possible to prove that (4.8) has at most $c_5(d) \times c_6^s$ solutions with $\mathcal{H}(\mathbf{y}) \geq c_7(d)(2A)^{c_8(d)}$ by techniques which are less powerful than Roth's (cf. [19], [21]).

§5. The rational case

We specialise the theorems of §§2, 3 to the case $K = \mathbf{Q}$. This paragraph can be read independently of §§1–4. The text between square brackets indicates the connection with the preceding paragraphs.

For any prime number p and any rational number α we define $|\alpha|_p = p^{-k}$ if $p^{-k}\alpha$ is the quotient of two rational integers both coprime to p. Let M denote the set of valuations on $\mathbf{Q}$ consisting of the absolute value $|.|$ and the valuation $|.|_p$ just defined for each prime number p. Then we have the *Product Formula*

$$\prod_{v \in M} |\alpha|_v = 1 \qquad \text{for all } \alpha \in \mathbf{Q}^*. \tag{5.1}$$

Here and elsewhere we put $V^* = V \setminus \{0\}$ for any set V. Furthermore, we define the height $h(\alpha)$ of α by

$$h(\alpha) = \prod_{v \in M} \max(1, |\alpha|_v). \tag{5.2}$$

Hence, if $\alpha = a/b$ with $a, b \in \mathbf{Z}$ and $\gcd(a, b) = 1$, then $h(\alpha) = \max(|a|, |b|)$. Note that, by (5.1),

$$h(\alpha) = h(\alpha^{-1}) = \prod_{v \in M} \min(1, |\alpha|_v)^{-1}. \tag{5.3}$$

Let $t \geq 2$. Let $T = \{p_1, \ldots, p_t\}$ be a set of prime numbers not exceeding P. Let S be the set of valuations $|.|, |.|_{p_1}, \ldots, |.|_{p_t}$. [Hence $s = t + 1$, the number of valuations in S.] By $v \notin S$ we mean $v \in M \setminus S$. For $\alpha \in \mathbf{Q}^*$ we write $\alpha = [\alpha]_S \{\alpha\}_S$ where $[\alpha]_S := \prod_{p \in T} |\alpha|_p^{-1}$ is the T-part of α and $\{\alpha\}_S := |\alpha| \prod_{p \in T} |\alpha|_p = \prod_{v \in S} |\alpha|_v$ is the T-free part of α. The set U_S of S-units consists of the rational numbers α with $\{\alpha\}_S = 1$. Hence every α in U_S is of the form $\pm p_1^{k_1} \ldots p_t^{k_t}$ with $k_1, \ldots, k_t \in \mathbf{Z}$. Put $\mathcal{S} = U_S \cap \mathbf{Z}$.

Since there are exactly two tuples $(x_0, \ldots, x_n)$ of rational integers with $\gcd 1$ which correspond with a given projective point in $\mathbf{P}^n(\mathbf{Q})$, we have the following consequence of Theorem 1.

Corollary 1.3. *Let Δ, δ be real constants with $\Delta > 0$, $0 \leq \delta < 1$. Then there are only finitely many tuples $x = (x_0, x_1, \ldots, x_n)$ of rational integers such that*

$$x_0 + x_1 + \ldots + x_n = 0, \tag{5.4}$$

$$x_{i_1} + \ldots + x_{i_k} \neq 0 \tag{5.5}$$

for each proper, non-empty subset $\{i_1, \ldots, i_k\}$ of $\{0, 1, \ldots, n\}$,

$$\gcd(x_0, x_1, \ldots, x_n) = 1, \tag{5.6}$$

and

$$\prod_{j=0}^{n} \{x_j\}_S \leq \Delta \big(\max_{j=0,\ldots,n} |x_j| \big)^{\delta}. \tag{5.7}$$

We express (5.5) succinctly by saying that no subsum of $x_0 + x_1 + \ldots x_n$ vanishes. Taking $\Delta = 1$, $\delta = 0$ means requiring that $x_0, x_1, \ldots, x_n$ are all elements of $\mathcal{S}$. Obviously Corollary 1.3 generalises the result of Schlickewei [70] and Dubois and Rhin [14], mentioned at the beginning of §2, who required pairwise coprimality instead of (5.5) and (5.6).

The subsequent results deal with the case $n = 2$. Let $\nu(a, b, c)$ denote the number of solutions of the equation

$$ax + by = cz \quad \text{in } x, y, z \in S \text{ with } \gcd(x, y, z) = 1, \qquad (5.8)$$

where $a, b, c \in \mathbf{Z}^*$. Since every solution $x_1, x_2 \in U_S$ of $ax_1 + bx_2 = c$ corresponds to exactly two solutions of (5.8), the following result follows from Theorem 2.

Corollary 2. $\nu(a, b, c) \leq 6 \times 7^{2t+3}$.

Erdős, Stewart and Tijdeman [17] have proved that there exist sets S of arbitrarily large cardinality t for which $\nu(1, 1, 1) > \exp(Ct^{1/2} \log t)$. Here and elsewhere C is a constant, but the constant may have a different value at each occurrence.

We call a triple $(a, b, c) \in (\mathbf{Z}^*)^3$ *S-normalised* if a, b, c, $p_1, \ldots, p_t$ are pairwise relatively prime and $0 < a \leq b \leq c$. [Then each S-equivalence class in $(\mathbf{Z}^*)^3$ contains exactly one S-normalised triple: Suppose $(a_0, a_1, a_2) \in (\mathbf{Z}^*)^3$. Put $\lambda = \gcd(a_0, a_1, a_2)$. Then (a_0, a_1, a_2) is S-equivalent with $(\{a_0/\lambda\}_S, \{a_1/\lambda\}_S, \{a_2/\lambda\}_S)$. Arrange the three numbers in the latter tuple in increasing order and call them a, b, c, respectively. Then (a, b, c) is the unique S-normalised triple in the S-equivalence class of (a_0, a_1, a_2).] The following result is an immediate consequence of Theorem 3.

Corollary 3. *There are only finitely many S-normalised triples $(a, b, c) \in (\mathbf{Z}^*)^3$ for which (5.8) has more than two solutions with positive z.*

Corollary 3 can be derived from Corollary 1.3 as follows. Let $(a, b, c) \in (\mathbf{Z}^*)^3$ be an S-normalised triple and suppose there are three distinct triples $(x_i, y_i, z_i) \in S^3$ satisfying (5.8) and $z_i > 0$ for $i = 1, 2, 3$. Then we obtain

$$\begin{vmatrix} x_1 & x_2 & x_3 \\ y_1 & y_2 & y_3 \\ z_1 & z_2 & z_3 \end{vmatrix} = 0. \qquad (5.9)$$

Note that the expression on the left-hand side does not change value if we permute x, y, z or subscripts 1, 2, 3 consistently. Furthermore, if $x_1 y_2 z_3 = x_2 y_1 z_3$, then $x_1 y_2 = x_2 y_1$, hence $x_1 = \pm x_2$, $y_1 = \pm y_2$ and therefore $z_1 = \pm z_2$. Since z_1 and z_2 are positive, we obtain

$$x_1 y_2 z_3 \neq x_2 y_1 z_3. \qquad (5.10)$$

We first prove that there are only finitely many possible values for $x_1 z_2 / x_2 z_1$ and $y_1 z_2 / y_2 z_1$. To do so we apply Corollary 1.3 with $\Delta = 1$, $\delta = 0$ in the following way. If all conditions with the possible exception of (5.5) are satisfied, then there are only finitely many possibilities for the quotients x_i / x_j $(0 \le i \le j \le n)$. If no (proper, non-empty) subsum of (5.9) vanishes, then Corollary 1.3 implies that there are only finitely many possibilities for $x_1 y_3 z_2 / x_2 y_3 z_1 = x_1 z_2 / x_2 z_1$ and $x_3 y_1 z_2 / x_3 y_2 z_1 = y_1 z_2 / y_2 z_1$. In this case our claim is correct, but we cannot be certain that no subsum vanishes. Suppose that there is a vanishing subsum. Then the complementary subsum vanishes too and we can apply Corollary 1.3 to both subsums. There are a great many cases to be considered, but, by using the symmetry and (5.10), their number can be brought down to five. The most difficult one is $x_1 y_2 z_3 + x_2 y_3 z_1 + x_3 y_1 z_2 = 0$, $x_2 y_1 z_3 + x_3 y_2 z_1 + x_1 y_3 z_2 = 0$. By Corollary 1.3 there are only finitely many possibilities for the quotients $x_3 y_1 z_2 / x_1 y_2 z_3$, $x_2 y_3 z_1 / x_1 y_2 z_3$, $x_1 y_3 z_2 / x_2 y_1 z_3$ and $x_3 y_2 z_1 / x_2 y_1 z_3$, hence for $x_1^2 y_2 z_2 / x_2^2 y_1 z_1$ and $x_1 y_2^2 z_1 / x_2 y_1^2 z_2$, whence for $x_1^3 z_2^3 / x_2^3 z_1^3$ and $y_1^3 z_2^3 / y_2^3 z_1^3$, whence for $x_1 z_2 / x_2 z_1$ and $y_1 z_2 / y_2 z_1$. The other cases can be treated similarly. We conclude that there are only finitely many possible values of $x_1 z_2 / x_2 z_1$ and $y_1 z_2 / y_2 z_1$, hence of $x_1 y_2 / x_2 y_1$. Since (x_1, y_1, z_1) and (x_2, y_2, z_2) satisfy (5.8), we have

$$\frac{a}{c} = \frac{y_1 z_2 - y_2 z_1}{x_2 y_1 - x_1 y_2}, \qquad \frac{b}{c} = \frac{x_1 z_2 - x_2 z_1}{x_1 y_2 - x_2 y_1},$$

hence

$$\frac{a x_1}{c z_1} = \frac{y_1 z_2 / y_2 z_1 - 1}{x_2 y_1 / x_1 y_2 - 1}, \qquad \frac{b y_1}{c z_1} = \frac{x_1 z_2 / x_2 z_1 - 1}{x_1 y_2 / x_2 y_1 - 1}.$$

Since $a, b, c, p_1, \ldots, p_t$ are pairwise coprime and x_1, y_1, z_1 are composed of $p_1, \ldots, p_t$, we obtain that there are only finitely many possible values for $a x_1 / c z_1$ and $b y_1 / c z_1$, whence for a, b and c. Thus there are only finitely many normalised triples $(a, b, c) \in (\mathbf{Z}^*)^3$ for which (5.8) has more than two solutions.

Corollaries 1.3, 2 and 3 do not provide any upper bounds for the solutions themselves. We shall show how such bounds can be derived from results on linear forms in logarithms of algebraic numbers.

Lemma 1. *Let* $\gamma_1, \ldots, \gamma_n \in \mathbf{Q}^*$ *with* $h(\gamma_i) \le \Gamma_i$ *where* $\Gamma_i \ge 3$ *for* $i = 1, \ldots, n$ *and* $n \ge 2$. *Let* $B \ge 2$ *and* $b_i \in \mathbf{Z}$ *with* $|b_i| \le B$ *for* $i = 1, \ldots, n$. *Put*

$$\Lambda = \gamma_1^{b_1} \ldots \gamma_n^{b_n} - 1,$$

$$\Omega = \prod_{i=1}^{n} \log \Gamma_i,$$

$$\Omega' = \prod_{i=1}^{n-1} \log \Gamma_i.$$

a) *Then either $\Lambda = 0$ or $|\Lambda| \geq \exp(-n^{Cn}\Omega \log \Omega' \log B)$.*
b) *Let p be any prime number. Then either $\Lambda = 0$ or*

$$|\Lambda|_p \geq \exp\big(-n^{Cn}p\Omega(\log B)^2\big).$$

The proofs of Lemma 1a) and 1b) can be found in Baker [3] and van der Poorten [65], respectively. For defects in the latter proof, see Yu [91], [92].

Győry [34] proved the Corollary below by applying a variation on Lemma 1.

Corollary 4. *If the triple $a, b, c \in (\mathbf{Z}^*)^3$ is S-normalised, then each solution (x, y, z) of (5.8) satisfies*

$$\max(|x|, |y|, |z|) < \exp(t^{Ct}P^{4/3}\log A)$$

where $A = \max(a, b, c, 3)$.

We shall give a simple proof of a slightly weaker assertion, namely with $\log A$ replaced by $\log A(\log \log A)^2$:

Let $(x, y, z) \in \mathcal{S}^3$ satisfy (5.8). Put $Z = \max(|x|, |y|, |z|)$. Then each of x, y, z is of the form $\pm p_1^{k_1} \ldots p_t^{k_t}$ with $k_1, \ldots, k_t \in \mathbf{Z}$. Observe that $|k_i| \leq C \log p_i^{k_i} \leq C \log Z$. Let $p \in T$. Suppose $|z|_p \neq 1$. Then $|x|_p = |y|_p = 1$ and

$$|z|_p = |cz|_p = |ax + by|_p = \left|-\frac{ax}{by} - 1\right|_p .$$

Hence, by Lemma 1b),

$$|z|_p \geq \exp\big(-t^{Ct}P(\log P)^t \log A(\log B)^2\big), \qquad (5.11)$$

where $B = \max(|k_1|, \ldots, |k_t|) \leq C \log Z$. Inequality (5.11) is also valid if $|z|_p = 1$. It follows that

$$|z| = \prod_{p \in T} |z|_p^{-1}$$

$$\leq \exp\big(t^{Ct}P(\log P)^t \log A(\log \log Z)^2\big).$$

By the symmetry of (5.8), the right-hand side is also an upper bound for $|x|$ and $|y|$. (We have not used the fact that $0 < a \le b \le c$.) Hence

$$\frac{\log Z}{(\log \log Z)^2} \le t^{Ct} P(\log P)^t \log A.$$

By transferring secondary factors we obtain

$$\log Z \le t^{Ct} P(\log P)^{t+3} \log A(\log \log A)^2.$$

If $\log P \le t^4$ then $(\log P)^{t+3} \le t^{Ct}$, otherwise

$$\begin{aligned}
(\log P)^{t+3} &\le (\log P)^{3t} \\
&\le \exp\bigl(3(\log P)^{1/4} \log \log P\bigr) \\
&\le C P^{1/3}.
\end{aligned}$$

Thus

$$\log Z \le t^{Ct} P^{4/3} \log A(\log \log A)^2$$

which is our claim.

An upper bound depending on P and t only can be given for the coefficients and the solutions of those S-normalised S-unit equations which have more than $t + 2$ solutions. This follows from the following consequence of Theorems 4 and 5. (Observe that for each triple $(a', b', c') \in (\mathbf{Q}^*)^3$ we have $\max\bigl(h(a'), h(b'), h(c')\bigr) \ge \max(|a|, |b|, |c|)$ where (a, b, c) is an S-normalised triple of non-zero integers which is S-equivalent to (a', b', c').)

Corollary 5. *Each S-normalised triple $(a, b, c) \in (\mathbf{Z}^*)^3$ such that (5.8) has at least $t + 3$ solutions (x, y, z) with $z > 0$ satisfies*

$$\max(|a|, |b|, |c|) \le \exp(t^{Ct} P^2)$$

and each solution of such an equation satisfies

$$\max(|x|, |y|, |z|) \le \exp(t^{Ct} P^3).$$

Proof. Put $\alpha = a/c$, $\beta = b/c$, $A = c = \max(a, b, c)$. Then $0 \le \alpha \le \beta \le 1$. If (5.8) has $t + 3$ solutions with $z > 0$, then the equation

$$\alpha x + \beta y = 1 \qquad\qquad (5.12)$$

has $t + 3$ solutions, $(x_0, y_0), (x_1, y_1), \ldots, (x_{t+2}, y_{t+2}) \in U_S^2$, say. Without loss of generality we may assume

$$h(x_0) \leq \ldots \leq h(x_{t+2}). \tag{5.13}$$

We shall prove that

$$\log A \leq t^{Ct} P^{4/3}. \tag{5.14}$$

By Corollary 4, this suffices to prove Corollary 5. In the sequel we shall assume $A > 2^{10}$.

First we prove that for $i = 1, \ldots, t + 2$ there exists a valuation $|\ |_v \in S = \{|\cdot|, |\cdot|_{p_1}, \ldots, |\cdot|_{p_t}\}$ such that

$$|\alpha x_i|_v \leq A^{-1/5(t+1)}. \tag{5.15}$$

We write $v \in S$. We distinguish two cases.

Case 1. $\alpha \leq A^{-1/4}$.

(This condition is equivalent to $a < c^{3/4}$. Essentially this is the case treated by Győry [34], cf. §3 above. In this case the solution (x_0, y_0) is not used so that the conclusion of the theorem can be reached when there are only $t + 2$ solutions.)

We have $[\alpha]_S = 1$ and $\{x_i\}_S = 1$. Hence

$$\prod_{v \in S} |\alpha x_i|_v = \prod_{v \in S} |\alpha|_v = \alpha \leq A^{-1/4}.$$

Since S has $t + 1$ elements, there is some v in S such that (5.15) holds.

Case 2. $\alpha > A^{-1/4}$. (This part contains the new argument.) We have $\beta \geq \alpha > A^{-1/4}$, $b \geq a > A^{3/4}$ and, by (5.12),

$$\alpha(x_i - x_0) = \beta(y_0 - y_i) \qquad \text{for } i = 1, \ldots, t + 2.$$

Since $x_0, x_i, y_0, y_i \in U_S$, we obtain for any prime $p \notin T$ that

$$|\alpha(x_i - x_0)|_p \leq \min(|\alpha|_p, |\beta|_p).$$

Hence, by (5.1) and $[\alpha]_S = [\beta]_S = 1$,

$$\prod_{v \in S} |\alpha(x_i - x_0)|_v = \prod_{p \notin T} |\alpha(x_i - x_0)|_p^{-1}$$

$$\geq \prod_{p \notin T} \left(\min(|\alpha|_p, |\beta|_p) \right)^{-1}$$

$$= \left(\prod_p \min(|\alpha|_p, |\beta|_p) \right)^{-1}.$$

By the coprimality condition on a, b, c we have $\min(|a|_p, |b|_p) = |\mathrm{lcm}(a,b)|_p = |ab|_p$. Hence

$$\min(|\alpha|_p, |\beta|_p) = \frac{\min(|a|_p, |b|_p)}{|c|_p} = \left|\frac{ab}{c}\right|_p.$$

By the Product Formula we obtain

$$\left(\prod_p \min(|\alpha|_p, |\beta|_p)\right)^{-1} = \frac{ab}{c}.$$

Therefore

$$\prod_{v \in S} |\alpha(x_i - x_0)|_v \geq ab/c > A^{3/4} A^{3/4} A^{-1} = A^{1/2}.$$

On the other hand, by $0 \leq \alpha \leq 1$ and (5.13),

$$\prod_{v \in S} |\alpha(x_i - x_0)|_v \leq \alpha \prod_{v \in S} |x_i - x_0|_v$$

$$\leq 2 \prod_{v \in S} \max(|x_0|_v, |x_i|_v)$$

$$\leq 2\left(\prod_{v \in S} \max(1, |x_0|_v)\right)\left(\prod_{v \in S} \max(1, |x_i|_v)\right)$$

$$= 2h(x_0)h(x_i) \leq 2(h(x_i))^2.$$

It follows that

$$h(x_i) \geq \left(\tfrac{1}{2}A^{1/2}\right)^{1/2} > A^{1/5}.$$

We infer from (5.2) and (5.3) that

$$\prod_{v \in S} \min(1, |x_i|_v) = \left(\prod_{v \in S} \max(1, |x_i|_v)\right)^{-1}$$

$$= (h(x_i))^{-1} < A^{-1/5}.$$

Since $0 \leq \alpha \leq 1$ and $[\alpha]_S = 1$, there is some v in S such that

$$|\alpha x_i|_v \leq |x_i|_v < A^{-1/(5(t+1))}.$$

This proves (5.15) in the second case.

Since $|S| = t + 1$, there are i, j in $\{1, \ldots, t + 2\}$ with $i < j$ such that, for the same valuation $v \in S$,

$$|\alpha x_i|_v \leq A^{-1/(5(t+1))}, \qquad |\alpha x_j|_v \leq A^{-1/(5(t+1))}.$$

Hence, by (5.12),

$$|\beta y_j|_v = |1 - \alpha x_j|_v \geq \frac{1}{2},$$

$$\left|\beta(y_i - y_j)\right|_v = \left|\alpha(x_j - x_i)\right|_v \leq 2A^{-1/(5(t+1))}.$$

We obtain

$$\left|\frac{y_i}{y_j} - 1\right|_v \leq 4A^{-1/(5(t+1))}. \tag{5.16}$$

We apply Lemma 1a) if v is the absolute value and Lemma 1b) otherwise. Note that $y_i/y_j = \pm p_1^{k_1} \ldots p_t^{k_t}$ with $|k_h| \leq C \log \max(h(y_i), h(y_j))$. Hence

$$\left|\frac{y_i}{y_j} - 1\right|_v \geq \exp\left(-t^{Ct} P(\log P)^{t+1} \left(\log \log \max(h(y_i), h(y_j))\right)^2\right). \tag{5.17}$$

Corollary 4 implies that

$$\log \max(h(y_i), h(y_j)) \leq t^{Ct} P^{4/3} \log A. \tag{5.18}$$

Combining (5.16), (5.17) and (5.18) we obtain

$$\log A \leq t^{Ct} P(\log P)^{t+3} (\log \log A)^2.$$

This implies, by transferring secondary factors,

$$\log A \leq t^{Ct} P(\log P)^{t+5}.$$

As in the proof of Corollary 4 we have $(\log P)^{t+5} \leq t^{Ct} P^{1/3}$. Hence $\log A \leq t^{Ct} P^{4/3}$ as claimed in (5.14). ∎

It is clear from the proofs that, in Corollaries 4 and 5, $P^{4/3}$ and P^2 can be replaced by $P^{1+\epsilon}$ and P^3 by $P^{2+\epsilon}$ for any positive ϵ, provided that the constants C are allowed to depend on ϵ.

§6. Applications to sums of products of given primes

Let $T = \{p_1, \ldots, p_t\}$ be a set of prime numbers not exceeding P (≥ 2). Let S be the set of rational integers of which each prime divisor belongs to T. We consider representations $x_1 + \ldots + x_n$ with $x_1, \ldots, x_n \in S$, so-called S-representations. We call two representations $x_1 + \ldots + x_n$ and $y_1 + \ldots + y_n$ distinct if $(x_1, \ldots, x_n)$ is not a permutation of $(y_1, \ldots, y_n)$. The representations $x_1 + \ldots + x_{n_1}$ and $y_1 + \ldots + y_{n_2}$ are called relatively prime if $\gcd(x_1, \ldots, x_{n_1}, y_1, \ldots, y_{n_2}) = 1$. They are called disjoint if $x_i \neq y_j$ for $i = 1, \ldots, n_1$ and $j = 1, \ldots, n_2$, and totally disjoint if there are no equal subsums, that means there are no non-empty proper subsets $\{i_1, \ldots, i_k\}$ of $\{1, \ldots, n_1\}$ and $\{j_1, \ldots, j_\ell\}$ of $\{1, \ldots, n_2\}$ such that $x_{i_1} + \ldots + x_{i_k} = y_{j_1} + \ldots + y_{j_\ell}$. If $n_1 = n_2 = 2$, then there is no difference between the notions distinct, disjoint and totally disjoint.

In this paragraph we give some applications of Corollaries 1.3 and 2–5. By C we shall denote absolute constants, by $C(T, n)$ numbers depending only on T and n, and so on.

Theorem 6. *Let n, n_1 and n_2 be positive integers.*

  a) *There is a number $C(T, n)$ such that every integer m has at most $C(T, n)$ representations as sums of n pairwise relatively prime elements from S.*

  b) *There are only finitely many integers which admit a representation $x_1 + \ldots + x_{n_1}$ of pairwise relatively prime elements of S and a representation $y_1 + \ldots + y_{n_2}$ of pairwise relatively prime elements of S such that the representations are disjoint.*

  c) *There are only finitely many integers which admit an S-representation $x_1 + \ldots + x_{n_1}$ and an S-representation $y_1 + \ldots + y_{n_2}$ such that the representations are relatively prime and totally disjoint.*

Proof. c) Suppose m admits the two described representations. Then $x_1 + \ldots + x_{n_1} - y_1 - \ldots - y_{n_2} = 0$. We may assume $m \neq 0$. We apply Corollary 1.3. Conditions (5.6) and (5.7) are satisfied (with $\Delta = 1$, $\delta = 0$). If (5.5) is not fulfilled, then $x_1 + \ldots + x_{n_1} - y_1 - \ldots - y_{n_2}$ has a vanishing subsum. The complementary subsum vanishes too. Since $m \neq 0$, one of both subsums involves both x's and y's. This leads to a contradiction with the supposition that the representations are totally disjoint. Thus (5.5) holds. By Corollary 1.3 we find that there exists a finite set of $(n_1 + n_2)$-tuples depending only on T, n_1 and n_2 to which $(x_1, \ldots, x_{n_1}, y_1, \ldots, y_{n_2})$ belongs. Thus $m = x_1 + \ldots + x_{n_1}$ satisfies $m < C(T, n_1, n_2)$.

b) Suppose m admits the two described representations. We may assume that $m \neq 0$ and that we do not have $x_i = -x_j = 1$ for some i, j in $\{1, \ldots, n_1\}$ or $y_i = -y_j = 1$ for some i, j in $\{1, \ldots, n_2\}$. Observe that $x_1 + \ldots + x_{n_1} - y_1 - \ldots - y_{n_2}$ can be split into a number of vanishing subsums so that none of these subsums has a vanishing subsum. The number of possible splittings is $C(n_1 + n_2)$. By the conditions of b) each subsum has at least three terms, hence involves at least two x's or two y's. By applying Corollary 1.3 to each of the possible splittings we obtain that there are only finitely many possibilities for the terms in each subsum, since these terms have no common prime factor. It follows that there is a finite set of $(n_1 + n_2)$-tuples depending only on T, n_1 and n_2 to which $(x_1, \ldots, x_{n_1}, y_1, \ldots, y_{n_2})$ belongs. Thus $|m| = |x_1 + \ldots + x_{n_1}| < C(T, n_1, n_2)$.

a) It suffices to prove that there are at most $C(T, n)$ distinct representations. If $m = 0$ then we apply the argument to the S-representations $x_1 + \ldots + x_n$ of m which we applied to $x_1 + \ldots + x_{n_1} - y_1 - \ldots - y_{n_2}$ in the proof of b). It follows that there is a finite set of n-tuples depending only on T and n to which $(x_1, \ldots, x_n)$ belongs.

Now assume $m \neq 0$. If m admits two representations of the described form, then, after permutation, they can be written as $x_1 + \ldots + x_n$ and $y_1 + \ldots + y_{n_1} + x_{n_1+1} + \ldots + x_n$ with $x_i \neq y_j$ for $i = 1, \ldots, n_1$ and $j = 1, \ldots, n_1$. According to b) there are only $C_1(T, n)$ integers m_1 which admit two disjoint representations $x_1 + \ldots + x_{n_1}$ and $y_1 + \ldots + y_{n_1}$ of pairwise relatively prime elements of S for some $n_1 \leq n$. Moreover, it follows from the proof of b) that there is a finite set of $2n_1$-tuples depending only on T and n_1 to which $(x_1, \ldots, x_{n_1}, y_1, \ldots, y_{n_1})$ belongs, that is, each m_1 admits at most $C_2(T, n_1)$ pairwise disjoint S-representations of pairwise relatively prime integers. Similarly $m - m_1$ admits at most $C_2(T, n - n_1)$ pairwise disjoint S-representations of pairwise relatively prime integers $x_{n_1+1} + \ldots + x_n$. Hence there are at most $C_1(T, n)C_2(T, n_1)$ possibilities for the distinct part $(x_1, \ldots, x_{n_1})$ and $C_2(T, n - n_1)$ possibilities for the remaining common part $(x_{n_1+1}, \ldots, x_n)$. Thus the total number of representations as sums of n pairwise relatively prime elements from S is bounded by $C(T, n) := C_1(T, n) \sum_{n_1=1}^{n} C_2(T, n_1)C_2(T, n - n_1)$. ∎

It is not hard to see that the restrictions in b) and c) cannot be omitted. Let $T = \{2, 3, 5\}$. In b) we require disjointness, since there are infinitely many integers with representations $5^k + 2^2 + 1 = 5^k + 3 + 2$. We require pairwise coprimality in view of the representations $2^{k+1} + 3 + 2$ and $2^k + 2^k + 5$. In c) we require total disjointness to exclude representations $2^{k+2} + 2^k + 2 + 1$ and $2^k \cdot 3 + 2^{k+1} + 3$. We require that the representations are relatively prime, since otherwise we may

have $2^{k+2} + 2^k$ and $2^k \cdot 3 + 2^{k+1}$ as representations. The restriction in
a) is necessary, since 1 has infinitely many representations of the form
$3 \cdot 2^k - 2^{k+1} - 2^k + 1$. However, Tijdeman and Wang [90] have proved that
there is a number $C_1(T, n)$ such that every integer has at most $C_1(T, n)$
representations as sums of n positive elements from S.

In a letter to one of us, P. Erdős drew our attention to the following
conjecture of D. Newman (cf. [16] p. 80). The number of representations
$m = 2^{\alpha}3^{\beta} + 2^{\gamma} + 3^{\delta}$ in non-negative integers α, β, γ, δ is bounded. Erdős
wondered whether this could be solved by Corollary 1.3. We show that
he is right, and that there are only finitely many integers which have
more than two disjoint representations. Any pair of coprime integers
greater than 1 could be taken in place of the bases 2 and 3.

Theorem 7. a) *There is a constant C such that every integer has at
most C representations of the form $2^{\alpha}3^{\beta} + 2^{\gamma} + 3^{\delta}$ with α, β, γ and δ
non-negative integers.*

b) *There are only finitely many integers which admit more than
two pairwise disjoint representations of the form $2^{\alpha}3^{\beta} + 2^{\gamma} + 3^{\delta}$ with α,
β, γ and δ non-negative integers.*

Proof. Put $T = \{2, 3\}$. By N-representation we shall mean a representa-
tion of the form $2^{\alpha}3^{\beta} + 2^{\gamma} + 3^{\delta}$. By Theorem 6c) there are only finitely
many integers m which admit two totally disjoint N-representations. Let
M_0 be the maximum of such numbers m.

We now assume that $m > M_0$ has two distinct N-representations,
$m = 2^{\alpha}3^{\beta} + 2^{\gamma} + 3^{\delta} = 2^{\epsilon}3^{\zeta} + 2^{\eta} + 3^{\theta}$. Hence these representations
are not totally disjoint. Since the representations are distinct, $2^{\alpha}3^{\beta} +
2^{\gamma} + 3^{\delta} - 2^{\epsilon}3^{\zeta} - 2^{\eta} - 3^{\theta}$ has exactly two vanishing subsums (which are
complementary).

Suppose first that each vanishing subsum has three terms. By using
the symmetry, also with respect to the bases 2 and 3, we may assume
without loss of generality that we have one of the following cases:

(i) $\qquad 2^{\alpha}3^{\beta} + 2^{\gamma} - 2^{\epsilon}3^{\zeta} = 0, \qquad 3^{\delta} - 2^{\eta} - 3^{\theta} = 0.$

(ii) $\qquad 2^{\alpha}3^{\beta} + 2^{\gamma} - 2^{\eta} = 0, \qquad 3^{\delta} - 2^{\epsilon}3^{\zeta} - 3^{\theta} = 0.$

(iii) $\qquad 2^{\alpha}3^{\beta} + 2^{\gamma} - 3^{\theta} = 0, \qquad 3^{\delta} - 2^{\epsilon}3^{\zeta} - 2^{\eta} = 0.$

(iv) $\qquad 2^{\gamma} + 3^{\delta} - 2^{\epsilon}3^{\zeta} = 0, \qquad 2^{\alpha}3^{\beta} - 2^{\eta} - 3^{\theta} = 0.$

We treat the cases separately and apply Corollary 1.3 to both subsums.

Case (i). It follows that there are only finitely many possibilities for δ, η and θ. The numbers $2^\alpha 3^\beta$, 2^γ and $2^\epsilon 3^\varsigma$ belong to a finite set apart from a common factor 2^k ($k \in \mathbf{Z}$, $k \geq 0$).

Case (ii). The numbers $2^\alpha 3^\beta$, 2^γ, 2^η belong to a finite set apart from a common factor 2^k. The numbers 3^δ, $2^\epsilon 3^\varsigma$, 3^θ belong to a finite set apart from a common factor 3^ℓ.

Cases (iii) and (iv). There are only finitely many possibilities for the exponents.

We conclude that m is both of the form $2^k(2^{k_1} 3^{\ell_1} + 2^{k_2}) + 3^{\ell + \ell_2}$ and of the form $2^{k+k_3} + 3^\ell(2^{k_4} 3^{\ell_3} + 3^{\ell_4})$ where $k_1, \ldots, k_4, \ell_1, \ldots, \ell_4$ belong to a certain finite set.

Suppose next that one vanishing subsum has two elements and the other four. This implies that the representations are not disjoint and hence we are not in situation b). Without loss of generality we may assume that we have one of the following cases:

(v) $2^\alpha 3^\beta - 2^\epsilon 3^\varsigma = 0,$ $2^\gamma + 3^\delta - 2^\eta - 3^\theta = 0.$

(vi) $2^\gamma - 2^\eta = 0,$ $2^\alpha 3^\beta + 3^\delta - 2^\epsilon 3^\varsigma - 3^\theta = 0.$

Case (v). By applying Corollary 1.3 to the second sum we obtain that $(\gamma, \delta, \eta, \theta)$ belongs to a finite set of quadruples.

Case (vi). By applying Corollary 1.3 to the second sum we find that $(2^\alpha 3^\beta, 3^\delta, 2^\epsilon 3^\varsigma, 3^\theta)$ belongs to a finite set of quadruples apart from a common factor 3^ℓ.

To prove a) we observe that each N-representation of m is either of the form $a2^k + b3^\ell$ or of the form $2^k 3^\ell + a + b$ where in each case (a, b) belongs to a finite set. Put $M_1 = \max(a + b)$ where (a, b) runs over this finite set. The number of representations of m of the form $a2^k + b3^\ell$ is bounded by Corollary 2. For representations of the form $2^k 3^\ell + a + b$ we remark that it follows from Corollary 1.3 that the distance between numbers of the form $2^k 3^\ell$ exceeds $2M_1$ when $2^k 3^\ell > M_2$. Hence if $m > M_1 + M_2$ and $m = 2^k 3^\ell + a + b$, then k and ℓ are uniquely determined by m. It follows that for $m > M_0 + M_1 + M_2$ the number of N-representations of the second type is also bounded. We conclude that the total number of N-representations of any number $m > M_0 + M_1 + M_2$ is bounded. It is obvious that the number of N-representations of numbers $m \leq M_0 + M_1 + M_2$ can be bounded. This proves a).

To prove b) we recall that if $m > M_0$ admits two disjoint N-representations then each representation is of the form $a \cdot 2^k + b \cdot 3^\ell$ where

(a, b) belongs to some finite set and either $a = 2^{k_1}3^{\ell_1} + 2^{k_2}$, $b = 3^{\ell_2}$ (the first type) or $a = 2^{k_3}$, $b = 2^{k_4}3^{\ell_3} + 3^{\ell_4}$ (the second type). We define T_1 as the set consisting of $2, 3$ and all the prime divisors of a and b where (a, b) runs over the finite set. Denote by S_1 the set of rational integers of which each prime divisor belongs to T_1. Let M_3 be so large that if m has two totally disjoint, relatively prime S_1-representations $x_1 + x_2$ and $y_1 + y_2$, then $m < M_3$. This number M_3 exists by Theorem 6c).

Suppose $m > \max(M_0, M_3)$ admits three pairwise disjoint N-representations. Then there are two disjoint representations of the same type. Since $m > M_3$, these representations as sum of two elements of S_1, are not totally disjoint. It follows that the corresponding N-representations are not totally disjoint. If the representations are of the first type,

$$2^k(2^{k_1}3^{\ell_1} + 2^{k_2}) + 3^{\ell+\ell_2}$$

and

$$2^{k'}(2^{k_3}3^{\ell_3} + 2^{k_4}) + 3^{\ell'+\ell_4}$$

say, then

$$2^k(2^{k_1}3^{\ell_1} + 2^{k_2}) = 3^{\ell'+\ell_4}$$

and

$$2^{k'}(2^{k_3}3^{\ell_3} + 2^{k_4}) = 3^{\ell+\ell_2},$$

since the N-representations are disjoint. It follows that k and ℓ are bounded, hence $m < M_4$. If the representations are of the second type, then we obtain similarly $m < M_5$. Thus no m greater than $\max(M_0, M_3, M_4, M_5)$ admits more than two disjoint N-representations. ∎

There exist infinitely many integers with two disjoint N-represent-ations in view of $2^a \cdot 3^1 + 2^a + 3^{b+2} = 2^3 3^b + 2^{a+2} + 3^b$ for any positive integers a, b. There are infinitely many integers m which admit four distinct N-representations, namely, for $a \geq 2$, $b \geq 2$,

$$(2^a + 3^b) = 2^{a-1}3^0 + 2^{a-1} + 3^b$$

$$= 2^{a-2}3^1 + 2^{a-2} + 3^b$$

$$= 2^1 3^{b-1} + 2^a + 3^{b-1}$$

$$= 2^3 3^{b-2} + 2^a + 3^{b-2}.$$

Tijdeman and Wang [90] have proved that apart from the numbers of the form $2^a + 3^b$ there are only finitely many positive integers m which admit at least four N-representations.

It follows from Corollary 2 that the number of representations of a non-zero integer m as difference of two elements of S is bounded (in terms of t). This was the clue to the solution of an old problem of Erdős and Turán.

Let $a_1 < a_2 < \ldots < a_k$ and $b_1 < b_2 < \ldots < b_\ell$ be positive integers. Assume that the prime factors of

$$P_{k,l} := \prod_{1 \leq i \leq k, 1 \leq j \leq \ell} (a_i + b_j)$$

are given by $p_1, \ldots, p_t$. Erdős and Turán (cf. [15] p. 36) conjectured that if $\ell = k$ and k tends to infinity then $t \to \infty$. They had settled the special case $b_j = a_j$ for $j = 1, \ldots, k$ in their first joint paper [18]. Győry, Stewart and Tijdeman [47] observed that a stronger assertion follows from Corollary 2. Since, for any non-zero c, the number of solutions of $x - y = c$ in integers x, y composed of given primes $p_1, \ldots, p_t$ is at most $6 \times 7^{2t+3}$, the number of positive integers a such that both $a + b_1$ and $a + b_2$ are composed of $p_1, \ldots, p_t$ is at most $6 \times 7^{2t+3}$. It follows that $\ell \geq 2$ already implies $t \geq \frac{1}{4} \log k - 2$, hence $t \to \infty$ as $k \to \infty$. An elementary solution of the problem of Erdős and Turán was presented in [88]. Erdős also posed the problem of investigating the number of distinct prime factors of $\prod(a_i + b_j)$ if the product extends over a given set of pairs (i, j). Results in this direction can be found in Győry, Stewart and Tijdeman [48].

There are several related results involving $P := \max_{i=1,\ldots,t} p_i$. We henceforth assume that P is fixed and that $a_1 + b_1, \ldots, a_1 + b_\ell$, $a_2 + b_1, \ldots, a_k + b_\ell$ have no prime factor in common. In [47] we showed that if $k \geq 2$, $\ell \geq 2$, then $a_k + b_\ell$ is bounded. This follows by applying Corollary 1.3 to $(a_1 + b_1) + (a_2 + b_2) - (a_1 + b_2) - (a_2 + b_1) = 0$. By applying Corollary 5 to $x - y = b_\ell - b_1$, we obtain the following refinement of [47, Theorem 2]:

If $k \geq t + 3$ and $\ell \geq 2$, then $a_k + b_\ell \leq \exp(t^{Ct} P^3)$ where C is some effectively computable absolute constant. Surveys on these and related results are given by Stewart [87] and Stewart and Tijdeman [88].

§7. Applications to finitely generated groups

In a letter to one of us A. Rhemtulla and S. Sidki asked to show that not every rational integer r is of the form $r_1 + r_2 + \ldots + r_{n'}$, $n' \leq n$, where each r_i belongs to G, where G is a fixed multiplicative group generated by algebraic integers $\alpha_1, \ldots, \alpha_t$. They needed this result for their study of ellipticity problems in group theory. An application of Theorem 1 yielded a positive answer. In fact, if n is a positive integer and $\alpha_1, \ldots, \alpha_t$ are non-zero algebraic numbers (not necessarily integers), then there exists a positive integer m which is not representable as the sum of at most n power products $\alpha_1^{b_1} \ldots \alpha_t^{b_t}$ $(b_1, \ldots, b_t \in \mathbf{Z})$. This assertion follows from the following theorem by letting G be the group generated by $\alpha_1, \ldots, \alpha_t$ and H the group generated by a prime number p of which no power belongs to G. We namely infer from the theorem that only finitely many powers of p can be represented as the sum of at most n power products of $\alpha_1, \ldots, \alpha_t$.

Theorem 8. *Let n_1 and n_2 be positive integers. Let G, H be finitely generated multiplicative subgroups of $\mathbf{C}^*$ with $G \cap H = \{1\}$. There are only finitely many complex numbers α which can be written both as*

$$\alpha = \epsilon_1 + \ldots + \epsilon_{n_1} \quad \text{with } \epsilon_1, \ldots, \epsilon_{n_1} \in G \qquad (7.1)$$

and as

$$\alpha = \eta_1 + \ldots + \eta_{n_2} \quad \text{with } \eta_1, \ldots, \eta_{n_2} \in H. \qquad (7.2)$$

Proof. We use induction on $n = n_1 + n_2$. We denote by $N(n)$ the number of complex numbers α which can be written both in the form (7.1) and in the form (7.2) with $n_1 + n_2 \leq n$. Further we denote by $N_0(n)$ the number of complex numbers which are of the forms (7.1) and (7.2) with $n_1 + n_2 \leq n$ such that no subsum of $\epsilon_1 + \ldots + \epsilon_{n_1}$ equals any subsum of $\eta_1 + \ldots + \eta_{n_2}$. Obviously $N(2) = 1$. Suppose $N(n-1) < \infty$. If α is of the forms (7.1) and (7.2) with $n_1 + n_2 = n$, then

$$\epsilon_1 + \ldots + \epsilon_{n_1} - \eta_1 \ldots - \eta_{n_2} = 0,$$

$$\epsilon_1, \ldots, \epsilon_{n_1} \in G, \ \eta_1, \ldots, \eta_{n_2} \in H. \qquad (7.3)$$

If no subsum of the left-hand side vanishes, then we obtain, by applying Theorem 1' to $G_0 = G^{n_1} \times H^{n_2}$, that there are only finitely many possibilities for $(\epsilon_1 : \ldots : \epsilon_{n_1} : \eta_1 : \ldots : \eta_{n_2})$. If $(\epsilon_1', \ldots, \epsilon_{n_1}', \eta_1', \ldots, \eta_{n_2}')$ and $(\epsilon_1'', \ldots, \epsilon_{n_1}'', \eta_1'', \ldots, \eta_{n_2}'')$ are two solutions corresponding to the same projective point, then $\epsilon_1'/\epsilon_1'' = \eta_1'/\eta_1'' \in G \cap H = \{1\}$. Thus there are only finitely many solutions of (7.3) without vanishing subsums, whence $N_0(n) < \infty$. If some subsum vanishes, then α can be written as $\beta + \gamma$ where both β and γ are of the forms (7.1) and (7.2), but in both cases $n_1 + n_2 \leq n - 1$. The number of numbers α representable in this way is at most $\left(N(n-1)\right)^2$. Thus $N(n) \leq N_0(n) + \left(N(n-1)\right)^2 < \infty$. This proves the induction hypothesis. ∎

§8. Applications to recurrence sequence of complex numbers

By a *recurrence sequence* we mean an infinite sequence of complex numbers $U = \{u_m\}_{m=0}^{\infty}$ satisfying a relationship of the type

$$u_{m+k} = c_1 u_{m+k-1} + \ldots + c_k u_m \qquad \text{for } m = 0, 1, 2, \ldots \qquad (8.1)$$

where $c_1, \ldots, c_k$ are complex numbers. The sequence U satisfies several recurrence relations of type (8.1); among these there is a unique recurrence relation for which k is minimal. Supposing that (8.1) is the recurrence with minimal k satisfied by U, we put

$$F(X) = X^k - c_1 X^{k-1} \ldots - c_{k-1} X - c_k. \qquad (8.2)$$

F is called the *companion polynomial* of U. Obviously $F(0) \neq 0$. Let

$$F(X) = (X - \omega_1)^{e_1} \ldots (X - \omega_r)^{e_r}$$

where $e_1, \ldots, e_r$ are positive integers and $\omega_1, \ldots, \omega_r$ non-zero distinct complex numbers. Then there are polynomials $f_1, \ldots, f_r \in \mathbb{C}[X]$, of degrees at most $e_1 - 1, \ldots, e_r - 1$ respectively, such that

$$u_m = f_1(m)\omega_1^m + \ldots + f_r(m)\omega_r^m \qquad \text{for } m = 0, 1, 2, \ldots. \qquad (8.3)$$

We call k the *order* and r the *rank* of U. We say that U is *non-degenerate* if ω_i/ω_j is not a root of unity for $1 \leq i < j \leq r$, and *degenerate* otherwise. If U is degenerate, then there exists a positive integer v such that each sequence $\{u_{\ell+mv}\}_{m=0}^{\infty}$ ($0 \leq \ell < v$) is either non-degenerate and of rank less than r or identically zero.

The following theorem can be derived from the Main Theorem on S-Unit Equations.

Theorem 9. *Let R be a subring of $\mathbb{C}$ which is finitely generated over $\mathbb{Z}$ and let $U = \{u_m\}_{m=0}^{\infty}$ be a non-degenerate recurrence sequence in $\mathbb{C}$ of rank at least 2. Then there are only finitely many pairs of integers (m, n) for which a $\zeta_{m,n} \in R \setminus \{0\}$ exists such that*

$$\zeta_{m,n} u_m = u_n, \qquad m > n \geq 0. \qquad (8.4)$$

This is a slight generalisation of Theorem 3 of Evertse [20] which gives the result when R is a finitely generated multiplicative subgroup of the field of algebraic numbers.

Before discussing the proof, we state some consequences of Theorem 9.

Corollary 9.1 (Skolem, Mahler, Lech [53]). *Let $U = \{u_m\}_{m=0}^{\infty}$ be a recurrence sequence in $\mathbb{C}$ for which the set $\mathcal{M} = \{m : u_m = 0\}$ is infinite. Then $\mathcal{M}$ is ultimately periodic (i.e. there are positive integers m_0 and v such that $m \in \mathcal{M}$ implies $m + v \in \mathcal{M}$ for all $m \geq m_0$).*

Proof. It is easy to check that $\mathcal{M}$ is finite if U has rank 1. Suppose U has rank at least 2 and $\mathcal{M}$ is infinite. Then U is degenerate by Theorem 9. Hence there exists an integer v such that each sequence $\{u_{\ell+mv}\}_{m=0}^{\infty}$ is either non-degenerate, whence has only finitely many zeros, or is identically zero.   ■

The following statement was made by Glass, Loxton and van der Poorten [27].

Corollary 9.2. *Let $U = \{u_m\}_{m=0}^{\infty}$ be a non-periodic non-degenerate recurrence sequence in $\mathbb{C}$. Then there are only finitely many pairs of integers m, n with $m > n \geq 0$ and $u_m = u_n$.*

Proof. If U has rank 1, then $u_m = u_n$ implies

$$f_1(m)\omega_1^m = f_1(n)\omega_1^n.$$

If f_1 is constant, then ω_1 is a root of unity and U is periodic, which contradicts our assumption. If f_1 is non-constant, then, by $m > n$, $|f_1(m)| > |f_1(n)|$ for m sufficiently large. Hence $|\omega_1| < 1$. Put $f_1(X) = a_0 X^{\ell} + a_1 X^{\ell-1} + \ldots + a_{\ell}$ with $a_0 \neq 0$. Then

$$1 < |\omega_1|^{-1} \leq |\omega_1|^{n-m} < \frac{|a_0| + \frac{1}{m}\sum_{j=1}^{\ell}|a_j|}{|a_0| - \frac{1}{n}\sum_{j=1}^{\ell}|a_j|} = \frac{1 + c/m}{1 - c/n}$$

for some $c > 0$. This implies that n is bounded. Since $u_m \neq 0$ for m large, we have $u_n \neq 0$, hence $|f_1(n)\omega_1^n|$ is bounded from below by a positive constant. We have $f_1(m)\omega_1^m \to 0$ as $m \to \infty$. Thus m is bounded.

If U has rank at least 2, then Corollary 9.2 follows at once from Theorem 9.   ■

The next result was proved by Pólya [64] in 1921 in the case that all the u_m are rational.

Corollary 9.3. *Let G be a finitely generated multiplicative subgroup of $\mathbf{C}^*$ and let $U = \{u_m\}_{m=0}^{\infty}$ be a recurrence sequence such that $u_m \in G \cup \{0\}$ for $m = 0, 1, 2, \ldots$. Then the formal power series $\sum_{m=0}^{\infty} u_m X^m$ is equal to*

$$\sum_{j=1}^{\ell} \frac{\beta_j X^{j-1}}{1 - \alpha_j X^{\ell}},$$

where $\ell \geq 1$ is an integer and $\alpha_1, \ldots, \alpha_\ell, \beta_1, \ldots, \beta_\ell$ are complex numbers with $\alpha_1, \ldots, \alpha_\ell \neq 0$.

Proof. We first prove Corollary 9.3 in the case when U has rank 1, that is $u_m = f(m)\alpha^m$ where $\alpha \in \mathbf{C}^*$ and $f \in \mathbf{C}[X]$. Suppose that $u_m \in G \cup \{0\}$ for $m = 0, 1, 2, \ldots$. Let G' be the multiplicative group generated by G and α. Then $f(m) \in G' \cup \{0\}$ for $m = 0, 1, 2, \ldots$. We shall prove that f is constant.

There exist complex numbers $c_1, \ldots, c_k$ such that

$$f(X + k) = c_1 f(X + k - 1) + \ldots + c_k f(X) \quad \text{identically in } X.$$

We suppose that k is minimal, hence $c_k \neq 0$. Choose m_0 such that $f(m_0) \neq 0$ for $m > m_0$. By assumption, we have for $m > m_0$ that $(f(m + k), f(m + k - 1), \ldots, f(m))$ is a solution of the equation

$$c_0 x_k + c_1 x_{k-1} + \ldots + c_k x_0 = 0$$

$$\text{in } x_0, \ldots, x_k \in G' \text{ (where } c_0 = -1). \quad (8.5)$$

For each proper non-empty subset J of $\{0, \ldots, k\}$, there are only finitely many m with $\sum_{j \in J} c_j f(m + k - j) = 0$, since the polynomial $\sum_{j \in J} c_j f(X + k - j)$ does not vanish identically in X. Hence there is an m_1 such that for $m \geq m_1$, $(f(m+k), \ldots, f(m))$ is a solution of (8.5) with $\sum_{j \in J} c_j x_{k-j} \neq 0$ for each proper non-empty subset J of $\{0, 1, \ldots, k\}$. We obtain from Theorem 1' that $f(m + k)/f(m)$ assumes only finitely many values for $m = 0, 1, 2, \ldots$. Since $f(m + k)/f(m) \to 1$ as $m \to \infty$, this implies that f is constant. Hence $u_m = \beta \alpha^m$ for $m = 0, 1, 2, \ldots$ and therefore

$$\sum_{m=0}^{\infty} u_m X^m = \frac{\beta}{1 - \alpha X}.$$

Now suppose that U has order at least 2. By Theorem 9, U is degenerate. Hence there is a v such that the sequences $U_\ell = \{u_{\ell+mv}\}_{m=0}^{\infty}$

$(0 \le \ell < v)$ are either non-degenerate or identically zero. But the non-degenerate sequences among the U_ℓ must have order 1. Now Corollary 9.3 follows by applying the result for recurrence sequences of rank 1. ∎

Recently, Bézivin [6] proved the following result by applying Theorem 1′.

Let G be a finitely generated multiplicative subgroup of $\mathbf{C}^*$ and let $F(X) = \sum_{m=0}^{\infty} u_m X^m \in \mathbf{C}[[X]]$ be a formal power series with the following properties:

(a) there are polynomials $f_0, \ldots, f_k \in \mathbf{C}[X]$ such that

$$\sum_{i=0}^{k} f_i(m) u_{m+i} = 0 \quad \text{for } m = 0, 1, 2, \ldots \tag{8.6}$$

and

(b) there are sequences $\{c_j(m)\}_{m=0}^{\infty}$ $(1 \le j \le \ell)$, with entries in $G \cup \{0\}$ such that

$$u_m = \sum_{j=1}^{\ell} c_j(m) \quad \text{for } m = 0, 1, 2, \ldots . \tag{8.7}$$

Then $F(X)$ is the Taylor expansion around the origin of a rational function with only simple zeros.

Using Bezivin's result, Pólya's result can be extended to a recurrence sequence $\{u_m\}_{m=0}^{\infty}$ satisfying a relationship of type (8.6). A relation of type (8.6) is satisfied if $F(X)$ is the Taylor expansion of an algebraic function around the origin.

We now turn to the proof of Theorem 9, which resembles the proof of Theorem 3 of [20]. We shall only work out in detail the new ideas. The lemma below is used to reduce Theorem 9 to the case that R is contained in the field of algebraic numbers A.

Lemma 2. *Let $R \subset \mathbf{C}$ be a ring which is finitely generated over $\mathbf{Z}$ and let V be a finite subset of R such that V does not contain 0 or a root of unity. Then there exists a ring homomorphism $\phi : R \to \mathbf{A}$ ("specialisation") such that ϕ is invariant on $R \cap \mathbf{Q}$ and, for each α in V, $\phi(\alpha) \ne 0$ and $\phi(\alpha)$ is not a root of unity.*

Proof. Let K be the quotient field of R. Then K is finitely generated over $\mathbf{Q}$, whence $K = \mathbf{Q}(\mathbf{x}; y)$ where $\mathbf{x} = (x_1, \ldots, x_t)$ is a tuple of numbers which are algebraically independent over $\mathbf{Q}$ and y is integral over the ring

$\mathbf{Z}[\mathbf{x}] = \mathbf{Z}[x_1, \ldots, x_t]$. Thus y is a zero of a polynomial $F(\mathbf{x}; Y)$ which is irreducible in $\mathbf{Z}[\mathbf{x}, Y]$ and has leading coefficient 1. Thus R can be written as

$$R = \mathbf{Z}\left[\frac{f_1(\mathbf{x}; y)}{p(\mathbf{x})}, \ldots, \frac{f_\ell(\mathbf{x}; y)}{p(\mathbf{x})}\right]$$

where $f_1, \ldots, f_\ell \in \mathbf{Z}[\mathbf{x}; Y]$ and $p \in \mathbf{Z}$. Hence R is contained in the ring

$$\tilde{R} = \left\{\frac{f(\mathbf{x}; y)}{p^n(\mathbf{x})} : f \in \mathbf{Z}[\mathbf{x}; Y], n \in \mathbf{Z}, n \geq 0\right\}.$$

Each pair $\tilde{\mathbf{x}}, \tilde{y}$ with $\tilde{\mathbf{x}} = (\tilde{x}_1, \ldots, \tilde{x}_t) \in \mathbf{Z}^t$ with $p(\tilde{\mathbf{x}}) \neq 0$ and $F(\tilde{\mathbf{x}}, Y)$ not identically zero in Y and $\tilde{y}$ a zero of $F(\tilde{\mathbf{x}}, Y)$, defines a ring homomorphism $\phi : \tilde{R} \to \mathsf{A}$ with

$$\phi\left(\frac{f(\mathbf{x}, y)}{p^n(\mathbf{x})}\right) = \frac{f(\tilde{\mathbf{x}}, \tilde{y})}{p^n(\tilde{\mathbf{x}})}.$$

The image of ϕ is contained in the algebraic number field $\mathbf{Q}(\tilde{y})$ of which the degree is bounded by $[K : \mathbf{Q}(\mathbf{x})]$. Hence there is an integer $m > 0$, independent of $\tilde{\mathbf{x}}$ and $\tilde{y}$, such that every root of unity ρ in $\phi(\tilde{R})$ satisfies $\rho^m = 1$.

Let $G_1, \ldots, G_v$ denote the minimal polynomials in $\mathbf{Z}[\mathbf{x}; Y]$ of the elements of V. Choose $\tilde{\mathbf{x}} \in \mathbf{Z}^t$ such that $p(\tilde{\mathbf{x}}) \neq 0$, $F(\tilde{\mathbf{x}}; Y)$ is not identically zero and $G_i(\tilde{\mathbf{x}}; \alpha) \neq 0$ for each i in $\{1, \ldots, v\}$ and $\alpha \in \{0\} \cup \{\rho : \rho^m = 1\}$. (Note that no $G_i(\mathbf{X}, \alpha)$ is identically zero). Let $\tilde{y}$ be a zero of $F(\tilde{\mathbf{x}}; Y)$. It is now obvious that ϕ, defined by $\tilde{\mathbf{x}}$ and $\tilde{y}$, satisfies the assertion of Lemma 2. ∎

Proof of Theorem 9. Let $U = \{u_m\}_{m=0}^\infty$ be a non-degenerate recurrence sequence of rank $r \geq 2$. Then there are non-zero polynomials $f_i \in \mathbf{C}[X]$ and $\omega_i \in \mathbf{C}^*$ for $i = 1, \ldots, r$ such that

$$u_m = \sum_{i=1}^r f_i(m)\omega_i^m \qquad \text{for } m = 0, 1, 2, \ldots$$

and ω_i/ω_j is not a root of unity for $1 \leq i < j \leq r$. Let $R \subset \mathbf{C}$ be a ring which is finitely generated over $\mathbf{Z}$, and let K be the smallest subfield of $\mathbf{C}$ which contains R, $\omega_1, \ldots, \omega_r$ and the coefficients of $f_1, \ldots, f_r$.

We first show that it suffices to show Theorem 9 in the algebraic case, i.e. when $K \subseteq \mathsf{A}$. So suppose that Theorem 9 holds in the algebraic case and that $K/\mathbf{Q}$ is transcendental. Let $\hat{R}$ be the ring generated by R, the coefficients of $f_1, \ldots, f_r$ and $\omega_1, \ldots, \omega_r, \omega_1^{-1}, \ldots, \omega_r^{-1}$. By Lemma 2 there is a specialisation $\phi : \hat{R} \to A$ such that ϕ maps non-zero coefficients of the f_i on non-zero numbers, and all quotients ω_i/ω_j $(i \neq j)$ on numbers different from 0 and roots of unity. Then ϕ maps U on the sequence $\tilde{U} := \{\tilde{u}_m\}_{m=0}^\infty := \{\phi(u_m)\}_{m=0}^\infty$ where

$$\tilde{u}_m = \sum_{i=1}^r \tilde{f}_i(m)\tilde{\omega}_i^m \qquad \text{for } m = 0, 1, 2, \ldots,$$

$\tilde{f}_i \in A[X]$ and $\tilde{\omega}_i = \phi(\omega_i) \in A^*$. Obviously $\tilde{U}$ is a non-degenerate recurrence sequence of rank $r \geq 2$. Let (m, n) be a pair of integers with

$$m > n \geq 0, \qquad \zeta_{m,n} u_m = u_n,$$

$$\zeta_{m,n} \in R \setminus \{0\}, \qquad \zeta_{m,n} = 1 \text{ if } u_n = 0. \tag{8.8}$$

Then, on putting $\tilde{R} = \phi(R)$, $\tilde{\zeta}_{m,n} = \phi(\zeta_{m,n})$, we obtain

$$\tilde{\zeta}_{m,n}\tilde{u}_m = \tilde{u}_n, \qquad \tilde{\zeta}_{m,n} \in \tilde{R}. \tag{8.9}$$

$\tilde{R}$ is finitely generated over $\mathbf{Z}$, but $\tilde{\zeta}_{m,n}$ may be 0. However, from Theorem 9 in the algebraic case it follows that there is an n_0 such that $\tilde{u}_n \neq 0$ for $n \geq n_0$, whence (8.9) is satisfied by at most finitely many pairs m, n with $m > n \geq n_0$. We infer that (8.8) is satisfied by only finitely many pairs m, n with $m > n \geq n_0$. To prove that (8.8) holds for only finitely many pairs with $n < n_0$, take a specialisation ϕ' having the same properties as ϕ and the additional property that $\phi'(u_n) \neq 0$ for each $n < n_0$ with $u_n \neq 0$, and repeat the arguments given above.

We shall now prove Theorem 9 in the algebraic case. Henceforth, the field K is an algebraic number field and S a finite set of places on K containing the infinite places, such that all non-zero coefficients of the polynomials f_i and all ω_i $(i = 1, \ldots, r)$ are S-units, and such that all elements of R are S-integers.

We shall need two other lemmas.

Lemma 3. *Let $p \in A(X)$ be a rational function with no poles outside the disc $\{z \in \mathbf{C} : |z| \leq A\}$ and let $\alpha \in A^*$. If there are infinitely many pairs of integers m, n with*

$$m > n \geq A, \qquad p(m)\alpha^m = p(n)\alpha^n$$

then p is constant and α is a root of unity.

Proof (cf. [77] pp. 84–85). We assume that

$$p(X) = \frac{X^k + a_{k-1}X^{k-1} + \ldots + a_1 X + a_0}{X^\ell + b_{\ell-1}X^{\ell-1} + \ldots + b_1 X + b_0}$$

which is no restriction. Let $\overline{p}(z)$ be the rational function obtained from $p(z)$ by replacing all coefficients of p by their complex conjugates. If $k \neq \ell$ we put $F(X) = p(X)\overline{p}(X)(\alpha\overline{\alpha})^X$ and if $k = \ell$ and α is a root of unity of order q say, we define $F(X)$ to be a non-constant function from $\{p(X)^q + \overline{p}(X)^q, i(p(X)^q - \overline{p}(X)^q)\}$. In both cases $p(m)\alpha^m = p(n)\alpha^n$ implies $F(m) = F(n)$, and there is an x_0 such that $F(x)$ is monotone for $x \geq x_0$. We conclude that $p(m)\alpha^m = p(n)\alpha^n$ for at most finitely many pairs of integers m, n with $m > n$.

We now consider the remaining case: $k = \ell$ and α is not a root of unity. We suppose $|\alpha| \leq 1$ which is no restriction. By $c_1, c_2, \ldots$ we denote (effectively computable) numbers depending only on α and p. We have $|p(m) - 1| \leq c_1/m$ for $m \geq m_0$. Hence, for $m > n \geq m_0$ with $p(m)\alpha^m = p(n)\alpha^n$,

$$|\alpha^{m-n} - 1| \leq |\alpha|^{m-n}|1 - p(m)| + |p(n) - 1| \leq \frac{c_2}{n}.$$

On the other hand, by Baker's theorem [3] (which is the analogue of Lemma 1a) for algebraic numbers) we have, noting that α is not a root of unity,

$$|\alpha^{m-n} - 1| \geq |m - n|^{-c_3}.$$

Hence

$$n \leq c_4(m - n)^{c_5}. \tag{8.10}$$

By assumption, α is not a root of unity. Hence there is a valuation $|.|_v$ on the smallest number field K containing α and the coefficients of p with $|\alpha|_v := c_6 > 1$. Therefore

$$c_6^{m-n} = |\alpha|_v^{m-n} = \left|\frac{p(n)}{p(m)}\right|_v \leq c_7 m^{c_8}$$

which implies $m - n \leq c_9 \log m$. Together with (8.10) this shows that

$$m = n + m - n \leq c_4(m - n)^{c_5} + m - n \leq c_{10}(\log m)^{c_{11}}.$$

Hence m is bounded.   ∎

The next lemma was already stated in van der Poorten and Schlickewei [67].

Lemma 4. *Let T be a finite set of places on K, containing S. Then for every ϵ with $0 < \epsilon < 1$ there is an m_0 depending only on ϵ and T such that for all $m \geq m_0$,*

$$\prod_{v \in T} |u_m|_v \geq \left\{ \prod_{v \in S} \max(|\omega_1|_v, \ldots, |\omega_r|_v) \right\}^{m(1-\epsilon)} > 0. \qquad (8.11)$$

Proof. We shall prove Lemma 4 by induction on r. For $r = 1$, Lemma 4 is obvious. Suppose that (8.11) holds for all non-degenerate recurrence sequences of rank less than r where $r \geq 2$ (induction hypothesis). We have the identity

$$u_m - \sum_{i=1}^{r} f_i(m)\omega_i^m = 0 \quad \text{for } m = 0, 1, 2, \ldots. \qquad (8.12)$$

By the induction hypothesis, there is an m_1 such that no proper, non-empty subsum of this sum vanishes for $m \geq m_1$. Let $0 < \epsilon < 1$. Let H_m be the projective height of the projective point $(u_m : -f_1(m)\omega_1^m : \ldots : -f_r(m)\omega_r^m)$. It is easy to show that there is an $m_2 = m_2(\epsilon, T) > m_1$ such that

$$H_m \geq \left\{ \prod_{v \in S} \max(|\omega_1|_v, \ldots, |\omega_r|_v) \right\}^{m(1-\epsilon/5)}$$

for all $m \geq m_2$. Moreover, there is an $m_3 = m_3(\epsilon, T) > m_2$ such that for all $m \geq m_3$,

$$\prod_{v \in T} \prod_{i=1}^{r} |f_i(m)\omega_i^m|_v \leq H_m^{\epsilon/5},$$

since all ω_i are S-units. If $m > m_3$ does not satisfy (8.11) then

$$\prod_{v \in T} \left(|u_m|_v \prod_{i=1}^{r} |f_i(m)\omega_i^m|_v \right) \leq H_m^{\frac{\epsilon}{5} + \frac{1-\epsilon}{1-\epsilon/5}} \leq H_m^{1-\epsilon/2}.$$

Recall that u_m and the $f_i(m)\omega_i^m$ are all T-integers. Together with the Main Theorem on S-Unit Equations this implies that there are only finitely many projective points $P_m = (u_m : f_1(m)\omega_1^m : \ldots : f_r(m)\omega_r^m)$ such that (8.11) is not satisfied. For each projective point $\underline{\alpha}$, there are only finitely many m with $P_m = \underline{\alpha}$. For if there were infinitely many such m, then there would be infinitely many pairs (m, n) with $m > n$ and

$$\frac{f_1(m)\omega_1^m}{f_1(n)\omega_1^n} = \frac{f_2(m)\omega_2^m}{f_2(n)\omega_2^n},$$

whence

$$\frac{f_1(m)}{f_2(m)}\left(\frac{\omega_1}{\omega_2}\right)^m = \frac{f_1(n)}{f_2(n)}\left(\frac{\omega_1}{\omega_2}\right)^n.$$

The latter is impossible by Lemma 3. This completes the proof of Lemma 4. ∎

Suppose that there are infinitely many pairs m, n with $m > n \geq 0$ for which a $\zeta_{m,n} \in R \setminus \{0\}$ exists such that $\zeta_{m,n} u_m = u_n$, i.e.

$$\sum_{i=1}^{r} \zeta_{m,n} f_i(m)\omega_i^m - \sum_{i=1}^{r} f_i(n)\omega_i^n = 0.$$

Then there are pairs of subsets (I_j, J_j) of $\{1,\ldots,r\}$, for $j = 1,\ldots,\ell$, such that $\bigcup_{j=1}^{\ell} I_j = \bigcup_{j=1}^{\ell} J_j = \{1,\ldots,r\}$, the sets I_j and the sets J_j are pairwise disjoint, at least one of I_j, J_j is non-empty, there is an infinite set V of pairs (m,n) for which

$$m > n \geq 0, \quad \sum_{i\in I_j} \zeta_{m,n} f_i(m)\omega_i^m + \sum_{i\in J_j}(-f_i(n)\omega_i^n) = 0 \qquad (8.13)$$

and no proper, non-empty subsum of this sum is 0.

We shall show that the cardinality of each I_j is at most 1. Suppose I_j contains two subscripts, which, for convenience, are taken equal to 1 and 2. Let $P_{m,n}$ be the projective point with entries $\zeta_{m,n} f_i(m)\omega_i^m$ ($i \in I_j$) and $-f_i(n)\omega_i^n$ ($i \in J_j$). Then the entries of $P_{m,n}$ are all S-integers. Moreover there are infinitely many different points among the $P_{m,n}$ with $(m,n) \in V$, for the sequence of m with $(m,n) \in V$ is unbounded, and by Lemma 3 there are only finitely many pairs (m_1, m_2) with $m_1 < m_2$, $(m_1, n_1) \in V$, $(m_2, n_2) \in V$ and

$$P_{m_1,n_1} = P_{m_2,n_2},$$

whence

$$\frac{f_1(m_1)}{f_2(m_1)}\left(\frac{\omega_1}{\omega_2}\right)^{m_1} = \frac{f_1(m_2)}{f_2(m_2)}\left(\frac{\omega_1}{\omega_2}\right)^{m_2}.$$

Let $H_{m,n}$ denote the height of $P_{m,n}$. It is easy to check that for m sufficiently large,

$$H_{m,n} \geq C^{4m/5} \qquad (8.14)$$

where

$$C = \prod_{v\in S} \max(|\omega_1|_v, |\omega_2|_v).$$

By enlarging S if necessary, we may assume that $\prod_{v \in S} |u_n|_v \leq C_1^{n(1+\kappa/10r)}$ for all $n \geq 0$ where $\kappa = \log C / \log C_1$ and $C_1 = \prod_{v \in S} \max(|\omega_1|_v, \ldots, |\omega_r|_v)$. By Lemma 4 we have for m sufficiently large, that

$$\prod_{v \in S} |u_m|_v \geq C_1^{m(1-\kappa/10r)}.$$

Hence

$$\prod_{v \in S} |\zeta_{m,n}|_v = \prod_{v \in S} \left| \frac{u_n}{u_m} \right|_v \leq C_1^{(m+n)\kappa/10r} \leq C^{m/5r}.$$

This shows that for m sufficiently large, in view of (8.14)

$$\prod_{v \in S} \left(\prod_{i \in I_j} |\zeta_{m,n} f_i(m)\omega_i^m|_v \prod_{i \in J_j} |f_i(n)\omega_i^n|_v \right) \leq C^{m/5} < H_{m,n}^{1/2}.$$

By Theorem 1, we obtain that there are only finitely many projective points $P_{m,n}$ and this yields a contradiction. Therefore no set I_j or J_j can have cardinality larger than 1.

We infer that there is a permutation σ of $(1, \ldots, n)$ such that for all (m, n) in V

$$\zeta_{m,n} f_i(m)\omega_i^m = f_{\sigma(i)}(n)\omega_{\sigma(i)}^n. \qquad (8.15)$$

If σ is the identity, then there are infinitely many pairs (m, n) with

$$\frac{f_1(m)}{f_2(m)} \left(\frac{\omega_1}{\omega_2} \right)^m = \frac{f_1(n)}{f_2(n)} \left(\frac{\omega_1}{\omega_2} \right)^n,$$

which is impossible in view of Lemma 3. If σ is not the identity, we derive a contradiction as follows (cf. [20] pp. 242–243). Let $i \in \{1, \ldots, n\}$ such that $i \neq \sigma(i)$, and put

$$\theta_\ell = \omega_{\sigma^\ell(i)}/\omega_{\sigma^{\ell+1}(i)}, \qquad q_\ell = f_{\sigma^{\ell+1}(i)}(n)/f_{\sigma^\ell(i)}(m).$$

Then, by (8.15),

$$\theta_\ell^m = \frac{q_\ell}{q_{\ell+1}} \theta_{\ell+1}^n \quad \text{for } \ell = 0, 1, 2 \ldots. \qquad (8.16)$$

Let μ be the order of σ. Then $\theta_\mu = \theta_0$, $q_\mu = q_0$. By starting with $\theta_0^{m^\mu}$ and applying (8.16) μ times, we find

$$\theta_0^{m^\mu - n^\mu} = q_0^{m^{\mu-1} - n^{\mu-1}} q_1^{m^{\mu-2}n - m^{\mu-1}} q_2^{m^{\mu-3}n^2 - m^{\mu-2}n} \ldots q_{\mu-1}^{n^{\mu-1} - mn^{\mu-2}}. \qquad (8.17)$$

All exponents are divisible by $m - n$. Choose a valuation $|.|_v$ with $|\theta_0|_v \geq c > 1$. Then the v-adic valuation of the left-hand side of (8.17) is bounded from below by $c^{m^{\mu-1}(m-n)}$, while the v-adic valuation of the right-hand side is bounded from above by $m^{c_1 m^{\mu-2}(m-n)}$ for some constant c_1. This shows that m is bounded. Thus the proof of Theorem 9 is complete. ∎

Lewis and Turk [55] studied the solubility of equation $u_m = a u_n$ in integers $m > n$ where $U = \{u_m\}_{m=0}^{\infty}$ is a non-degenerate recurrence sequence and a some complex number. They gave various upper bounds which should be treated with care, since not all results are well stated. The methods are however correct and some of them are quite interesting, but they do not involve S-unit equations.

§9. Applications to recurrence sequences of algebraic numbers

We shall use the notation of §8 and consider recurrence sequences of algebraic numbers $U = \{u_m\}_{m=0}^{\infty}$. We assume that U is non-degenerate. It then follows that the coefficients c_i of the minimal recurrence relation (8.2), the roots ω_i of the companion polynomial and the coefficients of the polynomials f_i are all algebraic (cf. [77] Ch. R). Let K be an algebraic number field which contains all these algebraic numbers. By an argument similar to that employed in the proof of Lemma 4 in §8, van der Poorten and Schlickewei [67] proved that for every positive ϵ there exists a positive number C_1 depending only on U and ϵ such that

$$|u_m| > C_1 \left(\max_{i=1,\ldots,r} |\omega_i| \right)^{(1-\epsilon)m} \tag{9.1}$$

for $m = 1, 2, \ldots$. This inequality should be compared with the opposite inequality

$$|u_m| < C_2 \left(\max_{i=1,\ldots,r} |\omega_i| \right)^{(1+\epsilon)m}$$

for some effectively computable number C_2 depending only on U and ϵ, which follows directly from (8.3). Inequality (9.1) says that cancellation of terms in the sum on the right-hand side of (8.3) can occur only to a very limited extent (in the algebraic case).

For $\alpha \in K^*$ we define $P_K(\alpha)$ ($P_K^+(\alpha)$, respectively) to be the maximum of the norms of prime ideals $\wp$ such that for valuations v corresponding to $\wp$ one has $|\alpha|_v \neq 1$ ($|\alpha|_v < 1$ respectively) if such prime ideals $\wp$ exist and $P_K(\alpha) = 1$ ($P_K^+(\alpha) = 1$, respectively) otherwise. Further we put $P_K(0) = P_K^+(0) = 0$. We assume that U is non-degenerate and of rank at least 2. Van der Poorten [66] noticed that Theorem 1 implies that $P_K(u_m) \to \infty$ as $m \to \infty$. Evertse [20] generalised this by proving

$$\lim_{\substack{m \to \infty \\ m > n \\ u_n \neq 0}} P_K(u_m/u_n) = \infty. \tag{9.2}$$

Theorem 9 implies the following stronger result.

Corollary 9.4.

$$\lim_{\substack{m \to \infty \\ m > n \\ u_n \neq 0}} P_K^+(u_m/u_n) = \infty.$$

Proof. It follows from Corollary 9.1 that $u_m u_n \neq 0$ for $m > n \geq n_0$. Put $\zeta_{m,n} = u_n/u_m$ for these values of m and n. Suppose there are infinitely many pairs (m, n) with $m > n \geq n_0$ such that $P_K^+(u_m/u_n)$ is bounded from above by M. Let S be the union of all infinite places on K and all finite places corresponding to prime ideals on K with norms at most M. Then the cardinality of S is finite. Further, $\zeta_{m,n}$ is an S-integer for infinitely many pairs m, n with $m > n \geq n_0$. The S-integers form a finitely generated subring of $\mathbf{C}$. Thus Theorem 9 implies that there are only finitely many pairs m, n with $m > n \geq n_0$ such that $\zeta_{m,n}$ is an S-integer, a contradiction. ∎

Since the proof of Theorem 1 is ineffective, it is impossible to derive lower bounds for $P_K(u_m)$ or $P_K^+(u_m/u_n)$ from the proof of Theorem 1.

If $k = r = 2$, then (8.3) becomes

$$u_m = \alpha \omega_1^m + \beta \omega_2^m \qquad (m = 0, 1, \ldots) \tag{9.3}$$

and it is possible to apply Theorems 2–5. We assume

$$\omega_1 \neq \omega_2, \qquad \omega = |\omega_1| \geq |\omega_2| > 0,$$

$$\omega_1/\omega_2 \text{ not a root of unity.} \tag{9.4}$$

Let S be the smallest set of places on K containing all infinite places such that α, β, ω_1 and ω_2 are all S-units. Theorem 2 implies that for any non-zero complex number A the equation $u_m = A$ has at most $3 \times 7^{d+2s}$ solutions where d is the degree of K and s the cardinality of S. Theorem 3 implies that for given roots ω_1 and ω_2 there are only finitely many equivalence classes of recurrence sequences such that $u_m = A$ has more than two solutions m. If $u_m \in \mathbf{Z}$ for $m = 0, 1, \ldots$, then results of Kubota

[49] and Beukers [4] imply that $u_m = A$ has at most four solutions m. Upper bounds for the number of subscripts m with $u_m = A$ in the case that (9.3) yields a sequence of rational numbers, or even just algebraic numbers, are contained in the papers of Kubota [49] and Beukers and Tijdeman [5].

Since the proofs of Theorems 4 and 5 are effective, it is possible to derive effective bounds for sequences (9.3). First we state some results in case $u_m \in \mathbf{Z}$ for all m. Stewart [85] proved by Baker's method that there exist effectively computable numbers C_3 and m_1 depending *only on α and β* such that

$$|u_m| \geq \omega^{m - C_3 \log m} \qquad (m \geq m_1).$$

Parnami and Shorey [62] used this result to prove that $u_m = u_n$ with $m > n$ implies that m is bounded by an effectively computable constant and Shorey [76] even derived lower bounds for $|u_m - u_n|$. In the latter paper Shorey also proved that

$$P^+(u_m/u_n) \geq C_4 \left(\frac{m}{\log m} \right)^{1/(d_1+1)}$$

where $d_1 = [\mathbf{Q}(\omega_1) : \mathbf{Q}]$ and C_4 is an effectively computable number depending only on U. Stewart [86] and Shorey [75] also considered lower bounds for the greatest squarefree factor of u_m.

It is possible to generalise most of the above mentioned results to arbitrary algebraic recurrence sequences in K. Further Mignotte, Shorey and Tijdeman [58] have extended some results to the case $r = 3$. Their main result is that there exist effectively computable numbers C_5 and m_2 such that

$$|u_m| \geq \omega^m \exp\left(-C_5 (\log m)^2\right) \qquad (m \geq m_2).$$

It seems impossible to prove such a result by Baker's method for r larger than 3. For these and related results, see Shorey and Tijdeman [77] Ch. R, 2, 3, 4.

§10. Applications to irreducible polynomials and arithmetic graphs

Let K be an algebraic number field, and S a finite set of places on K containing S_∞. Let N be a positive integer. For any finite subset

$\mathcal{A} = \{\alpha_1, \ldots, \alpha_m\}$ of $\mathcal{O}_S$ with $m \geq 3$, we denote by $\mathcal{G}_K(\mathcal{A}, S, N)$ the graph whose vertex set is $\mathcal{A}$ and whose edges are the unordered pairs α_i, α_j such that $N_S(\alpha_i - \alpha_j) = \prod_{v \in S} |\alpha_i - \alpha_j|_v^{[K:\mathbb{Q}]} > N$. If in particular $S = S_\infty$, then we shall denote this graph simply by $\mathcal{G}_K(\mathcal{A}, N)$. Many diophantine problems, for instance related to irreducibility of polynomials (see Theorem 12), decomposable form equations (see §11) or algebraic number theory (see §12), can be reduced to the study of connectedness properties of graphs $\mathcal{G}_K(\mathcal{A}, S, N)$. Such properties are stated in Theorems 10 and 11 below and these theorems can be used to solve the diophantine problems mentioned. Before stating Theorem 10, we introduce some terminology. If $\mathcal{G}$ is a graph, then, as usual, $|\mathcal{G}|$ and $\overline{\mathcal{G}}$ denote the order (the number of vertices) and the complement of $\mathcal{G}$, respectively. The *triangle hypergraph* $\mathcal{G}^T$ of $\mathcal{G}$ is that hypergraph whose vertices are the edges of $\mathcal{G}$ and whose edges are the triples of edges of $\mathcal{G}$ that form a triangle.

Theorem 10 (Győry [35], [42]). *Let $m \geq 3$ be a rational integer, $\mathcal{A} = \{\alpha_1, \ldots, \alpha_m\}$ a subset of $\mathcal{O}_S$ and $\mathcal{G}_1, \ldots, \mathcal{G}_\ell$ the connected components of $\mathcal{G} = \mathcal{G}(\mathcal{A}, S, N)$ such that $|\mathcal{G}_1| \geq |\mathcal{G}_2| \geq \ldots \geq |\mathcal{G}_\ell|$. Then at least one of the following cases holds:*

i) *$\ell = 1$ and $\overline{\mathcal{G}}$ or $\overline{\mathcal{G}}^T$ is not connected;*

ii) *$\ell = 2$, $|\mathcal{G}_2| = 1$ and $\overline{\mathcal{G}}_1$ is not connected;*

iii) *$\ell = 2$, $2 \leq |\mathcal{G}_2| \leq |\mathcal{G}_1|$ and both $\mathcal{G}_1$ and $\mathcal{G}_2$ are complete;*

iv) *there is an $\epsilon \in U_S$ and for every pair i, j with $1 \leq i < j \leq m$ there is an $\alpha_{ij} \in \mathcal{O}_S$ such that*

$$\alpha_i - \alpha_j = \epsilon \alpha_{ij}$$

and

$$\max_{i,j} h(\alpha_{ij}) \leq \exp\{(c_1 s)^{c_2 s} P^{2d} \log 2N\}$$

where c_1, c_2 are effectively computable numbers depending only on m, d and D_K.

Except for certain trivial situations, each of the cases i) to iv) can occur (cf. [35]). The graphs $\mathcal{G}_K(\epsilon \mathcal{A} + \beta, S, N)$ have obviously the same structure for every $\epsilon \in U_S$ and $\beta \in \mathcal{O}_S$. It follows from Theorem 10 that apart from translation by elements of $\mathcal{O}_S$ and multiplication by elements of U_S, there are only finitely many m-tuples $\mathcal{A} = \{\alpha_1, \ldots, \alpha_m\}$ for which $\mathcal{G}_K(\mathcal{A}, S, N)$ is not of the type i), ii) or iii) and all those $\mathcal{A}$ can be, at least in principle, effectively determined.

Theorem 10 is proved by repeatedly applying Theorem 4. We sketch some ideas behind the proof of Theorem 10:

Suppose i), ii) and iii) do not hold. Then one can prove that

a) $\overline{\mathcal{G}}$ and its triangle hypergraph $\overline{\mathcal{G}}^T$ are connected; or

b) $\mathcal{G}$ has two connected components of order ≥ 2 of which at least one is not complete.

We shall sketch the proof that iv) holds in case a). Let $\{\alpha_i, \alpha_j, \alpha_k\}$ be an edge of $\overline{\mathcal{G}}^T$. Then

$$N_S(\alpha_i - \alpha_j) \leq N, \quad N_S(\alpha_j - \alpha_k) \leq N, \quad N_S(\alpha_k - \alpha_i) \leq N.$$

This implies that $\alpha_i - \alpha_j$, $\alpha_j - \alpha_k$, $\alpha_k - \alpha_i$ are $\tilde{S}$-units where $\tilde{S}$ is a finite set of places which contains S and depends only on N, K and S. Moreover,

$$\frac{\alpha_i - \alpha_j}{\alpha_i - \alpha_k} + \frac{\alpha_j - \alpha_k}{\alpha_i - \alpha_k} = 1.$$

By Theorem 4 this implies that there are only finitely many possible values for the quotient $(\alpha_i - \alpha_j)/(\alpha_j - \alpha_k)$, which can all be effectively determined. If $\{\alpha_j, \alpha_k, \alpha_\ell\}$ is another edge of $\overline{\mathcal{G}}^T$, then $(\alpha_j - \alpha_k)/(\alpha_k - \alpha_\ell)$ belongs to a finite, effectively determinable set, and so $(\alpha_i - \alpha_j)/(\alpha_k - \alpha_\ell)$ must belong to such a set. By continuing this argument, it follows that for any two connected pairs (α_i, α_j), (α_p, α_q) in $\overline{\mathcal{G}}^T$, $(\alpha_i - \alpha_j)/(\alpha_p - \alpha_q)$ belongs to a finite effectively determinable set. But $\overline{\mathcal{G}}^T$ is connected, hence for each quadruple $(\alpha_i, \alpha_j, \alpha_p, \alpha_q)$ for which $[\alpha_i, \alpha_j]$ and $[\alpha_p, \alpha_q]$ are edges in $\overline{\mathcal{G}}$, the quotient $(\alpha_i - \alpha_j)/(\alpha_p - \alpha_q)$ can assume only finitely many values which can be effectively determined. Fix p and q. Since $\overline{\mathcal{G}}$ is connected, each pair (α_a, α_b) can be connected by a path in $\overline{\mathcal{G}}$. By summing over all terms $(\alpha_i - \alpha_j)/(\alpha_p - \alpha_q)$ for the edges in this path we obtain that for each pair (a, b) the quotient $(\alpha_a - \alpha_b)/(\alpha_p - \alpha_q)$ can assume only finitely many values which can be determined effectively. Since $N_S(\alpha_p - \alpha_q) \leq N$, we have $\alpha_p - \alpha_q = \alpha_{pq}\epsilon$ where $\epsilon \in U_S$ and α_{pq} belongs to a finite set which can be effectively determined. From these facts it follows easily that $\alpha_a - \alpha_b = \alpha_{ab}\epsilon$ for each pair (α_a, α_b), where $\epsilon \in U_S$, and each α_{ab} belongs to a finite set which can be effectively determined. This proves (iv).

If the order of $\mathcal{G} = \mathcal{G}_K(\mathcal{A}, S, N)$ is large enough then $\mathcal{G}$ cannot have property iii). This fact plays a crucial role in some applications to irreducible polynomials (see below and [41]) and polynomials of given discriminant (cf. §12 and [38]). Győry [35], [42] proved the following theorem but with a weaker estimate for $|\mathcal{G}|$ than (10.1), since he used a weaker version of Theorem 2.

Theorem 11. *Let $\mathcal{A}$ be a finite subset of $\mathcal{O}_S$ and let $\mathcal{G} = \mathcal{G}_K(\mathcal{A}, S, N)$. There exists an effectively computable positive number c_3, depending only on d and D_K, such that if*

$$|\mathcal{G}| > c_3 7^{2s} N^2 \qquad (10.1)$$

then $\mathcal{G}$ has at most two connected components, and one of them is of order at least $|\mathcal{G}| - 1$.

For certain more general (but ineffective) versions of Theorems 10 and 11, see Győry [40]. Theorems 10 and 11 are slightly modified versions of Theorems 1, 2 of [35].

Theorems 10 and 11 have applications to irreducible polynomials. Here we shall present a consequence of Theorem 11. I. Schur and later A. Brauer, R. Brauer and H. Hopf investigated the reducibility of polynomials of the form $g(f(X))$ where f, g are monic polynomials in $\mathbf{Z}[X]$, g is irreducible over $\mathbf{Q}$ and the zeros of f are distinct elements of $\mathbf{Z}$. For a survey of results in this direction, see [29], [41]. Győry [28], [29], [41] extended these investigations to the case that the zeros of f are in an arbitrary totally real algebraic number field K of degree d. Let $\mathcal{A} = \{\alpha_1, \ldots, \alpha_m\} \in \mathcal{O}_K^m$ be the set of zeros of such a monic polynomial $f \in \mathbf{Z}[X]$ and suppose that $g \in \mathbf{Z}[X]$ is an irreducible monic polynomial whose splitting field is a totally imaginary quadratic extension of a totally real number field. Consider the graph $\mathcal{G} = \mathcal{G}_K(\mathcal{A}, N)$ with the choice $N = 2^d |g(0)|^{d/\deg(g)}$. Győry [28] proved that if this graph $\mathcal{G}$ has a connected component with k vertices, then the number of irreducible factors of $g(f(X))$ over $\mathbf{Q}$ is not greater than $\deg(f)/k$. This estimate is in general best possible (cf. [29]). Therefore, Theorem 11 implies the following

Theorem 12. *Let $f, g \in \mathbf{Z}[X]$ with the properties specified above. There is an effectively computable number c_4, depending only on d, h_K and D_K, such that if*

$$\deg(f) > c_4 |g(0)|^{2/\deg(g)}$$

then $g(f(X))$ is irreducible over $\mathbf{Q}$.

This theorem was proved by Győry [41] with a slightly weaker but explicit lower bound for $\deg(f)$ and in the more general case that the ground field is an arbitrary totally real number field (instead of $\mathbf{Q}$).

§11. Applications to decomposable form equations

Decomposable form equations form a very important class of polynomial diophantine equations. Many problems in number theory can be reduced to such equations. The most important types of decomposable form equations are Thue equations, norm form equations, discriminant form equations and index form equations. There is an extensive literature of equations of these types, and this will be the theme of the next memoir [25] in these Proceedings. Here we shall restrict ourselves to the application of unit equations to decomposable form equations. As will be seen, finiteness problems for decomposable form equations are in fact equivalent to finiteness questions concerning unit equations.

Let $F(X_1, X_2)$ be a binary form with coefficients in $\mathcal{O}_K$ and splitting field G over K. Let $\beta \in \mathcal{O}_K \setminus \{0\}$. By using their results on approximations of algebraic numbers, Thue [89] (in the case $K = \mathbf{Q}$) and Siegel [78], [80] showed that if

(a) F has at least three pairwise linearly independent linear factors in its factorisation over G

then the equation

$$F(x_1, x_2) = \beta \qquad \text{in } x_1, x_2 \in \mathcal{O}_K \qquad (11.1)$$

has only finitely many solutions. Equation (11.1) is called a *Thue equation over K*. Further, as was (implicitly) pointed out by Siegel [79], [80], any unit equation in two variables (over K) can be reduced to a finite number of Thue equations (over K) and conversely, any Thue equation over K leads to a finite number of unit equations in two variables (over an appropriate extension of K). Indeed, since U_K is finitely generated, every solution of

$$\alpha_1 u + \alpha_2 v = 1 \qquad \text{in } u, v \in U_K \qquad (11.2)$$

(where $\alpha_1, \alpha_2 \in K^*$) can be written in the form $u = u'x_1^n$, $v = v'x_2^n$ where $n \geq 3$ is a given positive integer, $x_1, x_2 \in U_K$, and $u', v' \in U_K$ can assume only finitely many values. Hence (11.2) reduces to finitely many Thue equations

$$(\alpha_1 u')x_1^n + (\alpha_2 v')x_2^n = 1 \qquad \text{in } x_1, x_2 \in \mathcal{O}_K.$$

We shall now show how the finiteness of the number of solutions of (11.1) follows from the fact that any equation of the form (11.2) has only finitely many solutions. After multiplying (11.1) by an appropriate rational integer, (11.1) takes the form

$$\ell_1(\mathbf{x})\ldots\ell_n(\mathbf{x}) = \beta' \quad \text{in } \mathbf{x} = (x_1, x_2) \in \mathcal{O}_K^2 \qquad (11.3)$$

where the $\ell_i(\mathbf{X})$ $(i = 1,\ldots,n)$ are linear forms in X_1, X_2 with coefficients in the ring of integers $\mathcal{O}_G$ of G. For every solution $\mathbf{x}$ of (11.3), each $\ell_i(\mathbf{x})$ divides β' in $\mathcal{O}_G$, and hence lies in a finite number of cosets of G^* with respect to the unit group U_G. If now e.g. ℓ_1, ℓ_2, ℓ_3 are pairwise linearly independent, then

$$\lambda_1 \frac{\ell_1(\mathbf{X})}{\ell_3(\mathbf{X})} + \lambda_2 \frac{\ell_2(\mathbf{X})}{\ell_3(\mathbf{X})} = 1$$

for appropriate $\lambda_1, \lambda_2 \in G^*$. The numbers $\ell_1(\mathbf{x})/\ell_3(\mathbf{x})$ and $\ell_2(\mathbf{x})/\ell_3(\mathbf{x})$ are contained in a finite number of cosets of G^* with respect to U_G, hence (11.3) yields a finite number of unit equations

$$\lambda_1' u + \lambda_2' v = 1 \quad \text{in } u, v \in U_G.$$

For every solution u, v of this equation, $\ell_1(\mathbf{x})/\ell_3(\mathbf{x}) = u$, $\ell_2(\mathbf{x})/\ell_3(\mathbf{x}) = v$ determine $\ell_1(\mathbf{x})$, $\ell_2(\mathbf{x})$, $\ell_3(\mathbf{x})$ and hence $\mathbf{x}$, up to a proportional factor which can be determined from (11.3). There is a similar relationship between *Thue equations over $\mathcal{O}_S$*, i.e. equations of the type

$$F(x_1, x_2) = \beta \quad \text{in } x_1, x_2 \in \mathcal{O}_S \qquad (11.1')$$

and S-unit equations in two variables (with not necessarily the same ground field and set of places S). Cf. Mahler [56] and Parry [63].

Thanks to Baker [1] and others, it turns out that the above arguments can be made effective and Theorem 4 (as well as its other versions) can be applied to obtain effective results for Thue equations. Baker [1], [2] proved (implicitly) the first version of Theorem 4 (for ordinary units) and used it to make effective Thue's and Siegel's finiteness theorems mentioned above by giving explicit upper bounds for the heights of the solutions of (11.1). Coates [11], [12], in the case $K = \mathbf{Q}$, and Győry [37], [39], in the general case, extended these results to equation (11.1'). By using (a more explicit version of) Theorem 4, it was shown in [37], [39] that all solutions x_1, x_2 of (11.1') satisfy

$$\max\{h(x_1), h(x_2)\} < \exp\{(c_1 s)^{c_2 s} P^{d+1}\}$$

where c_1 and c_2 are positive numbers depending only on β, F and K (which were given explicitly in [39]).

By means of (a generalisation of) Theorem 2, Evertse [19] and later Evertse and Győry [22] derived explicit upper bounds for the numbers of solutions of (11.1) and (11.1′) which are independent of the coefficients of F. In [22], the bound

$$4n \times 7^{2g(d+s+w(\beta))}$$

has been obtained for the number of solutions of (11.1′) where $n = \deg(F)$, $g = [G : K]$ (hence $1 \leq g \leq n!$) and $w(\beta)$ denotes the number of distinct prime ideal divisors of (β).

As a generalisation of (11.1) and (11.1′), consider the *decomposable form equations*

$$F(x_1, \ldots, x_m) = \beta \qquad \text{in } x_1, \ldots, x_m \in \mathcal{O}_K \qquad (11.4)$$

or, more generally,

$$F(x_1, \ldots, x_m) = \beta \qquad \text{in } x_1, \ldots, x_m \in \mathcal{O}_S \qquad (11.4')$$

where $F(\mathbf{X}) = F(X_1, \ldots, X_m)$ is a *decomposable form* in $m \geq 2$ variables with coefficients in $\mathcal{O}_K$, i.e. a homogeneous polynomial which factorises into linear factors, $\ell_1(\mathbf{X}), \ldots, \ell_n(\mathbf{X})$ say, over some finite extension G of K.

In the case that F is a norm form and $K = \mathbf{Q}$, Schmidt [72] and Schlickewei [71] gave finiteness criteria for (11.4) and (11.4′), repectively. Their proofs are based on Schmidt's Subspace Theorem and its p-adic generalisation (cf. §4) and are ineffective. For generalisations to norm form equations over arbitrary finitely generated domains over $\mathbf{Z}$, see Laurent [52].

(11.4) and (11.4′) can be reduced to unit equations in a similar way as in the case $m = 2$ described above. Any linear relation $\lambda_{i_1}\ell_{i_1} + \ldots + \lambda_{i_r}\ell_{i_r} = 0$ with $\lambda_{i_1}, \ldots, \lambda_{i_r} \in G^*$ leads to finitely many inhomogeneous unit equations in $r - 1$ variables. But in contrast to the case $m = 2$, where one linear relation with $r = 3$ was enough, in general several linear relations are needed to prove the finiteness of the number of solutions of (11.4) and (11.4′). Győry (partly with Papp) extended the above method of reducing Thue equations to unit equations to all decomposable form equations in m (≥ 2) variables whose system of linear factors $\mathcal{L}_0 = \{\ell_1, \ldots, \ell_n\}$ satisfies the following conditions:

(b) rank $\mathcal{L}_0 = m$;

(c) $\mathcal{L}_0$ can be divided into subsets $\mathcal{L}_1, \ldots, \mathcal{L}_h$ such that if $|\mathcal{L}_j| \geq 2$ for some j, then for each i, i' with ℓ_i, $\ell_{i'} \in \mathcal{L}_j$ there exists a sequence

$\ell_i = \ell_{i_1}, \ldots, \ell_{i_r} = \ell_{i'}$ in $\mathcal{L}_j$ with the property that for $q = 1, \ldots, r-1$ there is a linear combination of ℓ_{i_q} and $\ell_{i_{q+1}}$, with coefficients in G^*, which also belongs to $\mathcal{L}_j$;

(d) There is a k with $1 \leq k \leq m$ such that X_k can be expressed as a linear combination of the forms from $\mathcal{L}_j$ for each j in $\{1, \ldots, h\}$.

By using Theorem 10, Győry [37], [39] showed that under assumptions (b), (c), (d), equation (11.4') has only finitely many solutions with $x_k \neq 0$ and he gave an effective bound for the heights of the solutions. The condition $x_k \neq 0$ is necessary in general, but if $h = 1$ in (c) then conditions (d) and $x_k \neq 0$ can be dropped. This is always the case for Thue equations. Then $h = 1$ and (b), (c) are equivalent to condition (a).

Important special types of decomposable form equations are the *discriminant form equation*

$$D_{M/K}(\alpha_1 x_1 + \ldots + \alpha_m x_m) = \beta \quad \text{in } x_1, \ldots, x_m \in \mathcal{O}_S, \qquad (11.5)$$

and the *norm form equation*

$$N_{M/K}(\alpha_0 x_0 + \ldots + \alpha_m x_m) = \beta \quad \text{in } x_0, \ldots, x_m \in \mathcal{O}_S, \qquad (11.6)$$

where $\alpha_0 = 1$, $M = K(\alpha_1, \ldots, \alpha_m)$ is a finite extension of K, $1, \alpha_1, \ldots, \alpha_m$ are linearly independent over K, and $D_{M/K}$ and $N_{M/K}$ denote the discriminant and norm over K, respectively. As an application of his result on decomposable form equations with properties (b),(c),(d), Győry [39] gave explicit upper bounds for the heights of the solutions of (11.5) and (11.6), where in (11.6) he assumed that

$$[K(\alpha_0, \ldots, \alpha_{i+1}) : K(\alpha_0, \ldots, \alpha_i)] \geq 3 \quad \text{for } i = 0, \ldots, m-1.$$

Győry [36], [39] derived several results on index form equations and algebraic number theory from his result on (11.5).

Recently, Evertse and Győry [25] replaced conditions (b),(c),(d) by the slightly weaker condition (e) of Theorem 13. To state this theorem we need some further notation. Let $\mathcal{L}^*$ be a maximal set of pairwise linearly independent linear factors of $\mathcal{L}_0$. For any subspace V of K^m, let $r(V, \mathcal{L}^*)$ denote the minimum of all positive integers r for which there are $\ell_{i_1}, \ldots, \ell_{i_r}$ in $\mathcal{L}^*$ whose restrictions to V are linearly dependent, but pairwise linearly independent. If this minimum exists, then $r(V, \mathcal{L}^*) \geq 3$. Otherwise we put $r(V, \mathcal{L}^*) = 2$. Let $\mathcal{L} \supseteq \mathcal{L}^*$ be a finite set of pairwise linearly independent linear forms in $X_1, \ldots, X_m$ with coefficients in G. By applying Theorem 4 the following result can be proved.

Theorem 13. (Evertse and Győry [25]). *Suppose that*

(e) *for every subspace V of K^m of dimension ≥ 2 on which none of the forms in $\mathcal{L}$ vanishes identically, we have $r(V, \mathcal{L}^*) = 3$.*

Then there exists an effectively computable number c_3, depending only on K, S, F and β, such that all solutions of the equation

$$F(\mathbf{x}) = F(x_1, \ldots, x_m) = \beta \ in \ \mathbf{x} \in \mathcal{O}_S^m$$

$$with \ \ell(\mathbf{x}) \neq 0 \ for \ all \ \ell \in \mathcal{L} \quad (11.7)$$

satisfy $\max_i h(x_i) < c_3$.

If in particular $\mathcal{L} = \mathcal{L}^*$, then equation (11.7) reduces to (11.4') and Theorem 13 provides an effective bound for the solutions of (11.4'). In [25], Theorem 13 is proved under a slightly weaker assumption which involves only a finite and effectively determinable collection of subspaces V of K^m.

The next result can be deduced from Theorem 2.

Theorem 14 (Evertse and Győry [22]). *With the above notation and under assumption (e), the number of solutions of (11.7) is at most*

$$n\left(4 \times 7^{2g(d+s+w(\beta))}\right)^{m-1}.$$

It is a remarkable fact that this bound is independent of the coefficients of F. As a consequence of Theorem 14, Evertse and Győry [22] derived also explicit bounds for the numbers of solutions of (11.5) and (11.6), with similar conditions as for the effective results.

An application of Corollary 1.2 led to the following general finiteness criterion for decomposable form equations.

Theorem 15 (Evertse and Győry [24]). *The following two statements are equivalent:*

(f) *For every subspace V of K^m of dimension ≥ 2 on which none of the forms in $\mathcal{L}$ vanishes identically, we have $r(V, \mathcal{L}^*) \geq 3$;*

(g) *For every $\beta \in K^*$ and every finite subset S of M_K containing all infinite places, (11.7) has only finitely many solutions.*

Condition (f) is obviously weaker than (e). Theorem 15 implies, in an ineffective form, all the above-mentioned finiteness results for decomposable form equations (cf. [24]). In [57] Mason gave an analogous result for decomposable form equations over function fields which he derived from his own effective analogue of Corollary 1.2 over function fields.

As was pointed out in [24], Corollary 1.2 is a consequence of implication (f)$\Longrightarrow$(g) of Theorem 15. Indeed, let $\alpha_1,\ldots,\alpha_m \in K^*$ and consider the unit equation

$$\alpha_1 x_1 + \alpha_2 x_2 + \ldots + \alpha_m x_m = 1 \quad \text{in } x_1, x_2, \ldots, x_m \in U_S \qquad (11.8)$$

with $\alpha_{i_1} x_{i_1} + \ldots + \alpha_{i_r} x_{i_r} \neq 0$ for all non-empty subsets $\{i_1,\ldots,i_r\}$ of $\{1,\ldots,m\}$. Put $F(\mathbf{X}) = X_1\ldots X_m(\alpha_1 X_1 + \ldots + \alpha_m X_m)$, $\mathcal{L}^* = \{X_1,\ldots,X_m,\alpha_1 X_1 + \ldots + \alpha_m X_m\}$ and let $\mathcal{L}$ be the set of linear forms of the type $\alpha_{i_1} X_{i_1} + \ldots + \alpha_{i_r} X_{i_r}$ where $\{i_1,\ldots,i_r\}$ is a non-empty subset of $\{1,\ldots,m\}$. Since $F(\mathbf{x}) \in U_S$ for every solution $\mathbf{x} = (x_1,\ldots,x_m) \in U_S^m$ of (11.8), we have $F(\mathbf{x}) = \beta\epsilon^{m+1}$ where β, $\epsilon \in U_S$ and β can assume only finitely many values. This means that, with the notation $\mathbf{x}' = \mathbf{x}/\epsilon$, (11.8) reduces to finitely many equations of the type

$$F(\mathbf{x}') = \beta \quad \text{in } \mathbf{x}' = (x_1',\ldots,x_m') \in U_S^m$$

$$\text{with } \ell(\mathbf{x}') \neq 0 \text{ for all } \ell \in \mathcal{L}. \qquad (11.9)$$

It was, however, shown in [24] that these $\mathcal{L}^*$ and $\mathcal{L}$ satisfy assumption (f) of Theorem 15. Therefore, by Theorem 15, equation (11.9) has only finitely many solutions $\mathbf{x}'$. This implies that (11.8) has indeed only finitely many solutions. In other words, Corollary 1.2 on unit equations is equivalent to the implication (f)$\Longrightarrow$(g) of Theorem 15. This contains as a special case the relationship observed by Siegel between Thue equations and unit equations in two variables.

Finally, we note that Theorems 14, 15 were proved in [22], [24], respectively, in the more general form when the ground ring is an arbitrary finitely generated extension ring of $\mathbf{Z}$. In the proofs the authors used Theorem 1' and the general version of Theorem 2, respectively. Győry's effective results on decomposable form equations in [37], [39] have been extended to this more general situation in [43], [44].

§12. Applications to algebraic number theory

Several diophantine problems in algebraic number theory can be reduced to the study of the equations

$$D_{K/\mathbf{Q}}(\alpha) = D_0 \quad \text{in } \alpha \in \mathcal{O}_K \qquad (12.1)$$

and

$$D(f) = D_0 \quad \text{in monic polynomials } f \in \mathbf{Z}[X] \qquad (12.2)$$

where K is now an algebraic number field of degree $d \geq 2$, $D(f)$ and $D_{K/\mathbf{Q}}(\alpha)$ denote the discriminant of f and α, respectively, and $D_0 \in \mathbf{Z} \setminus \{0\}$. If α satisfies (12.1) then its minimal defining polynomial over $\mathbf{Z}$ satisfies (12.2). Equation (12.2) can have, however, other (not necessarily irreducible) solutions without zeros in K. Hence (12.2) is more general than (12.1). If α is a solution of (12.1) then so is $\alpha + a$ for all $a \in \mathbf{Z}$. Elements α, α' of $\mathcal{O}_K$ with $\alpha - \alpha' \in \mathbf{Z}$ are called $\mathbf{Z}$-*equivalent*. Similarly, if f is a solution of (12.2), then so is $f^*(X) = f(X + a)$ for every $a \in \mathbf{Z}$. Polynomials f, f^* of this kind are called $\mathbf{Z}$-*equivalent*. By repeatedly applying an earlier version of Theorem 4, Győry proved, in 1973, the following

Theorem 16 (Győry [30]). *Every solution α of* (12.1) *is $\mathbf{Z}$-equivalent to a solution $\alpha' \in \mathcal{O}_K$ for which*

$$H(\alpha') < c_1$$

where c_1 is an effectively computable number depending only on d, D_K and D_0.

In other words, there are only finitely many pairwise $\mathbf{Z}$-inequivalent elements in $\mathcal{O}_K$ with discriminant D_0, and a full set of representatives of such elements can be, at least in principle, effectively determined. This finiteness assertion was independently proved in a non-effective form by Birch and Merriman [7] in 1972.

We shall now sketch how (12.1) can be reduced to a finite system of unit equations. Let G be the normal closure of $K/\mathbf{Q}$ with degree g (over $\mathbf{Q}$) and let $\alpha^{(1)} = \alpha$, $\alpha^{(2)}, \ldots, \alpha^{(d)}$ denote the conjugates of α with respect to $K/\mathbf{Q}$. If $d \geq 3$ then

$$\frac{\alpha^{(1)} - \alpha^{(i)}}{\alpha^{(1)} - \alpha^{(2)}} + \frac{\alpha^{(i)} - \alpha^{(2)}}{\alpha^{(1)} - \alpha^{(2)}} = 1 \quad \text{for } i = 3, \ldots, d. \qquad (12.3)$$

Further, the numbers $\alpha^{(1)} - \alpha^{(2)}$, $\alpha^{(1)} - \alpha^{(i)}$, $\alpha^{(i)} - \alpha^{(2)}$ divide D_0 in $\mathcal{O}_G$, whence they belong to finitely many cosets of G^* with respect to U_G. Thus (12.3) reduces indeed to finitely many unit equations in two variables and, by Theorem 4, $\alpha^{(1)} - \alpha^{(i)}$, $\alpha^{(2)} - \alpha^{(i)}$ and so $\alpha^{(i)} - \alpha^{(j)}$ can be determined up to the common factor $\alpha^{(1)} - \alpha^{(2)}$ which is however determinable from (12.1), and Theorem 16 follows.

In fact Theorem 16 is an immediate consequence of Theorem 10. Let $\mathcal{A} = \{\alpha^{(1)}, \ldots, \alpha^{(d)}\}$ and $N = |D_0|^g$. By (12.1) we have

$$\left| N_{G/\mathbf{Q}}(\alpha^{(i)} - \alpha^{(j)}) \right| \leq N \quad \text{for } 1 \leq i < j \leq d,$$

hence the graph $\mathcal{G}_G(\mathcal{A}, N)$ (cf. §10) has only isolated vertices. Therefore case iv) of Theorem 10 applies and the differences $\alpha^{(i)} - \alpha^{(j)}$ can assume only finitely many effectively determinable values, up to a common factor, while this common factor can be derived from (12.1).

If (12.1) is solvable then $D_K | D_0$. Denote by w the number of distinct prime factors of D_0/D_k. By means of Theorem 2 one can prove the following

Theorem 17 (Evertse and Győry [23]). *Equation* (12.1) *has at most* $7^{g(d-1)(2w+3)}$ *pairwise* $\mathbf{Z}$-*inequivalent solutions.*

We note that $d \leq g \leq d!$.

In view of a theorem of Minkowski d can be estimated from above explicitly in terms of D_K. Further, (12.1) implies $|D_K| \leq |D_0|$. Hence the dependence of c_1 on d and D_K in Theorem 16 can be dropped (cf. [30]). For irreducible polynomials $f \in \mathbf{Z}[X]$ this implies the following

Theorem 18 (Győry [30]). *Every solution* f *of* (12.2) *is* $\mathbf{Z}$-*equivalent to a solution* $f^* \in \mathbf{Z}[X]$ *for which*

$$\deg(f^*) \leq c_2, \qquad H(f^*) \leq c_3$$

where c_2, c_3 *are effectively computable numbers depending only on* D_0, *and* $H(f^*)$ *denotes the maximum absolute value of the coefficients of* f^*.

The 'reducible' case can be reduced to the 'irreducible' one by using the relation

$$D(F) = \prod_{i=1}^{k} D(f_i) \prod_{1 \leq i < j \leq k} \left(\text{Res}(f_i, f_j) \right)^2$$

where $f = \prod_{i=1}^{k} f_i$ in $\mathbf{Z}[X]$ and $\text{Res}(f_i, f_j)$ denotes the resultant of f_i and f_j. We note that in Theorem 18 an upper bound for $\deg(f^*)$ can also be derived by means of Theorem 11.

Theorem 18 implies that up to $\mathbf{Z}$-equivalence, there are only finitely many monic polynomials $f \in \mathbf{Z}[X]$ with discriminant $D_0 \neq 0$ and a full set of representatives of such polynomials can be effectively determined. For binary forms of given degree and given non-zero discriminant, a similar but ineffective finiteness theorem was independently proved by Birch and Merriman [7].

We present one consequence of Theorems 16 and 17 here. For other applications we refer to [32], [33], [36], [23]. As is known, there exist algebraic number fields K having *power integral bases* (i.e. integral bases of the form $\{1, \alpha, \ldots, \alpha^{d-1}\}$ where $d = [K : \mathbf{Q}])$, but this is not the case in general. For references to results concerning power integral bases, see [46]. It is known that $\alpha \in \mathcal{O}_K$ generates a power integral basis if and only if $D_{K/\mathbf{Q}}(\alpha) = D_K$. If α is a generator then so are all $\alpha' \in \mathcal{O}_K$ which are $\mathbf{Z}$-equivalent to α. By applying Theorem 16 with $D_0 = D_K$, we have

Corollary 16.1 (Győry [32]). *If $\{1, \alpha, \ldots, \alpha^{d-1}\}$ is an integral basis of K, then there is an $\alpha' \in \mathcal{O}_K$ which is $\mathbf{Z}$-equivalent to α such that*

$$H(\alpha') < c_4$$

where c_4 is an effectively computable number depending only on d and D_K.

Thus, up to $\mathbf{Z}$-equivalence, there are only finitely many elements in $\mathcal{O}_K$ which generate a power integral basis and they can be effectively determined. In particular, one can decide at least in principle, whether K has a power integral basis or not.

Corollary 17.1 (Evertse and Győry [23]). *Up to $\mathbf{Z}$-equivalence there are at most $7^{3g(d-1)}$ elements $\alpha \in \mathcal{O}_K$ for which $\{1, \alpha, \ldots, \alpha^{d-1}\}$ is an integral basis for K.*

Since $g \leq d!$, this implies an upper bound depending only on d.

For explicit expressions for c_1 to c_4 and for references, see Győry [31], [33]. The results presented above have various generalisations; for references see [45], [46], [23].

§13. Applications to transcendental number theory

Let $g(z) = \sum_{k=1}^{\infty} z^{k!}$. Let $\alpha_1, \ldots, \alpha_n$ be algebraic numbers with $0 < |\alpha_i| < 1$ for $i = 1, \ldots, n$. D. W. Masser conjectured that if α_i/α_j is not a root of unity for $1 \leq i < j \leq n$, then $g(\alpha_1), \ldots, g(\alpha_n)$ are algebraically independent. Nishioka [60] used Theorem 1 to prove the stronger assertion that under the above conditions all numbers $g^{(\ell)}(\alpha_i)$ $(1 \leq i \leq n, \ell \geq 0)$ are algebraically independent.

Nishioka generalised the above result to more general gap series f. Let K be an algebraic number field. Let $f(z) = \sum_{k=0}^{\infty} a_k z^{e_k}$ be a power series with non-zero coefficients $a_k \in K$, positive convergence radius R and increasing non-negative exponents e_k satisfying

$$\lim_{k \to \infty} (e_k + \log M_k + \log A_k)/e_{k+1} = 0$$

where $A_k = \max(1, \overline{|a_0|}, \ldots, \overline{|a_k|})$ and M_k is a positive integer such that $M_k a_0, \ldots, M_k a_k$ are algebraic integers. Cijsouw and Tijdeman [10] proved that $f(\alpha)$ is transcendental for any algebraic number α with $0 < |\alpha| < R$. Bundschuh and Wylegala [8] proved the remarkable result that $f(\alpha_1), \ldots, f(\alpha_n)$ are algebraically independent for any algebraic numbers $\alpha_1, \ldots, \alpha_n$ with $0 < |\alpha_1| < \ldots < |\alpha_n| < R$. There are several other papers on the algebraic independence of values of gap series, but nobody could handle the case of α_i of equal absolute values until Nishioka [61] applied Theorem 1. She proved the following general result.

Theorem 19. *Let* $\alpha_1, \ldots, \alpha_n$ *be algebraic numbers with* $0 < |\alpha_i| < 1$ *for* $i = 1, \ldots, n$. *Then the following three properties are equivalent.*

(i) *The numbers* $f^{(\ell)}(\alpha_i)$ $(1 \le i \le n,\ \ell \ge 0)$ *are algebraically dependent over the rationals.*

(ii) *The numbers* $1, f(\alpha_1), \ldots, f(\alpha_n)$ *are linearly dependent over the algebraic numbers.*

(iii) *There is a non-empty subset* $\{i_1, \ldots, i_m\}$ *of* $\{1, \ldots, n\}$, *a number* γ, *roots of unity* $\zeta_1, \ldots, \zeta_m$ *and algebraic numbers* $d_1, \ldots, d_m$, *not all zero, such that*

$$\alpha_{i_j} = \zeta_j \gamma \quad (1 \le j \le m) \qquad and \qquad \sum_{j=1}^{m} d_j \zeta_j^{e_k} = 0$$

for all sufficiently large k.

References

[1] A. Baker, Contributions to the theory of diophantine equations, *Philos. Trans. Roy. Soc. London* **A 263** (1967/68), 173–208.

[2] A. Baker, Bounds for the solutions of the hyperelliptic equation, *Proc. Cambridge Philos. Soc.* **65** (1969), 439–444.

[3] A. Baker, The theory of linear forms in logarithms, *Transcendence Theory: Advances and Applications*, Academic Press, London, 1977, pp. 1–27.

[4] F. Beukers, The multiplicity of binary recurrences, *Compositio Math.* **40** (1980), 251–267.

[5] F. Beukers and R. Tijdeman, On the multiplicities of binary complex recurrences, *Compositio Math.* **51** (1984), 193–213.

[6] J.-P. Bézivin, Sur un Théorème de G. Pólya, *J. Reine Angew. Math.* **364** (1986), 60–68.

[7] B. J. Birch and J. R. Merriman, Finiteness theorems for binary forms with given discriminant, *Proc. London Math. Soc.* (3) **24** (1972), 385–394.

[8] P. Bundschuh and F.-J. Wylegala, Über algebraische Unabhängigkeit bei gewissen nichtfortsetzbaren Potenzreihen, *Arch. Math.* **34** (1980), 32–36.

[9] J. W. S. Cassels, On a class of exponential equations, *Ark. Mat.* **4** (1961), 231–233.

[10] P. L. Cijsouw and R. Tijdeman, On the transcendence of certain power series of algebraic numbers, *Acta Arith.* **23** (1973), 301–305.

[11] J. Coates, An effective *p*-adic analogue of a theorem of Thue, *Acta Arith.* **15** (1968/69), 279–305.

[12] J. Coates, An effective *p*-adic analogue of a theorem of Thue II: The greatest prime factor of a binary form, *Acta Arith.* **16** (1969/70), 399–412.

[13] H. Davenport and K. F. Roth, Rational approximations to algebraic numbers, *Mathematika* **2** (1955), 160–167.

[14] E. Dubois and G. Rhin, Sur la majoration de formes linéaires à coefficients algébriques réels et *p*-adiques, Démonstration d'une conjecture de K. Mahler, *C. R. Acad. Sci. Paris* A **282** (1976), 1211–1214.

[15] P. Erdős, Problems in number theory and combinatorics, Proc. 6th Manitoba Conference on Numerical Math., *Congress Numer. 18, Utilitas Math.*, Winnipeg, Man., 1977.

[16] P. Erdős and R. L. Graham, Old and new problems and results in combinatorial theory: van der Waerden's theorem and related topics, *Enseign. Math.* (2) **25** (1979), 325–344.

[17] P. Erdős, C. L. Stewart and R. Tijdeman, On the number of solutions of the equation $x + y = z$ in *S*-units, *Compositio Math.*, to appear.

[18] P. Erdős and P. Turán, On a problem in the elementary theory of numbers, *Amer. Math. Monthly* **41** (1934), 608–611.

[19] J.-H. Evertse, On equations in *S*-units and the Thue-Mahler equation, *Invent. Math.* **75** (1984), 561–584.

[20] J.-H. Evertse, On sums of S-units and linear recurrences, *Compositio Math.* **53** (1984), 225–244.

[21] J.-H. Evertse, On equations in two S-units over function fields of characteristic 0, *Acta Arith.* **47** (1986), 233–253.

[22] J.-H. Evertse and K. Győry, On unit equations and decomposable form equations, *J. Reine Angew. Math.* **358** (1985), 6–19.

[23] J.-H. Evertse and K. Győry, On the number of polynomials and integral elements of given discriminant, *Acta Math. Hungar.*, to appear.

[24] J.-H. Evertse and K. Győry, Finiteness criteria for decomposable form equations, *Acta Math.*, to appear.

[25] J.-H. Evertse and K. Győry, Decomposable form equations, *New Advances in Transcendence Theory* (A. Baker ed.), Cambridge Univ. Press, 1988, Chapter 10.

[26] J.-H. Evertse, K. Győry, C. L. Stewart and R. Tijdeman, On S-unit equations in two unknowns, *Invent Math.*, to appear.

[27] J. P. Glass, J. H. Loxton and A. J. van der Poorten, Identifying a rational function, *C. R. Math. Rep. Acad. Sci. Canada* **3** (1981), 279–284. Corr. **4** (1982), 309–314.

[28] K. Győry, Sur l'irréductibilité d'une classe des polynômes I, *Publ. Math. Debrecen* **18** (1971), 289–307.

[29] K. Győry, Sur l'irréducibilité d'une classe des polynômes II, *Publ. Math. Debrecen* **19** (1972), 293–326.

[30] K. Győry, Sur les polynômes à coefficients entiers et de discriminant donné, *Acta Arith.* **23** (1973), 419–426.

[31] K. Győry, Sur les polynômes à coefficients entiers et de discriminant donné II, *Publ. Math. Debrecen* **21** (1974), 125–144.

[32] K. Győry, Sur les polynômes à coefficients entiers et de discriminant donné III, *Publ. Math. Debrecen* **23** (1976), 141–165.

[33] K. Győry, On polynomials with integer coefficients and given discriminant IV, *Publ. Math. Debrecen* **25** (1978), 155–167.

[34] K. Győry, On the number of solutions of linear equations in units of an algebraic number field, *Comment. Math. Helv.* **54** (1979), 583–600.

[35] K. Győry, On certain graphs composed of algebraic integers of a number field and their applications I, *Publ. Math. Debrecen* **27** (1980), 229–242.

[36] K. Győry, Résultats effectifs sur la représentation des entiers par des formes décomposables, *Queen's Papers in Pure and Applied Math.* No. 56, Kingston, Ont., 1980.

[37] K. Győry, On the representation of integers by decomposable forms in several variables, *Publ. Math. Debrecen* **28** (1981), 89–98.

[38] K. Győry, On discriminants and indices of integers of an algebraic number field, *J. Reine Angew. Math.* **324** (1981), 114–126.

[39] K. Győry, On S-integral solutions of norm form, discriminant form and index form equations, *Studia Sci. Math. Hungar.* **16** (1981), 149–161.

[40] K. Győry, On certain graphs associated with an integral domain and their applications to Diophantine problems, *Publ. Math. Debrecen* **29** (1982), 79–94.

[41] K. Győry, On the irreducibility of a class of polynomials III, *J. Number Theory* **15** (1982), 164–181.

[42] K. Győry, Effective finiteness theorems for Diophantine problems and their applications (in Hungarian), Academic doctor's thesis, Debrecen, 1983.

[43] K. Győry, Bounds for the solutions of norm form, discriminant form and index form equations in finitely generated domains, *Acta Math. Hungar.* **42** (1983), 45–80.

[44] K. Győry, On norm form, discriminant form and index form equations, *Coll. Math. Soc. J. Bolyai* **34**, Topics in Classical Number Theory, Budapest, 1981, North-Holland, Amsterdam etc., 1984, pp. 617–676.

[45] K. Győry, Effective finiteness theorems for polynomials with given discriminant and integral elements with given discriminant over finitely generated domains, *J. Reine Angew. Math.* **346** (1984), 54–100.

[46] K. Győry, Sur les générateurs des ordres monogènes des corps de nombres algébriques, *Sém. Théorie des Nombres* 1983–84, Univ. Bordeaux, No. 32, 1984, 12 pp.

[47] K. Győry, C. L. Stewart and R. Tijdeman, On prime factors of sums of integers I, *Compositio Math.* **59** (1986), 81–88.

[48] K. Győry, C. L. Stewart and R. Tijdeman, On prime factors of sums of integers III, *Acta Arith.*, to appear.

[49] K. K. Kubota, On a conjecture of Morgan Ward, *Acta Arith.* **33** (1977), 11–48, 99–109.

[50] S. Lang, Integral points on curves, *Publ. Math. I.H.E.S.* **6** (1960), 27–43.

[51] S. Lang, Fundamentals of Diophantine Geometry, *Springer-Verlag*, Berlin, Heidelberg, New York, 1983.

[52] M. Laurent, Équations diophantiennes exponentielles, *Invent. Math.* **78** (1984), 299–327.

[53] C. Lech, A note on recurring series, *Ark. Mat.* **2** (1953), 417–421.

[54] D. J. Lewis and K. Mahler, On the representation of integers by binary forms, *Acta Arith.* **6** (1960/61), 333–363.

[55] D. J. Lewis and J. Turk, Repetitiveness in binary recurrences, *J. Reine Angew. Math.* **356** (1985), 19–48.

[56] K. Mahler, Zur Approximation algebraischer Zahlen I: Über den grössten Primteiler binärer Formen, *Math. Ann.* **107** (1933), 691–730.

[57] R. C. Mason, Norm form equations IV; rational functions, *Mathematika* **33** (1986), 204–211.

[58] M. Mignotte, T. N. Shorey and R. Tijdeman, The distance between terms of an algebraic recurrence sequence, *J. Reine Angew. Math.* **349** (1984), 63–76.

[59] T. Nagell, Quelques problèmes relatifs aux unités algébriques, *Ark. Mat.* **8** (1969), 115–127.

[60] K. Nishioka, Proof of Masser's conjecture on the algebraic independence of values of Liouville series, *Proc. Japan Acad. Ser.* **A 62** (1986), 219–222.

[61] K. Nishioka, Conditions for algebraic independence of certain power series of algebraic numbers, *Compositio Math.*, **62** (1987), 53–61.

[62] J. C. Parnami and T. N. Shorey, Subsequences of binary recursive sequences, *Acta Arith.* **40** (1981/82), 193–196.

[63] C. J. Parry, The $\wp$-adic generalisation of the Thue-Siegel theorem, *Acta Math.* **83** (1950), 1–100.

[64] G. Pólya, Arithmetische Eigenschaften der Reihenentwicklungen rationaler Funktionen, *J. Reine Angew. Math.* **151** (1921), 1–31.

[65] A. J. van der Poorten, Linear forms in logarithms in the p-adic case, *Transcendence Theory: Advances and Applications*, Academic Press, London, 1977, pp. 29–57.

[66] A. J. van der Poorten, Some problems of recurrent interest, *Coll. Math. Soc. János Bolyai* **34**, Topics in Classical Number Theory, North-Holland, Amsterdam, 1984, pp. 1265–1294.

[67] A. J. van der Poorten and H. P. Schlickewei, The growth conditions for recurrence sequences, Macquarie Univ. Math. Rep. 82-0041, North Ryde, Australia, 1982.

[68] H. P. Schlickewei, Linearformen mit algebraischen Koeffizienten, *Manuscripta Math.* **18** (1976), 147–185.

[69] H. P. Schlickewei, The $\wp$-adic Thue-Siegel-Roth-Schmidt theorem, *Arch. Math.* **29** (1977), 267–270.

[70] H. P. Schlickewei, Über die diophantische Gleichung $x_1 + x_2 + \ldots + x_n = 0$, *Acta Arith.* **33** (1977), 183–185.

[71] H. P. Schlickewei, On norm form equations, *J. Number Theory* **9** (1977), 370–380.

[72] W. M. Schmidt, Linearformen mit algebraischen Koeffizienten II, *Math. Ann.* **191** (1971), 1–20.

[73] W. M. Schmidt, Simultaneous approximation to algebraic numbers by elements of a number field, *Monatsh. Math.* **79** (1975), 55–66.

[74] W. M. Schmidt, *Diophantine Approximation*, Lect. Notes Math. **785**, Springer-Verlag, Berlin, Heidelberg, New York, 1980.

[75] T. N. Shorey, The greatest square free factor of a binary recursive sequence, *Hardy Ramanujan J.* **6** (1983), 23–36.

[76] T. N. Shorey, Linear forms in members of a binary recursive sequence, *Acta Arith.* **43** (1984), 317–331.

[77] T. N. Shorey and R. Tijdeman, *Exponential Diophantine Equations*, Cambridge Univ. Press, 1986.

[78] C. L. Siegel, Approximation algebraischer Zahlen, *Math. Z.* **10** (1921), 173–213.

[79] C. L. Siegel, The integer solutions of the equation $y^2 = ax^n + bx^{n-1} + \ldots + k$, *J. London Math. Soc.* **1** (1926), 66–68.

[80] C. L. Siegel, Über einige Anwendungen diophantischer Approxima-
     tionen, *Abh. Preuss. Akad. Wiss., Phys.-Math. Kl.*, **1929**, No. 1.

[81] J. H. Silverman, Quantitative results in diophantine geometry,
     Preprint, M. I. T., Cambridge, Mass., 1983.

[82] J. H. Silverman, A quantitative version of Siegel's theorem: integral
     points on elliptic curves and Catalan curves, *J. Reine Angew. Math.*
     378 (1987), 60–100.

[83] Th. Skolem, A method for the solution of the exponential equation
     $A_1^{x_1} \ldots A_m^{x_m} - B_1^{y_1} \ldots B_n^{y_n} = C$ (Norwegian), *Norsk Mat. Tidsskr.*
     27 (1945), 37–51.

[84] V. G. Sprindzhuk, Effective estimates in 'ternary' exponential dio-
     phantine equations (Russian), *Dokl. Akad. Nauk BSSR* **13** (1969),
     777–780.

[85] C. L. Stewart, Divisor Properties of Arithmetical Sequences, Ph.D.
     Thesis, Univ. of Cambridge, Cambridge, 1976.

[86] C. L. Stewart, On divisors of Fermat, Fibonacci, Lucas and Lehmer
     numbers III, *J. London Math. Soc.* (2) **28** (1983), 211–217.

[87] C. L. Stewart, Some remarks on prime divisors of sums of inte-
     gers, Séminaire de Théorie des Nombres, Paris, 1984–85, *Progress
     in Mathematics* **63**, Birkhäuser, Boston etc., 1986, pp. 217–223.

[88] C. L. Stewart and R. Tijdeman, On prime factors of sums of integers
     II, in *Diophantine Analysis*, edited by J. H. Loxton and A. J. van
     der Poorten, Cambridge Univ. Press, 1986, pp. 83–98.

[89] A. Thue, Über Annäherungswerte algebraischer Zahlen, *J. Reine
     Angew. Math.* **135** (1909), 284–305.

[90] R. Tijdeman and L. Wang, Sums of products of powers of given
     prime numbers, *Pacific J. Math.* to appear.

[91] K. R. Yu, Linear forms in logarithms in the *p*-adic case, *New Ad-
     vances in Transcendence Theory* (A. Baker ed.), Cambridge Univ.
     Press, 1988, Chapter 26.

[92] K. R. Yu, *Linear forms in the p-adic logarithms*, Max-Planck-
     Institut für Mathematik MPI/87-20, Bonn, F. R. Germany, 1986.

10

DECOMPOSABLE FORM EQUATIONS

J.-H. Evertse and K. Győry[1]

Let $F(\underline{x}) = F(x_1, \ldots, x_m)$ be a form (homogeneous polynomial) in $m \geq 2$ variables with coefficients in $\mathbf{Z}$. Suppose that F is *decomposable* (that is that F factorizes into linear factors over the field $\overline{\mathbf{Q}}$ of algebraic numbers). For $m = 2$ every form is decomposable, but for $m > 2$ this is not always the case. Let $b \in \mathbf{Z} \setminus \{0\}$ and consider the *decomposable form equation*

$$F(\underline{x}) = F(x_1, \ldots, x_m) = b \tag{1}$$

in $\underline{x} = (x_1, \ldots, x_m) \in R^m$ where R is a subring of $\mathbf{Q}$ finitely generated over $\mathbf{Z}$. Equations of this type are of basic importance in the theory of Diophantine equations and have many applications to algebraic number theory. Important classes of such equations are *Thue equations*, when $m = 2$, *norm form equations*, *discriminant form equations* and *index form equations*. In the last twenty years much progress has been made in the study of decomposable form equations. By means of the Thue-Siegel-Roth-Schmidt method general finiteness criteria have been established which guarantee, under the most general conditions possible for F and R, the finiteness of the numbers of solutions for every b. These criteria do not provide, however, any procedure to solve the equations in question or decide the solvability and hence are *ineffective*. *Effective* finiteness theorems have been obtained for a restricted class of decomposable form equations, including Thue equations, discriminant form equations, index form equations and a class of norm form equations. By using Baker's method concerning linear forms in logarithms of algebraic numbers, explicit upper bounds have been derived for the absolute values (heights) of the solutions. These bounds make it possible, at least in principle, to determine all solutions. Finally, for the restricted class of decomposable form equations mentioned above, explicit upper bounds have been given for the numbers of solutions which are independent of the coefficients of the decomposable forms involved. The most important theorems have

---

[1] Research supported in part by the Hungarian National Foundation for Scientific Research, grant 273

been generalized to the case that the ground ring (cf. §1) is an arbitrary finitely generated integral domain over **Z**, and analogous results have been established over function fields. It turns out that the theory of decomposable form equations is in fact equivalent to the theory of unit equations (see [15], and [16] in this volume) and this close connection has proved very useful for decomposable form equations. The most general ineffective and effective finiteness results concerning decomposable form equations have been obtained via unit equations.

In §1, we shall give a historical survey on the advances in the study of decomposable form equations and their applications. We shall state results only over finitely generated subrings of **Q** and indicate extensions to the case of more general ground rings finitely generated over **Z**. Results over function fields will be discussed in Mason's paper in this volume. For the methods which have been used, related results, further applications and references we refer to [62], [10], [7], [4], [56], [28], [33], [67], [44], [58], [16].

In §2, two new results, Theorems 1′ and 2′, will be presented. Theorem 1′ is an effective finiteness result for a wide class of decomposable form equations over finitely generated subrings of algebraic number fields. Theorem 2′ provides, for the same class of equations, an explicit upper bound for the numbers of solutions which is independent of the coefficients of the decomposable forms involved. Theorems 1′ and 2′ are extensions of the previous results to a slightly larger class of equations. The proofs of Theorems 1′ and 2′ are given in §3.

§1. Historical Survey

In this section, we first give a brief survey of results obtained for decomposable form equations in two unknowns (Thue equations), and then discuss to what extent these results have been generalized to norm form equations, discriminant form and index form equations and in general to decomposable form equations in more than two unknowns. In the sequel, C_1, C_2, ... will denote effectively computable positive numbers which depend only on appropriate parameters of the equations under consideration. Unless otherwise stated, explicit expressions for these numbers have been given in the papers to which we shall refer.

Thue equations

Consider the equation
$$F(x_1, x_2) = b \tag{2}$$

in $x_1, x_2 \in \mathbf{Z}$ where $F \in \mathbf{Z}[X_1, X_2]$ is a binary form and $b \in \mathbf{Z} \setminus \{0\}$. If $n = \deg(F) \leq 2$, (2) may have infinitely many solutions and all these can be described. In 1909 Thue [70] proved the following.

Theorem A (Thue [70]). *If $F \in \mathbf{Z}[X_1, X_2]$ is an irreducible binary form of degree $n \geq 3$ then (2) has only finitely many solutions in $x_1, x_2 \in \mathbf{Z}$.*

After Thue, equations of this type are named *Thue equations*. It is easy to see that in Thue's theorem the "irreducibility" condition can be replaced by the weaker assumption that F is not a constant multiple of a linear form or of an irreducible quadratic form with positive discriminant.

Thue deduced his finiteness result from his approximation theorem. Theorem A was later improved and generalized by several authors. Siegel [60] gave a general finiteness criterion for polynomial Diophantine equations in two unknowns. In 1933, Mahler [46] extended Thue's theorem to the equation

$$F(x_1, x_2) = bp_1^{z_1} \ldots p_s^{z_s} \text{ in } x_1, x_2, z_1 \ldots, z_s \in \mathbf{Z}$$
$$\text{with } (x_1, x_2) = 1 \tag{3}$$

where $p_1, \ldots, p_s$ ($s \geq 0$) are distinct primes.

Theorem B (Mahler [46]). *Let $F \in \mathbf{Z}[X_1, X_2]$ be a binary form having at least three pairwise linearly independent linear factors in its factorization over $\overline{\mathbf{Q}}$. Then equation (3) has only finitely many solutions.*

Equations of the type (3) are called *Thue-Mahler equations*. An equivalent formulation of Mahler's theorem is that equation (2) has only finitely many solutions x_1, x_2 in the ring $\mathbf{Z}\left[\frac{1}{p_1}, \ldots, \frac{1}{p_s}\right]$. Since every subring of $\mathbf{Q}$ which is finitely generated over $\mathbf{Z}$ can be written in this form (with finitely many appropriate primes), Mahler's result implies that (2) has finitely many solutions in every finitely generated subring of $\mathbf{Q}$. We note that these results of Mahler do not remain valid in general if F has at most two pairwise linearly independent linear factors over $\overline{\mathbf{Q}}$.

Siegel [59], [60], Parry [48] extended the above-mentioned results of Thue and Mahler, respectively, to the so-called *relative case* when the ground ring (that is, the ring containing b, the coefficients of F and the values assumed by the unknowns) is the ring of integers of an arbitrary algebraic number field. Finally, Lang [44] gave a further generalization to the case of arbitrary finitely generated ground rings over $\mathbf{Z}$. By a recent result of Faltings [17] (see also [18]), even the number of "rational" solutions is finite, provided the degree $n \geq 4$. All these results are, however, *ineffective*, that is, their proofs do not provide an algorithm for deciding the solvability or determining the solutions.

The first general *effective* result on the Thue equations was proved by Baker [1] in 1968. By using his fundamental effective inequalities concerning linear forms in the logarithms of algebraic numbers, he showed the following.

Theorem C (Baker [1]). *If $F \in \mathbf{Z}[X_1, X_2]$ is an irreducible binary form of degree $n \geq 3$ then all solutions $x_1, x_2 \in \mathbf{Z}$ of (2) satisfy*

$$\max(|x_1|, |x_2|) < \exp\{n^{v^2} H^{vn^3} + (\log|b|)^{2n+2}\}$$

where $v = 128n(n + 1)$ and H denotes the maximum absolute value of the coefficients of F.

This made it possible, at least in principle, to solve Thue's equations. Baker's estimate was later improved by Feldman [19], Sprindžuk [65], Stark [68], Baker [3] and others. The best known upper bound, due to Győry and Papp [37] is of the form

$$\max(|x_1|, |x_2|) < \left(H \cdot |b|\right)^{(C_1 n)^{C_2 n} (R_K \log R_K^*)^2},$$

where C_1 and C_2 are effectively computable positive absolute constants, R_K is the regulator of the field K generated by a root of $F(X, 1) = 0$ and $R_K^* = \max(R_K, 3)$. The above bounds led to effective improvements of Liouville's approximation theorem (cf. [1], [19], [37]).

By proving and using a p-adic analogue of Baker's inequality concerning linear forms in logarithms, Coates [8], [9] made effective Mahler's theorem for irreducible binary forms F. Coates' estimate for the solutions was improved and generalized by Sprindžuk [63], [64], [66] and Shorey, van der Poorten, Tijdeman and Schinzel [57]. In [57] the authors gave an effective version of Mahler's theorem in full generality. The results of Baker, Coates and Shorey, van der Poorten, Tijdeman and Schinzel were later extended to the relative case by Baker [2], Kotov [39] and Győry [24], [26], respectively. Further extensions to the case of arbitrary finitely generated ground rings over $\mathbf{Z}$ were recently obtained by Győry [31], [33].

Several authors derived upper bounds for the numbers of solutions of the Thue and Thue-Mahler equations; for references see e.g. [11], [12].

In 1983, Evertse [11], [12] was the first to obtain (without any additional restriction concerning F or b) bounds for the numbers of solutions of (2) and (3) which are independent of the coefficients F. Let $\omega(b)$ denote the number of distinct prime factors of b.

Theorem D (Evertse [12]). *Under the assumption of Theorem B, equation* (3) *has at most*

$$2 \times 7^{n^3(2s+2\omega(b)+3)}$$

solutions.

Evertse [12] proved his theorem in a more general form, in the relative case, by using a modification of a method of Thue and Siegel. For further generalizations to equations over arbitrary finitely generated domains over $\mathbf{Z}$ see Evertse and Győry [14].

It follows from Theorem D that (2) has at most $2 \times 7^{n^3(2s+2\omega(b)+3)}$ solutions x_1, x_2 in the ring $\mathbf{Z}\left[\frac{1}{p_1}, \ldots, \frac{1}{p_s}\right]$. By restricting themselves to solutions x_1, x_2 in $\mathbf{Z}$ with $(x_1, x_2) = 1$, Bombieri and Schmidt [6] have recently improved this upper bound to $C_3 \times n^{\omega(b)+1}$, where C_3 is an absolute constant (which was not explicitly computed but smaller than 215 for n sufficiently large). For further recent related results we refer to the paper of Schmidt in this volume.

Norm form equations

Let K be an algebraic number field of degree $n \geq 2$ with $\mathbf{Q}$-isomorphisms $\sigma_1, \ldots, \sigma_n$ in $\mathbf{C}$. Put $\alpha^{(i)} = \sigma_i(\alpha)$ for any $\alpha \in K$. Let $\alpha_1 = 1, \alpha_2, \ldots \alpha_m$, $m \geq 2$, be linearly independent elements of K over $\mathbf{Q}$, (i.e. $m \leq n$) and suppose, for simplicity, that $K = \mathbf{Q}(\alpha_1, \ldots, \alpha_m)$. Put

$$L^{(i)}(\underline{X}) = \alpha_1^{(i)} X_1 + \ldots + \alpha_m^{(i)} X_m \qquad \text{for } i = 1, \ldots, n. \qquad (4)$$

Then

$$N(\underline{X}) = N(\alpha_1 X_1 + \ldots + \alpha_m X_m) = \prod_{i=1}^{n} L^{(i)}(\underline{X})$$

is a decomposable form with coefficients in $\mathbf{Q}$ which is called a *norm form*. In what follows, let[1] $b \in \mathbf{Q}^*$. An equation of the type

$$N(\underline{x}) = b \qquad (5)$$

---

[1] K^* will denote the set of non-zero elements of a field K. In general, for any integral domain R, R^+ and R^* will denote the additive group and the unit group (that is, the multiplicative group of invertible elements of R).

in $\underline{x} \in \mathbf{Z}^m$ is called *norm form equation* (over $\mathbf{Z}$). If in particular $m = 2$ and $n \geq 3$, then (5) is just a Thue equation.

For $m = n$, (5) is a generalization of the Pell-equation and then (5) can be completely solved (cf. [7]). For $m < n$, the problem is much more difficult. Let V denote the $\mathbf{Q}$-vector space generated by $\alpha_1, \ldots, \alpha_m$. By means of his powerful subspace theorem Schmidt [53] proved in 1971 the following general finiteness criterion.

Theorem E (Schmidt [53]). *The following two statements are equivalent:*

(i) *V has no subspace of the form $\mu K'$ where $\mu \in K^*$ and K' is a subfield of K different from $\mathbf{Q}$ and the imaginary quadratic fields;*

(ii) *(5) has finitely many solutions $\underline{x}$ in $\mathbf{Z}^m$ for all $b \in \mathbf{Q}^*$.*

For $m = 2$, Schmidt's theorem reduces to Thue's theorem. Later Schmidt [54] proved a more general theorem by showing that all solutions of an arbitrary norm form equation over $\mathbf{Z}$ belong to finitely many so-called families of solutions. In 1977, Schlickewei [51] generalized certain weaker versions of the results of Mahler and Schmidt. His results imply that if V has no subspace of the form $\mu K'$ with $\mu \in K^*$ and a subfield K' of K such that $K' \neq \mathbf{Q}$ then (5) has only finitely many solutions in every finitely generated subring of $\mathbf{Q}$. A further generalization has been recently obtained by Laurent [45] to the case when the ground ring is an arbitrary finitely generated integral domain over $\mathbf{Z}$.

The above-mentioned finiteness results concerning norm form equations are all ineffective. In 1978, Győry and Papp [35] succeeded in obtaining, as an immediate consequence of a more general result (cf. [35], Theorem 3), the following effective finiteness theorem for norm form equations. They used Baker's method concerning linear forms in logarithms of algebraic numbers.

Theorem F (Győry and Papp [35]). *Suppose that in (5) (i') α_{j+1} has degree ≥ 3 over $\mathbf{Q}(\alpha_1, \ldots, \alpha_j)$ for $j = 1, \ldots, m - 1$. Then all solutions $\underline{x} = (x_1, \ldots, x_m) \in \mathbf{Z}^m$ of (5) satisfy $\max(|x_1|, \ldots, |x_m|) \leq C_4$, where C_4 is an effectively computable number.*

For $m = 2$, Theorem F (with the explicit C_4) reduces to Baker's Theorem B with another bound. In (i'), the lower bound 3 for the degrees cannot be diminished in general. Condition (i') is, however, stronger than (i) in Theorem E, that is Theorem F did not make Schmidt's result effective. By a recent conjecture of Mignotte, Schmidt's theorem cannot be made effective in full generality.

Later another effective result on norm form equations was obtained independently by Kotov [41], [42] and Győry [29] which is not implied, even in ineffective form, by Schmidt's theorem.

Theorem G (Győry [29], Kotov[2] [42]). *Suppose that in* (5) α_m *is of degree* ≥ 3 *over* $\mathbf{Q}(\alpha_1, \ldots, \alpha_{m-1})$. *Then all solutions* $\underline{x} = (x_1, \ldots, x_m) \in \mathbf{Z}^m$ *of* (5) *with* $x_m \neq 0$ *satisfy* $\max(|x_1|, \ldots, |x_m|) \leq C_5$, *where* C_5 *is an effectively computable number.*

The restriction $x_m \neq 0$ and the condition concerning the degree of α_m cannot be dropped in general. We remark that Theorem F can be deduced from Theorem G.

Theorems F and G were established in [35], [29], [41] in the relative case. For generalizations to the case of finitely generated ground rings in number fields see Győry [27], [30] and Kotov [40], [41], and in arbitrary finitely generated extensions of $\mathbf{Q}$ see Győry [31], [33]. Under the assumptions of Theorems F and G, respectively, upper bounds for the numbers of solutions, independent of $\alpha_1, \ldots, \alpha_m$, were derived by Evertse and Győry [14].

Discriminant form and index form equations

We shall use the same notation as before. In particular, $L(\underline{X}) = \alpha_1 X_1 + \ldots + \alpha_m X_m$ and $L^{(1)}(\underline{X}), \ldots, L^{(n)}(\underline{X})$ are defined by (4). Here we do not assume, however, that $m \geq 2$ and $\alpha_1 = 1$. Then

$$D(\underline{X}) = D(\alpha_1 X_1 + \ldots + \alpha_m X_m) = \prod_{1 \leq i < j \leq m} \left(L^{(i)}(\underline{X}) - L^{(j)}(\underline{X}) \right)^2$$

is a decomposable form of degree $n(n-1)$ with coefficients in $\mathbf{Q}$ which is called *discriminant form* (cf. Kronecker [43], Hensel [38]). The equations of the type

$$D(\underline{x}) = b \tag{6}$$

in $\underline{x} \in \mathbf{Z}^m$, named *discriminant form equations* (over $\mathbf{Z}$), play an important rôle in algebraic number theory. After several special results, in 1976 the following general and effective finiteness criterion was established by Győry [21].

Theorem H (Győry [21]). *The following two statements are equivalent:*

---

[2] Kotov [42] made the stronger hypothesis that α_m is of degree ≥ 5 over $\mathbf{Q}(\alpha_1 \ldots, \alpha_{m-1})$.

(i) $1, \alpha_1, \ldots, \alpha_m$ are linearly independent over $\mathbf{Q}$;

(ii) *(6) has only finitely many solutions in $\underline{x} \in \mathbf{Z}^m$ for every $b \in \mathbf{Q}^*$.*
Further, if (i) holds, then all solutions $\underline{x} = (x_1 \ldots, x_m) \in \mathbf{Z}^m$ of (6)
satisfy $\max(|x_1|, \ldots, |x_m|) \leq C_6$, where C_6 effectively computable.

In fact, in [21] a more general result was proved which asserts that
all solutions of (6) belong to finitely many so-called families of solutions
and all these can be effectively found.

Of particular importance is the special case when $m = n - 1$ and
$\{\alpha_0 = 1, \alpha_1, \ldots, \alpha_{n-1}\}$ is a $\mathbf{Z}$-basis of O_K, the ring of integers of K.
Then Theorem H implies that up to the obvious translation by elements
of $\mathbf{Z}$, the equation

$$D(\alpha) = b \quad \text{in } \alpha \in O_K$$

has only finitely many solutions and all these can be effectively determined. This finiteness assertion was earlier proved by Birch and Merriman [5] in a non-effective form and, independently, by Győry [20] in an
effective form.

If $\alpha \in O_K$ and if $\alpha = x_0 + x_1\alpha_1 + \ldots + x_{n-1}\alpha_{n-1}$ is the representation
of α with $x_0, \ldots, x_{n-1} \in \mathbf{Z}$ then it is easy to verify that

$$D(\alpha_1 x_1 + \ldots + \alpha_{n-1} x_{n-1}) = I^2(x_1, \ldots, x_{n-1}) D_K \qquad (7)$$

where D_K denotes the discriminant of K and $I(X_1, \ldots, X_{n-1})$ is a decomposable form of degree $n(n-1)/2$ with coefficients in $\mathbf{Z}$. Further,
the index of α in O_K, defined by

$$I(\alpha) = [O_K^+ : (\mathbf{Z}[\alpha])^+] \quad \text{is equal to } |I(x_1, \ldots, x_{n-1})|,$$

see for example [28]. For that reason, $I(X_1, \ldots, X_{n-1})$ is called the *index
form* of the integral basis $\{\alpha_0, \ldots, \alpha_{n-1}\}$ in question, and the equations
of the type

$$I(x_1, \ldots x_{n-1}) = \pm b \quad \text{in } x_1, \ldots, x_{n-1} \in \mathbf{Z} \qquad (8)$$

are called *index form equations.* For cubic number fields, index form
equations were earlier extensively studied by Nagell [47], Delone and
Faddeev [10] and others. For further references, see [28]. In view of (7),
(8) can be reduced to a discriminant form equation. As a consequence
of his Theorem H, Győry obtained, in 1976, the following result.

Theorem I (Győry [21]). *All solutions $\underline{x} = (x_1, \ldots, x_{n-1}) \in \mathbf{Z}^{n-1}$ of (8)
satisfy $\max(|x_1|, \ldots, |x_{n-1}|) \leq C_7$, where C_7 is effectively computable.*

Of particular interest is the special case $b = \pm 1$ when (8) is equivalent to the equation

$$I(\alpha) = 1 \quad (\alpha \in O_K) \iff O_K = \mathbf{Z}[\alpha]$$
$$\iff \{1, \alpha, \ldots, \alpha^{n-1}\} \text{ is an integral basis.} \qquad (9)$$

The existence of such a power integral basis considerably facilitates the calculation in O_K and the study of arithmetical properties of O_K. It is known that, for example, quadratic and cyclotomic fields have such integral bases, but this is not the case in general. If α is a solution of (9), then so is $\alpha + a$ for all $a \in \mathbf{Z}$. It follows from Theorem I that, up to translation by elements of $\mathbf{Z}$, (9) has only finitely many solutions and all these can be effectively determined; cf. Győry [21].

For further applications of Theorems H and I see Győry [28]. Theorems H, I and their consequences mentioned above were later extended by Győry [22], [23], Trelina [71], [72] and Győry and Papp [36] to the relative and p-adic case, and recently by Győry [31], [33], [34] to the case of arbitrary finitely generated ground domains.

We derived in [14] explicit upper bounds for the numbers of solutions of (6) and (8) which are independent of the coefficients of the forms involved. As a consequence we showed that up to the translation by elements of $\mathbf{Z}$, the number of solutions of (9) is at most $2(4 \times 7^{3g})^{n-2}$ where g denotes the degree of the normal closure of $K/\mathbf{Q}$ (hence $n \leq g \leq n!$).

Decomposable form equations of general type

We return now to the general decomposable form equation (1). Under various restrictive conditions made for F, Schmidt [53], [55], [56] and Schlickewei [50], [52] obtained ineffective finiteness results for certain other special types of decomposable form equations. A system of linear forms with coefficients in $\overline{\mathbf{Q}}$ is called *symmetric* (cf. [53]) if every form in the system occurs as often among the forms as its complex conjugate. In (1) the system of linear factors of F over $\overline{\mathbf{Q}}$ can be chosen to be symmetric. The following theorem can be deduced from a more general result of Schmidt (cf. [53], Satz 1).

Theorem J (Schmidt [53]). *Suppose that $F(\underline{x}) \neq 0$ for all $\underline{0} \neq \underline{x} \in \mathbf{Z}^m$. Then the following two statements are equivalent:*

(i) *For every subspace V of $\mathbf{Q}^m$ of dimension $d \geq 1$ and for every symmetric subsystem φ of the linear factors of F over $\overline{\mathbf{Q}}$, the rank of φ on V is greater than*

$$\min\{dt/n, d-1\}$$

where $n = \deg(F)$ and t is the number of the linear forms in φ;

(ii) *For every $b \in \mathbf{Q}^*$, equation (1) has only finitely many solutions in $\underline{x} \in \mathbf{Z}^m$.*

Theorem J implies Theorem E. The condition $F(\underline{x}) \neq 0$ does not hold however for discriminant forms and index forms, hence the criterion above does not apply to decomposable form equations in full generality. This result of Schmidt was later extended by Schlickewei [52] to the case of finitely generated ground rings in $\mathbf{Q}$.

We shall now present a general finiteness criterion which guarantees the finiteness of the number of solutions of (1) for every $b \in \mathbf{Q}^*$ and every finitely generated subring R of $\mathbf{Q}$. Let G be the splitting field of F (i.e. the smallest extension of $\mathbf{Q}$ over which F factorizes into linear forms), and let $\mathcal{L}_0$ be a maximal set of pairwise linearly independent linear factors of F with coefficients in G. For every subspace V of $\mathbf{Q}^m$ of dimension ≥ 1, we denote by $r(V, \mathcal{L}_0)$ the minimum of all integers r for which there exist linear forms $L_{i_1}, \ldots, L_{i_r}$ in $\mathcal{L}_0$ whose restrictions to V are linearly dependent but pairwise linearly independent. If this minimum exists then $r(V, \mathcal{L}_0) \geq 3$. Otherwise we put $r(V, \mathcal{L}_0) = 2$. Let $\mathcal{L} \supset \mathcal{L}_0$ be a finite set of linear forms in $X_1, \ldots, X_m$ with coefficients in G. A subspace V of $\mathbf{Q}^m$ is called $\mathcal{L}$-*admissible* if no form in $\mathcal{L}$ vanishes identically on V.

Theorem K (Evertse and Győry [15]). *The following two statements are equivalent:*

(i) *For every $\mathcal{L}$-admissible subspace V of $\mathbf{Q}^m$ of dimension ≥ 2, we have $r(V, \mathcal{L}_0) \geq 3$;*

(ii) *For every $b \in \mathbf{Q}^*$ and every subring R of $\mathbf{Q}$ which is finitely generated over $\mathbf{Z}$, the equation*

$$F(\underline{x}) = b \text{ in } \underline{x} \in R^m \text{ with } L(\underline{x}) \neq 0 \text{ for all } L \in \mathcal{L} \setminus \mathcal{L}_0 \qquad (1')$$

has only finitely many solutions.

Further, we showed in [15] that for every $b \in \mathbf{Q}^*$ and for every finitely generated subring R of $\mathbf{Q}$, all solutions of (1') belong to finitely many $\mathcal{L}$-admissible subspaces V of $\mathbf{Q}^m$ with $r(V, \mathcal{L}_0) = 2$. Since every subspace of $\mathbf{Q}^m$ of dimension 1 can contain only finitely many solutions of (1'), the implication (i) $\Longrightarrow$ (ii) in Theorem K is an immediate consequence of this latter finiteness assertion. In [15] we proved these results in a more general form, over finitely generated subrings of an arbitrary finitely generated extension of $\mathbf{Q}$.

In the special case $\mathcal{L} = \mathcal{L}_0$ equation (1') reduces to equation (1) and Theorem K provides a finiteness criterion for (1). Theorem K implies, in an ineffective form, the finiteness assertions of Theorems A, B, E, F, G, H and I (cf. [15]). The finiteness result quoted after Theorem K, and therefore the implication (i) $\Longrightarrow$ (ii) in Theorem K can be deduced from the following finiteness theorem on S-unit equations which was established independently by van der Poorten and Schlickewei [49] and Evertse [13].

Let K be an algebraic number field, Γ a finitely generated subgroup of K^*, and $m \geq 2$ an integer. Then the equation

$$u_1 + u_2 + \ldots + u_m = 1 \quad in \ u_1, \ldots, u_m \in \Gamma$$

has only finitely many solutions such that $\sum_{i \in I} u_i \neq 0$ for each non-empty subset I of $\{1, 2, \ldots, m\}$.

In [15] (see also [16]) it has been pointed out that the implication (i) $\Longrightarrow$ (ii) of Theorem K is in fact equivalent to the above theorem on S-unit equations. Since this latter theorem has been deduced from the Schmidt-Schlickewei subspace theorem which is ineffective, Theorem K is also ineffective. Moreover, if Mignotte's conjecture is true, then it cannot be made effective in full generality. There does exist, however, an algorithm to decide whether condition (i) in Theorem K holds (cf. [15]).

For decomposable form equations of general type the first general effective finiteness result was obtained by Győry and Papp [35] in 1978. Later, their result was improved and generalized by Győry [29], [30] to Theorem L stated below. In the remainder of this section, let R be an arbitrary but fixed finitely generated subring of $\mathbf{Q}$ over $\mathbf{Z}$. Then $R = \mathbf{Z}\left[\frac{1}{p_1}, \ldots, \frac{1}{p_s}\right]$ with appropriate rational primes $p_1, \ldots, p_s$ ($s \geq 0$). For every $a \in \mathbf{Q}$ with $a = k/l$; $k, l \in \mathbf{Z}$, $(k, l) = 1$ we put $h(a) = \max(|k|, |l|)$.

Theorem L (Győry [30]). *Suppose that*

(i) *$\mathcal{L}_0$ has rank m;*

(ii) *$\mathcal{L}_0$ can be divided into subsets $\mathcal{L}_1, \ldots, \mathcal{L}_h$ with the following properties: if $\mathrm{Card}(\mathcal{L}_j) \geq 2$ for some j, then for each r, r' with $L_r, L_{r'} \in \mathcal{L}_j$, there exists a sequence $L_r = L_{r_1}, \ldots, L_{r_t} = L_{r'}$ in*

$\mathcal{L}_j$ such that, for $q = 1, \ldots, t - 1$, L_{r_q}, $L_{r_{q+1}}$ has a linear combination with coefficients in G^* which belongs to $\mathcal{L}_j$;

(iii) $\mathcal{L} = \mathcal{L}_0$ or $\mathcal{L} = \mathcal{L}_0 \cup \{X_k\}$ for some $k \in \{1, \ldots, m\}$ according as $h = 1$ or $h > 1$;

(iv) If $h > 1$, then X_k can be expressed as a linear combination of the forms from $\mathcal{L}_j$ for every $j \in \{1, \ldots, h\}$.

Then all solutions $\underline{x} = (x_1, \ldots, x_m)$ of $(1')$ satisfy $\max(h(x_1), \ldots, h(x_m)) \leq C_8$, where C_8 is effectively computable.

If $h = 1$ (this is the case e.g. for Thue equations), conditions (iii) and (iv) can be obviously dropped and Theorem L provides an effective finiteness result for (1). The discriminant forms, binary forms considered in Theorem B, and norm forms considered in Theorem G all satisfy the conditions of Theorem L. Therefore Theorem L implies (with the explicit C_8) Theorems B, C, F, G, H and I (cf. [30]). For extensions of Theorem L to the case of arbitrary ground rings which are finitely generated over $\mathbf{Z}$ we refer to Győry [31], [33]. Apart from the forms of the bounds, these general versions imply all the above-mentioned effective finiteness results for decomposable form equations (cf. [31], [33]). The proofs involve among other things Baker's method, the analogues over function fields of the results in question and some effective specialization argument.

In [14] we have recently shown that conditions (i), (ii), (iii), (iv) of Theorem L together imply the following condition

(i*) For every $\mathcal{L}$-admissible subspace V of $\mathbf{Q}^m$ of dimension ≥ 2 we have $r(V, \mathcal{L}_0) = 3$.

Since the number of subspaces V under consideration is in general infinite, it is hard to decide whether condition (i*) is satisfied or not. Let again $\mathcal{L}$ be as in Theorem K. We shall show that in Theorem L conditions (i) to (iv) can be replaced by a weaker version of (i*) which involves only finitely many and effectively determinable subspaces.

Theorem 1. *There is a finite, effectively determinable set of $\mathcal{L}$-admissible subspaces V of $\mathbf{Q}^m$ of dimension ≥ 2 such that if $r(V, \mathcal{L}_0) = 3$ for all V in this set, then all solutions $\underline{x} = (x_1, \ldots, x_m)$ of $(1')$ satisfy $\max(h(x_1), \ldots, h(x_m)) \leq C_9$, where C_9 is effectively computable.*

Theorem L is a consequence of Theorem 1. Further, Theorem 1 is easier to compare with Theorem K. We should, however, remark that in the most important special cases when $\mathcal{L} = \mathcal{L}_0 \cup \{X_k\}$ for some k, we do not know any equation to which Theorem 1 can be applied but Theorem L cannot. Furthermore, it is easier to check the more explicit conditions of Theorem L.

The proof of Theorem 1 will be based on an effective finiteness result of Győry [25] obtained for S-unit equations in two unknowns, which was proved by means of Baker's method and its p-adic analogue.

In what follows, we may suppose without loss of generality that in $(1')$, $b \in \mathbf{Z} \setminus \{0\}$. Under assumption (i^*), we derived in [14] the bound $n(4 \times 7^{g(2s+2\omega(b)+3)})^{m-1}$ for the number of solutions of $(1')$ where $n = \deg(F)$ and $g = [G : \mathbf{Q}]$. By using an upper bound of Evertse [12] established for the numbers of solutions of S-unit equations in two unknowns we shall here deduce almost the same bound subject to the weaker and effective assumption of Theorem 1.

Theorem 2. *There is a finite, effectively determinable set of $\mathcal{L}$-admissible subspaces V of $\mathbf{Q}^m$ of dimension ≥ 2 such that if $r(V, \mathcal{L}_0) = 3$ for all V in this set, then the number of solutions of $(1')$ is at most*

$$n(3 \times 7^{g(2s+2\omega(b)+3)})^{m-1}.$$

We note that $g \leq n!$. From Theorem 2 one can easily obtain bounds of a similar type for the numbers of solutions of the Thue equations, Thue-Mahler equations (cf. Theorem D), discriminant form and index form equations, and the norm form equations considered in Theorems F and G.

In §§2 and 3, we shall state and prove Theorems 1 and 2 in a more precise and more general form, for equations considered over the rings of S-integers of algebraic number fields.

§2. Some new results

Let K be an algebraic number field of degree d. Denote by M_K the set of places (that is, equivalence classes of multiplicative valuations) on K. Places in M_K are called *finite* if they contain non-archimedean valuations, and *infinite* otherwise. The field K has at most d infinite places. In every place p on $\mathbf{Q}$ we choose a valuation $| \cdot |_p$ normalized in the usual way (for elementary properties of places and heights which will be used in §§2 and 3 we refer to [16], §§1, 2, in this volume). Further, in every place v on K we normalize a valuation $| \cdot |_v$ in the following way: if v lies above $p \in M_{\mathbf{Q}}$ and if $\mathbf{Q}_p$, K_v denote the completions of $\mathbf{Q}$ at p and K at v respectively, then we choose $| \cdot |_v$ such that $|\alpha|_v = |\alpha|_p^{[K_v:\mathbf{Q}_p]/d}$ for each $\alpha \in \mathbf{Q}$. The set of valuations thus normalized satisfies the product formula.

Let S be a finite subset of M_K with cardinality s which contains all infinite places. Suppose that the finite places in S lie above rational primes not exceeding $P(\geq 2)$. By O_S we shall denote the ring

$$\{\alpha \in K : |\alpha|_v \leq 1 \text{ for all } v \in M_K \setminus S\}.$$

The elements of O_S and O_S^* are called *S-integers* and *S-units*, respectively. If S consists only of the infinite places, then O_S is just the ring O_K of integers in K. We note that the ring O_S is finitely generated over $\mathbf{Z}$ and every subring R of K which is finitely generated over $\mathbf{Z}$ is a subring of such a ring O_S. Moreover, if in particular $K = \mathbf{Q}$ then $R = O_S$ for an appropriate finite subset S of $M_{\mathbf{Q}}$.

For any integer $t \geq 1$, we define the *height* of $\underline{\alpha} = (\alpha_1,\ldots,\alpha_t) \in K^t$ by

$$h(\underline{\alpha}) = \prod_{v \in M_K} \max\{1, \max_{1 \leq j \leq t} |\alpha_j|_v\}.$$

In particular,

$$h(\alpha) = \prod_{v \in M_K} \max\{1, |\alpha|_v\}$$

will denote the height of $\alpha \in K$. For every positive number C there are only finitely many $\underline{\alpha}$ in K^t with $h(\underline{\alpha}) \leq C$ and these belong to an effectively determinable subset of K^t (cf. (11), (12)). We define the height of a polynomial

$$P(X_1,\ldots,X_t) = \sum_{i_1,\ldots,i_t} a(i_1,\ldots,i_t)X_1^{i_1}\ldots X_t^{i_t}$$

$$\text{in } K[X_1,\ldots,X_t]$$

by

$$h(P) = \prod_{v \in M_K} \max\{1, \max_{i_1,\ldots,i_t} |a(i_1,\ldots,i_t)|_v\}.$$

The heights $h(\underline{\alpha})$, $h(\alpha)$ and $h(P)$ depend only on $\underline{\alpha}$, α and P, respectively, and not on the choice of the number field K.

Let $F(\underline{X}) = F(X_1,\ldots,X_m) \in K[X_1,\ldots,X_m]$ be a decomposable form of degree $n \geq 3$ in $m \geq 2$ variables with height $\leq A$ which factorizes into linear factors over a finite extension G of K. Let $g = [G : K]$ and let D_G denote the absolute value of the discriminant of G (over $\mathbf{Q}$). Let $\mathcal{L}_0$ be a maximal set of pairwise linearly independent linear factors of F over G and let $\mathcal{L} \supseteq \mathcal{L}_0$ be a finite set of linear forms in $G[X_1,\ldots,X_m]$. For any subspace V of K^m of dimension ≥ 1 we define $r(V,\mathcal{L}_0)$ in the same way as in §1. Similarly, we shall say that V is $\mathcal{L}$-*admissible* if no form in $\mathcal{L}$ vanishes identically on V. For given $C \geq 1$, there are only finitely many linear forms $L \in K[X_1, \ldots, X_m]$ with $h(L) \leq C$. We shall denote by $\mathcal{W}(K, m, C)$ the collection of subspaces V of K^m of the type

$$V = \{\underline{x} \in K^m : L_1(\underline{x}) = L_2(\underline{x}) = \ldots = L_r(\underline{x}) = 0\}$$

where r can be any integer with $1 \leq r \leq m-1$, and $\{L_1, \ldots, L_r\}$ can be any set of linear forms from $K[X_1, \ldots, X_m]$ with heights $\leq C$. The set $\mathcal{W}(K, m, C)$ is finite and $K^m \in \mathcal{W}(K, m, C)$. Let $\beta \in K^*$ with height $\leq B$ and consider the equation

$$F(\underline{x}) = \beta \text{ in } \underline{x} \in O_S^m \text{ with } L(\underline{x}) \neq 0 \text{ for all } L \in \mathcal{L} \setminus \mathcal{L}_0. \qquad (10)$$

Theorem 1'. *There are effectively computable numbers C_1, C_2 depending only on d, g, D_G, s, P, n, A, m and B with the following property : if*

(i') $r(V, \mathcal{L}_0) = 3$ *for every $\mathcal{L}$-admissible subspace V of K^m of dimension ≥ 2 which belongs to $\mathcal{W}(K, m, C_1)$,*

then all solutions $\underline{x}$ of (10) satisfy $h(\underline{x}) \leq C_2$.

Theorem 1' implies that there are finitely many $\mathcal{L}$-admissible subspaces V of K^m of dimension ≥ 2 such that if $r(V, \mathcal{L}_0) = 3$ for all of these V then (10) has only finitely many solutions. Moreover, if K, S, β, G, n and the coefficients of F and of the linear forms in $\mathcal{L}$ are effectively given in the sense defined in [69] and [34], p. 59, then both the subspaces V in question and the solutions of (10) can be effectively determined. In the special case $K = \mathbf{Q}$, this gives Theorem 1 stated in §1.

There are only finitely many $v \in M_K \setminus S$ for which $|\beta|_v \neq 1$ or F has a coefficient with v-value > 1. In the sequel the number of these v will be denoted by $\omega_S(\beta, F)$.

Theorem 2'. *If the condition (i') holds with the C_1 specified in Theorem 1', then the number of solutions of (10) is at most*

$$n(3 \times 7^{g(d+2s+2\omega_S(\beta,F))})^{m-1}.$$

In the case $K = \mathbf{Q}$, Theorem 2' gives Theorem 2 formulated in §1.

As was mentioned in §1, the conditions (i) to (iv) of Theorem L together imply the assumption (i') of Theorems 1' and 2'. Therefore Theorem 1' provides, as a consequence, a generalization of Theorem

L to equations over O_S. Such a generalization was earlier proved by Győry [30] with an explicit upper bound for the heights of the solutions. It implied more general versions over O_S of Theorems C, F, G, H, I presented in §1 (cf. Győry [30]). This means that apart from the forms of the bounds, these more general versions of Theorems C, F, G, H, I can also be deduced from Theorem 1′. Similarly, Theorem 2′ implies (with slightly different bounds) the results of Evertse [12] and Evertse and Győry [14] on Thue equations, discriminant form equations and index form equations over S-integers of number fields.

§3. Proofs

We shall keep the notation of §2. It is important to note that if $H(\alpha)$ denotes the maximum absolute value of the coefficients of the minimal polynomial of an algebraic number α over $\mathbf{Z}$ and if d is the degree of α, then $H(\alpha)$ and $h(\alpha)$ (called sometimes the usual and absolute height of α, respectively) are related by

$$(d+1)^{1/2}(h(\alpha)) \leq H(\alpha) \leq 2^d(h(\alpha))^d \tag{11}$$

(cf. [44], Ch. 3, p. 54 and Theorem 2.8). Further, if $\alpha = a/b \in \mathbf{Q}$ with $a, b \in \mathbf{Z}$ and $(a, b) = 1$ then

$$h(\alpha) = H(\alpha) = \max(|a|, |b|).$$

Let M_G be the set of places on G and suppose that the valuation $|\cdot|_w$ in $w \in M_G$ is normalized in the same way as was indicated in §2. The height function can be extended to $q \times t$ matrices with entries in G. Let $\underline{A} = (\alpha_{jk})$ be such a matrix. Put

$$h(\underline{A}) = \prod_{w \in M_G} \max\{1, \max_{\substack{1 \leq j \leq q \\ 1 \leq k \leq t}} |\alpha_{jk}|_w\}.$$

The following lemma states a few elementary properties of height functions. Let $\mathcal{M}_{q,t}(G)$ denote the set of $q \times t$ matrices with entries in G.

Lemma 1. (i) If $\underline{A} = (\alpha_{jk}) \in \mathcal{M}_{q,t}(G)$ then

$$\max_{j,k} h(\alpha_{jk}) \leq h(\underline{A}) \leq \prod_{j,k} h(\alpha_{jk}). \tag{12}$$

(ii) *If $\underline{A} \in \mathcal{M}_{q,t}(G)$, $\underline{B} \in \mathcal{M}_{t,r}(G)$, $\alpha \in G^*$ then*

$$h(\underline{A}\,\underline{B}) \leq t h(\underline{A}) h(\underline{B}), \quad h(\alpha \underline{A}) \leq h(\alpha) h(\underline{A}) \tag{13}$$

(iii) *If $\underline{A} \in \mathcal{M}_{q,q}(G)$ and $\underline{A}$ is invertible then*

$$h(\underline{A}^{-1}) \leq h(\det \underline{A})(q-1)! h(\underline{A})^{q-1} \leq (q!)^2 h(\underline{A})^{2q-1}. \tag{14}$$

(iv) *Let $\omega_1, \ldots, \omega_q$ be K-linearly independent elements of G. There exist effectively computable numbers c_1 and c_2, depending only on q, such that for every $\gamma = \xi_1 \omega_1 + \ldots + \xi_q \omega_q$ with $\xi_1, \ldots, \xi_q \in K$ we have*

$$h(\xi_1, \ldots, \xi_q) \leq c_1 \big(h(\omega_1, \ldots, \omega_q) h(\gamma) \big)^{c_2}. \tag{15}$$

Proof. (i). Straightforward consequence of the definitions of the heights.

(ii), (iii). Straightforward application of the inequality

$$|\alpha_1 + \ldots + \alpha_t|_w \leq t \max\{|\alpha_1|_w, \ldots, |\alpha_t|_w\}$$
$$\text{for } \alpha_1, \ldots, \alpha_t \in G \text{ and } w \in M_G.$$

(iv). Let $\sigma_1 \ldots, \sigma_q$ be distinct K-isomorphisms of G for which $\Omega = \big(\sigma_j(\omega_k)\big)$, with $1 \leq j, k \leq q$, is invertible. Let[3] $\underline{x} = (\xi_1, \ldots, \xi_q)^T$, $\underline{b} = \big(\sigma_1(\gamma), \ldots, \sigma_q(\gamma)\big)^T$. Then $\underline{b} = \Omega \underline{x}$. Since $h(\sigma_j(\omega_k)) = h(\omega_k)$ for $1 \leq j, k \leq q$, we have $h(\Omega) \leq h(\omega_1, \ldots, \omega_q)^{c_4}$ where $c_4 = c_4(q)$ is effectively computable in terms of q. Now (15) follows by applying first (14) to Ω and then (13) to $\underline{x} = \Omega^{-1} \underline{b}$. ■

For every polynomial $P \in G[X_1, \ldots, X_m]$ and for every $w \in M_G$ we denote by $|P|_w$ the maximum of the w-values of the coefficients of P. It is known (cf. [44]) that if w is a finite place then for every $P, Q \in G[X_1, \ldots, X_m]$

$$|PQ|_w = |P|_w \cdot |Q|_w. \tag{16}$$

Put

$$h^*(P) = \prod_{w \in M_G} |P|_w.$$

Then, by the product formula, $h^*(\alpha P) = h^*(P)$ for any $\alpha \in G^*$ and

$$1 \leq h^*(P) \leq h(P). \tag{17}$$

---

[3] We denote by $\underline{B}^T$ the transposed matrix of a matrix $\underline{B}$.

Further, if at least one of the coefficients of P is equal to 1 then $h^*(P) = h(P)$.

Lemma 2. *Let $P, Q \in G[X_1, \ldots, X_m]$. Suppose that at least one of the coefficients of P is equal to 1 and that PQ has degree $\leq n$. Then*

$$h(P) \leq 4^{dg(n+1)^m} h(PQ).$$

Proof. We have

$$4^{-dg(n+1)^m} h^*(PQ) \leq h^*(P)h^*(Q)$$
$$\leq 4^{dg(n+1)^m} h^*(PQ) \tag{18}$$

(cf. [44], Ch. 3, Prop. 2.4). Now Lemma 2 follows from (17) and (18).

Let S' be the subset of M_G lying above the places in S, and let $O_{S'}$ be the ring of S'-integers in G. Then $O_{S'}$ contains as a subring the ring O_G of integers of G. We define the S'-*norm* by

$$N_{S'}(\alpha) = \left(\prod_{w \in S'} |\alpha|_w \right)^{dg} \qquad \text{for } \alpha \in G^*,$$

where $dg = [G : \mathbf{Q}]$. It is easily seen that the S'-norm is multiplicative. Further, if $\alpha \in O_{S'}$ then

$$1 \leq N_{S'}(\alpha) \leq \left(h(\alpha) \right)^{dg}.$$

The proof of Theorem 1′ is based on the following lemma which is an easy consequence of an effective result of Győry [25] on homogeneous S'-unit equations in three unknowns. We note that Győry [25] gave an explicit bound for the heights of the solutions. Following the arguments of the proofs below of Theorems 1′ and 2′ and using the explicit bound mentioned, it would not be difficult to derive explicit values for C_1 and C_2 in Theorems 1′ and 2′.

Lemma 3. *Let $N \geq 1$ and let $\gamma_1, \gamma_2, \gamma_3 \in O_G \setminus \{0\}$ with heights $\leq \Gamma$. Then all solutions of the equation*

$$\gamma_1 x_1 + \gamma_2 x_2 + \gamma_3 x_3 = 0 \text{ in } x_1, x_2, x_3 \in O_{S'} \setminus \{0\}$$
$$\text{with } \max_{1 \leq i \leq 3} N_{S'}(x_i) \leq N \tag{19}$$

satisfy $h(x_1/x_2) \leq c_5 N^{c_6}$ where c_5, c_6 are effectively computable positive numbers depending only on d, g, D_G, s, P and Γ.

In fact this lemma can be found, in a more explicit form, in the work [32] of Győry. Since [32] was written in Hungarian, we shall give here a complete proof.

Proof. Let $\wp_1, \ldots, \wp_{s'}$, be the prime ideals in O_G corresponding to the finite places in S'. Clearly $s' \leq g \cdot s$. Since $x_i \in O_{S'}$, the ideal (x_i) generated by x_i can be written as $v_i \wp_1^{a_{i1}} \ldots \wp_{s'}^{a_{is'}}$, where v_i is an integral ideal in G coprime with $\wp_1, \ldots, \wp_{s'}$, and $a_{i1}, \ldots, a_{is'}$ are non-negative rational integers. Obviously $N(v_i) \leq N$. Let h_G and R_G be the class number and regulator of G. Then

$$\max\{h_G, R_G\} < c_7(d, g, D_G),$$

where c_7 is effectively computable in terms of d, g, D_G (cf. [61]). Let b_{ij} be rational integers such that $b_{ij} \equiv a_{ij} \pmod{h_G}$ and $0 \leq b_{ij} < h_G$ for each i and j. Then $v_i \wp_1^{b_{i1}} \ldots \wp_{s'}^{b_{is'}}$ is a principal ideal, say (y_i), with $y_i \in O_G$ and $|N_{G/\mathbb{Q}}(y_i)| \leq c_8 N$, $i = 1, 2, 3$. Here c_8 and c_9, c_{10} below denote effectively computable positive numbers depending only on d, g, D_G, s, P and Γ. We have $x_i = y_i \delta_i$ with some $\delta_i \in O_{S'}^*$, $i = 1, 2, 3$. Putting $\wp_j^{h_G} = (\pi_j)$ with appropriate $\pi_j \in O_G$ for $j = 1, \ldots, s'$, there are positive integers $b_1, \ldots, b_{s'}$ such that $\rho_i := \delta_i \pi_1^{b_i} \ldots \pi_{s'}^{b_{s'}} \in O_G \cap O_{S'}^*$ for $i = 1, 2, 3$. For $x_i' := \pi_1^{b_1} \ldots \pi_{s'}^{b_{s'}} x_i = y_i \rho_i$ we have

$$\gamma_1 x_1' + \gamma_2 x_2' + \gamma_3 x_3' = 0$$

and Lemma 6 of Győry [25] gives

$$h(x_1/x_2) = h(x_1'/x_2') \leq c_9 N^{c_{10}}. \qquad \blacksquare$$

Denote by T the smallest subset of M_K containing S such that both $|\beta|_v = 1$ and the v-values of all coefficients of F are ≤ 1 for all $v \in M_K \setminus T$. Further, let T' be the subset of places of M_G lying above the places in T, $O_{T'}$ the ring of T'-integers in G and t' the cardinality of T'. Then $t' \leq g(s + \omega_S(\beta, F))$. Furthermore, we have $\beta \in O_{T'}^*$, $F \in O_{T'}[X_1, \ldots, X_m]$ and $O_S \subseteq O_{T'}$.

The main tool in the proof of Theorem 2' will be the following result of Evertse [12].

Lemma 4. *For every* γ, $\delta \in G^*$ *the equation*

$$\gamma x + \delta y = 1 \qquad in \ x, y \in O_{T'}^*$$

has at most $3 \times 7^{dg+2t'}$ *solutions.*

Proof. This is Theorem 1 in [12].

Before proving our theorems we show that we can make certain assumptions without loss of generality. We may assume that the coefficient of X_1^n in F, say a_0, is different from zero. Indeed, there is a rational integer a with $0 \le a \le mn$ such that $F(1, a, \ldots, a^{m-1}) \ne 0$. On applying the linear transformation

$$X_i = a^{i-1} X_1' + X_i', \qquad i = 1, \ldots, m,$$

to F, $\mathcal{L}_0$, $\mathcal{L}$ and (10), all conditions of our theorems will be satisfied together with the assumption required.

Further, replacing the linear forms in $\mathcal{L}$ by appropriate proportional factors if necessary, we may choose the factorization

$$F(\underline{X}) = L_1(\underline{X}) \ldots L_n(\underline{X}) \tag{20}$$

of F into linear forms $L_1, \ldots, L_n$ from $G[X_1, \ldots, X_m]$ such that

$$\max_{1 \le j \le n} h(L_j) \le c_{11} \tag{21}$$

where c_{11} and c_{12}, c_{13} below are effectively computable positive numbers depending only on d, g, n, A and m. Indeed, we may choose the coefficients of X_1 in $L_1, \ldots, L_n$ to be $a_0, 1, \ldots, 1$, respectively. Then by Lemma 2,

$$A \ge h(F) \ge c_{12} \max\{h(L_1/a_0), h(L_2), \ldots, h(L_m)\}. \tag{22}$$

Further, a_0 is a coefficient of F so that $h(a_0) \le h(F)$. Hence, by (13), $h(L_1) \le h(a_0)h(L_1/a_0) \le c_{13}$. Together with (22) this proves (21).

Finally, we show that if (10) is solvable then there are $\mu_1, \ldots, \mu_n \in G^*$ such that

$$\mu_j L_j(\underline{x}) \in O_{T'}^* \ (j = 1, \ldots, n) \tag{23}$$
$$\text{for all solutions } \underline{x} \text{ of (10)}$$

(see also [14]). Indeed, let $w \in M_G \setminus T'$. Then for every solution $\underline{x}$ we have

$$|L_j(\underline{x})|_w \le |L_j|_w \quad \text{for } j = 1, \ldots, n,$$

and, by (16), $|\beta|_w = 1$ and $|F|_w \le 1$,

$$\prod_{j=1}^{n} (|L_j(\underline{x})|_w / |L_j|_w) = |F(\underline{x})|_w / |F|_w = |\beta|_w / |F|_w \ge 1.$$

Hence

$$|L_j(\underline{x})|_w = |L_j|_w \qquad \text{for } j = 1, \ldots, n. \tag{24}$$

Let $\underline{x}_0$ be a fixed solution of (10) and put $\mu_1 = \beta L_1(\underline{x}_0)^{-1}$ and $\mu_j = L_j(\underline{x}_0)^{-1}$ for $j = 2, \ldots, n$. Then, by (24), $|\mu_j L_j(\underline{x})|_w = 1$ ($j = 1, \ldots, n$) and (23) holds.

Both Theorem 1' and Theorem 2' are easy to deduce from the next lemma.

Lemma 5. *Let q be a rational integer with $0 \le q \le m - 2$, let $C_q^* \ge 1$, and let V be an $\mathcal{L}$-admissible subspace of K^m of dimension $m - q$ with $V \in \mathcal{W}(K, m, C_q^*)$ and $r(V, \mathcal{L}_0) = 3$. Then there exists an effectively computable number C_{q+1}^* $(\ge C_q^*)$ depending only on d, g, D_G, s, P, A, n, m, B and C_q^* such that all solutions $\underline{x} \in V \cap O_S^m$ of (10) are contained in at most $3 \times 7^{dg+2t'}$ $\mathcal{L}$-admissible subspaces of V with dimension $m-q-1$ which belong to $\mathcal{W}(K, m, C_{q+1}^*)$.*

Proof. In what follows, $c_{14}, \ldots, c_{19}$ will denote effectively computable positive numbers depending only on d, g, D_G, s, P, A, n, m, B, C_q^*. Further, for convenience we put $N = 3 \times 7^{dg+2t'}$.

Suppose that (10) has a solution in $V \cap O_S^m$. Consider a fixed factorization of the form (20) of F with the property (21) and fix some $\mu_1, \ldots, \mu_n \in G^*$ for which (23) holds. In view of (11) and (12), the height of the least common multiple, say a, of the denominators of the coefficients of $L_1, \ldots, L_n$ occurring in (2) is at most c_{14}. Let $\underline{x} \in V \cap O_S^m$ be an arbitrary but fixed solution of (10). Then $aL_j(\underline{x}) \in O_{S'}$ and by (10) and (20), we have

$$a^n \beta \in O_{S'} \text{ and } aL_j(\underline{x}) \text{ divides } a^n \beta \text{ in } O_{S'} \text{ for } j = 1, \ldots, n. \tag{25}$$

By assumption, among $L_1, \ldots, L_n$ there are three linear forms, say L_1, L_2, L_3, whose restrictions to V are linearly dependent but pairwise linearly independent. Therefore there exist $\gamma_1, \gamma_2, \gamma_3 \in G^*$ such that $\gamma_1 L_1(\underline{X}) + \gamma_2 L_2(\underline{X}) + \gamma_3 L_3(\underline{X}) = 0$ identically on V. In view of $V \in \mathcal{W}(K, m, C_q^*)$, $\gamma_1, \gamma_2, \gamma_3$ can be chosen from $O_G \setminus \{0\}$ with heights at most c_{15}. For the solution $\underline{x}$ under consideration we have

$$\gamma_1 \big(aL_1(\underline{x})\big) + \gamma_2 \big(aL_2(\underline{x})\big) + \gamma_3 \big(aL_3(\underline{x})\big) = 0. \tag{26}$$

Further, by (25)

$$N_{S'}\big(L_j(\underline{x})\big) \le N_{S'}(a^n\beta) \le \big(h(a^n\beta)\big)^{dg} \le c_{16}.$$

By applying now Lemma 3 to (26) we obtain

$$h\big(L_1(\underline{x})/L_2(\underline{x})\big) \le c_{17}.$$

On the other hand, it follows from (26) that

$$\left(-\frac{\gamma_1\mu_1^{-1}}{\gamma_2\mu_2^{-1}}\right)\left(\frac{\mu_1 L_1(\underline{x})}{\mu_2 L_2(\underline{x})}\right) + \left(-\frac{\gamma_3\mu_3^{-1}}{\gamma_2\mu_2^{-1}}\right)\left(\frac{\mu_3 L_3(\underline{x})}{\mu_2 L_2(\underline{x})}\right) = 1.$$

Together with Lemma 4 and (23), this implies that $\big(\mu_1 L_1(\underline{x})/\mu_2 L_2(\underline{x})\big)$ and hence $L_1(\underline{x})/L_2(\underline{x})$ belongs to a subset of G^* of cardinality at most N which does not depend on $\underline{x}$. Consequently, there exist at most N elements $\lambda \in G^*$ with heights $\le c_{17}$ such that every solution $\underline{x} \in V \cap O_S^m$ of (10) is a zero of at least one of the linear forms

$$L_\lambda(\underline{X}) = L_1(\underline{X}) - \lambda L_2(\underline{X}).$$

By the assumption made on L_1, L_2, none of the forms L_λ vanishes identically on V.

As is known (see e.g. [28], p. 71), O_G has an integral basis $\{\omega_1, \ldots, \omega_{dg}\}$ such that

$$\begin{aligned}
h(\omega_1, \ldots, \omega_{dg}) &\le h(\omega_1)\ldots h(\omega_{dg}) \\
&\le \big(|\overline{\omega_1}|\ldots|\overline{\omega_{dg}}|\big)^{dg} \le c_{18},
\end{aligned} \tag{27}$$

where $|\overline{\omega_i}|$ denotes the maximum of the absolute values of the conjugates of ω_i, $i = 1, \ldots, dg$. We may assume that $\omega_1, \ldots, \omega_g$ are K-linearly independent. Each L_λ considered above can be written as

$$L_\lambda(\underline{X}) = \sum_{j=1}^{g} \omega_j l_{\lambda,j}(\underline{X}) \tag{28}$$

with linear forms $l_{\lambda,j} \in K[X_1, \ldots, X_m]$ which do not all identically vanish on V. By using Lemma 1, (iv) and (27) we obtain from (28) that $h(l_{\lambda,j}) \le c_{19}$ for $j = 1, \ldots, g$. Further, $l_{\lambda,j}(\underline{x}) = 0$ $(j = 1, \ldots, g)$ for every $\underline{x} \in V$ with $L_\lambda(\underline{x}) = 0$. Thus we conclude that all solutions $\underline{x} \in V \cap O_S^m$ are contained in at most N $\mathcal{L}$-admissible subspaces of V of dimension

$m - q - 1$ which belong to $\mathcal{W}(K, m, C_{q+1}^*)$ for $C_{q+1}^* = \max(C_q^*, c_{19})$. This completes the proof of the lemma. ∎

Proof of Theorem 1'. If assumption (i') holds with $C_1 \geq 1$ then $r(K^m, \mathcal{L}_0) = 3$. Hence we can apply Lemma 5 successively with $q = 0, 1, \ldots, m - 2$ and with $C_0^* = 1$. Let $1 \leq C_1^* \leq \cdots \leq C_{m-2}^* \leq C_{m-1}^*$ be the corresponding effectively computable numbers, specified in Lemma 5, which depend now only on d, g, D_G, s P, A, n, m and B. Put $C_1 = C_{m-2}^*$ and suppose that $r(V, \mathcal{L}_0) = 3$ for every $\mathcal{L}$-admissible subspace V of K^m of dimension ≥ 2 with $V \in \mathcal{W}(K, m, C_1)$. Then, by Lemma 5, it follows that all solutions of (10) are contained in subspaces of K^m of dimension 1 belonging to $\mathcal{W}(K, m, C_{m-1}^*)$. This means that every solution $\underline{x}$ of (10) can be written in the form $\underline{x} = \kappa \underline{x}'$ with some $\kappa \in K^*$ and $\underline{x}' \in K^m$ for which $h(\underline{x}') \leq c_{20}$, where c_{20} as well as c_{21}, c_{22} introduced below are effectively computable numbers depending only on on d, g, D_G, s, P, A, n, m and B. From (10) we obtain

$$\kappa^n F(\underline{x}') = \beta.$$

Together with (12) and (13) this implies $h(\kappa) \leq c_{21}$ whence $h(\underline{x}) \leq c_{22}$.

Proof of Theorem 2'. Apply again Lemma 5 successively in the same way as in the proof of Theorem 1'. Let $1 = C_0^* \leq C_1^* \leq \cdots \leq C_{m-2}^* \leq C_{m-1}^*$ be as above, and suppose again that $r(V, \mathcal{L}_0) = 3$ for every $\mathcal{L}$-admissible subspace V of K^m of dimension ≥ 2 which belongs to $\mathcal{W}(K, m, C_1)$ for $C_1 = C_{m-2}^*$. Then Lemma 5 implies that all solutions of (10) are contained in at most N^{m-1} subspaces of K^m of dimension 1. If $\underline{x}$ and $\rho\underline{x}$ are solutions of (10) with some $\rho \in K^*$ then

$$\rho^n = F(\rho\underline{x})/F(\underline{x}) = 1.$$

Therefore, every subspace of K^m of dimension 1 contains at most n solutions. Hence the number of solutions of (10) is at most

$$nN^{m-1} = n(3 \times 7^{dg+2t'})^{m-1}$$
$$\leq n(3 \times 7^{g(d+2s+2\omega_S(\beta,F))})^{m-1}.$$ ∎

References

[1] A. Baker, Contributions to the theory of Diophantine equations, *Philos. Trans. Roy. Soc. London,* A **263** (1968), 173–208.

[2] A. Baker, Bounds for the solutions of the hyperelliptic equation, *Math. Proc. Cambridge Philos. Soc.* **65** (1969), 439–444.

[3] A. Baker, A sharpening of the bounds for linear forms in logarithms II, *Acta Arith.* **24** (1973), 33–36.

[4] A. Baker, *Transcendental number theory*, 2nd ed., Cambridge University Press, 1979.

[5] B. J. Birch and J. R. Merriman, Finiteness theorems for binary forms with given discriminant, *Proc. London Math. Soc.* **25** (1972), 385–394.

[6] E. Bombieri and W. M. Schmidt, On Thue's equation, *Invent. Math.* **88** (1987), 69–81.

[7] Z. I. Borevich and I. R. Shafarevich, *Number theory*, 2nd ed., 1967.

[8] J. Coates, An effective p-adic analogue of a theorem of Thue, *Acta Arith.* **15** (1969), 279–305.

[9] J. Coates, An effective p-adic analogue of a theorem of Thue II, The greatest prime factor of a binary form, *Acta Arith.* **16** (1970), 399–412.

[10] B. N. Delone and D. K. Faddeev, *The theory of irrationalities of the third degree*, American Mathematical Society, Providence, 1964.

[11] J. H. Evertse, Upper bounds for the numbers of solutions of Diophantine equations, MC – tract 168, Amsterdam, 1983.

[12] J. H. Evertse, On equations in S-units and the Thue-Mahler equation, *Invent. Math.* **75** (1984), 561–584.

[13] J. H. Evertse, On sums of S-units and linear recurrences, *Compositio Math.* **53** (1984), 225–244.

[14] J. H. Evertse and K. Győry, On unit equations and decomposable form equations, *J. reine angew. Math.* **358** (1985), 6–19.

[15] J. H. Evertse and K. Győry, Finiteness criteria for decomposable form equations, *Acta, Arith.* to appear.

[16] J. H. Evertse, K. Győry, C. L. Stewart and R. Tijdeman, S-unit equations and their applications, This volume.

[17] G. Faltings, Endlichkeitssätze für abelsche Varietäten über Zahlkörpern, *Invent. Math.* **73** (1983), 349–366.

[18] G. Faltings, G. Wüstholz et al., *Rational points*, Vieweg 1984.

[19] N. I. Feldman, An effective refinement of the exponent in Liouville's theorem (Russian), *Izv. Akad. Nauk SSSR* **35** (1971). 973–990.

[20] K. Győry, Sur les polynômes à coefficients entiers et de discriminant donné, *Acta Arith.* **23** (1973), 419–426.

[21] K. Győry, Sur les polynômes à coefficients entiers et de discriminant donné III, *Publ. Math Debrecen* **23** (1976), 141–165.

[22] K. Győry, On polynomials with integer coefficients and given discriminant IV, *Publ. Math. Debrecen* **25** (1978), 155–167.

[23] K. Győry, On polynomials with integer coefficients and given discriminant V, $\wp$-adic generalizations, *Acta Math. Acad. Sci. Hungar.* **32** (1978), 175–190.

[24] K. Győry, On the greatest prime factors of decomposable forms at integer points, *Ann. Acad. Sci. Fenn. Ser. A I, Math.* **4** (1978/ 1979), 341–355.

[25] K. Győry, On the number of solutions of linear equations in units of an algebraic number field, *Comment. Math. Helv.* **54** (1979), 583–600.

[26] K. Győry, Explicit upper bounds for the solutions of some Diophantine equations, *Ann. Acad. Sci. Fenn., Ser. A, Math.* **5** (1980), 3–12.

[27] K. Győry, Sur certaines généralisations de l'équation de Thue-Mahler, *Enseign. Math.* **26** (1980), 247–255.

[28] K. Győry, Résultats effectifs sur la représentation des entiers par des formes décomposables, *Queen's Papers in Pure and Applied Math.*, No. 56, Kingston, Canada, 1980.

[29] K. Győry, On the representation of integers by decomposable forms in several variables, *Publ. Math. Debrecen* **28** (1981), 89–98.

[30] K. Győry, On S-integral solutions of norm form, discriminant form and index form equations, *Studia Sci. Math. Hungar.* **16** (1981), 149–161.

[31] K. Győry, Bounds for the solutions of norm form, discriminant form and index form equations in finitely generated integral domains, *Acta. Math. Hungar.* **42** (1983), 45–80.

[32] K. Győry, Effective finiteness theorems for Diophantine problems and their applications (in Hungarian), Academic doctor's thesis, Debrecen, 1983.

[33] K. Győry, On norm form, discriminant form and index form equations, *Coll. Math. Soc. J. Bolyai 34.* Topics in Classical Number Theory, Budapest, 1981. North-Holland 1984, pp. 617–676.

[34] K. Győry, Effective finiteness theorems for polynomials with given discriminant and integral elements with given discriminant over finitely generated domains, *J. reine angew. Math.* **346** (1984), 54–100.

[35] K. Győry and Z. Z. Papp, Effective estimates for the integer solutions of norm form and discriminant form equations, *Publ. Math. Debrecen* **25** (1978), 311–325.

[36] K. Győry and Z. Z. Papp, On discriminant form and index form equations, *Studia Sci. Math. Hungar.* **12** (1977), 47–60.

[37] K. Győry and Z. Z. Papp, Norm form equations and explicit lower bounds for linear forms with algebraic coefficients, *Studies in Pure Mathematics* (To the memory of Paul Turán), Budapest, 1983. pp. 245–257.

[38] K. Hensel, *Theorie der algebraischen Zahlen*, Leipzig und Berlin, 1908.

[39] S. V. Kotov, The Thue-Mahler equation in relative fields (Russian), *Acta Arith.* **27** (1975), 293–315.

[40] S. V. Kotov, On Diophantine equations of norm form type II (Russian), *Inst. Math. Akad. Nauk BSSR*, Preprint No. 10, Minsk, 1980.

[41] S. V. Kotov, Effective bound for a linear form with algebraic coefficients in the archimedean and *p*-adic metrics, *Inst. Math. Akad. Nauk BSSR*, Preprint No. 24, Minsk, 1981.

[42] S. V. Kotov, Effective bound for the values of the solutions of a class of Diophantine equations of norm form type (Russian), *Mat. Zametki* **33** (1983), 801–806.

[43] L. Kronecker, Grundzüge einer arithmetischen Theorie der algebraischen Größen, *J. reine angew. Math.* **92** (1882), 1–122.

[44] S. Lang, *Fundamentals of Diophantine geometry*, Springer 1983.

[45] M. Laurent, Equations diophantiennes exponentielles, *Invent. Math.* **78** (1984), 299–327.

[46] K. Mahler, Zur Approximation algebraischer Zahlen I, Über den größten Primteiler binärer Formen, *Math. Ann.* **107** (1933), 691–730.

[47] T. Nagell, Zur Theorie der kubischen Irrationalitäten, *Acta Math.* **55** (1930), 33-65.

[48] C. J. Parry, The *P*-adic generalization of the Thue-Siegel theorem, *Acta Math.* **83** (1950), 1–100.

[49] A. J. van der Poorten and H-P. Schlickewei, The growth conditions for recurrence sequences, *Macquarie Univ. Math. Rep.* **82–0041**, North Ryde, Australia, 1982.

[50] H-P. Schlickewei, Inequalities for decomposable forms, *Astérisque* **41–42** (1977), 267–271.

[51] H-P. Schlickewei, On norm form equations, *J. Number Theory* **9** (1977), 370–380.

[52] H-P. Schlickewei, On linear forms with algebraic coefficients and Diophantine equations, *J. Number Theory* **9** (1977), 381–392.

[53] W. M. Schmidt, Linearformen mit algebraischen Koeffizienten II, *Math. Ann.* **191** (1971), 1–20.

[54] W. M. Schmidt, Norm form equations, *Annals of Math.* **96** (1972), 526–551.

[55] W. M. Schmidt, Inequalities for resultants and for decomposable forms, *Proc. Conf. Diophantine approximation and its applications,* Washington, 1972. New York and London, 1973, pp. 235–253.

[56] W. M. Schmidt, *Diophantine approximation, Lecture Notes in Math.* **785**, Springer 1980.

[57] T. N. Shorey, A. J. van der Poorten, R. Tijdeman and A. Schinzel, *Applications of the Gelfond-Baker method to Diophantine equations, Transcendence Theory: Advances and Applications,* 59–77. Academic Press, 1977.

[58] T. N. Shorey and R. Tijdeman, *Exponential Diophantine Equations,* Cambridge University Press, 1986.

[59] C. L. Siegel, Approximation algebraischer Zahlen, *Math. Z.* **10** (1921), 173–213.

[60] C. L. Siegel, Über einige Anwendungen diophantischer Approximationen, *Abh. Preuss. Akad. Wiss.* (1929), 1–41.

[61] C. L. Siegel, Abschätzung von Einheiten, *Nachr. Göttingen* (1969), 71–86.

[62] T. Skolem, *Diophantische Gleichungen,* Berlin, 1938.

[63] V. G. Sprindžuk, Rational approximations to algebraic numbers (Russian), *Izv. Akad. Nauk SSSR* **35** (1971), 991–1007.

[64] V. G. Sprindžuk, The greatest prime divisor of a binary form (Russian), *Dokl. Akad. Nauk BSSR* **15** (1971), 389–391.

[65] V. G. Sprindžuk, On an estimate for solutions of Thue's equation (Russian), *Izv. Akad. Nauk SSSR* **36** (1972), 712–741.

[66] V. G. Sprindžuk, On the structure of numbers representable by binary forms (Russian), *Dokl. Akad. Nauk BSSR* **17** (1973), 685–688.

[67] V. G. Sprindžuk, *Classical Diophantine equations in two variables* (Russian), Moscow, 1982.

[68] H. M. Stark, Effective estimates of solutions of some Diophantine equations, *Acta Arith.* **24** (1973), 251–259.

[69] K. B. Stolarsky, *Algebraic numbers and Diophantine approximation*, New York, 1974.

[70] A. Thue, Über Annäherungswerte algebraischer Zahlen, *J. reine angew. Math.* **135** (1909), 284–305.

[71] L. A. Trelina, On algebraic integers with discriminants having fixed prime factors (Russian), *Mat. Zametki* **21** (1977), 289–296.

[72] L. A. Trelina, On the greatest prime factor of an index form (Russian), *Dokl. Akad. Nauk BSSR* **21** (1977), 975–976.

11

A NOTE ON GELFOND'S METHOD

N. I. Feldman

For the 80th anniversary of A. O. Gelfond's birth

In this note we look at some possible improvements of Gelfond's method. For simplicity of presentation, we restrict ourselves to the standard problem of the measure of transcendence of π and $\log \alpha$.

1. In the construction of the auxiliary forms

$$f(z) = \sum_{\kappa=0}^{q_0-1} \sum_{\ell=0}^{q-1} C_{\kappa,\ell} z^\kappa e^{\ell z}, \tag{1}$$

the numbers $C_{\kappa,\ell}$ are chosen using Dirichlet's principle. If the approximating algebraic numbers ζ have increasing degree n, then one usually takes

$$C_{\kappa,\ell} = \sum_{t=0}^{n-1} C_{\kappa,\ell,t} \zeta^t, \qquad C_{\kappa,\ell,t} \in \mathbf{Z}, \tag{2}$$

so that the equations $C_{\kappa,\ell} = 0$ imply

$$C_{\kappa,\ell,0} = \ldots = C_{\kappa,\ell,n-1} = 0.$$

At the same time, the numbers

$$\varphi_{s,x} = \sum_{\kappa,\ell} \sum_{\sigma=0}^{s} C_s^\sigma C_\kappa^\sigma \kappa! x^{\kappa-\sigma} \zeta^{\kappa-\sigma} \ell^{s-\sigma} \alpha^{\ell x}, \qquad x \in \mathbf{Z}, \ s \in \mathbf{N}_0, \tag{3}$$

arising in the course of the proof have a larger degree n_1 in ζ (at the present state of the method $n_1 \sim n \log n$ for $\alpha = -1$ and is of order n^2 in other cases). Thus the subsequent reasoning would also apply in the case when the coefficients $C_{\kappa,\ell}$ are given by polynomials in ζ of degree n_1. It is necessary only to ensure that one of these coefficients is different from zero. The question naturally arises in connection with this argument as to the number of polynomials $\varphi(z) \in \mathbf{Z}[z]$, $H(\varphi) \le H_0$, $\deg \varphi(z) \le n_1$,

$n_1 \geq n$, incongruent modulo $p(z)$, where $p(z)$ is the minimum polynomial ζ. It is clear that all polynomials (see (2))

$$\sum_{t=0}^{n-1} C_{\kappa,\ell,t} z^t$$

for different choices of $C_{\kappa,\ell,t}$ are incongruent modulo $p(z)$, but there are possibly still many others with degrees between n and n_1. However, there cannot be very many. In fact, suppose there exist N polynomials of degree at most n_1 mutually incongruent modulo $p(z)$. Supposing for simplicity that $\zeta \in \mathbf{R}$, we can prove with the help of Dirichlet's principle that there exists a polynomial $Q(z) \in \mathbf{Z}[z]$, $H(Q) \leq 2H_0$, $\deg Q \leq n_1$, for which

$$0 < |Q(\zeta)| \leq 4H_0 (1 + |\zeta|)^{n_1} N^{-1}.$$

But, as is well-known, the following inequality must be satisfied

$$|Q(\zeta)| \geq ((2n_1 + 2)H_0)^{1-n} L(\zeta)^{-n_1},$$

so

$$N \leq L(\zeta)^{n_1} H_0^n 2^{n+1} (1 + |\zeta|)^{n_1} (n_1 + 1)^{n-1} \leq L(\zeta)^{n_1} H_0^n C_0(n_1),$$

as we may suppose that $|\zeta|$ is bounded. Only the first two factors play an essential rôle in connection with the problems mentioned above.

By estimating from above the number of elements in each class of polynomials congruent modulo $p(z)$ (of course, their degrees and heights must satisfy the above conditions), it is possible to show that

$$N \geq C(n_1) L(\zeta)^{n_1} H_0^n.$$

By this method we obtain the "supplementary magnitude" $L(\zeta)^{n_1}$. Unfortunately, for our problems the sizes of $L(\zeta)^{n_1}$ and H_0^n are approximately equal, so this strengthening does not lead to an improvement on the sharpness of the estimates relative to n. It is, however, advantageous in the calculation of the constant $c \exp(-c n^a \log^b n)$.

This example shows that apparently it is worthwhile seeking other possibilities than Dirichlet's principle.

In several cases it is profitable to replace the z^k in (1) by polynomials of the type

$$(z + k)(z + k + 1) \ldots (z + k + q_0), \tag{4}$$

or analogous ones, for example, powers of such polynomials. This technique enabled us to replace the factor $\log \log H$ that occurred in the estimates for the measures of transcendence for π and $\log \alpha$, following Gelfond's method, by $\log n$. It also worked successfully in the deduction of degree estimates associated with the height and in a number of other cases. At the bottom of success here lay the fact that, for *arbitrary* κ, z in $\mathbf{Z}$, the set of values of the polynomials (4) and their derivatives of not very high order consists of numbers having a very large common divisor. Unfortunately, this common divisor quickly decreases with the growth of the order of the differentiation. At the same time, the polynomials (4) have a redundant property (for our purposes): for us, not all integers κ, z are needed, but only some. In this connection the question arises about a search for other polynomials having similar properties. Possibly, of course, the polynomials (4) or their powers, are optimal here, but a proof of this assertion is not known to the author. If there are better polynomials, then this would allow us to obtain more precise bounds.

3. For the solution of our problems we have to estimate the non-zero magnitudes (3) from below. They are polynomials over $\mathbf{Z}$ in the algebraic numbers α and ζ (in other cases we may come across polynomials in a larger number of independent variables). Usually one uses estimates of the form

$$\delta = \left| P(\beta_1, \ldots, \beta_s) \right| \geq L(P)^{1-n} \prod_{j=1}^{s} L(\beta_j)^{\frac{-nN_j}{n_j}}, \qquad (5)$$

where $P \in \mathbf{Z}[z_1, \ldots, z_s]$, $n = \deg \mathbf{Q}(\beta_1, \ldots, \beta_s)$, $n_j = \deg \beta_j$, $N_j = \deg_{z_j} P$, $\delta \neq 0$.

The estimate for the measure of transcendence of π is better than that for $\log \alpha$ only because in (3) we have ± 1 in place of $\alpha^{\ell x}$. Making use of (5) for the estimation of $\varphi_{s,x}$ we may suppose that $N_\alpha = qx$. At the same time, the polynomial $\varphi_{s,x}$ does not contain all powers of α, but only multiples of x. Hence the question naturally arises about the possibility of strengthening the estimates (5) for polynomials of the special form

$$P(\zeta, \alpha^x).$$

Of course, to derive an estimate it is possible to make use of an estimate for the linear form $x \log \alpha - \log \beta$, where β is the closest root to α^x of the polynomial $P(\zeta, z)$. This allows us to "transfer" x from the exponent to the base in the estimate; but for our problems this gives nothing, as the degree and height of the number β turn out too large, although these will not depend on x. Nevertheless, the existence of estimates better in their dependence on x for the numbers (6) shows that it is worthwhile looking for other methods to deduce similar bounds.

To conclude, we note that what has been said above relates to the method of A. Baker concerning bounds for linear forms in several logarithms of algebraic numbers.

12

ON EFFECTIVE BOUNDS FOR CERTAIN LINEAR FORMS

A. I. Galochkin

To get bounds for linear forms in the values of analytic functions satisfying linear differential equations with coefficients in the field $C(z)$, it is often necessary to establish beforehand the linear independence of these functions over the field $C(z)$ (see [1] Chapter 7, §2). Under certain conditions, the constants in these bounds may be effectively computed if arbitrary non-zero solutions of the corresponding differential equations are linearly independent (see [2], [3]).

In the present paper we show the linear independence over $C(z)$ of the solutions of linear differential equations of a sufficiently general type. These results may be used to obtain effective bounds for the corresponding numerical linear forms.

Let $a_j(x)$, $b_j(x)$ be polynomials with complex coefficients $m_j = \deg b_j(x) \geq \deg a_j(x)$, $m_j \geq 1$;

$$\psi_j(z) = 1 + \sum_{\nu=1}^{\infty} z^\nu \prod_{x=1}^{\nu} \frac{a_j(x)}{b_j(x)}, \tag{1}$$

$j = 1, \ldots, t$; $b_j(x) \neq 0$ for $x = 1, 2, \ldots$.

The functions $\psi_j(z)$ satisfy the linear differential equations

$$U_j : b_j(\delta)y_j = a_j(\delta)zy_j + b_j(0), \quad \delta = z\frac{d}{dz}. \tag{2}$$

Theorem 1. Suppose that the following conditions are satisfied.

(A) For each index j $(1 \leq j \leq t)$ and any $c \in \mathbf{Z}$ the two polynomials $a_j(x)$ and $b_j(x + c)$ are coprime.

(B) For any two polynomials $a_k(x)b_\ell(x) = c(x + \lambda_1)\ldots(x + \lambda_N)$ and $a_\ell(x)b_k(x)$ $(1 \leq k < \ell \leq t)$ and any set $c_1, \ldots, c_N$ of integers

$$c(x + \lambda_1 + c_1)\ldots(x + \lambda_N + c_N) \not\equiv a_\ell(x)b_k(x).$$

Further, let $y_j(z)$ be an arbitrary solution of U_j in (2), $j = 1, \ldots, t$ not belonging to the ring $\mathbf{C}\left[z, z^{-1}\right]$. Then the set of functions

$$1, y_j^{(s)}(z), \qquad j = 1, \ldots, t;\ s = 0, 1, \ldots, m_j - 1, \tag{3}$$

is linearly independent over the field $\mathbf{C}(z)$.

Theorem 2. Let $m_j \geq 1$, $a_j(x) \neq 0$ for $x = 1, 2, \ldots$; $j = 1, \ldots, t$. Suppose, further, that condition (B) of Theorem 1 is satisfied, and that

(C) For each index j and any $c \in \mathbf{Z}$, $c \geq 0$, the two polynomials $a_j(x)$ and $b_j(x + c)$ are coprime.

Then the functions

$$1, \psi_j^{(s)}(z), \qquad j = 1, \ldots, t;\ s = 0, 1, \ldots, m_j - 1, \tag{4}$$

are linearly independent over the field $\mathbf{C}(z)$.

Note that it is impossible to substitute condition (A) of Theorem 1 by the weaker condition (C). Let us take an example. The equation

$$\delta(\delta + \lambda)y = (\delta + \lambda - 1)zy, \qquad \lambda \notin \mathbf{Z},$$

is satisfied by the function $z^{-\lambda} \notin \mathbf{C}(z)$ and this is linearly dependent, together with its derivative, over $\mathbf{C}$.

Theorem 3. Let I be the field of rationals or an imaginary quadratic field, $a_j(x) = 1$, $\deg b_j(x) = m \geq 1$, $b_j(x) = f_j(x)g(x) \in I[x]$, $j = 1, \ldots, t$, where all the roots of the polynomials $f_j(x)$ are rationals. Let condition (B) be satisfied. Then for an arbitrary choice $\{h_0, h_{js}\}$ of integers from I with $H = \max_{j,s}(|h_0|, |h_{js}|) \geq 3$ we have

1) For any positive ϵ and for $H > H_0(\epsilon, g(x), f_1(x), \ldots, f_t(x))$,

$$\left| h_0 + \sum_{j=1}^{t} \sum_{s=0}^{m-1} h_{js}\psi_j^{(s)}(1) \right| > H^{(-mt+\tau+\epsilon)/(1-\tau)} \tag{5}$$

where $\tau = m^{-1}v - (mx_1)^{-1} - \ldots - (mx_v)^{-1}$, $v = \deg g(x)$, and $x_1, \ldots, x_v$ are the degrees of the algebraic numbers which are the roots of the polynomial $g(x)$.

2) If $g(0) = 0$, then

$$\left| \sum_{j=1}^{t} \sum_{s=0}^{m-1} h_{js}\psi_j^{(s)}(1) \right| > H^{1-mt-\gamma/\sqrt{\ln \ln H}} \tag{6}$$

The constants H_0 and γ may be effectively computed.

Let

$$a_{j0}(x) = b_{j0}(x) = 1,$$

$$a_{jn}(x) = \prod_{\ell=0}^{n-1} a_j(x - \ell),$$

$$b_{jn}(x) = \prod_{\ell=0}^{n-1} b_j(x - \ell),$$

$$A_n(x) = a_{1n}(x) \ldots a_{tn}(x), \quad B_n(x) = b_{1n}(x) \ldots b_{tn}(x), \qquad (7)$$

$$A_{jn}(x) = \frac{A_n(x)}{a_{jn}(x)},$$

$$B_{jn}(x) = \frac{B_n(x)}{b_{jn}(x)}.$$

Write $f(z) \asymp g(z)$ to denote that

$$f(z) - g(z) \in \mathbf{C}\left[z, z^{-1}\right].$$

It is easy to check that the operator $\delta = z\frac{d}{dz}$ has the following property:

$$P(\delta)z^n y = z^n P(\delta + n)y, \qquad P(x) \in \mathbf{C}[x]. \qquad (8)$$

By using induction on k and using (2), (7) and (8) it is easy to prove that if y_j is a solution of U_j from (2), then

$$b_{jk}(\delta)y_j \asymp a_{jk}(\delta)z^k y_j. \qquad (9)$$

Lemma 1. Suppose that condition (C) of Theorem 2 is satisfied and that $y_j(z)$ is an arbitrary solution of equation U_j in (2), $m_j \geq 1$, $j = 1, \ldots, t$. Further, let the functions (3) be linearly dependent over $\mathbf{C}(z)$. Then there are polynomials $Q_j(x) \in \mathbf{C}[x]$, $j = 1, \ldots, t$, not all identically zero, such that

$$Q = Q_1(\delta)y_1(z) + \ldots + Q_z(\delta)y_t \asymp 0. \qquad (10)$$

Proof. If the functions (3) are linearly dependent over $\mathbf{C}(z)$ then the functions 1, $\delta^s y_j(z)$, $j = 1, \ldots, t$; $s = 0, 1, \ldots, m_j - 1$ are linearly dependent over $\mathbf{C}[z]$, and so for some natural number n

$$G = \sum_{j=1}^{t} \sum_{k=0}^{n} z^k g_{jk}(\delta) y_j \asymp 0,$$

(11)

$$g_{jk}(x) \in \mathbf{C}[x], \quad \deg g_{jk}(x) < m_j.$$

Without loss of generality we may suppose that not all the polynomials $g_{j0} \equiv 0$. Suppose that $g_{10} \not\equiv 0$. From (7), (8) and (9) we obtain:

$$0 \asymp A_n(\delta)G$$

$$= \sum_{j=1}^{t} \sum_{k=0}^{n} g_{jk}(\delta - k) A_{jn}(\delta) \frac{a_{jn}(\delta)}{a_{jk}(\delta)} \left(a_{jk}(\delta) z^k y_j \right)$$

$$\asymp \sum_{j=1}^{t} \left[\sum_{k=0}^{n} g_{jk}(\delta - k) A_{jn}(\delta) \frac{a_{jn}(\delta)}{a_{jk}(\delta)} b_{jk}(\delta) \right] y_j.$$

Thus we obtain a relationship of the form (10). It remains to show that the polynomial in δ in square brackets for $j = 1$ is not identically zero. Suppose the contrary. Then, in view of (7), we obtain

$$g_{10}(x) \equiv - \sum_{k=1}^{n} g_{1k}(x - k) \frac{b_1(x)}{a_1(x)} \prod_{\ell=1}^{k-1} \frac{b_1(x - \ell)}{a_1(x - \ell)}.$$

(12)

The polynomial $b_1(x)$ is, by condition (C), relatively prime to the polynomial $a_1(x)a_1(x-1)\ldots a_1(x-n+1)$. Therefore the right-hand side of (12) has at least $m_1 = \deg b_1(x)$ zeros (counting multiplicity). At the same time, $\deg g_{10}(x) < m_1$ in equalities (10) and (11). Consequently, $g_{10}(x) \equiv 0$ contradicting the earlier assertion. This proves the lemma.

Lemma 2. Let $y = y(z)$ be a function, analytic in some neighbourhood of the point $z = 0$ and satisfying the following differential equation:

$$Q(\delta)y = R(z), \quad 0 \not\equiv Q(x) \in \mathbf{C}[x], \; R(z) \in \mathbf{C}\left[z, z^{-1}\right].$$

(13)

Then $y(z)$ is a polynomial.

To prove the lemma it is sufficient to substitute into (13) the Taylor series expansion of $y(z)$ in powers of z and to equate the coefficients for the same powers of z in the left and right hand sides of this equality.

Lemma 3. Let $y = y_j(z)$ be a solution of U_j in (2), such that

$$Q_0 = q_0(\delta)y \asymp 0, \quad 0 \not\equiv q_0(x) \in C[x], \ \deg q_0(x) = n. \qquad (14)$$

Then for any integer k, $0 \leq k \leq n$, we have

$$Q_k = a_{jk}(\delta + k)q_k(\delta)y \asymp 0, \quad 0 \not\equiv q_k(x) \in C[x],$$
$$\deg q_k(x) = n - k. \qquad (15)$$

The proof of the Theorem proceeds by induction on k. For $k = 0$ its assertion follows from (14) and the equality $a_{j0}(x) \equiv 1$. Suppose relation (15) holds for the case Q_k. We shall show it for Q_{k+1}. From (2), (7), (8) and the induction hypothesis, we obtain

$$0 \asymp \left[z^{-1}b_j(\delta) - a_j(\delta + k + 1) \right] Q_k \asymp$$
$$\asymp a_{j,k+1}(\delta + k + 1) \left[q_k(\delta + 1) - q_k(\delta) \right] y,$$

from which the assertion of the Lemma follows for Q_{k+1}. This proves the lemma.

Lemma 4. If condition (A) of Theorem 1 and the conditions of Lemma 3 are satisfied, then $y_j(z) \in C\left[z, z^{-1} \right]$.

Proof. Applying Lemma 3 for $k = n$, we obtain

$$a_{jn}(\delta + n)y_j(z) \asymp 0. \qquad (16)$$

By virtue of (8), (9) and (16)

$$b_{jn}(\delta)y_j(z) \asymp z^n a_{jn}(\delta + n)y_j(z) \asymp 0. \qquad (17)$$

From (7) and condition (A) it follows that the polynomial $a_{jn}(x+n)$ and $b_{jn}(x)$ are coprime. Therefore there exist polynomials $u_n(x)$ and $v_n(x)$ such that

$$a_{jn}(x + n)u_n(x) + b_{jn}(x)v_n(x) \equiv 1$$

Consequently, in view of (16) and (17),

$$y_j(z) = \left[a_{jn}(\delta + n)u_n(\delta) + b_{jn}(\delta)v_n(\delta) \right] y_j(z) \asymp 0.$$

This proves the lemma.

Lemma 5. Let $y_1(z)$, ..., $y_t(z)$ be functions, analytic in some domain, such that

$$P = P_1(\delta)y_1(z) + \ldots + P_t(\delta)y_t(z) \asymp 0,$$

$$P_j(x) \in C[x], \qquad j = 1, \ldots, t, \tag{18}$$

where for any polynomials $R_2(x)$, $R_3(x)$, ..., $R_t(x)$, not all identically zero,

$$R_2(\delta)y_2(z) + \ldots + R_t(\delta)y_t(z) \notin C\left[z, z^{-1}\right]. \tag{19}$$

Further, let (10) be one of the relations of the form (18), in which the polynomial $Q_1(x) \not\equiv 0$ has smallest possible degree. Then there exists a polynomial $S(x) \in C[x]$, such that

$$P_j(x) \equiv Q_j(x)S(x), \qquad j = 1, \ldots, t.$$

Proof. Let us divide with remainder the polynomial $P_1(x)$ by $Q_1(x)$:

$$P_1(x) \equiv Q_1(x)S(x) + R_1(x), \qquad \deg R_1 < \deg Q_1.$$

Then by virtue of (10) and (18)

$$0 \asymp P - S(\delta)Q = R_1(\delta)y_1(z) + \ldots + R_t(\delta)y_t(z), \tag{20}$$

$$R_j(x) = P_j(x) - S(x)Q_j(x) \qquad j = 1, \ldots, t.$$

From the inequality $\deg R_1 < \deg Q_1$ and the choice of relation (10) in the condition of the lemma, it follows that $R_1(x) \equiv 0$, and then from (19) and (20) we get that all $R_j(x) \equiv 0$. This proves the lemma.

Lemma 6. Suppose that condition (B) of Theorem 1 and condition (C) of Theorem 2 are satisfied and that $m_j \geq 1$, $j = 1, \ldots, t$. Let $y_j(z)$ be an arbitrary solution of U_j in (2), and suppose that none of these $y_j(z)$ satisfy (13). Then the functions (3) are linearly independent over $C(z)$.

Proof. Suppose that the functions (3) are linearly dependent over $C(z)$. Then by Lemma 1 the equality (10) holds. Without loss of generality we may suppose that the functions (3) and the polynomial $Q_1(z)$ satisfy the conditions of Lemma 5 and that $Q_j(x) \not\equiv 0$ for $j = 1, \ldots, t$: for otherwise we can, if necessary, reduce the number of functions and rename them as $y_1, \ldots, y_t$. By the condition of the Lemma, $t \geq 2$. Using (7), (8), (9) and (10), we get

$$0 \asymp B_n(\delta)Q \asymp z^n \sum_{j=1}^{t} Q_j(\delta + n)B_{jn}(\delta + n)a_{jn}(\delta + n)y_j. \tag{21}$$

Applying Lemma 5 to (10) and (21), we establish that

$$\frac{Q_1(x+n)B_{1n}(x+n)a_{1n}(x+n)}{Q_2(x+n)B_{2n}(x+n)a_{2n}(x+n)} \equiv \frac{Q_1(x)}{Q_2(x)},$$

from which, using the notation of (7), we find

$$\frac{f(x+1)\ldots f(x+n)}{g(x+1)\ldots g(x+n)} \equiv \frac{Q_1(x)Q_2(x+n)}{Q_2(x)Q_1(x+n)}, \tag{22}$$

where $f(x) = a_1(x)b_2(x)$, $g(x) = a_2(x)b_1(x)$.

From (22) in the case $n = 1$ we get that the leading coefficients of the polynomials $f(x)$ and $g(x)$ are the same. Therefore, it follows from condition (B) of Theorem 1 that these polynomials can be factorised thus:

$$f(x) = f_1(x)f_2(x), \qquad g(x) = g_1(x)g_2(x),$$

$$u_i = \deg f_i \geq 0, \qquad v_i = \deg g_i \geq 0, \qquad u_1 \neq v_1,$$

that all the roots of the polynomial $f_1(x)g_1(x)$ will be congruent to one another modulo 1, and at the same time none of the roots of the polynomial $f_2(x)g_2(x)$ will be congruent modulo 1 to one of the roots of $f_1(x)g_1(x)$.

Denote the left hand side of (22) by $R_n(x)$. Suppose that

$$R_{in}(x) = \frac{f_i(x+1)\ldots f_i(x+n)}{g_i(x+1)\ldots g_i(x+n)}, \qquad R_n(x) = R_{1n}(x)R_{2n}(x).$$

For the rational function $R(x)$ let $d(R)$ be the sum of the degrees of the polynomials, where the functions are represented so that the numerator and denominator are irreducible over $C[x]$. From (22) and the definition of the polynomials $f_i(x)$, $g_i(x)$ it follows that

$$d(R_n) = d(R_{1n}) + d(R_{2n}) \geq$$

$$\geq d(R_{1n}) \geq |u_1 - v_1|n \geq n.$$

At the same time the value of the function d from the right-hand side of (22) remains bounded as n grows. Consequently, this equation is contradicted for a sufficiently large value of n. This proves the lemma.

The proof of Theorem 1 follows from Lemmas 4 and 6, and of Theorem 2 follows from Lemmas 2 and 6.

The assertion of Theorem 3 follows from Theorem 1 and a result in [4]. As in [3], it is possible to substantiate the possibility of computing effective constants in the bounds (5) and (6).

It is easy to check that if $\psi_1(z), \ldots, \psi_t(z)$ are integral functions, then the conditions of Theorem 2 are not only sufficient but also necessary for their linear independence over the field $C(z)$. An analogous assertion holds if $\psi_j(z) = \psi(\alpha_j z)$, where $\alpha_1, \ldots, \alpha_t$ are distinct non-zero complex numbers, and $\psi(z)$ is a function of the form (1). In this case condition (B) is always satisfied.

References

[1] A. B. Shidlovsky, *Diophantine approximations and transcendental numbers*, Moscow University Press, 1982.

[2] A. B. Shidlovsky, On the transcendence and algebraic independence of the values of integral functions of certain classes, *Academic Notes of Moscow State University*, **186**, 1959, 11–70, (*Matematika*), **9**.

[3] A. B. Shidlovsky, On the estimates of the algebraic independence measure of the values of E-functions, *J. Austral. Math. Soc.* (Ser A), **27**, 1979, 385–407.

[4] A. I. Galochkin, On an analogue of Siegel's method, *Vest. Moscow Univ., Math-Mechan.*, 1986, No. 2, 30–34.

13

AUTOMATA AND TRANSCENDENCE

J. H. Loxton

1. Introduction

In [15], Kurt Mahler describes how he began his studies of transcendental numbers in 1926. While very ill and in bed, he set himself to prove that the number $\sum_{n=0}^{\infty} z^{2^n}$ is irrational for rational z with $0 < |z| < 1$. In fact, he proved that this number is transcendental for algebraic z with $0 < |z| < 1$. Mahler's method can be used to prove results on the transcendence and algebraic independence of the values at algebraic points of functions satisfying a rather general type of functional equation. The achievements of the first fifty years of the method are described in [11]. It is an interesting chapter of transcendence theory, but in 1976 the ideas did not seem as widely applicable as those of other methods in the subject.

Mahler's method took on a new significance when Mendès France and his co-workers observed that some of Mahler's examples arose in their studies of finite automata. (See [3] and [17].) This connection between finite automata and functional equations was known some years earlier to Cobham [4], but was apparently never published. The following is an account of Cobham's analysis of finite automata via functional equations and of the application of transcendence theory to this area. This work is a joint project with Alf van der Poorten and much of this report is his.

The gripping drama "Folds!" [7] proves that finite automata and automatic sequences are ubiquitous. I give just two famous examples. The Thue sequence (see below) played a part in Thue's construction of a sequence without square and in later work of Morse and Hedlund in topological dynamics. The Rudin-Shapiro sequence, which arose in the solution of a problem in harmonic analysis, has links with dragon curves, fractals and the Ising model in statistical mechanics. (See [1].) More general classes of automata have now begun to emerge from the linguists' domain and in [2] they appear in the study of quasicrystals. Much remains to be discovered and some of the open problems bearing on transcendence theory will be discussed in what follows.

2. Automata and Substitutions.

Let $A = \{a_0, a_1, \ldots, a_{p-1}\}$ be a finite non-empty set of symbols and let $\Omega = A^{\mathbb{N}}$ be the set of sequences $\alpha = \alpha_0 \alpha_1 \alpha_2 \ldots$ of symbols from the alphabet A. A substitution is a map $w : A \to \Omega$ taking each symbol a_i to a word $w_i = w(a_i)$ in Ω. The map w has a natural extension to Ω by concatenation: $w(\alpha_0 \alpha_1 \alpha_2 \ldots) = w(\alpha_0) w(\alpha_1) w(\alpha_2) \ldots$. If the initial symbol of $w(\alpha_0)$ is α_0, then there is a unique shortest sequence α which begins with α_0 and is a fixed point of w, meaning that $w(\alpha) = \alpha$.

Suppose, in particular, that each w_i has the same length r with $r \geq 2$. Let B be another set of symbols and let $h : A \to B$ be any map. Extend h to a length-preserving map on Ω by $h(\alpha_0 \alpha_1 \alpha_2 \ldots) = h(\alpha_0) h(\alpha_1) h(\alpha_2) \ldots$. Any sequence $h(\alpha)$ constructed from a fixed point α of w and a length-preserving map h is called a regular sequence in base r. The regular sequences in base r are exactly those sequences which can be recognized by finite automata in base r. Indeed, given a regular sequence as described here, we can build an equivalent finite automaton with set of states A and input symbols $\{0, 1, \ldots, r-1\}$ as follows. The transition function δ is determined by taking $\delta(a, j)$ to be the $(j+1)$-st symbol of $w(a)$ for a in A and $0 \leq j \leq r-1$. The output discriminates between the states a and a' if $h(a) \neq h(a')$. As an example to illustrate how such an automaton works, consider the Thue sequence. This sequence can be defined by the finite automaton in base 2 with 2 states, a_0 and a_1, and 2 inputs, 0 and 1, and transition function given by $\delta(a_j, 0) = a_j$ and $\delta(a_j, 1) = a_{1-j}$ for $j = 0, 1$. The output is $h(a_j) = j$ if the final state is a_j. Now let α_n be the output obtained when the binary digits of n are used as successive inputs, starting the machine in the initial state a_0. This gives the Thue sequence

$$0\,1\,1\,0\,1\,0\,0\,1\,1\,0\,0\,1\,0\,1\,1\,0 \ldots.$$

Alternatively, this sequence arises as the fixed point of the substitution $0 \to 0\,1$, $1 \to 1\,0$. In fact, α_n is the parity of the number of ones in the binary representation of n. This is typical: a regular sequence in base r can recognize patterns in the digits of numbers written in base r. The equivalence between finite automata and regular sequences is described in detail in [6], Theorem 3.

In general, if α is a fixed point of a substitution w on A and $h : A \to B$ is a length-preserving map, we call $h(\alpha)$ a semi-regular sequence. These sequences also arise from automata. Take the states of the automaton to be A together with an extra dead state and take the input symbols as $\{0, 1, \ldots, r-1\}$, where r is the maximum of the lengths of the w_j. For the transition function, take $\delta(a_i, j)$ to be the $(j+1)$-st symbol of w_i, if there is one, and the dead state otherwise. Again, the output discriminates between the states a and a' if $h(a) \neq h(a')$. To recover the sequence, let $S_j = \{k : h(\delta(a_0, (k)_r)) = b_j\}$ be the set of integers k for which the automaton starting in state a_0 and reading in the string of digits $(k)_r$ of k in base r as successive inputs, arrives at a state a with $h(a) = b_j$. These S_j are regular in base r. Now let ϵ enumerate the set $T = \bigcup S_j$ in increasing order. The k-th symbol in the semi-regular sequence $h(\alpha)$ is b_j if and only if $\epsilon(k)$ is in S_j. The regular sequences correspond to the case in which the union of the S_j is all of $\mathbf{N}$, so that $\epsilon(k) = k$.

As an example, we show that the Fibonacci numbers are semi-regular. Take the automaton with input symbols 0 and 1 and six states, $a_0, a_1, \ldots, a_4$ and $-$ (the dead state), and transition function given by

$$\delta(a_0, 0) = a_0, \qquad \delta(a_1, 0) = \delta(a_2, 0) = a_2,$$

$$\delta(a_3, 0) = \delta(a_4, 0) = a_4, \qquad \delta(-, 0) = -,$$

$$\delta(a_0, 1) = a_1, \qquad \delta(a_1, 1) = \delta(a_3, 1) = -,$$

$$\delta(a_2, 1) = \delta(a_4, 1) = a_3, \qquad \delta(-, 1) = -.$$

With acceptance states $a_0, a_1, \ldots, a_4$, this automaton recognizes the set T of all numbers whose binary representations contain no adjacent ones. With acceptance states a_0, a_1, a_2, it recognizes the set consisting of 0 and the powers of 2. These sets are regular in base 2. If ϵ enumerates T and $F = \{k : \epsilon(k) \text{ is in } S\}$, then F is the set of Fibonacci numbers. Alternatively, the substitution

$$a_0 \to a_0 a_1, \quad a_1 \to a_2, \quad a_2 \to a_2 a_3, \quad a_3 \to a_4, \quad a_4 \to a_4 a_3,$$

derived from the transition function in the obvious way, generates the fixed point

$$a_0 a_1 a_2 a_2 a_3 a_2 a_3 a_4 a_2 a_3 a_4 a_4 a_3 a_2 a_3 \ldots$$

and the characteristic function of the Fibonacci numbers is the image of this sequence under the length-preserving map defined by $h(a_3) = h(a_4) = 0$ and $h(a_0) = h(a_1) = h(a_2) = 1$. This example depends on the fact that every natural number can be expressed uniquely as a sum of non-adjacent Fibonacci numbers: the automaton recognizes digit patterns in Fibonacci notation. In general, the recovery of the substitution from the automaton is more complicated and may require conversion of the automaton from base r to base r^k. The details are given in [4].

3. Automata and functional equations

Let w be a uniform substitution on the alphabet $A = \{a_0, a_1, \ldots, a_{p-1}\}$ for which each word $w_i = w(a_i)$ has length $r \geq 2$ and let $h : A \to B = \{0, 1, \ldots, q-1\}$ be a length-preserving map. Let $\alpha = \alpha_0\alpha_1 \ldots$ be a fixed point of w and let $\beta = \beta_0\beta_1 \ldots = h(\alpha)$ be the associated regular sequence. For each symbol a_i in A, define the generating function

$$F_i(z) = \sum_{n=0}^{\infty} f_{in} z^n$$

where

$$f_{in} = \begin{cases} 1 & \text{if } \alpha_n = a_i \\ 0 & \text{otherwise.} \end{cases}$$

Observe that, for any n, the string $\alpha_{rn}\alpha_{rn+1} \ldots \alpha_{rn+r-1}$ depends only on α_n and, in fact, coincides with w_j if $\alpha_n = a_j$. If we set

$$g_{ijk} = \begin{cases} 1 & \text{if } a_i \text{ is the } (k+1)\text{-st symbol of } w_j \\ 0 & \text{otherwise,} \end{cases}$$

then we have

$$f_{i,rn+k} = \sum_{j=0}^{p-1} g_{ijk} f_{jn}.$$

We can now obtain a system of functional equations for the functions $F_i(z)$ because

$$\sum_{m=0}^{\infty} f_{im} z^m = \sum_{n=0}^{\infty} \sum_{k=0}^{r-1} f_{i,rn+k} z^{rn+k}$$

$$= \sum_{j=0}^{p-1} \left(\sum_{k=0}^{r-1} g_{ijk} z^k \right) \left(\sum_{n=0}^{\infty} f_{jn} z^{rn} \right),$$

that is,

$$F_i(z) = \sum_{j=0}^{p-1} p_{ij}(z) F_j(z^r) \qquad (0 \leq i \leq p-1),$$

where the $p_{ij}(z) = \sum g_{ijk} z^k$ are polynomials. We can interpret the regular sequence β as the expansion of a real number in base q (with the decimal point after β_0 for convenience). Note that

$$\beta_0 \cdot \beta_1 \beta_2 \ldots = h(\alpha_0) \cdot h(\alpha_1) h(\alpha_2) \ldots = \sum_{n=0}^{\infty} h(\alpha_n) q^{-n}$$

$$= \sum_{n=0}^{\infty} \sum_{i=0}^{p-1} f_{in} h(a_i) q^{-n} = \sum_{i=0}^{p-1} h(a_i) F_i(q^{-1}).$$

Thus, if we set

$$G(z) = \sum_{i=0}^{p-1} h(a_i) F_i(z),$$

then the output of the automaton is $G(q^{-1})$.

For the Thue sequence, the associated generating functions are

$$F_0(z) = 1 + z^3 + z^5 + z^6 + z^9 + z^{10} + \ldots,$$

$$G(z) = F_1(z) = z + z^2 + z^4 + z^7 + z^8 + z^{11} + \ldots$$

and the functional equations take the form

$$F_0(z) = F_0(z^2) + z F_1(z^2), \qquad F_1(z) = z F_0(z^2) + F_1(z^2).$$

Since $F_0(z) + F_1(z) = (1 - z)^{-1}$, we can simplify the system to

$$G(z) = (1 - z) G(z^2) + \frac{z}{1 - z}.$$

Consider now a semi-regular sequence with the same notation as before except that the substitution w is no longer required to be regular. We can again arrive at a system of functional equations characterizing the sequence, but with added complications to keep track of the varying lengths of the words of the substitution. This involves functions of p complex variables $z_0, \ldots, z_{p-1}$ and we use the standard abbreviations with multi-indices: $\mu = (\mu_0, \mu_1, \ldots, \mu_{p-1})$,

$$z^\mu = z_0^{\mu_0} z_1^{\mu_1} \ldots z_{p-1}^{\mu_{p-1}}, \qquad |\mu| = |\mu_0| + |\mu_1| + \ldots + |\mu_{p-1}|.$$

With the symbol a_i, we associate the generating function

$$F_i(z) = \sum_{\mu} f_{i\mu} z^\mu,$$

where the sum is taken over all p-tuples $\mu = (\mu_0, \mu_1, \ldots, \mu_{p-1})$ of non-negative integers, $f_{i\mu} = 1$ whenever $\alpha_{|\mu|} = a_i$ and for each j there are exactly μ_j occurrences of a_j in the string $\alpha_0 \alpha_1 \ldots \alpha_{|\mu|-1}$ and $f_{i\mu} = 0$ otherwise. As before, we set

$$G(z) = \sum_{i=0}^{p-1} h(a_i) F_i(z),$$

so that the output of the automaton is the number $G(q^{-1}, \ldots, q^{-1})$.

Now, suppose the term z^μ occurs in the series $F_i(z)$. Imagine the fixed point α being constructed by applying the substitution successively to α_0, α_1, $\ldots$. When we reach $\alpha_{|\mu|}$, we must have examined the symbol a_i exactly μ_i times and so we must have written out the word w_i exactly μ_i times. Let t_{ij} denote the number of occurrences of the symbol a_j in the word w_i. Then the part of the sequence constructed by the time the substitution reaches $\alpha_{|\mu|}$ contains the symbol a_j exactly $\sum_{i=0}^{p-1} \mu_i t_{ij}$ times and altogether $\sum_{i,j=0}^{p-1} \mu_i t_{ij}$ symbols have been written down. The next symbol to be written will be the first symbol, say b_k, of w_i, so $F_k(z)$ must contain the term z^ν with $\nu_j = \sum_{i=0}^{p-1} \mu_i t_{ij}$. If b_ℓ, say, is the second symbol of w_i, then $F_\ell(z)$ contains the term z^λ with $\lambda_j = \nu_j$ $(j \neq k)$, $\lambda_k = \nu_k + 1$, and so on. We introduce the $p \times p$ matrix $T = (t_{ij})$ and define the transformation $w = Tz$ on $\mathbf{C}^p$ by

$$w_i = \prod_{j=0}^{p-1} z_j^{t_{ij}}.$$

With this notation, $z^\nu = z^{\mu T} = (Tz)^\mu$, so we can expect that each $F_i(z)$ will be expressible by means of the $F_j(Tz)$. This works as before. Set $g_{ij\kappa} = 1$ if a_i is the $(|\kappa| + 1)$-st symbol of w_j and is preceded by exactly κ_ℓ occurrences of the symbol a_ℓ for each ℓ, and set $g_{ij\kappa} = 0$ otherwise. Then

$$f_{i,\mu T + \kappa} = \sum_{j=0}^{p-1} g_{ij\kappa} f_{j\mu}.$$

Here, and in what follows, κ runs through a suitable finite set of non-negative integer vectors so that every non-negative integer vector can be uniquely expressed as $\nu = \mu T + \kappa$ where μ also has non-negative integer entries. Now

$$\sum_\nu f_{i\nu} z^\nu = \sum_\nu \sum_\kappa f_{i,\mu T + \kappa} z^{\mu T + \kappa} = \sum_{j=0}^{p-1} \left(\sum_\kappa g_{ij\kappa} z^\kappa \right) \left(\sum_\mu f_{j\mu} z^{\mu T} \right),$$

that is

$$F_i(z) = \sum_{j=0}^{p-1} p_{ij}(z) F_j(Tz) \qquad (0 \le i \le p-1)$$

where the $p_{ij}(z)$ are certain polynomials whose coefficients are 0 and 1.

For the Fibonacci sequence, we take T to be the 5×5 matrix

$$T = \begin{pmatrix} 1 & 1 & 0 & 0 & 0 \\ 0 & 0 & 1 & 0 & 0 \\ 0 & 0 & 1 & 1 & 0 \\ 0 & 0 & 0 & 0 & 1 \\ 0 & 0 & 0 & 1 & 1 \end{pmatrix}$$

and we have the system of functional equations

$$F_0(z) = F_0(Tz)$$
$$F_1(z) = z_0 F_0(Tz)$$
$$F_2(z) = F_1(Tz) + F_2(Tz)$$
$$F_3(z) = z_2 F_2(Tz) + z_4 F_4(Tz)$$
$$F_4(z) = F_3(Tz) + F_4(Tz).$$

In fact, $F_0(z) = 1$, $F_1(z) = z_0$ and the output function is $G(z) = 1 + z_0 + F_2(z)$ which satisfies the functional equation $G(z) = z_0 + G(Tz)$. The output number is $G(\frac{1}{2}, \ldots, \frac{1}{2}) = \sum 2^{-f_n}$, where f_n is the n-th Fibonacci number.

4. A transcendence problem

It is a truth universally acknowledged that the decimal expansion of an algebraic irrational must be complicated. One interpretation of this maxim is that an algebraic irrational should be a normal number, but this seems quite out of reach. A more promising approach is to investigate the computational complexity of the decimal expansion. In all methods of computing algebraic numbers so far devised, the rate of digit production decreases with the number of digits produced, that is the number is not real-time computable. This question was addressed by Hartmanis and Stearns [9] who asked whether there is any algebraic irrational which is real-time computable on a Turing machine. As stated, this too seems out of reach, but the problem becomes more approachable for some restricted classes of machines. The finite automata discussed above always perform their computations in real-time, so we are led as in [4] to the following hypothesis: the decimal expansion of an algebraic irrational is not semi-regular.

We shall formulate a transcendence statement which would resolve the problem posed above. We then explain how the transcendence proof by Mahler's method necessitates some technical hypotheses which prevent us from solving the problem completely.

Let T be a $p \times p$ matrix with non-negative integer entries and define the multiplicative action of T on C^p as before. Let $f_1(z), \ldots, f_m(z)$ be locally analytic functions of the p complex variables $z = (z_0, \ldots, z_{p-1})$, that is functions which are regular in some neighbourhood of the origin, and suppose they have algebraic Taylor coefficients. Further, suppose that the vector $f(z) = (f_1(z), \ldots, f_m(z))^t$ satisfies the system of functional equations $f(z) = A(z)f(Tz)$, where $A(z)$ is an $m \times m$ matrix whose entries are rational functions with algebraic coefficients. Let ζ be a proper algebraic point of C^p, that is a point with non-zero algebraic coordinates. We aim to show that if the functions $f_{i_1}(z), \ldots, f_{i_k}(z)$ are algebraically independent over $C(z)$, then the numbers $f_{i_1}(\zeta), \ldots, f_{i_k}(\zeta)$ are algebraically independent over Q. Given such a theorem, we could apply it to the functions $G(z)$, $F_0(z)$, $\ldots$, $F_{p-1}(z)$ of Section 3 and the point $\zeta = (q^{-1}, \ldots, q^{-1})$. Since $G(z, \ldots, z)$ is a power series with integer coefficients and radius of convergence equal to 1, it is either a rational function, or has the unit circle as a natural boundary and so is transcendental. Thus $G(q^{-1}, \ldots, q^{-1})$ is either rational or transcendental and we see that the decimal expansion of an algebraic irrational cannot be semi-regular.

Now for the caveats. An essential feature of the transcendence proof is the so-called vanishing theorem. We need to impose conditions on the matrix T and on the point ζ so that a non-trivial locally analytic function $f(z)$ cannot satisfy $f(T^k\zeta) = 0$ for all sufficiently large integers k. This problem has been completely solved by Masser [16]. His theorem allows for substantial extension of earlier work in this area, but also shows that we must make the following assumptions:

(1) the matrix T is non-singular and none of its eigenvalues is a root of unity,

(2) $T^k\zeta \to 0$ as $k \to \infty$, and

(3) there is no non-trivial integer p-tuple μ such that $(T^k\zeta)^\mu = 1$ for all integers k in some arithmetic progression.

From (1), it seems that Mahler's method cannot give results on the transcendence of theta functions, such as

$$\theta(z, w) = \sum_{n=1}^{\infty} w^n z^{n(n-1)}.$$

This function satisfies the functional equation $\theta(z, w) = w\theta(z, w^2 z) + w$, corresponding to the matrix $T = \begin{pmatrix} 1 & 0 \\ 1 & 2 \end{pmatrix}$. Since $\theta(\frac{1}{2}, \frac{1}{2}) = \sum 2^{-n^2}$, it is the generating function for the sequence of squares which we see to be a semi-regular sequence. This appears to be one of the functions which is not accessible to any of the current transcendence methods.

The transcendence argument itself comes down to a comparison of the algebraic size and absolute value of certain auxiliary functions evaluated at the points $T^k \zeta$. To make this work, we need a quantitative version of (2), as follows. Let $r = r(T)$ denote the maximum of the absolute values of the eigenvalues of T, so that r itself is an eigenvalue of T. Then we require

(4) $r > 1$,

(5) there is a positive eigenvector belonging to the eigenvalue r, and

(2 bis) $\log |T^k \zeta| \leq -cr^k$ for some positive constant c and all sufficiently large k.

In the application to automata with $\zeta = (q^{-1}, \dots, q^{-1})$, it is clear that (2 bis) follows from (4) and (5).

Finally, we need some assumptions on the system of functional equations $f(z) = A(z)f(Tz)$ to ensure that we can pass backwards and forwards between $f(z)$ and $f(T^k z)$. From the given functional equation, we obtain by iteration

$$f(z) = A^{(k)}(z)f(T^k z), \qquad A^{(k)}(z) = A(z)A(Tz) \dots A(T^{k-1} z).$$

In [14], we assume

(6) $A^{(k)}(\zeta)$ is defined and non-singular for each $k \geq 1$, and

(7) $A(0)$ is defined and non-singular.

The main theorem of [14] is the natural algebraic independence theorem obtained under the assumptions (1), (2 bis), (3)–(7). Of these assumptions, (1), (4) and (5) are apparently unavoidable restrictions on the class of automata accessible to the method, (2 bis) is automatic and (3) and (6) can very likely be realized by modifying the translation from the automaton to the functional equations to remove unnecessary variables and functions from the system. (Unfortunately, I do not see how to guarantee this in general.) In [14], we need (7) to obtain a fairly explicit description of the dependence of $A^{(k)}(z)$ on k and this is probably a defect in the argument. If we take the more modest aim to prove that an algebraic irrational is not regular, then we have generating functions of a single variable and the list of requirements reduces to (6) and (7). In this case, I believe that I have an argument which will eliminate the need for the hypothesis (7).

5. Base dependence and more on automata.

Digit patterns which are recognizable in one base may look quite random in another. Consider, for example, the digits of the successive powers of 2 in base 2 and in base 10. Cobham [5] shows that this is a characteristic of regular sequences. More precisely, suppose that s and t are multiplicatively independent positive integers, that is $s^m t^n \neq 1$ for integers m and n not both zero, so that the bases s and t are different. Then any sequence which is regular in both base s and base t must be periodic and hence regular in every base.

Cobham's proof in [5] is elementary, but curiously uncommunicative. The many extensions which come to mind do not seem to yield to the same methods. For example, we can ask to what extent a semi-regular sequence determines its associated transformation T. In this direction, we make the following conjecture. Let S and T be $p \times p$ matrices of non-negative integers satisfying condition (1) of Section 4. Suppose S and T are independent in the sense that if S' and T' are irreducible components of S and T then $S'^m T'^n \neq I$ for integers m and n not both zero. Suppose $f(z)$ is a locally analytic function satisfying the functional equations

$$f(z) = a_0(z) + \sum_{i=1}^{m} a_i(z) f(S^i z), \qquad f(z) = b_0(z) + \sum_{j=1}^{n} b_j(z) f(T^j z)$$

with rational coefficients $a_i(z)$ and $b_j(z)$. Then $f(z)$ must be a rational function. (Note that functional equations of this shape are equivalent to systems of functional equations for the vectors

$$(1, f(z), f(Sz), \ldots, f(S^{m-1}z)) \quad \text{and} \quad (1, f(z), f(Tz), \ldots, f(T^{n-1}z)).)$$

If the matrices S and T and the functional equations satisfy the extra requirements (2 bis), (4), (5) and (7) of Section 4, this conjecture can be attacked by transcendence theory as in [13]. We introduce the $2p$ variables $z = (z_S, z_T)$, where z_S and z_T are p-tuples, and the transformation $Ц = \begin{pmatrix} S & 0 \\ 0 & T \end{pmatrix}$ defined by $Цz = (Sz_S, Tz_T)$. The functions $\left(1, f(z_S), \ldots, f(S^{m-1}z_S), f(z_T), \ldots, f(T^{n-1}z_T)\right)$ satisfy a system of functional equations involving the transformation $Ц$. The transcendence theory does not apply directly to $Ц$ because it does not have the mixing property (5) of Section 4. However, we can find a sequence of transformations

$$Ц_k = \begin{pmatrix} S^{\sigma_k} & 0 \\ 0 & T^{\tau_k} \end{pmatrix}$$

where σ_k and τ_k are positive integers chosen so that

$$\frac{\sigma_k}{\tau_k} \sim \frac{\log r(T)}{\log r(S)}.$$

(Here, $r(T)$ is the spectral radius of T, as before.) The transcendence theory of Section 4 can be generalized to deal with such sequences of transformations with $Ц_k$ replacing the T^k of the earlier work. By the assumed independence of S and T, we can choose $z_S = z_T = \zeta$, say, to be a suitable algebraic point and meet requirement (3) of Section 4. The values of the functions $f(z_S)$ and $f(z_T)$ when $z_S = z_T = \zeta$ are clearly dependent, so the functions themselves must be dependent, that is $f(z)$ must be an algebraic function. In most examples, it is now easy to argue that $f(z)$ must be rational, as expected. In general, this raises the interesting problem: are algebraic solutions of the functional equations considered here always rational? There are some results in this direction in the work of Kubota [10] and Nishioka [20].

We can hope to push the transcendence theory further along the lines suggested above to deal with chains of systems of functional equations. For example, suppose $T_0, T_1, \ldots$ are $p \times p$ matrices of non-negative integers and $f_0(z), f_1(z), \ldots$ are m-tuples of locally analytic functions satisfying functional equations $f_i(z) = A_i(z)f_{i+1}(T_i z)$ of the same shape as those of Section 4. The results of [12] should extend to this situation, yielding a transcendence theorem for the functions in the m-tuple $f_0(z)$ at suitable algebraic points. Such a theorem should have applications to push-down automata.

6. Transcendence measures

One of the nice features of Mahler's method is that it does not require any difficult estimation. This is true, in particular, in the construction of the auxiliary function whose coefficients play only a secondary rôle in the later work and do not need to be estimated at all. On the other hand, with the aid of appropriate estimates, the whole construction can be made effective, leading to reasonably explicit transcendence measures.

The best transcendence measure for functions of several variables is that of Galochkin [8]. Suppose $f(z)$ is a transcendental function of the p complex variables $z = (z_0, z_1, \ldots, z_{p-1})$ and satisfies $f(z) = a(z)f(Tz) + b(z)$. Suppose, in addition to the hypotheses of Section 4, that the characteristic polynomial of T is irreducible over $\mathbf{Q}$ and $a(z)$ and $b(z)$ are polynomials. Let P be a non-trivial polynomial with rational coefficients, degree at most d and height at most H. Then

$$\left|P\big(f(\zeta)\big)\right| > H^{-cd^{2p-1}},$$

with $c = c(f, \zeta) > 0$, providing $H > H_0(d)$.

Algebraic independence measures present a bigger challenge. Nesterenko [18] has applied ideas from commutative algebra together with his estimates for the heights of ideals to obtain a measure for functions of one variable. Suppose, with the usual hypotheses, that the functions $f_1(z), \ldots, f_m(z)$ are algebraically independent and satisfy $f_i(z) = a_i(z)f_i(z^r) + b_i(z)$. Let P be a non-trivial polynomial with rational coefficients, degree at most d and height at most H. Then

$$\left|P\big(f_1(\zeta), \ldots, f_m(\zeta)\big)\right| > H^{-cd^m},$$

with $c = c(f_i, \zeta) > 0$, providing $H > H_0(d)$. It would be interesting to know whether these measures can reflect any properties of the automata associated with the functional equations.

In another direction, Nishioka [19] has extended Mahler's method to solutions of the functional equation $P\big(z, f(z), f(z^r)\big) = 0$, where $P(z, u, v)$ is a polynomial with algebraic coefficients. Here it is also necessary to invoke a quantitative form of Siegel's lemma in the construction of the auxiliary function and this leads to an extra hypothesis on the coefficients of the solution $f(z) = \sum a_n z^n$, say. If the height h_n of the vector $(a_0, a_1, \ldots, a_n)$ satisfies $\log h_n \ll n^\lambda$ and $\deg_v P < r^{1/\lambda}$, then $f(\zeta)$ is transcendental for algebraic ζ with $0 < |\zeta| < 1$. Unfortunately, the extra hypothesis means that the result cannot be applied to such interesting functions as $j(\log z/2\pi i)$, where j is the modular invariant.

References

[1] J.-P. Allouche and M. Mendès France, Suite de Rudin-Shapiro et modèle d'Ising, *Bull. Soc. Math. France* **113** (1985), 273–283.

[2] E. Bombieri and J. E. Taylor, Quasicrystals, tilings and algebraic number theory: some preliminary connections, (to appear)

[3] C. Christol, T. Kamae, M. Mendès France and G. Rauzy, Suites algébriques, automates et substitutions, *Bull. Soc. Math. France* **108** (1980), 401–419.

[4] A. Cobham, On the Hartmanis-Stearns problem for a class of tag machines, *Technical report RC 2178, IBM Research Centre*, Yorktown Heights, New York, 1968.

[5] A. Cobham, On the base dependence of sets of numbers recognizable by finite automata, *Math. Systems Theory* **3** (1969), 186–192.

[6] A. Cobham, Uniform tag sequences, *Math. Systems Theory* **6** (1972), 164–192.

[7] M. Dekking, M. Mendès France and A. van der Poorten, Folds! *Math. Intelligencer* **4** (1982), 130–138, 173–181, 190–195.

[8] A. I. Galochkin, Transcendence measures for the values of functions satisfying certain functional equations, *Mat. Zametki* **27** (1980), 175–183.

[9] J. Hartmanis and R. E. Stearn, On the computational complexity of algorithms, *Trans. Amer. Math. Soc* **117** (1965), 285–306.

[10] K. K. Kubota, On the algebraic independence of holomorphic solutions of certain functional equations and their values, *Math. Ann.* **227** (1977), 9–50.

[11] J. H. Loxton and A. J. van der Poorten, Transcendence and algebraic independence by a method of Mahler, in A. Baker and D. W. Masser (eds.), *Transcendence theory: advances and applications*, (Academic Press, 1977).

[12] J. H. Loxton and A. J. van der Poorten, Arithmetic properties of certain functions in several variables, III, *Bull. Austral. Math. Soc.* **16** (1977), 15–47.

[13] J. H. Loxton and A. J. van der Poorten, Algebraic independence properties of the Fredholm series, *J. Austral. Math. Soc* (A) **26** (1978), 31–45.

[14] J. H. Loxton and A. J. van der Poorten, Arithmetic properties of the solutions of a class of functional equations, *J. reine angew. Math.* **330** (1982), 159–172.

[15] K. Mahler, Fifty years as a mathematician, *J. Number Theory* **14** (1982), 121–155.

[16] D. W. Masser, A vanishing theorem for power series, *Invent. Math.* **67** (1982), 275–296.

[17] M. Mendès France, Nombres algébriques et théorie des automates, *L'Ens. Math.* (2) **26** (1980), 193–199.

[18] Ju. V. Nesterenko, On the estimates for the measures of algebraic independence of numbers and functions, (to appear).

[19] K. Nishioka, On a problem of Mahler for transcendency of function values, *J. Austral. Math. Soc.* (A) **33** (1982), 386–393.

[20] K. Nishioka, Algebraic function solutions of a certain class of functional equations, *Arch. Math.* **44** (1985), 330–335.

14

THE STUDY OF DIOPHANTINE EQUATIONS
OVER FUNCTION FIELDS

R. C. Mason

1. Introduction

In recent years there has been important progress in the study of Diophantine equations over function fields. This has aroused increasing interest in the subject, and has led to the re-examination of some old results, as well as the realisation that there is now a wide variety of areas to which the subject is applicable. We shall spend a little time on the history and present state of the subject before proving a specific result concerning decomposable form equations. This result forms the completion of a series of works which attack several general classes of equations by means of an important inequality on solutions of the multivariate unit equation.

There are now three distinct analytical approaches to the study of Diophantine equations over function fields. The first is that of algebraic geometry, the second that of differential equations, and the third that of Diophantine approximation.

The celebrated theorem of Manin and Grauert established the analogue of Mordell's conjecture (now Faltings' theorem) for function fields, on the finiteness of the number of solutions of equations in two variables. This confirmed the great progress made in algebraic geometry since the Second World War. Grauert's approach was heavily dependent on esoteric algebraic geometry, and his work led to further advances in that subject by Shafarevitch and Paršin. Although the Manin-Grauert result was extremely general, for many years it was believed that their methods would not lead to effective bounds on the actual solutions of equations, nor would it allow the solutions themselves to be determined explicitly. It may also be remarked that effectiveness was not then considered quite so important as now, at least in certain influential circles. However in 1973 in his work on the development of the approach of Grauert and Shafarevitch, Paršin showed that it was indeed possible to calculate in principle effective bounds on the solutions. The bounds he calculated were extremely large, being multiply exponential. Recently, as effective methods have reasserted themselves in importance, Beukers has developed the work of Grauert and Paršin to produce quite explicit bounds for many general classes of equations in two variables. His bounds are not multiply exponential, although they are too large for numerical computation of solutions.

The second approach to the study of Diophantine equations over function fields is that of differential equations. By developing work of Kolchin, Osgood and later Schmidt were able to establish important results giving explicit bounds on the solutions of various families of equations. This approach has been developed extensively, and the techniques have been found to have more general applications, to meromorphic functions and Nevanlinna theory. The drawback in this approach is that although the functional dependence of the bounds is good, the numerical constants tend to be quite large. This defect makes the actual computation of solutions impractical. Furthermore, the most recent advances have not produced numerical constants as actual numbers but 'effective constants which may be determined in principle', a coded phrase for very large numbers indeed.

The third approach to the subject, the one developed further in this paper, is that of Diophantine approximation. In 1982 the author established a fundamental inequality concerning Diophantine approximation in function fields. The result may be described as a loose analogue of Baker's celebrated results on linear forms in the logarithms of algebraic numbers, from which the vast majority of work presented at this conference traces its source. The present author's inequality of 1982 has since led to a series of applications to general classes of equations over function fields, such as the Thue equation, the hyperelliptic equation, equations of genera zero and one, and so on. The principal distinguishing features of these results from those of previous authors were first, explicit bounds were readily produced, with very small numerical constants and secondly, the sharpness of the bounds, coupled with the directness of the method, permitted many equations to be solved explicitly, and surprisingly without any machine computation. This was an important advance. More recently a generalization of the fundamental inequality of 1982 has been established, concerning equations in several variables. This multivariate form has had a number of important applications to equations for which effective bounds are not known, and believed by some to be impossible, in the analogous case of number fields. The object of the second part of this paper is to illustrate one of these applications with the solutions of the central class of decomposable form equations over function fields.

Finally in this introduction we must mention two important signs of progress in the dissemination of our approach. The first concerns the so-called 'ABC' conjecture of Oesterlé. Based on his extensive study of elliptic curves, in 1985 Oesterlé formulated a conjecture relating the additive and multiplicative behaviour of numbers. If true, his conjecture would resolve many outstanding questions, such as the Fermat problem and the Cassels conjecture. The conjecture was sharpened by Masser, to a familiar form involving an arbitrarily small positive ϵ and a constant depending on ϵ. It is a remarkable fact that the analogue of this conjecture for function fields has already been established; indeed it is a special case of our fundamental inequality proved in 1982. Moreover, the analogue has been established in the best possible form, with ϵ zero and the constant equal to unity. The second development concerns the application of our methods to the study of exponential Diophantine equations. In a contribution to these proceedings, Brindza recounts how the combination of our fundamental inequality together with the linear forms results of Baker and its p-adic analogue has enabled him to succeed in establishing bounds on various classes of exponential Diophantine equations. This combination of Diophantine approximation over function fields with that over number fields had earlier been employed by Győry, but these new results are particularly interesting, and augur well for the future health of the subject.

2. Preliminaries

The rest of this paper aims to provide a complete resolution of the general decomposable form equation over function fields. Recently a study was made [5] of the general norm form equation

$$\text{Norm}_{K/L}(x) = c$$

over function fields, where x lies in a given module $\mathcal{M}$. Here K and L denote finite extensions of the rational function field $k(z)$: k is an algebraically closed field of characteristic zero presented explicitly, c is a non-zero element of L and $\mathcal{M}$ is a free $\mathcal{O}_L$-module in K, where $\mathcal{O}_L$ is the integral closure of $k[z]$ in L. The principal element in the analysis in the norm form equation was a bound on the solutions of the general unit equation

$$u_1 + \ldots + u_n = 1$$

in which the poles and zeros of $u_1, \ldots, u_n$ are constrained to lie in some fixed finite set. The analysis yielded an algorithm for the construction of all solutions x in $\mathcal{M}$ and a bound on their heights, provided only that $\mathcal{M}$ is non-degenerate (see [5]). Degenerate modules were studied in an earlier paper [8].

This paper is concerned with the more general equation

$$F(x_1, \ldots, x_n) = c, \qquad (*)$$

where F is a decomposable form, that is, it factorises

$$F(x_1, \ldots, x_n) = \prod_{i=1}^{m} L_i(x_1, \ldots, x_n),$$

where $m > n$ and each L_i is a linear form

$$L_i(x_1, \ldots, x_n) = \sum_{j=1}^{n} a_{ij} x_j,$$

and the coefficients a_{ij} lie in the finite extension K of $k(z)$. We shall be concerned with the solutions of $(*)$ with $x_1, \ldots, x_n$ in $\mathcal{O}(= \mathcal{O}_K)$. There are three obvious circumstances when these solutions are infinite in number. First, this occurs if each a_{ij} and c lies in the ground field k, as then $(*)$ is defined over the algebraically closed field k, and so has infinitely many solutions there. Secondly, if there exist $y_1, \ldots, y_n$ in K, not all zero, with $L_j(y_1, \ldots, y_n) = 0$ for $1 \leq j \leq m$, then $x_i = z_i + t y_i$, $1 \leq i \leq n$, forms a solution of $(*)$ whenever $z_1, \ldots, z_n$ does, so that the solutions will have unbounded height. Thirdly, suppose that $(*)$ splits, so that $L_1, \ldots, L_a$ involve $x_1, \ldots, x_b$ only, and $L_{a+1}, \ldots, L_m$ involve $x_{b+1}, \ldots, x_n$ only. Then, whenever $x_1, \ldots, x_n$ is a solution of $(*)$, and η is any non-trivial unit in $\mathcal{O}$, $\eta^{m-a} x_1, \ldots, \eta^{m-a} x_b, \eta^{-a} x_{b+1}, \ldots, \eta^{-a} x_n$ is also a solution of $(*)$: as η varies, we obtain solutions of unbounded height. We will show that these three provide the only types of circumstances in which the solutions of $(*)$ in $\mathcal{O}^n$ are infinite in number, and that if none obtains, so that there are only finitely many solutions, then these solutions may be determined effectively; moreover, we shall construct a bound on the heights of the solutions. In the proof we again exploit the inequality on solutions of the unit equation mentioned above. However we require a slightly more general result than previously, so we prove it anew (see Lemma below). It may be helpful in this context to recall an important special case of $(*)$, when $n = 2$: in [3] a general algorithm was given for the resolution of the Thue equation

$$(X - \alpha_1 Y) \ldots (X - \alpha_m Y) = \mu$$

over function fields. It was proved that if at least three of $\alpha_1, \ldots, \alpha_m$ are distinct then there are only finitely many solutions X, Y in $\mathcal{O}$ provided that there is no bilinear substitution with coefficients in $\mathcal{O}$ which transforms the Thue equation into an equation $f(x, y) = 1$ with coefficients in the ground field k. This theorem on the Thue equation was established by means of an inequality concerning the heights of the solutions of the unit equation

$$u_1 + u_2 = 1.$$

We demonstrated in [3] that this inequality for algebraic functions is analogous to Baker's celebrated result on linear forms in the logarithms of algebraic numbers, which led to his effective bounds for the solutions of the Thue equation over number fields [1]. It is an important unsolved Diophantine problem to provide effective bounds for the general norm form and decomposable form equations over number fields: the ineffective result is in the famous paper of Schmidt [9]. The most general known effective result in Győry's [2]: his analysis succeeds precisely for those equations which give rise to a unit equation $u_1 + u_2 = 1$, which may be treated using Baker's method.

Let us recall briefly the notions of valuation, derivation and height on function fields. The most convenient reference for this, which also contains the full background to the Thue equation and others over function fields, is [4] Chapter I, §2. We remember that k is an algebraically closed field of characteristic zero presented explicitly, $k(z)$ is the rational function field over k and K is a finite extension of $k(z)$, of genus g say. Associated with K there is a projective curve $\mathcal{C}$, whose points correspond to valuations on K. Each valuation v is given as the order of vanishing of the Laurent expansion in powers of a local parameter z_v. Thus the valuations are written additively with value group $\mathbf{Z}$. Now

$$\mathcal{O} = \{f \in K : v(f) \geq 0 \text{ whenever } v(z) \geq 0\}.$$

The valuations v with $v(z) < 0$ are termed *infinite*; the number of such is denoted by r. We also define

$$\mathcal{O}_\mathcal{V} = \{f \in K : v(f) \geq 0 \text{ for } v \notin \mathcal{V}\}$$

and

$$\mathcal{O}_\mathcal{V}^* = \{f \in K : v(f) = 0 \text{ for } v \notin \mathcal{V}\}$$

for any finite subset $\mathcal{V}$ of valuations; the latter is termed the group of $\mathcal{V}$-units of K. The height of an element f in K is given by

$$H(f) = -\sum_{v \in \mathcal{C}} \min\bigl(0, v(f)\bigr);$$

that is, the number of poles of f counted according to multiplicity. If $\mathcal{F}$ is a finite subset of K then we define

$$H(\mathcal{F}) = -\sum_{v \in \mathcal{C}} \min\big(0, v(\mathcal{F})\big),$$

where $v(\mathcal{F}) = \min\{v(f) : f \in \mathcal{F}\}$. If $\underline{x} = (x_1, \ldots, x_n)$ is a vector in K^n then we define $H(\underline{x}) = H\big(\{x_1, \ldots, x_n\}\big)$.

The mapping $f \mapsto f'$ on $k(z)$, termed differentiation with respect to z, extends uniquely to a global derivation on K. Moreover, since each f may be expanded locally in terms of the local parameter z_v, we obtain local derivations $\frac{df}{dv}$ by differentiating with respect to z_v. The genus formula states that

$$2g - 2 = \sum_v v\left(\frac{df}{dv}\right) \tag{1}$$

for any f in $K \setminus k$. The following results on differentiation of a power series are noted:

$$v\left(\frac{df}{dv}\right) = v(f) - 1 \quad \text{if } v(f) \neq 0,$$

and

$$v\left(\frac{df}{dv}\right) \geq 0 \qquad \text{if } v(f) \geq 0. \tag{2}$$

We now wish to state our main theorem. However let us first formulate precisely the three exceptional cases mentioned above which allow (∗) to have infinitely many solutions. The first of these we shall include in the statement of the theorem. The second is easy to formulate: for each i, $1 \leq i \leq m$, let K_i denote the subspace of K^n consisting of those $(x_1, \ldots, x_n)$ for which $L_i(x_1, \ldots, x_n) = 0$; the required condition is then $\bigcap_{i=1}^m K_i = \{0\}$. The third is more complicated to express, as the splitting exhibited may occur only after some linear transformation, or when some of the variables are predetermined. Here we state the precise assumption made: its necessity will be derived in §5. For each subspace W of K^n with $W \not\subseteq \bigcup_{i=1}^m K_i$, the linear forms $L_1, \ldots, L_m$ induce non-zero linear maps $f_1, \ldots, f_m$ from W to K. We introduce a graph G_W with vertices $\{1, \ldots, m\}$, and an edge $\{i, j\}$ whenever there is a *minimal* linear relation $\sum_{p \in I} \alpha_p f_p = 0$ valid on W, where the non-zero coefficients α_p lie in K, and I contains both i and j. The assumption is that the graph G_W is connected for every such subspace W. If we assume that $\mathcal{O}^*$ contains a non-trivial unit, then this involved criterion is in fact necessary for the solutions of (∗) to have bounded height for all c. The theorem shows that it is also sufficient.

Theorem. Consider the equation $(*)$ above, in which we assume that $\bigcap_{i=1}^{m} K_i = \{0\}$, and that for each subspace W of K^n not contained in any K_i, the graph G_W is connected. Then all the solutions of $(*)$ with $x_1, \ldots, x_n$ in $\mathcal{O}$ may be determined effectively. Furthermore, their heights are bounded:

$$H(x_1, \ldots, x_n) \le 4^{n^2} \left(H(F) + H(c) + g + r \right). \qquad (3)$$

Finally, there are an infinite number of solutions if and only if there exists a substitution of the form

$$x_i = \sum_{j=1}^{p} b_{ij} \lambda_j, \quad b_{ij} \in \mathcal{O},$$

such that $(*)$ is satisfied whenever $\lambda_1, \ldots, \lambda_p$ satisfy $F^*(\lambda_1, \ldots, \lambda_p) = 1$, where F^* is a non-zero decomposable form with coefficients in the ground field k.

In the next section we shall state and prove our lemma on the solutions of the general unit equation. In §3 we shall derive the algorithm by which all the solutions of $(*)$ may be determined effectively, and in §4 this algorithm will be explored to yield the requisite bound (3) on the heights of the solutions. Finally, in §5 we debate the necessity of the hypotheses in the theorem.

3. The Unit Equation

Suppose that $\underline{x} = (x_1, \ldots, x_n)$ is a solution in $\mathcal{O}^n$ of $F(\underline{x}) = c$. Since the graph G_K is connected by assumption, we obtain some linear relations

$$\sum_{i \in I} \alpha_i L_i(\underline{x}) = 0$$

for all $\underline{x}$ in K^n. Since $\underline{x}$ lies in $\mathcal{O}^n$, the poles of each $L_i(\underline{x})$ lie in a fixed finite set; moreover, since their product is equal to c, the zeros of each $L_i(\underline{x})$ also lie in a fixed finite set. We may therefore rewrite the equation above as

$$\sum_{i=1}^{p} u_i = 0,$$

where $v(u_i) = 0$ for every valuation v outside a fixed set $\mathcal{V}$. This equation is called the general unit equation. In [5] we showed that, provided the equation $\sum_{i=1}^{p} u_i = 0$ is a minimal linear relation over k, the height of $H(u_2/u_1,\ldots,u_p/u_1)$ is bounded above by $4^{p-2}(|\mathcal{V}|+2g)$, where $|\mathcal{V}|$ denotes the cardinality of $\mathcal{V}$. Here we shall weaken the assumption that the relation is a minimal linear dependence over k, and show that the bound remains valid.

Lemma. Suppose that $u_1,\ldots,u_n$ are elements of K such that $\sum_{i=1}^{n} u_i = 0$, but that $\sum_{i\in I} u_i$ is non-zero for every proper non-empty subset I of $\{1,\ldots,n\}$. Suppose further that there is a finite set of valuations $\mathcal{V}$ on K such that $v(u_i) = 0$ for each v outside $\mathcal{V}$, $i = 1,\ldots,n$. Then

$$H(u_2/u_1,\ldots,u_n/u_1) \le 4^{n-2}(|\mathcal{V}|+2g).$$

We remark that it is necessary to assume that for no proper subset I is $\sum_{i\in I} u_i = 0$. For otherwise, choosing $u_i^* = u_i$ for i in I, and $u_i^* = uu_i$ for i outside I yields a new set of solutions with $\sum_{i=1}^{n} u_i^* = 0$, and as u varies over the $\mathcal{V}$-units, $H(u_2^*/u_1^*,\ldots,u_n^*/u_1^*)$ is unbounded if $\mathcal{O}_{\mathcal{V}}^*$ is non-trivial. We shall say that $u_1,\ldots,u_n$ is an *irreducible* solution of $\sum_{i=1}^{n} u_i = 0$ if $\sum_{i\in I} u_i$ is non-zero for each proper non-empty subset I of $\{1,\ldots,n\}$.

Proof of Lemma. The result plainly holds for $n = 1$ or 2: we shall prove it for remaining values by induction. Let us denote by $\mathcal{V}_i$, $2 \le i \le n$, the set of valuations v or K such that $v(u_i/u_1) \ne 0$, so each $\mathcal{V}_i$ is contained in $\mathcal{V}$, and if $N = \max_{2\le i\le n} |\mathcal{V}_i|$, then $N \le |\mathcal{V}|$. We shall actually prove that, if $w_i = u_i/u_1$, $i = 1,\ldots,n$,

$$H(w_2,\ldots,w_n) \le A_n N + B_n g \qquad (4)$$

for some functions A_n and B_n of n; our inductive hypothesis is that (4) holds for all irreducible equations in K with fewer that n terms: A_n and B_n will be determined by a recurrence relation, and so the proof of the lemma will be complete when we establish (4), and that $A_n \le 4^{n-2}$, $B_n \le 2.4^{n-2}$. We now divide the equation $\sum_{i=1}^{n} u_i = 0$ by u_1, and differentiate to obtain

$$w_2' + \ldots + w_n' = 0. \qquad (5)$$

If (5) is not irreducible, then we may rearrange $u_2,\ldots,u_n$ so that, for some s, $2 \le s < n$,

$$w_2' + \ldots + w_s' = w_{s+1}' + \ldots + w_n' = 0.$$

We deduce that there is some λ in k such that

$$\lambda u_1 + u_2 + \ldots + u_s = (1-\lambda)u_1 + u_{s+1} + \ldots + u_n = 0.$$

Since the original equation $\sum_{i=1}^{n} u_i = 0$ is irreducible, we deduce that each of these two equations is irreducible and that $\lambda \neq 0, 1$. Hence by the inductive hypothesis we have

$$H(w_2, \ldots, w_s) \leq A_s N + B_s g$$

and

$$H(w_{s+1}, \ldots, w_n) \leq A_{n-s+1} N + B_{n-s+1} g,$$

and thus we obtain the desired inequality (4), provided that $A_s + A_{n-s+1} \leq A_n$ for each s, $2 \leq s \leq n$, and that the same inequality holds for the sequence B_s.

The result (4) has thus been proved when the differentiated equation (5) is reducible, so let us suppose instead that (5) is irreducible. By the inductive hypothesis we have

$$H(w_3'/w_2', \ldots, w_n'/w_2') \leq A_{n-1} N' + B_{n-1} g,$$

where $N' = \max_{3 \leq i \leq n} |\mathcal{V}_i'|$ and $\mathcal{V}_i'$ is the set of valuations v with $v(w_i') \neq v(w_2')$. The inductive step is completed in two stages: we shall derive from this last inequality an upper bound for $H(w_2, \ldots, w_n)$ in terms of N', and also establish an upper bound for N'. To establish the upper bound for N' we denote by $\mathcal{W}_i$, $2 \leq i \leq n$, the set of valuations v such that $v\left(\frac{dw_i}{dv}\right) \neq 0$. Now from inequalities (2) and the genus formula (1) we obtain

$$|\mathcal{V}_i| + 2g - 2 = \sum_{v \in \mathcal{W}_i \setminus \mathcal{V}_i} v\left(\frac{dw_i}{dv}\right), \qquad (6)$$

and each term in the sum on the right is a positive integer. We deduce that

$$|\mathcal{W}_i| \leq 2|\mathcal{V}_i| + 2g.$$

Since $\mathcal{V}_i' \subset \mathcal{W}_i \cup \mathcal{W}_2$ for each i, we deduce that

$$N' \leq |\mathcal{W}_2| + \max_{3 \leq i \leq n} |\mathcal{W}_i| \leq 4N + 4g. \qquad (7)$$

In order to derive an inequality for $H(w_2, \ldots, w_n)$, we first obtain an upper bound on the difference

$$H(w_3/w_2, \ldots, w_n/w_2) - H(w_3'/w_2', \ldots, w_n'/w_2') = \sum_v n_v,$$

where

$$n_v = \max_{3 \leq i \leq n} \left(0, v(w_2) - v(w_i)\right) + \min_{3 \leq i \leq n} \left(0, v\left(\frac{dw_i}{dv}\right) - v\left(\frac{dw_2}{dv}\right)\right).$$

We examine each n_v as v ranges over the valuations on K. If $v(w_2) \le v(w_i)$ for each i, then $n_v \le 0$. If $v(w_2) > v(w_i)$ for some i, then we may select i with $v(w_i)$ minimal. If in addition $v(w_i) \ne 0$ then from (2)

$$v\left(\frac{dw_i}{dv}\right) = v(w_i) - 1 \quad \text{and} \quad v\left(\frac{dw_2}{dv}\right) \le v(w_2) - 1,$$

so again $n_v \le 0$. If on the other hand $v(w_i) = 0$, then $v(w_2) > 0$ so v lies in $\mathcal{V}_2 \setminus \mathcal{V}_i$, and again from (2)

$$n_v \le 1 + \left(\frac{dw_i}{dv}\right).$$

Hence

$$\sum_v n_v \le |\mathcal{V}_2| + \sum_{i=3}^{n} \sum_{v \in W_i \setminus \mathcal{V}_i} v\left(\frac{dw_i}{dv}\right) \le N + (n-2)(N+2g),$$

using (6) again. Finally, from [4] p. 9 we have

$$H(w_2,\ldots,w_n) = H(1,w_2,\ldots,w_n)$$
$$= H(1/w_2, 1, w_3/w_2, \ldots, w_n/w_2) = H(w_3/w_2, \ldots, w_n/w_2)$$

since $1 + (w_3/w_2) + \ldots + (w_n/w_2) = -1/w_2$. We conclude that

$$H(w_2,\ldots,w_n) \le (4A_{n-1} + n - 1)N + (4A_{n-1} + B_{n-1} + 2n - 4)g.$$

Solving the recurrence relations $A_n = 4A_{n-1} + n - 1$, $B_n = 4A_{n-1} + B_{n-1} + 2n - 4$ with $A_2 = B_2 = 0$, it may be verified that $A_n \le 4^{n-2}$, $B_n \le 2 \cdot 4^{n-2}$, $A_s + A_{n-s+1} \le A_n$ and $B_s + B_{n-s+1} \le B_n$ for $2 \le s < n$, as required. The proof of the lemma is thus complete.

We conclude this section with some remarks on the coefficient 4^{n-2} in the statement of the lemma.* The actual values of A_n and B_n given by the recurrences are asymptotic to $\frac{7}{9}4^{n-2}$ and $\frac{7}{27}4^{n-1}$ respectively. As in [5] it may be seen that a more naïve proof, using an inductive hypothesis involving $|\mathcal{V}|$ rather than N, would yield A_n growing as fast as $n!$. Finally, we remark that the best possible values of A_n and B_n may

---

* This coefficient has recently been improved to $(n-1)(n-2)/2$. This improvement leads to a sharpening of the bound in our Theorem. See W. D. Brownawell and D. W. Masser, Vanishing Sums in Function Fields, *Math. Proc. Camb. Phil. Soc.* 1986.

be conjectured to be of order n by writing $u_1, \ldots, u_n$ in terms of a basis of the unit group $\mathcal{O}_\mathcal{V}^*$, and comparing the number of unknowns in this representation, approximately $n(N + g)$, with the number of equations over k to be satisfied, approximately H: these equations correspond to the vanishing of the negative terms in the Laurent expansions of $\sum_{i=1}^n u_i$ at points in $\mathcal{V}$. It is evident A_n and B_n are at least of order n: this can be seen by considering the solution $u_i = \binom{n}{i}(1-f)^i f^{n-i}$ of $\sum_{i=1}^n u_i = 1$ for any non-constant f in K. In the present paper the loss in A_n and B_n serve only to affect the final bound (3) on the solutions of $(*)$, and this bound is still linear with respect to H. However, in [7] we proved that the solutions $\underline{x}$ in *vector space* V of

$$\mathrm{Norm}_{K/L}(\underline{x}) = c$$

are of bounded height. The condition that V must satisfy depends on the value of A_n in the inequality above; in other words, an improvement on A_n will achieve bounds on the solutions of more equations, not merely better bounds as here.

4. The Algorithm

We shall now utilise the lemma proved above to derive the algorithm for the construction of the solutions $x_1, \ldots, x_n$ in $\mathcal{O}$ of $F(x_1, \ldots, x_n) = c$ as requested in our theorem. Let $\mathcal{M}$ denote the $\mathcal{O}$-module $\mathcal{O}^n$, so $\mathcal{M}$ is a free $\mathcal{O}$-module in K^n of rank n. We wish to construct an algorithm for determining the solutions of $(*)$ in $\mathcal{M}$, and this will be achieved by induction on the rank of $\mathcal{M}$, so we shall assume that such an algorithm is known for free $\mathcal{O}$-modules of smaller rank. We must first deal with the case of free modules of rank one, of the form $\mathcal{O}\underline{y}$. The equation $(*)$ now becomes

$$\alpha^m F(y_1, \ldots, y_n) = c,$$

where $\underline{x} = \alpha \underline{y}$ with α in $\mathcal{O}$. The equation has at most m solutions for α, which are readily determined. Now let us suppose that $\mathcal{M}$ is a free $\mathcal{O}$-module in K^n of rank t: let W denote the t dimensional subspace KM generated over K by $\mathcal{M}$. If $W \subset K_i$ for some i, then L_i vanishes on W and so $(*)$ has no solutions in $\mathcal{M}$. If on the other hand $W \not\subset K_i$ for each i, then by assumption the graph G_W introduced in §1 is connected. Let us now consider a finite set of minimal linear relations over W

$$\sum_{p \in I_q} \alpha_{pq} f_p(\underline{w}) = 0, \qquad q = 1, \ldots, Q, \ \underline{w} \in W, \tag{8}$$

which together imply the connectedness of G_W: that is, any pair of suffixes from $1, \ldots, m$ is linked by a chain of edges $\{i, j\}$, where i and j lie in I_q for some q. We now fix attention on a particular solution $\underline{w} = \underline{x}$: either all relations (8) remain irreducible, or at least one splits. Let us suppose first that at least one splits, so that $\underline{x}$ satisfies a linear relation

$$L(\underline{x}) = 0$$

which is not satisfied on the whole of $\mathcal{M}$. Since there are only Q relations in (8), and since each can split in at most $2^{|I_q|-1} - 1$ ways, L has only finitely many possibilities which may be determined effectively. For each of these possibilities for L we may construct a free $\mathcal{O}$-module of rank $t - 1$ containing the solutions of $\underline{x}$ in $\mathcal{M}$ of $L(\underline{x}) = 0$ as follows. Let $\underline{x}_1, \ldots, \underline{x}_t$ denote a basis of $\mathcal{M}$, so that each $\underline{x}$ in $\mathcal{M}$ may be written as $\sum_{i=1}^{t} \gamma_i \underline{x}_i$ with each γ_i in $\mathcal{O}$. We may assume that L involves $\underline{x}_1$, and so on taking coefficients of $\underline{x}_1$, we obtain

$$\gamma_1 + \sum_{i=2}^{t} \gamma_i \beta_i = 0$$

with $\beta_2, \ldots, \beta_t$ in K. The $\mathcal{O}$-module $\mathcal{M}_L$ with basis $\underline{x}_2 - \beta_2 \underline{x}_1, \ldots, \underline{x}_t - \beta_t \underline{x}_1$ contains all the solutions $\underline{x}$ in $\mathcal{M}$ of $L(\underline{x}) = 0$ as required. By the inductive hypothesis, the solutions $\underline{x}$ of (*) in each of the modules $\mathcal{M}_L$ may be determined effectively, and thus all the solutions of (*) in $\mathcal{M}$ such that at least one of the relations (8) is reducible may be constructed. Henceforth we assume that $\underline{x}$ is a solution of (*) in $\mathcal{M}$ such that the relations (8) are all irreducible when $\underline{w} = \underline{x}$. Let $\mathcal{V}$ denote the set of valuations v on K such that $v|\infty$, $v(c) > 0$, $v(\underline{x}_i) < 0$, $v(a_{ij}) < 0$ for some i, j, or $v(\alpha_{pq}) \neq 0$ for some p, q: thus $\mathcal{V}$ is a fixed finite set. Now let us show that the hypotheses of the lemma in §2 are satisfied for each equation in (8), and this $\mathcal{V}$: let v denote any valuation on K outside $\mathcal{V}$. Since $\underline{x} = \sum_{i=1}^{t} \gamma_i \underline{x}_i$ with $\gamma_1, \ldots, \gamma_t$ in $\mathcal{O}$, $v(\underline{x}) \geq 0$ and so $v(L_j(\underline{x})) \geq 0$, $1 \leq j \leq m$. However, $\prod_{j=1}^{m} L_j(\underline{x}) = c$ and $v(c) \leq 0$, so $v(L_j(\underline{x})) = 0$ and hence $v(\alpha_{pq} f_p(\underline{x})) = 0$ for each p, q. We deduce that the hypotheses of the lemma in §2 are satisfied for each equation in (8), and so conclude from the lemma that the height of $f_i(\underline{x})/f_j(\underline{x})$ is bounded explicitly for i, j in any I_q. Since $f_i(\underline{x})/f_j(\underline{x})$ also has all its poles and zeros inside the fixed set $\mathcal{V}$, we deduce (see [4, Lemma 1, p. 11]) that, to within a factor in k^*, it has only finitely many possibilities which are effectively determinable. Since any pair of suffixes in $\{1, \ldots, m\}$ is linked by a chain of such pairs i, j, the conclusion of the last sentence holds for any pair of suffixes. Finally, since $f_1(\underline{x}) \ldots f_m(\underline{x}) = c$, we deduce that, to within a factor in k^*, each $f_i(\underline{x})$ has only finitely many possibilities which may be determined effectively. Taking each of these possibilities in turn, we write

$$f_i(\underline{x}) = \lambda_i \delta_i, \qquad 1 \le i \le m,$$

where $\delta_1, \ldots, \delta_m$ are fixed elements of K and $\lambda_1, \ldots, \lambda_m$ lie in k with

$$\lambda_1 \ldots \lambda_m = 1.$$

It remains to determine the range of possibilities for $\lambda_1, \ldots, \lambda_m$. Substituting in (8) we obtain

$$\sum_{p \in I_q} \alpha_{pq} \delta_p \lambda_p = 0.$$

Let i denote any suffix in I_q; dividing by $\alpha_{iq}\delta_i$ and differentiating we obtain

$$P(\underline{x}) = \sum_{i \ne p \in I_q} \left(\frac{\alpha_{pq}\delta_p}{\alpha_{iq}\delta_i} \right)' \frac{1}{\delta_p} f_p(\underline{x}) = 0.$$

There are now two possibilities: if, for some p in I_q, $(\alpha_{pq}\delta_p/\alpha_{iq}\delta_i)'$ is non-zero, then the equation $P(\underline{w}) = 0$ cannot be satisfied for all $\underline{w}$ in W by the minimality of each relation in (8). As for L above, we are led to a free $\mathcal{O}$-module $\mathcal{M}_P$ of $\mathcal{M}$ of rank $t-1$ containing all the solutions $\underline{x}$ in $\mathcal{M}$ of $P(\underline{x}) = 0$: the solutions of (*) in this module may be constructed using the inductive hypothesis. We thus assume that the alternative case holds, so that each ratio $\alpha_{pq}\delta_p/\alpha_{iq}\delta_i$ lies in k for i,p in I_q. If we denote this ratio by μ_{pq}, then the equations in (8) may be rewritten as

$$\sum_{p \in I_q} \mu_{pq} \lambda_p = 0, \qquad q = 1, \ldots, Q$$

with μ_{pq} in k for all p, q. We may rearrange $\lambda_1, \ldots, \lambda_m$ so that these equations are equivalent to

$$\lambda_i = \sum_{p=1}^{s} v_{ip} \lambda_p, \qquad s < i \le m$$

for some $s \le t$. The equations $f_i(\underline{x}) = \lambda_i \delta_i$, $1 \le i \le m$ with $\underline{x} = \sum_{p=1}^{t} \gamma_p \underline{x}_p$ for $\underline{x}$ in W may be inverted using $\bigcap_{i=1}^{m} K_i = \{0\}$ to yield

$$\gamma_p = \sum_{i=1}^{s} \lambda_i b_{pi}, \qquad 1 \le p \le t. \tag{9}$$

Here the coefficients b_{pi} lie in K and $\lambda_1, \ldots, \lambda_s$ are elements of k satisfying

$$\lambda_1 \ldots \lambda_s \prod_{i=s+1}^{m} \left(\sum_{p=1}^{s} \nu_{ip}\lambda_p \right) = 1. \tag{10}$$

The only remaining question to settle is when $\gamma_1, \ldots, \gamma_t$ actually lie in $\mathcal{O}$. If the coefficients b_{pi} in (9) do not all lie in $\mathcal{O}$, then for some valuation v on K we have $v(b_{pi}) < 0$. By considering the negative Laurent coefficients of each b_{qj} in the expansion in powers of z_v, we deduce, as in [4] p. 21, that $\gamma_1, \ldots, \gamma_t$ lie in $\mathcal{O}$ if and only if a set of linear equations over k are satisfied by $\lambda_1, \ldots, \lambda_s$. This leads to a reduction in the number of variables $\lambda_1, \ldots, \lambda_s$, and once this reduction has been made the coefficients in the equations corresponding to (9) will all lie in the $\mathcal{O}$. We may therefore assume from the start that all the coefficients b_{pi} lie in $\mathcal{O}$, and then $\lambda_1, \ldots, \lambda_s$ may assume any values in k consistent with the equations (10): each possibility leads to a solution of $F(\underline{x}) = c$ with $\underline{x}$ in $\mathcal{M}$, from $\underline{x} = \sum_{p=1}^{t} \gamma_p \underline{x}_p$. We have now shown that if we are given a free $\mathcal{O}$-module $\mathcal{M}$ of rank t, we can construct a finite family of free $\mathcal{O}$-modules of rank $t - 1$, which together comprise all the solutions $\underline{x}$ in $\mathcal{M}$ of (*), apart from those derived from a possible substitution (9) with coefficients in $\mathcal{O}$, corresponding to the solutions in k^s of (10). This result, together with the inductive hypothesis, which deals with the modules of rank $t - 1$, completes the proof of our theorem, save of the bound (3).

5. The Bound

It is the object of this section to complete the proof of our theorem by establishing the bound (3) on the heights of all the solutions in $\mathcal{O}^n$ of (*). We shall accomplish this by induction of the rank t of an $\mathcal{O}$-module $\mathcal{M}$ in which the solution $\underline{x}$ is supposed to lie. Accordingly, let $\mathcal{M}$ denote any free $\mathcal{O}$-module in K^n of rank t, with basis $\underline{x}_1, \ldots, \underline{x}_t$ say. We shall prove that any solution $\underline{x}$ in $\mathcal{M}$ of $F(\underline{x}) = c$ satisfies

$$H(\underline{x}) \leq 4^{t^2} \big(H(F) + H(\underline{x}_1, \ldots, \underline{x}_t) + H(c) + g + r \big). \tag{11}$$

The assertion (11) will be proved by induction: accordingly, we shall assume its truth for modules of rank smaller than t. We must first check the assertion for modules of rank one: then $\underline{x} = \alpha \underline{x}_1$ with α in $\mathcal{O}$, so $\alpha^m F(\underline{x}_1) = c$, so $mH(\alpha) \leq H(c) + H(F) + mH(\underline{x}_1)$, and

$$H(\underline{x}) \leq H(F) + 2H(\underline{x}_1) + H(c),$$

which confirms (11). In the general case let us write $\underline{x} = \sum_{i=1}^{t} \gamma_i \underline{x}_i$ with $\gamma_1, \dots, \gamma_t$ in $\mathcal{O}$. We shall show that, given $\underline{x}$ in $\mathcal{M}$ satisfying $F(\underline{x}) = c$, there exist $\beta_1, \dots, \beta_t$ in K, not all zero, such that $\sum_{i=1}^{t} \gamma_i \beta_i = 0$. We may suppose that $\beta_1 \neq 0$: thus $\underline{x}$ lies in the $\mathcal{O}$-module of rank $t-1$ with basis $\underline{y}_2 = \underline{x}_2 - (\beta_2/\beta_1)\underline{x}_1, \dots, \underline{y}_t = \underline{x}_t - (\beta_t/\beta_1)\underline{x}_1$, so by the inductive hypothesis

$$H(\underline{x}) \leq 4^{(t-1)^2}\left(H(F) + H(\underline{y}_2, \dots, \underline{y}_t) + H(c) + g + r\right).$$

Since

$$H(\underline{y}_2, \dots, \underline{y}_t) \leq H(\underline{x}_1, \dots, \underline{x}_t) + H(\beta_1, \dots, \beta_t),$$

we will have completed the proof when we show that $\beta_1, \dots, \beta_t$ may be chosen with

$$H(\beta_1, \dots, \beta_t) \leq (4^{2t-1} - 1)\left(H(F) + H(\underline{x}_t) + H(c) + g + r\right). \quad (12)$$

We first observe that if we replace each linear form $L_i(\underline{x})$ by a multiple $L_i^*(\underline{x})$ which has unity as its first non-zero coefficient, then $\underline{x}$ is a solution of $\prod_{i=1}^{m} L_i^*(\underline{x}) = c^*$, where c^* is non-zero and

$$\sum_{i=1}^{m} H(L_i^*) \leq H(F) \quad \text{and} \quad H(c^*) \leq H(F) + H(c)$$

(see [4] p. 9). As in §3, we may suppose that there is some minimal linear relation on W,

$$\sum_{p \in I} \gamma_p L_p^*(\underline{w}) = 0, \qquad \underline{w} \in W. \quad (13)$$

We observe that at least one such relation exists with $|I| \geq 3$. For otherwise the linear forms $L_i^*(\underline{w})$ with i in I are proportional, and by the connectedness of G_W, all the linear forms $L_i^*(\underline{w})$, $1 \leq i \leq n$, are proportional. Since the dimension of W exceeds one, this implies that there is a non-zero $\underline{x}$ in W with $L_i^*(\underline{x}) = 0$ for $1 \leq i \leq n$, in contradiction to our initial assumption on the form F. Hence $|I| \geq 3$ for some minimal relation (13) as required. We observe that we can choose γ_p to be equal to the determinant of the matrix with entries $L_q^*(\underline{x}_i)$, where q ranges through the elements of I other than p, and i ranges through a certain subset I' of $\{1, \dots, t\}$, where I' is chosen such that this determinant is non-zero for at least one value of p. Thus

$$H(\gamma_p; p \in I) \leq \sum_{p \in I} H(L_p^*) + H(\underline{x}_1, \dots, \underline{x}_t).$$

Now let us suppose that $\underline{x}$ is a particular solution of $F(\underline{x}) = c$ in $\mathcal{M}$. If the relation (13) is reducible for $\underline{w} = \underline{x}$, then we have $\sum_{p \in J} \gamma_p L_p^*(\underline{x}) = 0$ for some proper non-empty subset J of I. Substituting $\underline{x} = \sum_{i=1}^{t} \alpha_i \underline{x}_i$, we obtain a non-trivial relation $\sum_{i=1}^{t} \alpha_i \beta_i = 0$ with

$$H(\beta_1, \ldots, \beta_t) \leq H(\gamma_p; p \in J) + H(F) + H(\underline{x}_1, \ldots, \underline{x}_t)$$

and hence (12) is satisfied. We may therefore suppose instead that (13) is irreducible for $\underline{w} = \underline{x}$, and hence we may apply the lemma in §2. If we denote by $\mathcal{V}$ the set of valuations v on K such that $v|\infty$, $v(\underline{x}_1, \ldots, \underline{x}_t) < 0$, $v(c^*) > 0$, $v(L_p^*) < 0$ or $v(\gamma_p) \neq 0$ for some $p \in I$, then as in §3 each $\gamma_p L_p^*(\underline{x})$ is a $\mathcal{V}$-unit; moreover

$$|\mathcal{V}| \leq r + H(\underline{x}_1, \ldots, \underline{x}_t) + H(c^*) + H(F) + (t+1)H(\gamma_p; p \in I).$$

From the lemma in §2 we deduce that

$$H\big(\gamma_i L_i^*(\underline{x})/\gamma_j L_j^*(\underline{x})\big) \leq 4^{|I|-1}(|\mathcal{V}| + 2g)$$

for any i, j in I. Choosing $i \neq j$ in I, we have $L_i^*(\underline{x})/L_j^*(\underline{x}) = y$, where y in K satisfies

$$H(y) \leq 4^t(|\mathcal{V}| + 2g) + H(\gamma_p; p \in I)$$

since $|I| \leq t+1$: in the particular case $t = 2$ the term 4^t may be replaced by 2 using the formulae for A_3 and B_3 in §2. Moreover, since $|I| \geq 3$, the relation $L_i^*(x) = yL_j^*(x)$ is non-trivial on $\mathcal{M}$, so substituting $\underline{x} = \sum_{i=1}^{t} \alpha_i \underline{x}_i$ we deduce that $\sum_{i=1}^{t} \alpha_i \beta_i = 0$ for some $\beta_1, \ldots, \beta_t$ in K, not all zero, and with

$$H(\beta_1, \ldots, \beta_t) \leq H(y) + H(F) + H(\underline{x}_1, \ldots, \underline{x}_t).$$

Combining the inequalities above yields the desired result. It is to be observed that although y has bounded height, it may still have infinitely many possibilities *a priori*, due to the fact that k is infinite. Thus whilst this bound on the height of y suffices for the inductive step in proving the bound on the height of each solution of (*), it alone will not allow the actual construction of the solutions. This is the reason for the more delicate analysis in §3.

The multiplier 4^{n^2} in the final bound could be reduced easily to C^{n^2} for any $C > 2$, valid for $n \geq n_0(C)$, using the above method of proof. Of course, any improvement on the bound in the lemma in §2 would also yield an immediate improvement here. Nevertheless, it should be emphasised that the dependence on $H(F)$ and $H(c)$ is only linear, continuing our experience with equations over function fields (see [4]). Moreover, the algorithm for the construction of solutions in §3 is more efficient than the presence of the factor 4^{n^2} might suggest. For at each stage of the algorithm one solves certain associated unit equations, and those with no solutions are discarded. Thus although the bounds given on the eventual height may not be optimal, the algorithm does not use this bound, but only that in the lemma which refers to the simpler unit equation.

6. Necessity

In this final section we discuss the necessity in our theorems for the assumptions that $\bigcap_{i=1}^{m} K_i = \{0\}$ and that the graph G_W is always connected. Let us first suppose, then, that $\bigcap_{i=1}^{m} K_i \neq \{0\}$, so that there is some non-zero y in every K_i: after taking a multiple we may suppose that y lies in $\mathcal{O}^n$. Now if x is a solution in $\mathcal{O}^n$ of $F(x) = c$, then $x + z.y$ is also a solution in $\mathcal{O}^n$ for each z in $\mathcal{O}^n$. We deduce that for some non-zero c there is a family of solutions in $\mathcal{O}^n$ of $(*)$ of unbounded height. Secondly we suppose that $\bigcap_{i=1}^{m} K_i = \{0\}$, but the graph G_W is disconnected for some subspace W, not contained in any K_i: thus we may assume, after rearrangement, that each of the linear forms $f_1, \ldots, f_s$ is not connected by any minimal relation to any of $f_{s+1}, \ldots, f_m$ for some s, $1 \leq s < m$. Now, since $\bigcap_{i=1}^{m} K_i = \{0\}$, we may invert the linear map $w \mapsto \big(f_1(w), \ldots, f_m(w)\big)$ from W to K^m to obtain

$$w = \sum_{i=1}^{m} a_i f_i(w) \qquad (w \in W)$$

for some a_i in K^n. We now fix x_0 such that $a_i f_i(z_0)$ lies in $\mathcal{O}^n$ for each i, and then choose $c = F(x_0)$. Now let η be any non-trivial unit of $\mathcal{O}$. If we multiply each of $f_1(x_0), \ldots, f_s(x_0)$ by η^{m-s}, and $f_{s+1}(x_0), \ldots, f_m(x_0)$ by η^{-s}, then the corresponding x, given by

$$x = \eta^{m-s} \sum_{i=1}^{s} a_i f_i(x_0) + \eta^{-s} \sum_{i=s+1}^{m} a_i f_i(x_0)$$

satisfies all the minimal linear equations on $f_1, \ldots, f_m$, as no such relation involves both f_i and f_j with $1 \leq i \leq s < j \leq m$. Furthermore, x lies in $\mathcal{O}^n$ and satisfies $F(x) = c$, so as η varies in $\mathcal{O}^*$ we obtain a family of solutions of unbounded height as required. We conclude that, provided $\mathcal{O}$ contains a non-trivial unit, the requirement that G_W be connected is necessary in order that the solutions in $\mathcal{O}^n$ of $F(\underline{x}) = c$ are of bounded height for each non-zero c in K.

We conclude with the remark that the method of proof in §3 and §4 applies to the solutions of (∗) in an arbitrary free $\mathcal{O}$-module $\mathcal{M}$ in K^n, not just the module $\mathcal{O}^n$. Accordingly we obtain the following theorem, in which t denotes the rank of $\mathcal{M}$, and $H(\mathcal{M})$ denotes the minimal height of an $\mathcal{O}$-basis $\underline{x}_1, \ldots, \underline{x}_t$ of $\mathcal{M}$.

Theorem. Suppose that $\bigcap_{i=1}^{m} K_i = \{O\}$ and, for each subspace W of $\mathcal{M}K$, not contained in any K_i, the graph G_W, defined in §2, is connected. Then all the solutions of $F(\underline{x}) = c$ with $\underline{x}$ in $\mathcal{M}$ may be determined effectively, and their heights are bounded

$$H(\underline{x}) \leq 4^{t^2} \left(H(F) + H(\mathcal{M}) + H(c) + g + r \right).$$

There are infinitely many solutions in $\mathcal{M}$ if and only if there exists a substitution

$$\underline{x} = \sum_{j=1}^{p} \underline{b}_j \lambda_j, \qquad \underline{b}_j \in \mathcal{O}^n$$

such that $\underline{x}$ lies in $\mathcal{M}$ with $F(\underline{x}) = c$ whenever $\lambda_1, \ldots, \lambda_p$ are chosen in k with $f(\underline{\lambda}) = 1$, where f is some non-zero decomposable form with coefficients in k.

Finally, if $\bigcap_{i=1}^{m} K_i$ is non-zero then (∗) has solutions in $\mathcal{M}$ of unbounded height for some c, and the same conclusion obtains if G_W is disconnected for some W as above, provided $\mathcal{O}^*$ is non-trivial.

The only reason for not stating our original theorem in this more general form is the possibility of accusations of abstruseness or inelegance.

References

[1] Baker, A. Contributions to the theory of Diophantine equations: I On the representation of integers by binary forms, *Philos. Trans. Roy. Soc. London Ser.* **A263** (1968), 173–191.

[2] Győry, K. Résultats effectifs sur la répresentation des entiers par des formes décomposables, *Queen's Papers in Pure Appl. Maths.* **56** (1980).

[3] Mason, R. C. On Thue's equation over function fields, *J. London Math. Soc. Ser.* 2 **24** (1981), 414–426.

[4] Mason, R. C. *Diophantine equations over function fields*, London Math. Soc. Lecture Notes **96**, Cambridge University Press. (1984).

[5] Mason, R. C. Norm Form Equations I, *J. Number Theory* **22** (1986), 190–207.

[6] Mason, R. C. Norm Form Equations III: Positive Characteristic, *Math. Proc. Camb. Philos. Soc.* **99** (1986), 409–423.

[7] Mason, R. C. Norm Form Equations IV: Rational Functions, *Mathematika* **33** (1986), 204–211.

[8] Mason, R. C. Norm Form Equations V: Degenerate Modules *J. Number Theory* **25** (1987), 239–248.

[9] Schmidt, W. M. Norm Form Equations, *Ann. Math.* **96** (1972), 526–551.

15

LINEAR RELATIONS ON ALGEBRAIC GROUPS

D. W. Masser

1. Introduction

In this article we shall describe some aspects of the following type of problem, in particular its relationship with transcendence theory. Given non-zero algebraic numbers $\alpha_1, \ldots, \alpha_n$, how can we decide if they are multiplicatively dependent; that is, if there exist integers $m_1, \ldots, m_n$, not all zero, such that $\alpha_1^{m_1} \ldots \alpha_n^{m_n} = 1$?

The analogous question can be asked for an elliptic curve E defined over the field of algebraic numbers. Given points $P_1, \ldots, P_n$ of E also defined over this field, how can we decide if they are linearly dependent in the sense that there exist integers $m_1, \ldots, m_n$, not all zero, such that $m_1 P_1 + \ldots + m_n P_n$ is the origin O of E?

We shall see that there are elementary arguments from the geometry of numbers yielding certain estimates which enable the above problems to be solved fairly efficiently. In principle all the constants in these estimates are effectively computable. But it turns out that the fine structure of the constants can often be greatly improved by applying techniques from the theory of transcendental numbers. We present a short survey of such applications, without going into the proofs at all.

As is usual in the modern theory of transcendence, we generalize and unify the results and methods by stating them in terms of algebraic groups. Thus our setting is as follows. Let k be a number field, and let G be a commutative connected algebraic group, defined over k, with origin O. For a field K containing k denote by $G(K)$ the group of points in G defined over K. Let $\overline{k}$ be an algebraic closure of k, and suppose we are given points $P_1, \ldots, P_n$ in $G(\overline{k})$. We want to know if there exist integers $m_1, \ldots, m_n$, not all zero, such that

$$m_1 P_1 + \ldots + m_n P_n = O. \qquad (1)$$

In fact we shall show how to find all $m = (m_1, \ldots, m_n)$ in $\mathbf{Z}^n$ such that (1) holds, in the sense that we shall find estimates for generators of the additive group of all m in $\mathbf{Z}^n$ satisfying (1). For brevity we refer to this group as the "relation group" of $P_1, \ldots, P_n$. We will estimate the norms

$$|m| = \max(|m_1|, \ldots, |m_n|)$$

of such generators in terms of G and $P_1, \ldots, P_n$.

We lose no generality in supposing that $P_1, \ldots, P_n$ lie in $G(K)$ for some number field K. We will need a height function on $G(K)$, and for this we assume G is embedded in some projective space $\mathbf{P}_N$ by means of a very ample divisor $\mathcal{D}$ and a set of basis elements, defined over k, of the associated linear system. It follows that G may be identified with a quasi-projective variety in $\mathbf{P}_N$ whose defining equations are homogeneous polynomials in the corresponding variables $X_0, \ldots, X_N$ with coefficients in k. Then we have the usual logarithmic absolute Weil height h on $\mathbf{P}_N(K)$, defined as follows. For P in $\mathbf{P}_N(K)$ let $\xi_0, \ldots, \xi_N$ be projective coordinates in K of P, and put

$$h(P) = h(\xi_0, \ldots, \xi_N) = D^{-1} \sum_v \log \max(|\xi_0|_v, \ldots, |\xi_N|_v).$$

Here D is the degree of K and the sum is extended over all valuations v of K, normalized to extend the standard valuations on $\mathbf{Q}$ and to satisfy the sum formula $\sum_v \log |\xi|_v = 0$ for all non-zero ξ in K.

This function therefore induces a height on $G(K)$, which we also denote by h. Our estimates for the $|m|$ in (1) will involve an upper bound for $h(P_1), \ldots, h(P_n)$.

The plan of this article is as follows. In section 2 we prove a general result about generators of certain groups. This section could be omitted on a first reading. In sections 3, 4, 5 and 6 we discuss various special algebraic groups, namely $G = \mathbf{G}_a$ the additive group, $G = \mathbf{G}_m$ the multiplicative group, $G = E$ an elliptic curve, and $G = A$ an abelian variety of arbitrary dimension. For the last three of these we obtain quite precise estimates for the solutions of (1); however, the first is somewhat anomalous and we do not consider it in as much detail. Then in section 7 we show how to put all these results together to treat a general commutative algebraic group.

2. Geometry of numbers

Here we prove a proposition about bases of certain groups. This type of result has already appeared several times in the literature (see for example [11] pp. 93, 96, 98 and [12] pp. 127, 203, 206), but we give a more general treatment. We say that a real-valued function f on Z^n is a convex distance function if it satisfies the following conditions:

(a) $f(m) \geq 0$ for all m in Z^n,

(b) $f(tm) = |t| f(m)$ for all m in Z^n and all t in Z,

(c) $f(m_1 + m_2) \leq f(m_1) + f(m_2)$ for all m_1, m_2 in Z^n.

For such a function it is plain that the set $\Gamma = \Gamma(f)$ of all m in Z^n with $f(m) = 0$ is an additive group.

Proposition. Let f be a convex distance function on Z^n. Further let E be such that $f(m) \leq E$ for each of the standard basis elements m of Z^n, and suppose there exists $\epsilon > 0$ such that $f(m) \geq \epsilon$ for all m in Z^n with $f(m) \neq 0$. Finally let Γ_0 be a subgroup of $\Gamma(f)$ of finite index ν. Then Γ_0 is generated by elements m of Z^n with

$$|m| \leq n^{n-1} \nu (E/\epsilon)^{n-1}.$$

More precisely, if Γ_0 has rank $r \geq 1$, it has basis elements $m_1, \ldots, m_r$ satisfying

$$|m_1| \ldots |m_r| \leq n^{n-1} \nu (E/\epsilon)^{n-r}.$$

Proof. We work in $\mathbf{R}^n$ with Lebesgue measure μ_n and its induced measures μ_r on subspaces of dimension r. It is easy to see that f has a (unique) continuous extension to $\mathbf{R}^n$, which satisfies

$$f(x) \leq nE|x| \tag{2}$$

for all x in $\mathbf{R}^n$. Also the kernel of V of f is the space spanned over $\mathbf{R}$ by Γ; its dimension is the rank r of Γ.

If $r = 0$ the Proposition is trivial. So henceforth we assume $1 \leq r \leq n$. Now Γ is a lattice in V; we are going to prove that its determinant $\Delta(\Gamma)$ satisfies

$$\Delta(\Gamma) \leq (nE/\epsilon)^{n-r}. \tag{3}$$

It will suffice to show that every convex symmetric set S in V with

$$\mu_r(S) > 2^r (nE/\epsilon)^{n-r} \tag{4}$$

contains a non-zero point of Γ; for then (3) follows on taking S as a fundamental region for Γ, slightly reduced in size.

Thus let S be convex symmetric satisfying (4). Pick $\delta > 0$ with

$$\delta^{-1} > nE/\epsilon \tag{5}$$

$$\mu_r(S) > 2^r(\delta^{-1})^{n-r}. \tag{6}$$

We thicken S in V to S_δ in $\mathbf{R}^n$ as follows. Let W be the space perpendicular to V, and let C_W be the section cut out by W of the cube C defined by $|x| \leq 1$. Define S_δ as the sum $S + \delta C_W$. By Vaaler's cube-slicing theorem [25] we have $\mu_{n-r}(C_W) \geq 2^{n-r}$, and so by orthogonality

$$\mu_n(S_\delta) = \mu_r(S)\mu_{n-r}(\delta C_W) \geq 2^{n-r}\delta^{n-r}\mu_r(S)$$

which by (6) exceeds 2^n. Clearly S_δ is convex symmetric, and so by Minkowski's First Theorem ([8] p. 71) it contains a non-zero point m in $\mathbf{Z}^n$. Now $m = x + \delta y$ for x in S, y in C_W; and using (2) we obtain

$$f(m) \leq f(x) + f(\delta y) = f(\delta y) \leq n\delta E$$

which by (5) is strictly less than ϵ. Consequently $f(m) = 0$. Therefore m is in Γ, so in V, and so also in S. As remarked above, this establishes (3).

Since Γ_0 has index ν in Γ, we deduce

$$\Delta(\Gamma_0) \leq \nu(nE/\epsilon)^{n-r}. \tag{7}$$

Let now $\lambda_1, \ldots, \lambda_r$ be the successive minima of Γ_0 with respect to the distance function on V induced by $|x|$. By Minkowski's Second Theorem ([8] p. 203) we have $\lambda_1 \ldots \lambda_r \leq 2^r \Delta(\Gamma_0)/v$, where v is the volume of the section of C cut by V. Again by cube-slicing $v \geq 2^r$; so by (7) we find that

$$\lambda_1 \ldots \lambda_r \leq \nu(nE/\epsilon)^{n-r}. \tag{8}$$

Let $m'_1, \ldots, m'_r$ be linearly independent points of Γ_0 with $|m'_i| \leq \lambda_i$, $1 \leq i \leq r$. Then Lemma 8 (p. 135) of [8] provides us with basis elements $m_1, \ldots, m_r$ of Γ_0 satisfying

$$|m_i| \leq \max\{|m'_i|, \tfrac{1}{2}(|m'_1| + \ldots + |m'_i|)\}$$

$$\leq \max(1, \tfrac{1}{2}i)\lambda_i, \qquad 1 \leq i \leq r.$$

We conclude using (8) that

$$|m_1| \ldots |m_r| \leq r! 2^{-r+1}\nu(nE/\epsilon)^{n-r},$$

and since $r!2^{-r+1} \le r^r 2^{-r+1} \le r^{r-1} \le n^{r-1}$ the inequality of the Proposition follows at once.

Later we shall use this for $f = \varphi^{1/2}$ where φ is a positive semidefinite quadratic form; the condition (c) above is just the Cauchy-Schwarz inequality. In this case the Proposition resembles Satz 1 (p.36) of a recent paper of Schlickewei [19], with his integrality hypothesis replaced by the hypothesis that either $\varphi \ge \epsilon^2$ or $\varphi = 0$. However, Schlickewei's result applies more significantly to forms that are not definite.

3. The additive group

We identify the complex points of $\mathbf{G}_a$ with the additive group of complex numbers, and we embed this as usual in $\mathbf{P}_1$ by taking x to $(1, x)$. So $\mathbf{G}_a$ is $\mathbf{P}_1$ minus the single point $(0, 1)$, and we can suppose $k = \mathbf{Q}$. The points $P_1, \ldots, P_n$ of $G(K)$ correspond to algebraic numbers $\alpha_1, \ldots, \alpha_n$ of K, and the relation (1) reads

$$m_1 \alpha_1 + \ldots + m_n \alpha_n = 0. \tag{9}$$

It is not difficult to prove that the relation group of all $m = (m_1, \ldots, m_n)$ in $\mathbf{Z}^n$ satisfying (9) is generated by m with

$$|m| \le \exp(Ch), \tag{10}$$

where h is an upper bound for the heights $h(1, \alpha_1), \ldots, h(1, \alpha_n)$, and C depends only on the degree D of K. We leave the proof to the reader; he could for example use Lemma 4 (p. 442) of [16] together with suitable trace arguments.

Now select

$$\alpha_1 = 1, \quad \alpha_2 = g, \ldots, \alpha_n = g^{n-1}$$

for a positive integer g. Then well-known properties of g-adic expansions show that every solution $m \ne 0$ of (9) must satisfy $|m| \ge g$. Since we can take $h \le (n-1) \log g$, it follows that the exponential dependence on h in (10) is necessary. As we shall soon see, the additive group is anomalous in this respect.

4. The multiplicative group

We identify the complex points of $\mathbf{G}_m$ with the multiplicative group of non-zero complex numbers, and we embed this again in $\mathbf{P}_1$ as in section 3. Then $\mathbf{G}_m$ is identified with $\mathbf{P}_1$ minus the points $(0,1)$ and $(1,0)$. Now $P_1, \ldots, P_n$ correspond to non-zero algebraic numbers $\alpha_1, \ldots, \alpha_n$ of K, and the relation (1) becomes

$$\alpha_1^{m_1} \ldots \alpha_n^{m_n} = 1.$$

In this case, estimates for $|m|$ were first obtained by Baker [2], [3] using the theory of linear forms in logarithms. A more elementary method was found by Stark [23] and van der Poorten and Loxton [17]. It is convenient for our purposes to introduce the quantities

$$\eta = \eta(\mathbf{G}_m, K) = \inf h(P),$$

where the infimum is taken over all non-torsion points P of $\mathbf{G}_m(K)$, and

$$\omega = \omega(\mathbf{G}_m, K)$$

the cardinality of the torsion part of $\mathbf{G}_m(K)$.

Theorem $\mathbf{G}_m$. Suppose $P_1, \ldots, P_n$ on $\mathbf{G}_m(K)$ have heights at most $h \geq \eta$. Then the relation group of $P_1, \ldots, P_n$ is generated by m with

$$|m| \leq n^{n-1} \omega (h/\eta)^{n-1}.$$

Proof. Let f be the function on $\mathbf{Z}^n$ defined by

$$f(m) = h(m_1 P_1 + \ldots + m_n P_n).$$

Since the height $h(1, \alpha)$ satisfies $h(1, \alpha) \geq 0$, $h(1, \alpha^t) = |t| h(1, \alpha)$ and $h(1, \alpha_1 \alpha_2) \leq h(1, \alpha_1) + h(1, \alpha_2)$, we obtain a convex distance function. The set $\Gamma = \Gamma(f)$ then consists of all m such that $m_1 P_1 + \ldots + m_n P_n$ is a torsion point of $\mathbf{G}_m(K)$. We apply the Proposition to the relation group Γ_0 of $P_1, \ldots, P_n$. Clearly the index ν of Γ_0 in Γ is at most ω. And we can take $E = h$, $\epsilon = \eta$. The theorem follows immediately.

In particular we see that the dependence on h is not exponential, in contrast to the additive case $G = \mathbf{G}_a$. We may also see that the exponent $n - 1$ of h is best possible, by choosing $n - 1$ independent points $Q_1, \ldots, Q_{n-1}$ on $\mathbf{G}_m(K)$ and taking

$$P_1 = -gQ_1, \quad P_2 = Q_1 - gQ_2, \ldots, P_{n-1} = Q_{n-2} - gQ_{n-1}, \quad P_n = Q_{n-1};$$

compare [17] (p. 300).

It thus remains to estimate the quantities η, ω in terms of K. A simple way of doing this is as follows. Let B be the number of points in $\mathbf{P}_1(K)$ with height at most 1, say. By considering the multiples bP, $0 \leq b \leq B$, and noting that $h(bP) = bh(P)$, it is easy to see that

$$\eta \geq B^{-1}, \qquad \omega \leq B.$$

Actually these give a very poor dependence on the degree D of K (at least exponential); and many authors have tried to do better. A classical question of Lehmer amounts to asking if $\eta \geq c^{-1}D^{-1}$, where c (and all the c's of this section) is a positive absolute constant; and the celebrated result of Dobrowolski [10] asserts that

$$\eta \geq c^{-1}D^{-1}(\mathcal{L}/\log \mathcal{L})^{-3}, \tag{11}$$

where for convenience we have written

$$\mathcal{L} = \log(D + 2).$$

This was proved by techniques associated with transcendence theory, which were first applied in such a context by Stewart [24].

The analogous estimate for ω is much more elementary. Since the torsion part of $\mathbf{G}_m(K)$ is cyclic, and the field of Nth roots of unity has degree equal to Euler's function $\varphi(N)$, which satisfies $\varphi(N) \geq c^{-1}N/\log\log N$, $N \geq 3$, we deduce

$$\omega \leq cD \log \mathcal{L}. \tag{12}$$

Substituting (11) and (12) into Theorem $\mathbf{G}_m$, we obtain something like Theorem 1 (p. 84) of [17]; but for all relations m, not just one. On the other hand, [17] distinguishes between the heights of $\alpha_1, \ldots, \alpha_n$, and we do not. For some interesting refinements, which amount to working on the tangent space of $\mathbf{G}_m^n$ rather than $\mathbf{G}_m^n$ itself, see Waldschmidt [26], Bijlsma [6], and Bijlsma and Cijsouw [7].

5. Elliptic curves

It is convenient to embed our elliptic curve E in $\mathbf{P}_2$ using the Weierstrass form

$$y^2 z = 4x^3 - g_2 x z^2 - g_3 z^3$$

Thus the invariants g_2, g_3 lie in the number field k. In this case the analogue of Theorem $\mathbf{G}_m$ requires the Néron-Tate height q on $E(K)$ defined by

$$q(P) = \lim_{t\to\infty} 2^{-2t} h(2^t P).$$

We put

$$\eta = \eta(E, K) = \inf q(P),$$

where the infimum is taken over all non-torsion P in $E(K)$, and we write

$$\omega = \omega(E, K)$$

for the cardinality of the torsion group of $E(K)$.

Theorem E. Suppose $P_1, \ldots, P_n$ on $E(K)$ have Néron-Tate heights at most $q \geq \eta$. Then the relation group of $P_1, \ldots, P_n$ is generated by m with

$$|m| \leq n^{n-1} \omega (q/\eta)^{\frac{1}{2}(n-1)}.$$

Proof. Let f be the function on $\mathbf{Z}^n$ defined by

$$f(m) = \left\{ q(m_1 P_1 + \ldots + m_n P_n) \right\}^{1/2}.$$

Since q is a positive definite quadratic form, we obtain a convex distance function. Again the set $\Gamma = \Gamma(f)$ consists of all m such that $m_1 P_1 + \ldots + m_n P_n$ is a torsion point of $E(K)$, and we take Γ_0 as the relation group of $P_1, \ldots, P_n$. We can choose this time $E = q^{1/2}$, $\epsilon = \eta^{1/2}$, and the theorem follows immediately.

Once again the dependence on heights is not exponential, and an example exactly as in the previous section shows that the exponent $\frac{1}{2}(n-1)$ of q cannot be improved, at least if the rank of $E(K)$ is no less than $n - 1$.

To apply the result in practice we must first estimate q in terms of the Weil height h. But in fact it is well-known that their difference is bounded on $E(\overline{k})$; thus

$$\delta = \delta(E) = \sup|q(P) - h(P)| \tag{13}$$

is independent of K, where the supremum is taken over all P in $E(K)$. So if h is an upper bound for $h(P_1), \ldots, h(P_n)$ we can take

$$q = \max(\eta, h + \delta).$$

Also since $q(P) \leq \delta$ implies $h(P) \leq 2\delta$, the simple counting argument of section 4 gives

$$\eta \geq \delta B^{-2}, \qquad \omega \leq B, \qquad (14)$$

where B is the number of points P in $\mathbf{P}_2(K)$. But as before this leads to an exponential dependence on the degree D of K, whereas one expects $\eta \geq C^{-1}D^{-1}$, where C (and all the C's of this section) is a positive constant depending only on E. The techniques of Dobrowolski were ingeniously adapted by Laurent [13] in the case of complex multiplication, and he obtained exactly the same bound

$$\eta \geq C^{-1}D^{-1}(\mathcal{L}/\log \mathcal{L})^{-3}. \qquad (CM\ 1)$$

In this case, best possible bounds are known for the order of torsion and it is not difficult to show (compare [4] p. 23) that

$$\omega \leq CD(\log \mathcal{L})^{1/2}. \qquad (CM\ 2)$$

When there is no hypothesis of complex multiplication, the method of Stewart has to be used. Thus it was proved in [1] that $\eta \geq C^{-1}D^{-10}\mathcal{L}^{-6}$; but I recently sharpened this to

$$\eta \geq C^{-1}D^{-3}\mathcal{L}^{-2} \qquad (15)$$

using an idea of Paula Cohen [9]. By rather different techniques Silverman [20] obtained $\eta >> D^{-2}$ when K is an abelian extension. As regards the torsion, Paula Cohen showed that every point of $E(K)$ has order at most $CD\mathcal{L}$; but her argument can be adapted to yield even

$$\omega \leq CD\mathcal{L}. \qquad (16)$$

When there is no complex multiplication it can be proved that $\omega \ll (D \log \mathcal{L})^{1/2}$ (compare $(CM\ 2)$); but the implied constant is not always effective. The proof is a straightforward deduction from Serre's deep description [18] of the Galois group of the complete division fields $k(E_n)$ of level n, using the fact that if a subgroup of $(\mathbf{Z}/n\mathbf{Z})^2$ has index ν, then its stabilizer in $GL_2(\mathbf{Z}/n\mathbf{Z})$ has cardinality at most ν^2.

In all of the above estimates the constants C depend on the elliptic curve E. In her thesis [9] Paula Cohen showed how this dependence could be made explicit. In fact nothing is lost by taking a rather crude measure of the size of E; ignoring moduli space subtleties, we define

$$s = s(E) = \max(1, h(1, g_2, g_3)).$$

Thus for example the work of Zimmer [27] implies that for (13) we have

$$\delta \leq cs, \tag{17}$$

where now c depends only on k; and this is essentially best possible (see [21] p. 209). Even so, using (17) to calculate B in (14) we obtain only exponential dependence on s. Once again transcendence methods greatly improve matters, and we find for (15) and (16) the inequalities

$$\eta \geq c^{-1}s^{-1}D^{-3}(s+\mathcal{L})^{-2} \tag{18}$$

$$\omega \leq cs^{1/2}D(s+\mathcal{L}), \tag{19}$$

again for c depending only on k (and in fact c depends only on the degree d of k).

Actually, for fixed D at any rate, these fall somewhat short of what is predicted. A conjecture of Lang [11] (p. 92) implies that η should be bounded below independently of E (and should even tend to infinity as the logarithm of the discriminant). Silverman has made important progress on these and related questions: see [20], and also [22] for a more general account. And it is generally believed that ω should be bounded above independently of E (as in Mazur's famous result for $k = \mathbf{Q}$).

6. Abelian varieties

Up to now we have been able to find simple canonical embeddings of our algebraic group in projective space. For an abelian variety A this is usually not possible, and all our constants must depend on the embedding, or equivalently, on the divisor $\mathcal{D}$ that gives rise to the embedding. We suppress this dependence in our notation $\eta = \eta(A, K), \omega = \omega(A, K)$; these are defined as in the previous section now with reference to $A(K)$ and the corresponding Néron-Tate height q.

Theorem A. Suppose $P_1, \ldots, P_n$ on $A(K)$ have Néron-Tate heights at most $q \geq \eta$. Then the relation group of $P_1, \ldots, P_n$ is generated by m with

$$|m| \leq n^{n-1}\omega(q/\eta)^{\frac{1}{2}(n-1)}.$$

Proof. Exactly as for Theorem E.

Most of the remarks made in section 5 apply more generally here; thus, provided the embedding corresponds to a symmetric divisor, the quantity

$$\delta = \delta(A) = \sup|q(P) - h(P)| \tag{20}$$

is independent of K, where the supremum is taken over all P in $A(K)$. And the inequalities (14) hold, provided B now counts the points P in $\mathbf{P}_N(K)$ with $h(P) \le 2\delta$.

Let g be the dimension of A. This time one expects $\eta \ge C^{-1}D^{-1/g}$, where C depends only on A. But if $g > 1$ no-one so far has been able to adapt Dobrowolski's work. Recently, building on earlier work for myself [14] and Bertrand [4], I proved that

$$\eta \ge C^{-1}D^{-(2g+1)}\mathcal{L}^{-2g}$$

(which in the case of complex multiplication can be sharpened to $\eta \ge C^{-1}D^{-2}\mathcal{L}^{-1}$). I also proved that

$$\omega \le CD^g\mathcal{L}^g;$$

probably equally far from the truth, which is now connected with deep results of Bogolomov and Serre (see the discussion and references in [4]).

There has been some start in calculating the dependence of these constants C on the abelian variety A, at least for fixed D. It is convenient to express this in terms of families as follows. We take a variety V, defined over k, and an abelian variety A_{gen} defined over the function field $k(V)$. We pick a symmetric very ample divisor $\mathcal{D}_{\text{gen}}$ on A_{gen} also defined over $k(V)$, and also basis elements for the corresponding linear system again defined over $k(V)$. Throwing away a proper Zariski closed subset of V if necessary, we can suppose that for each v in $V(k)$ the corresponding specialization on V yields an abelian variety A_v and a symmetric very ample divisor $\mathcal{D}_v$ both defined over $k(v) = k$. We can now measure the size of A_v simply by the height $h_V(v)$ of v in some fixed projective embedding of V; thus

$$s = s(A_v) = \max(1, h_V(v)).$$

For example, we can get every elliptic curve in this way by taking V as affine space $\mathbf{A}^2$, with coordinates t_2, t_3, minus the discriminant locus defined by $t_2^3 = 27t_3^2$. Define E_{gen} already in $\mathbf{P}_2$ by

$$y^2 z = 4x^3 - t_2 xz^2 - t_3 z^3$$

(so that $\mathcal{D}_{\text{gen}}$ is three times the divisor defined by the origin). For v in $V(k)$ with coordinates g_2, g_3 the specialized curve E_v is just that of section 5, of course; hence with the obvious embedding of V in $\mathbf{P}_2$ the size $s(E_v)$ agrees with the size defined there.

In general Silverman and Tate [21] (p. 201) proved that for (20) we have

$$\delta \leq cs$$

for c depending only on k, and it was shown in [15] that

$$\eta \geq C^{-1}s^{-(2g+1)}$$

$$\omega \leq Cs^g,$$

this time for C depending only on D and k (we may have to throw away another closed subset of V). As in the case of elliptic curves, these are probably very far from the truth. See [22] for an interesting discussion and some partial results.

7. The general case

Let G be a commutative connected algebraic group defined over a number field k. We say that the "dependence problem" can be solved for G if, given any points $P_1, \ldots, P_n$ on $G(\overline{k})$ we can find the relation group of all m satisfying (1), in the sense, for example, of the estimates of the previous sections. We do not make this definition very precise, because of the anomalous nature of the additive group in section 4 (however, Bertrand in [5] has suggested an interesting way round this difficulty). At any rate we regard this problem as having been solved for $G = G_a$, G_m, E and more generally an abelian variety A.

To deduce its solution for arbitrary G, it is convenient to store the relation group as a "relation matrix" whose rows are basis elements of the relation group. Suppose first that we have an exact sequence

$$0 \longrightarrow G' \longrightarrow G \longrightarrow G'' \longrightarrow 0$$

of commutative connected algebraic groups G', G, G'' defined over k, such that the maps α from G' to G and π from G to G'' are defined over k. Assume we can solve the dependence problem for G' and G''. We solve it for G as follows. Take points $P_1, \ldots, P_n$ in $G(\overline{k})$; then the images $P_i'' = \pi(P_i)$, $1 \leq i \leq n$, are in $G''(\overline{k})$. If these are independent then so are $P_1, \ldots, P_n$; otherwise we can find a relation matrix M'' for $P_1'', \ldots, P_n''$ with rows $m_j'' = (m_{j1}'', \ldots, m_{jn}'')$, $1 \leq j \leq \ell$, for some $\ell \geq 1$. Then the points

$$Q_j = m_{j1}'' P_1 + \ldots + m_{jn}'' P_n, \qquad 1 \leq j \leq \ell$$

satisfy $\pi(Q_j) = O$; so that we can write $Q_j = \alpha(P_j')$ for P_j' in $G'(\overline{k})$, $1 \leq j \leq \ell$. If $P_1', \ldots, P_\ell'$ are independent then so are $P_1, \ldots, P_n$; otherwise we can find a relation matrix M' for $P_1', \ldots, P_\ell'$. Now it is easily seen that the product $M'M''$ is a relation matrix for the original points $P_1, \ldots, P_n$.

In particular, if we can solve the dependence problem for G' and G'', then we can solve it for $G' \times G''$. Since every commutative linear group is a product of G_a's and G_m's, this solves the dependence problem for every such linear group.

Finally every G occurs in an exact sequence

$$O \longrightarrow L \longrightarrow G \longrightarrow A \longrightarrow O$$

where L is linear and A is an abelian variety. It follows that the dependence problem can be solved for any commutative connected algebraic group.

References

[1] M. Anderson and D. W. Masser, Lower bounds for heights on elliptic curves, *Math. Z.* **174** (1980), 23–34.

[2] A. Baker, Linear forms in the logarithms of algebraic numbers IV, *Mathematika* **15** (1968), 204–216.

[3] A. Baker, A sharpening of the bounds for linear forms in logarithms III, *Acta Arith.* **27** (1975), 247–252.

[4] D. Bertrand, Galois orbits on abelian varieties and zero estimates, *Diophantine analysis* (eds. J. H. Loxton and A. J. van der Poorten), London Math. Soc. Lecture Notes 109, Cambridge 1986, pp. 21–35.

[5] D. Bertrand, Galois representations and transcendental numbers, *New Advances in Transcendence Theory* (ed. A. Baker), Cambridge Univ. Press, 1988, Chapter 3.

[6] A. Bijlsma, Simultaneous approximations in transcendental number theory, Math. Centre Tracts 94 (Mathematisch Centrum, Amsterdam 1978).

[7] A. Bijlsma and P. Cijsouw, Degree-free bounds for dependence relations, *J. Australian Math. Soc.* **31** (1981), 496–507.

[8] J. W. S. Cassels, *An introduction to the geometry of numbers*, Springer, Berlin Göttingen Heidelberg 1959.

[9] P. Cohen, Explicit calculation of some effective constants in transcendence proofs, Ph.D. Thesis, University of Nottingham 1985 (Chapter 3).

[10] E. Dobrowolski, On a question of Lehmer and the number of irreducible factors of a polynomial, *Acta Arith.* **34** (1979), 391–401.

[11] S. Lang, *Elliptic curves: Diophantine analysis*, Springer, Berlin Heidelberg New York 1978.

[12] S. Lang, *Fundamentals of Diophantine geometry*, Springer, Berlin Heidelberg New York 1983.

[13] M. Laurent, Minoration de la hauteur de Néron-Tate, *Progress in Math. 38,* Birkhäuser, Boston Basel Stuttgart 1983 (pp. 137–151).

[14] D. W. Masser, Small values of the quadratic part of the Néron-Tate height, *Compositio Math.* **53** (1984), 153–170.

[15] D. W. Masser, Small values of heights on families of abelian varieties, *Proceedings of the Bonn Conference* (ed. G. Wüstholz), Springer Lecture Notes (to appear).

[16] D. W. Masser and G. Wüstholz, Fields of large transcendence degree generated by values of elliptic functions, *Inventiones Math.* **72** (1983), 407–464.

[17] A. J. van der Poorten and J. H. Loxton, Multiplicative relations in number fields, *Bull. Australian Math. Soc.* **16** (1977), 83–98 and ibid. **17** (1977), 151–155; see also a similar title in *Acta Arith.* **42** (1983), 291–302.

[18] J-P. Serre, Propriétés galoisiennes des points d'ordre fini des courbes elliptiques, *Inventiones Math.* **15** (1972), 259–331.

[19] H. P. Schlickewei, Kleine Nullstellen homogener quadratischer Gleichungen, *Monats. Math.* **100** (1985), 35–45.

[20] J. H. Silverman, Lower bound for the canonical height on elliptic curves, *Duke Math. J.* **48** (1981), 633-648.

[21] J. H. Silverman, Heights and the specialization maps for families of abelian varieties, *J. reine angew. Math.* **342** (1983), 197–211.

[22] J. H. Silverman, A quantitative version of Siegel's Theorem: integral points on elliptic curves and Catalan curves, *J. reine angew. Math.* **378** (1987), 60–100.

[23] H. Stark, Further advances in the theory of linear forms in logarithms, *Diophantine approximation and its applications*, Academic Press, New York London 1973 (pp. 255-293).

[24] C. L. Stewart, Algebraic integers whose conjugates lie near the unit circle, *Bull. Soc. Math. France* **106** (1978), 169–176.

[25] J. Vaaler, A geometric inequality with applications to linear forms, *Pacific J. Math.* **83** (1979), 543–553.

[26] M. Waldschmidt, A lower bound for linear forms in logarithms, *Acta Arith.* **37** (1980), 257–283.

[27] H. G. Zimmer, On the difference of the Weil height and the Néron-Tate height, *Math. Z.* **147** (1976), 35–51.

16

ESTIMATES FOR THE NUMBER OF ZEROS
OF CERTAIN FUNCTIONS

Yu. V. Nesterenko

In this paper we sketch a proof of the following assertion.

Theorem 1. Let $\xi_1, \ldots, \xi_q$ be distinct complex numbers and let $f_0(z)$, $\ldots, f_m(z)$ be the set of solutions of the differential equations

$$y_j' = F_j(y_0, \ldots, y_m), \qquad j = 0, 1, \ldots, m,$$

where the non-zero F_js are homogeneous polynomials with complex coefficients of the same degree. Suppose that these functions are analytic at the points $\xi_1, \ldots, \xi_q$ and that the maximal number of homogeneous algebraically independent $f_i(z)$'s $(0 \leq i \leq m)$ over $\mathbf{C}$ is $\kappa + 1$, $\kappa \geq 1$. Then there is a constant $\gamma_1 = \gamma_1(\overline{f}) > 0$ such that

$$\sum_{j=1}^{q} \operatorname*{ord}_{\xi_j} P(\overline{f}) \leq \gamma_1^{\kappa-1} \sum_{j=1}^{\kappa} a_{k-j} \mathcal{D}^j,$$

for any homogeneous polynomial $P \in \mathbf{C}[X_0, \ldots, X_m]$ with $P(\overline{f}) \not\equiv 0$, where $\mathcal{D} = \deg P$ and $a_{\kappa-j} = a_{\kappa-j}(\gamma_1^{j-1} \mathcal{D}^j)$.

Here and in what follows $a_i(T)$ denotes the maximum number of points $\overline{f}(\xi_\ell)$, $1 \leq \ell \leq q$, lying in an irreducible manifold in $\mathbf{P}^m$ of dimension i and degree at most T. In particular, if no two $\overline{f}(\xi_\ell)$ are proportional, then $a_0 = 1$. This theorem strengthens several results of W. D. Brownawell and D. W. Masser [5], and W. D. Brownawell [3], [4]. Theorem 1 follows from Theorem 2 below.

Theorem 2. Suppose that the conditions of Theorem 1 are satisfied. Then there exists a constant $\gamma_\ell = \gamma_\ell(\overline{f}) > 0$ such that

$$\sum_{j=1}^{q} \operatorname*{ord}_{\xi_j} \mathcal{I}(\overline{f}) \leq \gamma_2^{\tau-1} \sum_{j=0}^{\tau-1} a_j N(\mathcal{I})^{\frac{\kappa-j}{\kappa-\tau+1}},$$

for any homogeneous unmixed ideal $\mathcal{I} \subset C[x_0, \ldots, x_m]$, of height $h(\mathcal{I}) = m + 1 - \tau$, $1 \leq \tau \leq \kappa$, where $a_j = a_j\left(\gamma_2^{\tau-j-1} N(\mathcal{I})^{\frac{\kappa-j}{\kappa-\tau+1}}\right)$.

We define $\operatorname{ord}_\xi \mathcal{I}(\bar{f})$ for unmixed homogeneous ideals $\mathcal{I} \subset C[X_0, \ldots, X_m]$ just as $|\mathcal{I}(\bar{\omega})|$ is defined in [2] for the numerical case, except that the valuation ord_ξ replaces the absolute value on the field of formal power series $C((z - \xi))$.

Denote by $\mathcal{E}$ the prime ideal in $C[x_0, \ldots, x_m]$ generated by the homogeneous polynomials vanishing on the vector $\bar{f}(z)$. To deduce Theorem 1 from Theorem 2 it is sufficient to employ the functional analogue of Proposition 3 in [2] for the polynomial P and the ideal $\mathcal{E}$.

We move now to the proof of Theorem 2. Suppose that there are ideals satisfying the conditions of the theorem but for which its conclusion is false. Let $\mathcal{I}$ be one of these ideals with maximal height, and put $\tau = m + 1 - h(\mathcal{I}) \geq 1$. By using the functional analogue of Proposition 2 in [2], in just the same way as Theorem 2 follows from Proposition 7 in [2], it is easy to prove the existence of a prime homogeneous ideal $\wp \subset C[x_0, \ldots, x_m]$, $h(\wp) = m + 1 - \tau$ such that

$$\sum_{j=1}^{q} \operatorname{ord}_{\xi_j} \wp(\bar{f}(z)) > \gamma_\ell^{\tau-1} \sum_{j=0}^{\tau-1} a_j N(\wp)^{\frac{\kappa-j}{\kappa-\tau+1}}, \tag{1}$$

where $a_j = a_j\left(\gamma_2^{\tau-j-1} N(\wp)^{\frac{\kappa-j}{\kappa-\tau+1}}\right)$. For this we use the fact that the functions $a_j(T)$ increase with T. Without loss of generality, we may suppose that $x_0 \notin \wp$.

Suppose, first, that $\tau = 1$. Then $N(\wp) = 1$ and there is a vector $\bar{\alpha} = (\alpha_0, \alpha_1, \ldots, \alpha_m) \in C^{m+1}$ such that the ideal $\wp$ is generated by the polynomials $\alpha_i x_j - \alpha_j x_i$. In this case, as is easily proved,

$$\operatorname{ord} \wp(\bar{f}) = \min_{i \neq j} \operatorname{ord}_\xi \left(\alpha_i f_j(z) - \alpha_j f_i(z)\right)$$

and in the conditions of the theorem, the inequality $\operatorname{ord}_\xi \wp(\bar{f}) \leq 1$ is satisfied. The number of indices j such that $\operatorname{ord}_{\xi_j} \wp(\bar{f}) > 0$ does not exceed $a_0(1)$, which, together with inequality (1), gives a contradiction. Thus, $\tau \geq 2$.

Suppose now that $\mathcal{E} \not\subset \wp$. Then there is a homogeneous polynomial Q (one of the basis polynomials of the ideal $\mathcal{E}$) such that $Q \notin \wp$, $Q(\bar{f}(z)) = 0$. It follows from the functional analogue of Proposition 3 in [2] that there exists a homogeneous unmixed ideal $\mathcal{J} \subset C[x_0, \ldots, x_m]$, $h(\mathcal{J}) = m - \tau + 2$, satisfying the inequalities

$$N(\mathcal{J}) \leq N(\wp) \cdot \deg Q, \tag{2}$$

$$\operatorname{ord}_\xi \mathcal{J}(\bar{f}) \geq \min\left(\operatorname{ord}_\xi \wp(\bar{f}), \operatorname{ord}_\xi Q(\bar{f})\right) = \operatorname{ord}_\xi \wp(\bar{f}). \tag{3}$$

Since $h(\mathcal{J}) > h(\mathcal{I})$ the assertion of Theorem 2 is satisfied for the ideal $\mathcal{I}$. In view of (2) and (3) this leads to a contradiction with inequality (1) if the constant γ_2 is chosen sufficiently large so that $\mathcal{E}$ has a basis consisting of polynomials of degree no greater than $\gamma_2^{\frac{1}{\kappa}}$. Thus $\mathcal{E} \subset \wp$.

Define the differential operator

$$\mathcal{D} = \sum_{i=0}^{m} F_i(x_0, \ldots, x_m) \frac{\partial}{\partial x_i},$$

acting on $\mathbf{C}[x_0, \ldots, x_m]$.

Lemma. Suppose that the conditions of Theorem 2 are satisfied, $\wp \subset \mathbf{C}[x_0, \ldots, x_m]$ is a prime homogeneous ideal, $\mathcal{E} \subset \wp$, $\tau = m + 1 - h(\wp) \geq 1$ and

$$\sum_{j=1}^{q} \operatorname*{ord}_{\varepsilon_j} \wp(\overline{f}) > 0.$$

Then there exists a homogeneous polynomial $R \in \wp$ such that $Q = \mathcal{D}R \notin \wp$ and

$$\deg Q \leq \lambda^{2^{\kappa - \tau + 2}} N(\wp)^{\frac{1}{\kappa - \tau + 1}} \tag{4}$$

where $\lambda = \lambda(\overline{f}) > 0$ is a sufficiently large constant.

Proof. Define a sequence $\mathcal{X}_\ell$ by $\mathcal{X}_0 = \lambda$, $\mathcal{X}_{\ell+1} = \lambda \mathcal{X}_\ell^2$, $\ell \geq 0$.

Let P be a homogeneous polynomial of minimal degree satisfying the conditions $P \in \wp$, $P \notin \mathcal{E}$. From Theorem 1 and Proposition 1 in [1] applied to the ideals $\wp$ and $\mathcal{E}$, it follows that for sufficiently large λ

$$L = \deg P \leq N(\wp)^{\frac{1}{\kappa - \tau + 1}}. \tag{5}$$

For each $\ell = 0, 1, \ldots$, let $\mathcal{J}_\ell$ be the ideal in the ring $\mathbf{C}[x_0, \ldots, x_m]$, generated by polynomials from the ideal $\mathcal{E}$ and the polynomials

$$\mathcal{D}^i P, \qquad 0 \leq i \leq \mathcal{X}_\ell.$$

If $\mathcal{J}_0 \not\subset \wp$, then the smallest number i for which $\mathcal{D}^i P \notin \wp$ satisfies $i \leq \mathcal{X}_0 = \lambda$ and it is easily seen that the assertion of the lemma is satisfied by the polynomial $R = \mathcal{D}^{i-1} P$. We shall therefore also suppose that $\mathcal{J}_0 \subset \wp$.

Let d be the largest integer, $d \geq 0$, such that $\mathcal{J}_d \subset \wp$ and there exist homogeneous polynomials $Q_0, \ldots, Q_{m-\kappa+d}$ satisfying the conditions

1)

$$\deg Q_i \leq \lambda, \qquad i = 0,1,\ldots,m-\kappa-1,$$

$$\deg Q_{m+\kappa+i} \leq \mathcal{X}_i L, \qquad i = 0,1,\ldots,d,$$

2) The ideal $\mathcal{A}_d = (Q_0,\ldots,Q_{m-\kappa+d})$ is contained in $\mathcal{J}_d$.

3) All the primary components of $\mathcal{A}_d$, contained in $\wp$, have height $m - \kappa + d + 1$, and if the intersection of these components is the unmixed ideal $\mathcal{U}_d$,

$$N(\mathcal{U}_d) \leq \lambda^{-2} \mathcal{X}_{d+1} L^{d+1}.$$

For $d = 0$ we can take $Q_{m-\kappa} = P$ and choose $Q_0, \ldots, Q_{m-\kappa-1}$ from $\mathcal{E}$ in the required way. From condition 3) it follows that $d \leq \kappa - \tau$. Suppose that $\mathcal{J}_{d+1} \subset \wp$. In what follows we shall derive a contradiction from this assertion.

Let $\mathcal{B}$ be a primary component of the ideal $\mathcal{A}_d$ contained in $\wp$, let $\mathcal{Q} = \sqrt{\mathcal{B}}$ and ℓ be the exponent of the ideal $\mathcal{B}$. We shall prove that

$$\ell \leq (2\rho)^{d+1}, \qquad \rho = \mathcal{X}_d^{\frac{2}{d+1}}. \tag{6}$$

We have the inequality

$$\ell.N(\mathcal{Q}) \leq N(\mathcal{U}_d) \leq \lambda^{-2} \mathcal{X}_{d+1} L^{d+1},$$

from which, since

$$\nu = 1 + \left[\rho L \ell^{-\frac{1}{d+1}}\right],$$

it follows that

$$\nu^{d+1} N(\mathcal{E}) \geq \nu^{d+1} > \rho^{d+1} \ell^{-1} L^{d+1} \geq$$

$$\mathcal{X}_d^2 \lambda^2 \mathcal{X}_{d+1}^{-1} N(\mathcal{Q}) = \lambda N(\mathcal{Q}). \tag{7}$$

By applying Theorem 1 and Proposition 1 in [1] to the ideals $\mathcal{Q}$ and $\mathcal{E}$ respectively, we conclude that there exists a homogeneous polynomial $P_1 \in \mathcal{Q} \subset \wp$, $P_1 \notin \mathcal{E}$ such that $\deg P_1 = \nu$. From (7) we get $\rho L \ell^{-\frac{1}{d+1}} \geq 1$ and thus

$$\nu \leq 2\rho L \ell^{-\frac{1}{d+1}}.$$

From the definition of the polynomial P it follows that

$$L = \deg P \leq \deg P_1 \leq 2\rho L \ell^{-\frac{1}{d+1}}$$

and we also get inequality (6).

From the inclusion $Q \subset \wp$ it easily follows that there exists an index j satisfying $\mathrm{ord}_{\xi_j} Q > 0$. If $\mathcal{D}Q \subset Q$ this inequality leads to the conclusion that $Q \subset \mathcal{E}$, which is impossible, since $h(Q) = m - \kappa + d + 1 > m - k = h(\mathcal{E})$. Consequently $\mathcal{D}Q \not\subset Q$.

The ideal Q is isolated in the set of prime ideals associated with $\mathcal{A}_d$. Therefore a polynomial $H \notin q$ can be found satisfying $E^\ell \cdot H \in \mathcal{A}_d$ for any $E \in Q$. From this it follows immediately that there exist indices $i, j, 0 \leq i \leq m - \kappa + d, 0 \leq j \leq (2\rho)^{d+1}$ such that $\mathcal{D}^j Q_i \notin Q$. Since $\mathcal{X}_d + (2\rho)^{d+1} \leq \mathcal{X}_{d+1}$, we have that $\mathcal{D}^j Q_i \in \mathcal{J}_{d+1}$. Further,

$$\deg \mathcal{D}^j Q_i \leq \mathcal{X}_d L + (2\rho)^{d+1} \max \deg F_i \leq \mathcal{X}_{d+1} L.$$

Combining the polynomials $\mathcal{D}^j Q_i$ constructed for all ideals Q with the above properties, we can construct a homogeneous polynomial $Q_{m-\kappa+d+1} \in \mathcal{J}_{d+1}$, not lying in one of the ideals Q associated with $\mathcal{A}_d$ and contained in $\wp$, $\deg Q_{m-\kappa+d+1} \leq \mathcal{X}L$. It is not hard to check that the ideal $\mathcal{A}_{d+1} = (Q_0, \ldots, Q_{m-\kappa+d+1})$ satisfies condition 3).

Thus the supposition $\mathcal{J}_{d+1} \subset \wp$ leads to a contradiction with the definition of d. Hence $\mathcal{J}_{d+1} \not\subset \wp$ and since $d \leq \kappa - \tau$, we deduce that $\mathcal{J}_{\kappa-\tau+1} \not\subset \wp$. From this it follows that there exists an index $i, 0 \leq i < \mathcal{X}_{\kappa-\tau+1}$, such that $R = \mathcal{D}^i P \in \wp$, but $Q = \mathcal{D}R \notin \wp$. The bound (4) is satisfied in view of (5) and the equality $\mathcal{X}_{\kappa-\tau+1} = \lambda^{2^{\kappa-\tau+2}-1}$.

We continue the proof of Theorem 2. For any $\xi \in C$ denote the algebraic closure of the field of formal power series $C((z - \xi))$ by $\mathcal{K}_\xi$. We extend the valuation ord_ξ to the field $\mathcal{K}_\xi$ and for any two vectors $\overline{\varphi} = (\varphi_0, \ldots, \varphi_m) \in \mathcal{K}_\xi^{m+1}$, $\overline{\psi} = (\psi_0, \ldots, \psi_m) \in \mathcal{K}_\xi^{m+1}$, define

$$|\overline{\varphi}|_\xi = \min_{0 \leq j \leq m} (\mathrm{ord}_\xi \varphi_j),$$

$$\|\overline{\varphi} - \overline{\psi}\|_\xi = \min_{i \neq j} \mathrm{ord}_\xi(\varphi_i \psi_j - \varphi_j \psi_i) - |\overline{\varphi}|_\xi - |\overline{\psi}|_\xi.$$

Let R and Q be the polynomials, the existence of which is proved in the lemma. By the functional variant of the proof of Proposition 3, [2], we can establish the existence of zeros

$$\overline{\beta}_{\ell_j} \in \mathcal{K}_{\xi_\ell}^{m+1}, \quad \ell = 1, \ldots, q, \quad j = 1, 2, \ldots, N(\wp),$$

of the ideal $\wp$ such that for $\ell = 1, \ldots, q$

$$\mathrm{ord}_{\xi_\ell} \tau_\ell(a) + \sum_{j=1}^{N(\wp)} |\overline{\beta}_{\ell_j}|_{\xi_\ell} \geq 0, \tag{8}$$

$$\operatorname*{ord}_{\xi_\ell} \tau_\ell(a) + \sum_{j=1}^{N(\wp)} \left(\|\overline{f} - \overline{\beta}_{\ell j}\|_{\xi_\ell} + |\overline{\beta}_{\ell j}|_{\xi_\ell} \right) \geq \operatorname*{ord}_{\xi_\ell} \wp(\overline{f}) \qquad (9)$$

(see inequalities (22) and (24) in [2]), and, further, the existence of a homogeneous unmixed ideal $\mathcal{J} \subset \mathbf{C}[x_0, \ldots, x_m]$, satisfying the conditions $h(\mathcal{J}) = m - \tau + 2$,

$$N(\mathcal{J}) \leq N(\wp) \deg Q \leq \lambda^{2^\kappa} N(\wp)^{\frac{\kappa - \tau + 2}{\kappa - \tau + 1}}, \qquad (10)$$

$$\sum_{\ell=1}^{q} \operatorname*{ord}_{\xi_\ell} \mathcal{J}(\overline{f}) = \deg Q \sum_{\ell=1}^{q} \operatorname*{ord}_{\xi_\ell} \tau_\ell(a) + \sum_{\ell=1}^{q} \sum_{j=1}^{N(\wp)} \operatorname*{ord}_{\xi_\ell} Q(\overline{\beta}_{\ell j}). \qquad (11)$$

Denote the set of indices ℓ, $1 \leq \ell \leq q$, for which $\overline{f}(\xi_\ell)$ is a zero of the ideal $\wp$ by Γ. If $\ell \notin \Gamma$, then

$$\|\overline{f} - \overline{\beta}_{\ell j}\|_{\xi_\ell} = 0, \qquad j = 1, \ldots, N(\wp)$$

and hence

$$\operatorname*{ord}_{\xi_\ell} Q(\overline{\beta}_{\ell j}) \geq \|\overline{f} - \overline{\beta}_{\ell j}\|_{\xi_\ell} + |\overline{\beta}|_{\xi_\ell} \deg Q. \qquad (12)$$

If $\ell \in \Gamma$, then since $R \in \wp$, we have

$$\operatorname*{ord}_{\xi_\ell} Q(\overline{f}) = \operatorname*{ord}_{\xi_\ell} R(\overline{f}) - 1 \geq \|\overline{f} - \overline{\beta}_{\ell j}\|_{\xi_\ell} - 1$$

and, consequently,

$$\operatorname*{ord}_{\xi_\ell} Q(\overline{\beta}_{\ell j}) \geq \|\overline{f} - \overline{\beta}_{\ell j}\|_{\xi_\ell} + |\overline{\beta}_{\ell j}|_{\xi_\ell} \deg Q - 1. \qquad (13)$$

The number of elements of Γ does not exceed $a_{\tau-1}(N(\wp))$, so by (8), (9), (11), (12), (13), it follows that

$$\sum_{\ell=1}^{q} \operatorname*{ord}_{\xi_\ell} \mathcal{J}(\overline{f}) \geq \deg Q \sum_{\ell=1}^{q} \operatorname*{ord}_{\xi_\ell} \tau_\ell(a) +$$

$$+ \sum_{\ell=1}^{q} \sum_{j=1}^{N(\wp)} \left(\|\overline{f} - \overline{\beta}_{\ell j}\|_{\xi_\ell} + |\overline{\beta}_{\ell j}|_{\xi_\ell} \deg Q \right) - a_{\tau-1}(N(\wp)) N(\wp) \geq$$

$$\geq \sum_{\ell=1}^{q} \operatorname*{ord}_{\xi_\ell} \wp(\overline{f}) - a_{\tau-1}(N(\wp)) N(\wp). \qquad (14)$$

Since $h(\mathcal{J}) = m - \tau + \ell > h(\mathcal{I})$, the ideal $\mathcal{J}$ satisfies the assertion of Theorem 2. Hence and by (14), (10), we arrive at a bound for the sum of the multiplicities of the zeros of the ideal $\wp$, contradicting inequality (2) for $\gamma_2 = \lambda^{2^\kappa \cdot \kappa}$. This contradiction proves Theorem 2.

References

[1] Yu. V. Nesterenko, Estimates for the characteristic functions of a prime ideal, *Mat. Sbornik*, 1984, **123** (165), 11–34, (translation in *Math. USSR Sbornik* **51** (1985), No. 1).

[2] Yu. V. Nesterenko, On the algebraic independence of algebraic powers of algebraic numbers, *Mat. Sbornik*, 1984, **123** (165), 435–459, (translation in *Math. USSR Sbornik* **51** (1985), No. 2).

[3] W. D. Brownawell, On the orders of zero of certain functions, *Bull. Soc. Mat. France*, 2e serie, Memoire No. 2, 1980, 5–20.

[4] W. D. Brownawell, Zero estimates for solutions of differential equations, *Progress in Math.*, **31**, 1983, 67–94, Birkhäuser.

[5] W. D. Brownawell, D. W. Masser, Multiplicity estimates for analytic functions, *Duke Math. J.*, **47**, 1980, 273–295.

17

AN APPLICATION OF THE S-UNIT THEOREM
TO MODULAR FORMS ON $\Gamma_0(N)$

R. W. K. Odoni

1. Introduction

The Fourier coefficients of modular forms of type (k, ϵ) on $\Gamma_0(N)$ have many interesting arithmetical properties, with important applications in algebraic number theory — see [2], [9]. Serre [8] raised a number of interesting questions about such coefficients, some of which we answered in [6]. A particularly difficult problem of this type is the following: given $F(z)$, a form of type $(1, \epsilon)$ on $\Gamma_0(N)$, and an $\alpha \in \mathbb{C}$, how often do we have $a(n) = \alpha$? (Here $F(z)$ has the Fourier expansion $\sum_{n \geq 0} a(n) q^n$, $q = e^{2\pi i z}$, $\operatorname{Im} z > 0$). We shall describe in this article a method which yields the solution of this problem in the form of an asymptotic for

$$\sharp\{n : 1 \leq n \leq x, a(n) = \alpha\} \qquad (A1)$$

as $x \to \infty$. By an application of the S-unit theorem of van de Poorten-Schlickewei [7] and Evertse [3] (see also the survey [4] in this volume) we shall reduce the problem to a form amenable to the methods developed in [6]. In that paper a special case of the problem was treated; the general case was not tractable at that time, since the S-unit theorem was not available.

2. Notation and Terminology

We shall adhere to the conventions of [2], [9]; these are convenient sources for various results quoted without proof in this section.

Let $\mathcal{H}$ be the complex upper half-plane $\{z \in \mathbb{C} : \operatorname{Im} z > 0\}$. Then $\mathrm{SL}(2, \mathbb{Z})$ acts discontinuously on $\mathcal{H}$ if we associate with $\begin{pmatrix} a & b \\ c & d \end{pmatrix}$ in $\mathrm{SL}(2, \mathbb{Z})$ the linear fractional transformation $z \mapsto (az + b)/(cz + d)$. Now let $N \geq 1$, $N \in \mathbb{Z}$. We define

$$\Gamma_0(N) = \left\{ \begin{pmatrix} a & b \\ c & d \end{pmatrix} \in \mathrm{SL}(2, \mathbb{Z}) : c \in N\mathbb{Z} \right\}. \qquad (2.1)$$

Suppose that $k \in \mathbf{Z}$, $k \geq 1$, and that ϵ is a Dirichlet character (mod N). We say that $F(z)$ is a *modular form of type* (k, ϵ) *on* $\Gamma_0(N)$ if

(i) $F(z)$ is analytic on $\mathcal{H}$;

(ii) $(cz + d)^{-k} F((az + b)/(cz + d)) = \epsilon(d) F(z)$ for all $z \in \mathcal{H}$ and all $\begin{pmatrix} a & b \\ c & d \end{pmatrix}$ in $\Gamma_0(N)$;

$$(2.2)$$

(iii) for each cusp $t = T(\infty)$, $T \in \Gamma_0(1)$, $F(z)$ tends to a limit (written $F(t)$) as $z \to t$.

If, moreover, $F(z)$ vanishes at each cusp it is called a *cusp form* of type (k, ϵ) on $\Gamma_0(N)$. Let $\mathcal{M}(\Gamma_0(N), k, \epsilon)$ be the set of all $F(z)$ satisfying (2.2). Then $\mathcal{M}(\Gamma_0(N), k, \epsilon)$ is a finite-dimensional C-vector space which reduces to 0 unless $\epsilon(-1) = (-1)^k$; from now onwards we shall assume that the latter holds.

By mapping $F(z)$ to $F(T(\infty))$ we see that the cusp forms yield a subspace $\mathcal{M}^0(\Gamma_0(N), k, \epsilon)$ of codimension $\leq c$, the number of $\Gamma_0(N)$-inequivalent cusps. In fact the codimension is c since one can find "generalized Eisenstein series" in $\mathcal{M}(\Gamma_0(N), k, \epsilon)$ whose values separate the cusps.

Since $\begin{pmatrix} 1 & 1 \\ 1 & 0 \end{pmatrix} \in \Gamma_0(N)$ it follows from (2.2) that $F(z) = F(z + 1)$ for all $z \in \mathcal{H}$ and $F \in \mathcal{M}(\Gamma_0(N), k, \epsilon)$, and thus that

$$F(z) = \sum_{n \geq 0} a(n) q^n, \qquad q = e^{2\pi i z}, \ z \in \mathcal{H}. \qquad (2.3)$$

The $a(n)$ are called the *Fourier coefficients* of F. It is possible to normalize each Eisenstein series mentioned above so that, for it, $a(1) = 1$; when this is done the $a(n)$ $(n \geq 1)$ are multiplicative. This normalized Eisenstein series is, in fact, the Mellin transform of the product of two Dirichlet L-functions, at least one of which has conductor N.

There is a very important canonical decomposition of $\mathcal{M}^0(\Gamma_0(N), k, \epsilon)$, due to Atkin and Lehner [1] for $\epsilon = 1$, and Li [5] for general ϵ. Suppose that d, N^* are natural numbers such that $dN^* | N$, and that $G^*(z) \in \mathcal{M}^0(\Gamma_0(N^*), k, \epsilon)$, where ϵ reduces to ϵ^* (mod N^*). The forms $G(z)$ obtained in this way from proper divisors N^* of N span a subspace $\mathcal{M}^{\text{old}}(\Gamma_0(N), k, \epsilon)$, called the space of *old* (cusp-) *forms*. Now $\mathcal{M}^{\text{old}}(\Gamma_0(N), k, \epsilon)$ is equipped with a natural inner product (generalizing the Petersson inner producer for $N = 1$, $\Gamma_0(1) = \mathrm{SL}(2, \mathbf{Z})$). The orthogonal complement of $\mathcal{M}^{\text{old}}(\Gamma_0(N), k, \epsilon)$ with respect to this inner product is denoted by $\mathcal{M}^{\text{new}}(\Gamma_0(N), k, \epsilon)$, and called the space of *new* (cusp-) *forms*. The space $\mathcal{M}^{\text{new}}(\Gamma_0(N), k, \epsilon)$ has a natural basis consisting of simultaneous eigenfunctions for certain operators T_p, U_p, p prime, called *Hecke operators*, which constitute an obvious generalization of the classical Hecke operators for $N = 1$.

From now onwards we shall assume that $k = 1$. A deep result of Deligne and Serre [2] shows that these simultaneous eigenfunctions, normalized so that $a(1) = 1$, are precisely the Mellin transforms of Artin L-functions corresponding to irreducible continuous two-dimensional representations: $\text{Gal}(\overline{\mathbf{Q}}/\mathbf{Q}) \to \text{GL}(2, \mathbf{C})$, where $\overline{\mathbf{Q}}$ is the algebraic closure of $\mathbf{Q}$, with Artin conductor N. If we drop the adjective "irreducible" from the above, each normalized Eisenstein series discussed earlier can be regarded as the Mellin transform of an Artin L-function for a continuous (reducible) two-dimensional representation, plus a constant.

From all this it follows that every $F \in \mathcal{M}(\Gamma_0(N), I, \epsilon)$ is expressible as a finite sum

$$F(z) = \sum_{j \in J} c_j F_j(zd_j), \qquad (2.4)$$

where the c_j are complex constants and $F_j(z)$ is the Mellin transform of an Artin L-function of a two-dimensional representation of conductor N_j, where $d_j N_j | N$.

Our main concern in this paper is the behaviour of the $a(n)$ of (2.3) when $k = 1$; in particular we wish to determine (A1) asymptotically. The case $\sharp J = 1$ was treated in detail in [6]. We showed there that, for $\sharp J = 1$, the function

$$G(s) = \sum_{\substack{n \geq 1 \\ a(n) = \alpha}} n^{-s}, \qquad \sigma = \operatorname{Re} s > 1 \qquad (2.5)$$

is expressible as an integral

$$\int_C H(s, \mathbf{z}) \Psi(z_1^{-1}, \dots, z_k^{-1}) \prod_{r=1}^k \frac{dz_r}{z_r} \qquad (2.6)$$

over some cycle C in $\mathbf{C}^k$, k finite, where $\Psi(\mathbf{z})$ is a rational function of known singularities and $H(s, z)$ is a certain Euler product in s with known singularities and an analytic continuation into a (cut) region extending to the left of $\sigma = 1$. From the integral representation (2.6) we were then able to give a qualitative description of the types of asymptotic behaviour possible for (A1). Our aim is to show that an analogous result of the type (2.6) is still valid if $\sharp J > 1$; to obtain it we shall have to apply the S-unit theorem.

3. Preliminary reductions

We can always assume that $n \geq 1$. If $n \geq 1$ and $a(n) = \alpha$, we see from (2.3) and (2.4) that

$$\alpha = a(n) = \sum_{j \in J} c_j a_j(nd_j^{-1}), \qquad (3.1)$$

where we define $a_j(y) = 0$ unless $y \in \mathbb{N}$, the natural numbers $(0 \notin \mathbb{N})$.

As a first step we separate off the terms in (3.1) corresponding to divisors of the various N_j. There are only finitely many prime divisors of the various N_j, and they all divide N. Let $\mathcal{D}$ be the set of natural numbers all of whose prime factors divide N; then every $n \in \mathbb{N}$ is uniquely expressible as dm where $d \in \mathcal{D}$ and $(m, N) = 1$. Since the a_j $(j \in J)$ are multiplicative we thus have

$$\alpha = a(n) = \sum_{j \in J} c_j a_j(dd_j^{-1}) a_j(m). \qquad (3.2)$$

We now consider the values taken by the various $a_j(m)$. Since $F_j(z)$ is the Mellin transform of an Artin L-function of a two-dimensional representation we have (see [2])

$$\sum_{(m,N)=1} a_j(m)m^{-s} = \prod_{p \nmid N} \det\{I_2 - p^{-s}R_j(p)\}^{-s}, \qquad \sigma = \operatorname{Re} s > 1. \quad (3.3)$$

Here $R_j(p)$ is the matrix representing some element of the Frobenius class of p in $\operatorname{Gal} K_j/\mathbb{Q}$, where $\operatorname{Gal} \mathbb{Q}/K_j$ is the kernel of R_j. In particular $K_j/\mathbb{Q}$ is a finite Galois extension unramified outside the set of prime factors of N, so that the Frobenius classes are well-defined. Moreover we shall have $\det R_j(p) = \epsilon_j(p)$ where $F_j \in \mathcal{M}^0(\Gamma_0(N_j), 1, \epsilon_j)$. Let $\lambda_j(p)$ and $\mu_j(p)$ be the eigenvalues of $R_j(p)$. They are clearly roots of unity of order dividing $[K_j : \mathbb{Q}]$, while $\lambda_j(p)\mu_j(p) = \epsilon_j(p)$. We also have

$$\det\{I_2 - p^{-s}R_j(p)\}^{-1} = (1 - \lambda_j(p)p^{-s})^{-1}(1 - \mu_j(p)p^{-s})^{-1}, \quad (3.4)$$

from which we see that

$$a_j(p^k) = \sum_{r=0}^{k} \lambda_j(p)^r \{\mu_j(p)\}^{k-r} \quad (\forall k \geq 1). \qquad (3.5)$$

The value of the general $a_j(m)$ is then obtainable from (3.5) once all the various $\lambda_j(p)$, $\mu_j(p)$ have been determined. The latter is accomplished as follows. Let K be the compositum of all the K_j, $j \in J$. Then $K/\mathbf{Q}$ is finite Galois and unramified outside $\{p : p|N\}$. If $p\nmid N$ and its Frobenius class in $K/\mathbf{Q}$ is known then its class in each $K_j/\mathbf{Q}$ is also known and hence, up to conjugacy, so is each $\mathbf{R}_j(p)$. Thus, for all $j \in J$, the set $\{\lambda_j(p), \mu_j(p)\}$ is known and hence, by (3.5), the values $a_j(p^k)$ are known for all $k \geq 1$ and all $j \in J$. It is clear that, for $(m, N) = 1$, the values of $a_j(m), j \in J$ are integers in the cyclotomic field L generated by all $[K : \mathbf{Q}]$th roots of unity.

In order to be able to apply the S-unit theorem it is necessary to subdivide the $p\nmid N$ into two types. We first observe from (3.5) that if $\lambda_j(p) \neq \mu_j(p)$ then $a_j(p^k)$ is periodic in k, so that the set $\{a_j(p^k) : k \geq 1\}$ is finite. On the other hand if $\lambda_j(p) = \mu_j(p)$ the set is infinite. We call $p\nmid N$ a *good prime* if $\lambda_j(p) \neq \mu_j(p)$ for all $j \in J$; otherwise it is a *bad prime*.

We now refine the decomposition used in (3.2). Let $t \geq 10$ be arbitrary but fixed. Let $\mathcal{E}$ be the submonoid of N generated by all $p\nmid N$ and all powers p^k, $k > t$, of bad primes, and let $\mathcal{F}$ be the submonoid generated by all good p and all powers p^k, $k \leq t$, of bad primes. Then every $n \in N$ is uniquely expressible in the form $n = ef$ with $e \in \mathcal{E}$, $f \in \mathcal{F}$ and $(e, f) = 1$. The condition $t \geq 10$ is, to some extent, arbitrary but convenient for our purpose; it ensures that $\mathcal{E}$ is a thin subset of N, in that one can easily show that

$$\#\mathcal{E} \cap [1, x] = O_\epsilon(x^{\epsilon + \frac{1}{t+1}}) \tag{3.6}$$

for arbitrary $\epsilon > 0$, and hence that the Dirichlet series $\sum_{e \in \mathcal{E}} e^{-s}$ is analytic and uniformly bounded for $\sigma \geq \epsilon + \frac{1}{t+1}$.

Roughly speaking our decomposition $n = ef$ is intended to serve the following purpose; we should like to enumerate $(A1)$ by fixing $e \in \mathcal{E}$, taking $n = ef$, $f \in \mathcal{F}$, $(e, f) = 1$, counting the appropriate $f \leq x/e$ giving $a(n) = \alpha$, and then summing over all $e \leq x$ in $\mathcal{E}$. It is not possible to do this directly, but the processes which we carry out (with Dirichlet series) do embody this idea implicitly.

We rewrite (3.1) in the form

$$\alpha = a(n) = a(ef) = \sum_{j \in J} \gamma_j(e) a_j(f), \tag{3.7}$$

where $\gamma_j(e) = c_j a_j(ed_j^{-1})$, and write $\underline{\gamma}(e)$ for the vector $\{\gamma_j(e)\}_{j \in J}$ and $\mathbf{a}(f)$ for the vector $\{a_j(f)\}_{j \in J}$. Let Γ be the set of all $\underline{\gamma}(e)$; it is clearly countable. We introduce, for $\sigma > 1$, the Dirichlet series

$$G(s) = \sum_{\substack{n \geq 1 \\ a(n) = \alpha}} n^{-s}. \qquad (3.8)$$

We shall aim to express $G(s)$ in terms of Dirichlet series involving only e's and f's separately. With this in view our first task is to consider a fixed $e \in \mathcal{E}$ and examine $f \in \mathcal{F}$, with $(e, f) = 1$, satisfying (3.7).

4. Application of the S-unit theorem

Let $e \in \mathcal{E}$ be fixed and let $\gamma(e) = \gamma$; we consider the equation

$$\alpha = \sum_{j \in J} \gamma_j a_j(f) \qquad (4.1)$$

with $f \in F$, $(e, f) = 1$. Let us first dispose of the case $\alpha = 0$. In this case the decomposition (3.2) will suffice. It is known [2], [9] that complex conjugation in $\operatorname{Gal} K_j/\mathbf{Q}$ corresponds to an element τ such that trace $R_j(\tau) = 0$. Hence if m in (3.2) has, for each $j \in J$, at least one prime factor $P_j \| m$ whose Frobenius class in $K_j/\mathbf{Q}$ contains complex conjugation then $a_j(m) = 0$ for all $j \in J$. One can easily show (see e.g. [8]) that the number of $m, 1 \leq m \leq x$, with $(m, N) = 1$, not satisfying the above condition is asymptotically $c_1 (\log x)^{c_2 - 1}$ with $c_1, c_2 > 0$ and $c_2 < 1$; in fact this result can be refined to a full asymptotic expansion in the scale of functions $x(\log x)^{c_2 - n}$, $n \geq 1$. On the other hand the total number of m, $1 \leq m \leq x$, $(m, N) = 1$, is $\{\varphi(N)/N\}x + O(1)$. Thus we have that $a(n)$ is "almost always" 0. We could also have $a(n) = 0$ without all $a_j(m)$ zero; the enumeration of such m is easily reduced to a problem similar to that occurring when $\alpha \neq 0$, and we now turn to that case.

We recall that the values of all $a_j(f)$, $f \in F$, $j \in J$, are integers in the cyclotomic field L of §3. Moreover the multiplicative subgroup of L^* generated by all non-zero $a_j(f)$ is finitely generated, by §3. We denote it by A.

Starting from (4.1) we choose an arbitrary L-basis for the vector space $L\alpha + \sum_{j \in J} L\gamma_j$. On expressing α and the γ_j in terms of this basis we arrive at a finite family of equations

$$\theta_s = \sum_{j \in J} \varphi_{sj} a_j(f), \qquad \text{for all } s \in S, \qquad (4.2)$$

with θ_s and the φ_{sj} elements of L uniquely determined by γ. For at least one s not all of θ_s and the φ_{sj} are 0. We choose such an s and proceed as follows. We may delete from J all those j with $\varphi_{sj} = 0$, obtaining a subset $J_1 = J_1(s)$. Next we are faced with various alternatives, depending on those $j \in J_1$ for which $a_j(f) = 0$. This leads to the union of a finite number of systems of the type

$$\left.\begin{aligned}
\theta_s &= \sum_{j \in J_2} \varphi_{sj} a_j(f), \quad \text{no summand } 0 \\
0 &= a_j(f), \qquad\qquad \text{for all } j \in J_1 \setminus J_2
\end{aligned}\right\} \tag{4.3}$$

for all $J_2 \subseteq J_1$.

At this stage we can apply the S-unit theorem. Consider the system (4.3). Let G be the subgroup of L^* generated by A and all non-zero numbers amongst θ_s and the φ_{sj}, $j \in J_2$. Then G is finitely generated. The S-unit theorem can be expressed as follows: for any $k > 1$ the equation $g_1 + \ldots + g_k = 0$, with no proper subsum zero, has only finitely many solutions $\mathbf{g} \in G^k$, up to equivalence. Here $\mathbf{g}$ and $\mathbf{g}^*$ are regarded as equivalent if $\mathbf{g} = h\mathbf{g}^*$ for some $h \in G$.

Obviously the upper equation in (4.3) is expressible in the form $g_1 + \ldots + g_k = 0$, but certain proper subsums may vanish. The general system $g_1 + \ldots + g_k = 0$ clearly decomposed as the union of a finite number of simultaneous equations $\sum_{t \in T_r} g_t = 0$, no proper subsum 0, with T_r, $r \in R$, a subset of $\{1, \ldots, k\}$. Accordingly a single non-vacuous equation chosen from (4.2) gives rise to a finite collection of simultaneous equations of the type

$$\left.\begin{aligned}
a_j(f) &= 0, \qquad\qquad j \in J_1 \setminus J_2 \\
a_j(f)\varphi_{sj}b_i &= a_i(f)\varphi_{si}b_j \neq 0, \quad \text{for all } i, j \in T_r
\end{aligned}\right\} \tag{4.4}$$

with $\mathbf{b} \in B_r$, a finite set in G^{T_r}, together with an additional system

$$a_j(f)\varphi_{sj} = \theta_s \beta_j \neq 0, \quad \text{for all } j \in U, \tag{4.5}$$

with finitely many choices of $\beta \in G^U$, in case $\theta_s \neq 0$.

On recalling how the system (4.4) (or (4.5)) was obtained from (4.1), we see that the f occurring in (4.1) must, together with the condition $f \in \mathcal{F}$, $(e, f) = 1$, satisfy the intersection, over all $s \in S$, of a finite union of systems of the type (4.4) or (4.5), as the T_r vary over all consistent decompositions. The form of these systems allows us to employ the method of Frobenian functions [6] in order to obtain the required decomposition of $G(s)$ of (3.8). We shall carry this out in the next section.

5. Use of Frobenian functions

Now let p be a prime in $\mathcal{F}$. Then p is unramified in $K/\mathbb{Q}$ and the values $a_j(p^k)$, $j \in J$, $k \geq 1$ are determined by the Frobenius class of p in $K/\mathbb{Q}$. If p is bad we have $p^k \in \mathcal{F}$ only for $k \leq t$, by §3. Thus the total set of values taken by the vectors $\mathbf{a}(p^k)$ for prime powers p^k in $\mathcal{F}$ is finite. We denote this set by M. For each $\mathbf{m} \in M$ we introduce a complex variable $z(\mathbf{m})$. For $p \in \mathcal{F}$ let

$$H_p(\mathbf{z}, s) = 1 + \sum_{\mathbf{m} \in M} z(\mathbf{m}) \bigg\{ \sum_{\substack{k \geq 1, p^k \in \mathcal{F} \\ a(p^k) = \mathbf{m}}} p^{-ks} \bigg\}. \tag{5.1}$$

Certainly $H_p(\mathbf{z}, s)$ is determined by the Frobenius class of p in $K/\mathbb{Q}$. Now let $\sigma > 1$; we put

$$H(\mathbf{z}, s) = \prod_{p \in \mathcal{F}} H_p(\mathbf{z}, s). \tag{5.2}$$

Then $H(\mathbf{z}, s)$ is entire in $\mathbf{z}$ and analytic $\sigma > 1$.

Now let $f \in \mathcal{F}$; we denote by $\Omega(\mathbf{m}, f)$ the number of prime powers $p^k \in \mathcal{F}$ such that $p^k \| f$ and $a(p^k) = \mathbf{m}$, and by $\underline{\Omega}(f)$ the vector of all $\Omega(\mathbf{m}, f)$ ($\mathbf{m} \in M$). We abbreviate $\prod_{\mathbf{m} \in M} z(\mathbf{m})^{\Omega(\mathbf{m},f)}$ to $\mathbf{z}^{\underline{\Omega}(f)}$. Then, for $\sigma > 1$, we can expand (5.2) as a Dirichlet series in s, obtaining

$$H(\mathbf{z}, s) = \sum_{f \in \mathcal{F}} f^{-s} \mathbf{z}^{\underline{\Omega}(f)}. \tag{5.3}$$

Now consider a fixed $\gamma = \gamma(e)$ in (4.1). Ignoring for the moment the condition $(e, f) = 1$, we consider all $f \in F$ which satisfy (4.1) with the given γ. We assume first that $\|\mathbf{z}\| = \max_{\mathbf{m}} |z(\mathbf{m})| < 1$. Let $Q(\gamma, \mathbf{z})$ be the sum of $\mathbf{z}^{\underline{\Omega}}$, taken over all vectors $\underline{\Omega}$ expressible as $\underline{\Omega}(f)$ for some $f \in \mathcal{F}$ satisfying (4.1). It is easily seen that the conditions imposed on $\underline{\Omega}(f)$ by the associated systems (4.4), (4.5) imply that $Q(\gamma, \mathbf{z})$ is a rational function of $\mathbf{z}$ whose poles, other than ∞, are in the region $\|\mathbf{z}\| \geq 1$. Now let $\rho > 1$, and let $\mathcal{C}$ be the product of the anticlockwise circles $C(\mathbf{m}) : |z(\mathbf{m})| = \rho$. Then it is easily seen that

$$(2\pi i)^{-\sharp M} \int_{\mathcal{C}} \cdots \int \frac{H(\mathbf{z}, s)}{\prod_{p|e} H_p(\mathbf{z}, s)} Q(\gamma, \mathbf{z}^{-1}) \prod_{\mathbf{m} \in M} \frac{dz(\mathbf{m})}{z(\mathbf{m})} \tag{5.4}$$

exists for $\sigma > 1$ and coincides with the sum $\sum' f^{-s}$, taken over all $f \in \mathcal{F}$ with $(e, f) = 1$ satisfying (2.1). (Here $\mathbf{z}^{-1}$ means the vector with components $z(m)^{-1}$.)

To proceed further we need to consider the sum

$$\sum_{\substack{e \in \mathcal{E} \\ \gamma(e)=\gamma}} e^{-s} \prod_{p|e} H_p(\mathbf{z}, s)^{-1}. \tag{5.5}$$

The factor $H_p(\mathbf{z}, s)$ is taken to be 1 when $p|N$. Thus the only interesting H_p occurring in (5.5) are the bad primes, while $H_p(\mathbf{z}, s) = 1 + \sum_{m \in M} z(\mathbf{m}) \sum_{\substack{k \leq t \\ a(p^k)=m}} p^{-ks}$ in this case. If we drop the condition $\gamma(e) = \gamma$ for the moment, we see without difficulty that

$$\sum_{e \in \mathcal{E}} e^{-s} \prod_{p|e} H_p(\mathbf{z}, s)^{-1} \tag{5.6}$$

is expressible as an Euler product

$$\prod_{p|N}(1 - p^{-s})^{-1} \prod_{p \text{ bad}} \left\{ 1 + \frac{p^{-s(1+t)}}{(1 - p^{-s})H_p(\mathbf{z}, s)} \right\}, \tag{5.7}$$

and this expression is valid for $\sigma > (1+t)^{-1}$, provided that the radius ρ of the circles $C(\mathbf{m})$ is sufficiently close to 1. It follows that (5.5) converges absolutely for $\sigma > (1 + t)^{-1}$. Moreover the $Q(\gamma, \mathbf{z}^{-1})$ are uniformly bounded for $\|\mathbf{z}^{-1}\| \leq \rho_0^{-1}$ when $\rho_0 > 1$, independently of γ. It is thus permissible to multiply (5.4) by $\sum_{\gamma(e)=\gamma} e^{-s}Q(\gamma, \mathbf{z}^{-1})$ and then to sum over all $\gamma \in \Gamma$, when $\sigma > 1$, obtaining

$$G(s) = \int_C H(\mathbf{z}, s)\Psi(\mathbf{z}, s) \prod_{m \in M} \frac{dz(\mathbf{m})}{z(\mathbf{m})} \tag{5.8}$$

for $\sigma > 1$, where $\Psi(\mathbf{z}, s)$ is analytic for $\sigma > (1 + t)^{-1}$ and $\|\mathbf{z}^{-1}\| < 1$. The representation (5.8) is of an appropriate type to yield the required asymptotic expansion of $(A1)$. The analysis required is virtually the same as that used in [6]; an application of a suitably strong form of the Čebotarev density theorem allows us to continue $H(\mathbf{z}, s)$ analytically to the left of $s = 1$, at the cost of a branch cut leftwards from $s = 1$ along the real axis, and a logarithmic singularity at $s = 1$. The continuation is valid in a suitably common zero-free region for a certain finite family of

Artin L-functions, and the growth of H in this region is sufficiently slow to allow an inverse Mellin transform to be applied to (3.8). The reader should be able to fill in the details after examining the corresponding arguments in [6].

References

[1] A. O. L. Atkin, J. Lehner, Hecke Operators on $\Gamma_0(N)$, *Math. Annalen* **185** (1970), 134–160.

[2] P. Deligne, J-P. Serre, Formes modulaires de poids 1, *Ann. Sci. École Norm. Sup.* (4) **7** (1974), 507–530.

[3] J-H. Evertse, On sums of S-units and linear recurrences, *Compositio Math.* **53** (1984), 225–244.

[4] J-H. Evertse, K. Győry, C. L. Stewart, R. Tijdeman, S-unit equations and their applications, *New Advances in Transcendence Theory* (ed. A. Baker), Cambridge Univ. Press, 1988, Chap. 9.

[5] W. Li, New forms and functional equations, *Math. Annalen* **212** (1975), 285–315.

[6] R. W. K. Odoni, Notes on the method of Frobenian functions, with applications to Fourier coefficients of modular forms, *Elementary and analytic theory of numbers* (Banach Center Publications No. 17), PWN, Warsaw, 1985, 371–403.

[7] A. van der Poorten, H-P. Schlickewei, The growth conditions for recurrent sequences, *MacQuarie Math. Reports* 82–0041, 1982.

[8] J-P. Serre, Divisibilité de certaines fonctions arithmétiques, *L'Enseignement Math.* **22** (1976), 227–260.

[9] J-P. Serre, Modular forms of weight 1 and Galois representations, *Algebraic Number Fields* (ed. A. Fröhlich), Academic Press, 1977, 193–268.

18

LOWER BOUNDS FOR LINEAR FORMS IN LOGARITHMS

P. Philippon and M. Waldschmidt

1. Introduction

Let $\alpha_1, \ldots, \alpha_n$ be non-zero algebraic numbers and $\beta_0, \ldots, \beta_n$ be algebraic numbers. For $1 \le j \le n$, let $\log \alpha_j$ be any determination of the logarithm of α_j. Assume that the number

$$\Lambda = \beta_0 + \beta_1 \log \alpha_1 + \ldots + \beta_n \log \alpha_n$$

does not vanish. Our aim is to provide a new lower bound for $|\Lambda|$. For a history of the subject we refer to [1]. See also [11] and [6].

Our estimates improve previously known results on this subject.

If one pays special attention to the dependence on the degree of the algebraic numbers, it is more efficient to work with the absolute logarithmic height. We will do that in the next sections, but here we first give a few corollaries of our main results in terms of the usual height: for an algebraic number α, we denote by $H(\alpha)$ the maximum of the absolute values of the coefficients of the minimal polynomial of α over $\mathbf{Z}$.

Let D be a positive integer and A, A_1, $\ldots$, A_n be positive real numbers satisfying

$$D \ge [\mathbf{Q}(\alpha_1, \ldots, \alpha_n, \beta_0, \ldots, \beta_n) : \mathbf{Q}],$$
$$A_j \ge \max\{H(\alpha_j), e\}, \qquad 1 \le j \le n,$$

and

$$A = \max\{A_1, \ldots, A_n, e^e\}.$$

Theorem 1.1. *Let* $B = \max\{H(\beta_j),\ 0 \le j \le n\}$. *Assume*

$$A_j \ge \max\{\exp|\log \alpha_j|, e^n\}, \qquad 1 \le j \le n.$$

Then

$$|\Lambda| \geq e^{-U},$$

where

$$U = C_{11}(n) \cdot D^{n+2} \cdot \log A_1 \ldots \log A_n \cdot (\log B + \log \log A)$$

and

$$C_{11}(n) \leq 2^{8n+53} \cdot n^{2n}.$$

The main difference with Theorem 1.1 of [11] is that we omit a factor $\log \log A_{n-1}$ when, say, $A_n \geq A_{n-1} \geq \ldots \geq A_1 \geq e^n$ (with $A_{n-1} = e^e$ if $n = 1$). The cost is a factor 4 for the constant, but also the assumption $A_1 \geq e^n$.

We turn now to the so-called rational case, where $\beta_0 = 0$ and $\beta_1, \ldots, \beta_n$ are rational integers. In this case we write $\beta_i = b_i$, $1 \leq i \leq n$.

Theorem 1.2. *Let* $b_1, \ldots, b_n$ *be rational integers such that*

$$\alpha_1^{b_1} \ldots \alpha_n^{b_n} \neq 1.$$

Let B be a positive real number satisfying

$$B \geq \max\{|b_i|, \ 1 \leq i \leq n\} \quad and \quad B \geq e.$$

Then

$$|\alpha_1^{b_1} \ldots \alpha_n^{b_n} - 1| \geq B^{-C_{12}\Omega}$$

with $\Omega = \log A_1 \ldots \log A_n$, and C_{12} is a positive effectively computable constant depending only on n and on the degree of $\mathbf{Q}(\alpha_1, \ldots, \alpha_n)$ over $\mathbf{Q}$.

Finally, here is a variant of Theorem 1.2 which is useful for instance in the study of Diophantine equations.

Theorem 1.3. *In the situation of Theorem 1.2 assume* $e^e \leq A_1 \leq \ldots \leq A_{n-1} \leq A_n$, *and*

$$0 < |\alpha_1^{b_1} \ldots \alpha_n^{b_n} - 1| < e^{-\epsilon B}$$

for some $\epsilon > 0$. Then there exists a positive effectively computable constant C_{13}, depending only on n, on the degree of $\mathbf{Q}(\alpha_1, \ldots, \alpha_n)$ over $\mathbf{Q}$, *and on ϵ, such that*

$$B < C_{13} \log A_1 \ldots \log A_n \log \log A_n.$$

Moreover, if $b_n = 1$, then

$$B < C_{13} \log A_1 \ldots \log A_n \log \log A_{n-1}.$$

Here is the plan of this paper. In §2 we state two results (Theorem 2.1 for the "general" case, Theorem 2.2 for the "rational" case), and we deduce from them Theorems 1.1, 1.2 and 1.3.

Compared with previous proofs of similar results, the new feature here is that we apply the zero estimate of [9] (the previous zero estimates, which are referred to in [9], would not be sufficient for our purpose). The main result of [9] is stated in a much more general context, involving commutative algebraic groups. In §3 we state the special case which is needed here in terms of a lower bound for the rank of certain matrices. We need also to know that this zero estimate is not far from the best possible one, which means that we need an upper bound for the rank of these matrices. The proof of the zero estimate is postponed to the appendix. Indeed, for this proof, it is convenient to use the language of algebraic groups, and a reader who wishes to avoid this language can do so provided he takes for granted the Proposition 3.4. The proof of Theorem 2.1 is given in §4, and the proof of Theorem 2.2 in §5. A discussion on the explicit values of the constants is given in §2.

2. The two main results

Apart from the explicit dependence of the constant in terms of n, the two following statements include all previously known lower bounds for linear forms in logarithms which have been obtained so far by Baker's method.

When α is an algebraic number, we denote by $h(\alpha)$ the absolute logarithmic height of α (see for instance [11]).

We consider a non-zero linear form in logarithms of algebraic numbers with algebraic coefficients

$$\Lambda = \beta_0 + \beta_1 \log \alpha_1 + \ldots + \beta_n \log \alpha_n,$$

where $\alpha_1, \ldots, \alpha_n$ are non-zero algebraic numbers, $\beta_0, \ldots, \beta_n$ are algebraic numbers, and $\log \alpha_1, \ldots, \log \alpha_n$ are any non-zero determinations of the logarithms of $\alpha_1, \ldots, \alpha_n$. Let K be a number field containing $\alpha_1, \ldots, \alpha_n, \beta_0, \ldots, \beta_n$, of degree D over $\mathbb{Q}$.

Let $V_1, \ldots, V_n, V, E$ be positive real numbers satisfying

$$V_j \geq \max\{h(\alpha_j), \, |\log \alpha_j|/D, \, n/D\},$$

$$V = \max\{V_1, \ldots, V_n, \, 1\},$$

and

$$1 < E \leq \min\{e^{DV_j/n}, \ eDV_j/|\log \alpha_j|\}, \qquad 1 \leq j \leq n.$$

Our first main result deals with the "general case".

Theorem 2.1. *Let W be a positive real number satisfying*

$$W \geq \max_{0 \leq j \leq n} \{h(\beta_j)\}.$$

Then

$$|\Lambda| > \exp\{-C_{21}(n) . D^{n+2} . V_1 \ldots V_n.$$
$$(W + \log(EDV)) . (\log(ED)) . (\log E)^{-n-1}\}$$

with

$$C_{21}(n) \leq 2^{8n+51} . n^{2n}.$$

An improvement of our numerical value for $C_{21}(n)$ has been obtained recently by J. Blass, A. M. W. Glass, D. B. Meronk and R. P. Steiner [2].

Our second main result deals with the so-called "rational case".

Theorem 2.2. *Assume $\beta_0 = 0$ and $\beta_i = b_i \in \mathbf{Z}$ for $1 \leq i \leq n$. Let B, B_n, W be positive real numbers satisfying*

$$B \geq \max_{1 \leq j \leq n-1} |b_j|, \qquad B_n \geq |b_n|,$$

and

$$W \geq \max\left\{\log\left[\frac{B_n}{V_1} + \frac{B}{V_n} + 1\right]; \ \frac{1}{D} . \log E; \ 1\right\},$$

where we assume

$$V_1 \leq V_2 \leq \ldots \leq V_n.$$

Then

$$|\Lambda| > \exp\{-C_{22}(n) . D^{n+2} . V_1 \ldots V_n . W . (\log(ED)) . (\log E)^{-n-1}\},$$

where $C_{22}(n)$ is an effectively computable constant which depends only on n.

If one tries to compute an explicit value for $C_{22}(n)$ just by working out the proof in §5 below, one finds $C_{22}(n) \leq C_{23}.n^{2n^2}$ for some absolute constant C_{23}. However, assuming that the field $\mathbf{Q}(\sqrt{\alpha_1}, \ldots \sqrt{\alpha_n})$

has degree 2^n over $\mathbf{Q}$, Blass, Glass, Meronk and Steiner find $C_{22}(n) \le C_{24}^n \cdot n^n$ (with $C_{24} \le 2^{59}$). From the final descent of [11], one deduces $C_{12}(n) \le C_{25}^n \cdot n^{2n}$ for the constant C_{12} of Theorem 1.2, and one expects that $C_{22}(n) \le C_{26}^n \cdot n^{2n}$ without any assumption on $\mathbf{Q}(\sqrt{\alpha_1}, \ldots \sqrt{\alpha_n})$ (here C_{25} and C_{26} denote effectively computable absolute constants). However, as C.L. Stewart pointed out to the authors of [6], a further argument is necessary in order to achieve such an estimate. Also it was suggested by C. L. Stewart that the zero estimate of [9] could enable one to remove the $\log \log A_{n-1}$ as we did in Theorem 1.1.

Concerning the definition of E, it may be useful to notice that

$$eDV_j / |\log \alpha_j| \le 2^D \cdot e^{2DV_j};$$

indeed, for each non-zero algebraic number $\alpha \ne 0$, of degree $\le d$, and for $\log \alpha \ne 0$, we have

$$|\log \alpha| \ge 2^{-d} \cdot e^{-dh(\alpha)},$$

(using $|e^z - 1| \le |z| \cdot e^{|z|}$ with the Liouville inequality).

Proof of Theorem 1.1. We deduce Theorem 1.1 from Theorem 2.1, with $C_{11}(n) \le 4 \cdot C_{21}(n)$. We assume that the hypotheses of Theorem 1.1 hold, and we choose

$$V_j = \log A_j, \qquad W = \max\{1, \log B\}, \qquad E = eD.$$

Since

$$h(\alpha) \le \frac{1}{d}(\log H(\alpha) + \log d)$$

with $d = [\mathbf{Q}(\alpha) : \mathbf{Q}]$ (cf. [11], p. 260), we check that $V_j \ge h(\alpha_j)$. We use Theorem 2.1, with the fact that $A \ge e^e$ and $n \ge 1$, hence

$$(W + 1 + 2\log D + \log\log A)(1 + 2\log D)$$
$$< 4(1 + \log D)^{n+1}(\log B + \log\log A).$$

Therefore

$$C_{11}(n) \le 4C_{21}(n) \le 2^{8n+53} \cdot n^{2n}.$$

Proof of Theorem 1.2. We now deduce Theorem 1.2 from Theorem 2.2. We will use the following simple lemma.

Lemma 2.3. *Let $t \in \mathbf{C}$ and $r \in \mathbf{R}$ satisfy*

$$0 < r < 1 \quad and \quad |e^t - 1| \le r.$$

Then there exists $\kappa \in \mathbf{Z}$ *such that*

$$|t - 2i\kappa\pi| \leq \frac{1}{r} \cdot |\log(1 - r)| \cdot |e^t - 1|.$$

Proof of Lemma 2.3. The principal value of the logarithm satisfies

$$\sup_{|z|=r} |\log(1 + z)| = |\log(1 - r)|.$$

From Schwarz' lemma we get for $|z| \leq r$

$$|\log(1 + z)| \leq |z| \cdot \frac{1}{r} \cdot |\log(1 - r)|.$$

We use this inequality for $z = e^t - 1$, and we define

$$\kappa = \frac{1}{2i\pi} \cdot \left(t - \log(e^t)\right).$$

This proves Lemma 2.3.

For the proof of Theorem 1.2, we assume, as we may without loss of generality, $A_1 \leq A_2 \leq \ldots \leq A_{n-1} \leq A_n$ (with $A_{n-1} = e$ if $n = 1$). From Lemma 2.3 we deduce that as soon as

$$|\alpha_1^{b_1} \ldots \alpha_n^{b_n} - 1| \leq 1/3,$$

we have, for the principal value of the logarithm,

$$|b_1 \log \alpha_1 + \ldots + b_n \log \alpha_n - 2i\pi\kappa| \leq \frac{3}{2} \cdot |\alpha_1^{b_1} \ldots \alpha_n^{b_n} - 1|$$

for some $\kappa \in \mathbf{Z}$. Considering the imaginary parts, we get

$$|\kappa| < (nB + 1)/2.$$

We now choose

$$V_j = 7 \log A_j, \quad (1 \leq j \leq n), \quad V_0 = 2\pi n/D,$$

$$E = e, \quad \text{and} \quad W = \log(DB + 1).$$

Notice that

$$|\log \alpha_j| \leq \pi + \log |\alpha_j| \leq \pi + \log(A_j + 1) \leq \frac{9}{2} \cdot \log A_j.$$

We use Theorem 2.2 with n replaced by $n+1$:

$$\left| 2\kappa\pi i - \sum_{j=1}^{n} b_j \log \alpha_j \right|$$

$$> \exp\left\{ -C_{22}(n+1) . D^{n+3} 7^n \log A_1 \ldots \log A_n (2\pi n/D) W(1 + \log D) \right\}$$

$$> \exp\left\{ -C_{28}(n, D) \log A_1 \ldots \log A_n \log B \right\}.$$

Theorem 1.2 follows.

Proof of Theorem 1.3. Finally we deduce Theorem 1.3 from Theorem 2.2.

1) From the assumptions of Theorem 1.3 and from Theorem 1.2 we deduce

$$\epsilon B < C_{29}(n, D) . \log A_1 \ldots \log A_n \log B,$$

which yields

$$B < C_{13}(n, D, \epsilon) . \log A_1 \ldots \log A_n \log \log A_n.$$

2) If $b_n = 1$, we follow the proof of Theorem 1.2 above with the definition of W replaced by $W = \log(e + nB/V_n)$. There is no loss of generality to assume $B \geq e . \log A_n$. Using Theorem 2.2 we obtain

$$\exp(-\epsilon B) > \left| 2\kappa\pi i - \sum_{j=1}^{n} b_j \log \alpha_j \right|$$

$$> \exp\left\{ -C_{29}(n, D) \log A_1 \ldots \log A_n \log(B/\log A_n) \right\},$$

for some $\kappa \in \mathbf{Z}$ such that $|\kappa| < ((n-1)B + 2)/2$. Hence

$$B < C_{13}(n, D, \epsilon) . \log A_1 \ldots \log A_n \log \log A_{n-1}.$$

3. On the rank of certain matrices

In this section $\beta_0, \ldots, \beta_n$, $\ell_1, \ldots, \ell_n$ denote arbitrary complex numbers with $\beta_n = -1$. We write $\alpha_j = e^{\ell_j}$, $1 \leq j \leq n$.

The transcendence proofs will involve auxiliary functions of the form

$$F(z_0, z_1, \ldots, z_n) = P(z_0, e^{z_1}, \ldots, e^{z_n})$$

where $P \in \mathbf{C}[X_0, \dots, X_n]$.

We define derivations $D_0, \dots, D_{n-1}$ on the ring $R = \mathbf{C}[X_0, \dots X_n]$ by setting, for $0 \le i \le n$,

$$D_i X_0 = \delta_{i,0} \qquad \text{(Kronecker symbol),}$$
$$D_i X_j = \delta_{i,j} . X_j, \qquad 1 \le j \le n - 1,$$

and

$$D_i X_n = \beta_i . X_n.$$

Therefore

$$(\partial/\partial z_0)^{\tau_0} \dots (\partial/\partial z_{n-1})^{\tau_{n-1}} F(z_0, \dots, z_{n-1}, \beta_0 z_0 + \dots + \beta_{n-1} z_{n-1})$$
$$= D_0^{\tau_0} \dots D_{n-1}^{\tau_{n-1}} P(z_0, e^{z_1}, \dots, e^{z_{n-1}}, e^{\beta_0 z_0 + \dots + \beta_{n-1} z_{n-1}}).$$

We say that F has a zero of order $\ge T$ in the direction of the hyperplane W:

$$z_n = \beta_0 z_0 + \dots + \beta_{n-1} z_{n-1}$$

at a point $(u_0, \dots, u_n) \in \mathbf{C}^{n+1}$ if

$$D_0^{\tau_0} \dots D_{n-1}^{\tau_{n-1}} P(u_0, \dots, u_n) = 0$$

for all non-negative integers $\tau_0, \dots, \tau_{n-1}$ with $\tau_0 + \dots + \tau_{n-1} \le T$. This means that the function of n variables

$$F(u_0 + z_0, \dots, u_{n-1} + z_{n-1}, u_n + \beta_0 z_0 + \dots + \beta_{n-1} z_{n-1})$$

has a zero order of $\ge T$ at the point $(0, \dots, 0) \in \mathbf{C}^n$.

We need to know whether there exists a non-zero polynomial P of degrees say

$$\deg_{X_i} P \le L_i, \qquad 0 \le i \le n$$

such that F has a zero of order $\ge T$ in the direction of W at all points

$$(s, s\ell_1, \dots, s\ell_n), \qquad 0 \le s < S$$

for some given non-negative integers $L_0, L_1, \dots, L_n, T, S$.

By linear algebra, a sufficient condition for the existence of such a P is

$$L_0 \dots L_n > \binom{T + n}{n} . S.$$

The following easy remark will be used several times.

Remark 3.1. Assume $L_0 < TS$, $L_i < T$, $1 \leq i \leq n-1$, and $L_n = 0$. Then there is no non-zero polynomial P of degrees $\leq L_i$ such that

$$D_0^{\tau_0} \ldots D_{n-1}^{\tau_{n-1}} P(s, \alpha_1^s, \ldots, \alpha_n^s) = 0$$

for $0 \leq \tau_i < T$, $(0 \leq i \leq n-1)$, $0 \leq s < S$; (see [1], [11], [10]).

We denote by $\mathcal{A}$ the matrix whose entries are

$$\sum_{\tau_0' + \tau_0'' = \tau_0} \frac{\tau_0!}{\tau_0'! \cdot \tau_0''!} \cdot (\lambda_n \beta_0)^{\tau_0'} \cdot s^{\lambda_0 - \tau_0''} \cdot \frac{\lambda_0!}{(\lambda_0 - \tau_0'')!} \prod_{i=1}^{n-1} (\lambda_i + \lambda_n \beta_i)^{\tau_i} \prod_{i=1}^{n} e^{\lambda_i \ell_i s},$$

where the index of row is

$$(\tau_0, \ldots, \tau_{n-1}, s), \quad \text{with } \tau_0 + \ldots + \tau_{n-1} < T, 0 \leq s < S,$$

and the index of column is

$$(\lambda_0, \ldots, \lambda_n), \quad \text{with } 0 \leq \lambda_i \leq L_i, 0 \leq i \leq n.$$

The existence of a non-zero polynomial P as above amounts to saying that $\mathcal{A}$ has rank $< (L_0 + 1) \ldots (L_n + 1)$.

We begin by giving upper bounds for the rank of $\mathcal{A}$, and then we will explain that these upper bounds are essentially best possible.

We need to introduce a few notations. Let r be an integer, $0 \leq r \leq n$, and let $\lambda^{(1)}, \ldots, \lambda^{(r)}$ be linearly independent elements of $\mathbf{Z}^n$, with

$$\lambda^{(\rho)} = (\lambda_1^{(\rho)}, \ldots, \lambda_n^{(\rho)}), \quad 1 \leq \rho \leq r.$$

Let us write

$$\mathcal{L} = (\lambda^{(1)}, \ldots, \lambda^{(r)}),$$

and let $\varphi_{r,n}$ denote the set composed of all increasing sequences of r elements from $\{1, \ldots, n\}$; for each θ in $\varphi_{r,n}$, we will denote by $\mathcal{L}_\theta$ the minor of $\mathcal{L}$ whose columns are indexed by $\theta_1, \ldots, \theta_r$. Define

$$\sigma_0(\mathcal{L}) = \sigma_0 = \begin{cases} r & \text{if } (\beta_1, \ldots, \beta_n) \in \mathbf{C}\lambda^{(1)} + \ldots + \mathbf{C}\lambda^{(r)}, \\ r+1 & \text{otherwise}, \end{cases}$$

and

$$H(\mathcal{L}; x_1, \ldots, x_n) = (n-r)! \sum_{\theta \in \varphi_{r,n}} |\det \mathcal{L}_\theta| \cdot \prod_{i \notin \theta} x_i.$$

In this paragraph we will denote by $\mathcal{I}$ the ideal, associated to the matrix $\mathcal{L}$, generated by the r polynomials

$$\prod_{i=1}^{n} X_i^{\lambda_i^{(j)}} - 1; \qquad 1 \le j \le r$$

in $\mathbf{C}[X_1, \ldots, X_n]$. We will need an upper estimate for the maximal number of monomials in $X_1, \ldots, X_n$ of given degrees which are linearly independent modulo the above ideal $\mathcal{I}$. The following lemma will do the job.

Lemma 3.2. *The maximal number of monomials in $X_1, \ldots, X_n$ of degrees $\le L_1, \ldots, L_n$ which are linearly independent modulo $\mathcal{I}$, is bounded above by*

$$((4\sqrt{n})^{n-r}/(n-r)!)H(\mathcal{L}; L_1, \ldots, L_n).$$

Proof. Put $\Lambda = \mathbf{Z}(1/L_1, 0, \ldots, 0) + \ldots + \mathbf{Z}(0, \ldots, 0, 1/L_n)$ in $\mathbf{R}^n$,

$$\mathcal{C} = \{z \in \mathbf{R}; \ |z_i| \le 1, \ 1 \le i \le n\},$$

and $\Lambda' = \mathbf{Z}\tilde{\lambda}^{(1)} + \ldots + \mathbf{Z}\tilde{\lambda}^{(r)}$ where $\tilde{\lambda}^{(j)} = (\lambda_1^{(j)}/L_1, \ldots, \lambda_n^{(j)}/L_n)$. It is clear that two monomials $X_1^{\mu_1} \ldots X_n^{\mu_n}$ and $X_1^{\nu_1} \ldots X_n^{\nu_n}$ are congruent modulo $\mathcal{I}$ as soon as $\mu - \nu \in \Lambda'$, where $\mu = (\mu_1/L_1, \ldots, \mu_n/L_n)$ and $\nu = (\nu_1/L_1, \ldots, \nu_n/L_n)$. Let p be the orthogonal projection of $\mathbf{R}^n$ on the orthogonal of $\Lambda' \otimes_{\mathbf{Z}} \mathbf{R}$; then $p(\Lambda)$ is a discrete subgroup of $\mathbf{R}^n$. If $\tilde{\lambda}^{(r+1)}, \ldots, \tilde{\lambda}^{(n)}$ is a basis of $p(\Lambda)$, it follows from Lemma 3 of [3] that

$$\det {}^t\mathcal{M}\mathcal{M} = [\overline{\Lambda'} : \Lambda']^2/(L_1 \ldots L_n)^2 \det {}^t\tilde{\mathcal{L}}\tilde{\mathcal{L}},$$

where $\tilde{\mathcal{L}} = (\tilde{\lambda}^{(1)}, \ldots, \tilde{\lambda}^{(r)})$, $\mathcal{M} = (\tilde{\lambda}^{(r+1)}, \ldots, \tilde{\lambda}^{(n)})$ and $\overline{\Lambda'} = (\Lambda' \otimes_{\mathbf{Z}} \mathbf{R}) \cap \Lambda$. The maximal number, say $\mathcal{N}$, of monomials in $X_1, \ldots, X_n$ of degrees $\le L_1, \ldots, L_n$ which are linearly independent modulo $\mathcal{I}$ is at most $[\overline{\Lambda'} : \Lambda']$ times the number of points of $p(\Lambda)$ in $p(\mathcal{C})$. Since $p(\mathcal{C})$ is contained in the ball $\{z \in \mathbf{R}^{n-r}; \ |z| \le \sqrt{n}\}$, and since the diameter of a fundamental parallelogram of $p(\Lambda)$ is $\le \sqrt{n}$, we get the upper bound

$$\mathcal{N} \le (4\sqrt{n})^{n-r}[\overline{\Lambda'} : \Lambda']/(\det {}^t\mathcal{M}\mathcal{M}^{\frac{1}{2}}) \le (4\sqrt{n})^{n-r}L_1 \ldots L_n(\det {}^t\tilde{\mathcal{L}}\tilde{\mathcal{L}})^{\frac{1}{2}}.$$

It remains to compare $\det {}^t\tilde{\mathcal{L}}\tilde{\mathcal{L}}$ with $H(\mathcal{L}; L_1, \ldots, L_n)$. But

$$\det {}^t\tilde{\mathcal{L}}\tilde{\mathcal{L}} = \sum_{\theta \in \varphi_{r,n}} (\det \tilde{\mathcal{L}}_\theta)^2 = \sum_{\theta \in \varphi_{r,n}} \left(\frac{\det \mathcal{L}_\theta}{L_{\theta_1} \ldots L_{\theta_r}} \right)^2,$$

by the Cauchy-Binet formula (cf. [5], pp. 23–24). Finally

$$L_1 \ldots L_n (\det {}^t\tilde{\mathcal{L}}\tilde{\mathcal{L}})^{1/2} = L_1 \ldots L_n \left[\sum_{\theta \in \varphi_{r,n}} \left(\frac{\det \mathcal{L}_\theta}{L_{\theta_1} \ldots L_{\theta_r}} \right)^2 \right]^{1/2}$$

$$\leq L_1 \ldots L_n \sum_{\theta \in \varphi_{r,n}} \frac{|\det \mathcal{L}_\theta|}{L_{\theta_1} \ldots L_{\theta_r}}$$

$$\leq H(\mathcal{L}; L_1, \ldots, L_n)/(n-r)!,$$

which conclude the proof of Lemma 3.2.

Remark. We could also deduce Lemma 3.2 with $(4\sqrt{n})^{n-r}/(n-r)!$ replaced by 4^{n-r} using Nesterenko's result in [8] (compare with [10]).

Lemma 3.3. *For each $\mathcal{L}$ as above, the rank of the matrix $\mathcal{A}$ is at most*

$$\frac{(4\sqrt{n})^{n-r}}{(n-r)!} \cdot \binom{T+\sigma_0}{\sigma_0} \cdot S \cdot H(\mathcal{L}; L_1, \ldots, L_n).$$

Proof. The linear system associated to the matrix $\mathcal{A}$ can be written

$$(*) \qquad D_0^{\tau_0} \circ \ldots \circ D_{n-1}^{\tau_{n-1}} P(s, e^{s\ell_1}, \ldots, e^{s\ell_n}) = 0$$

$$(\tau_0 + \ldots + \tau_{n-1} < T; 0 \leq s \leq S),$$

where P stands for the general polynomial of degrees $L_0, \ldots, L_n$.

We will use the following remark. Let $Q \in C[X_1, \ldots, X_n]$ and $\alpha = (\alpha_1, \ldots, \alpha_n) \in C^n$ be such that

$$Q(x_1 \alpha_1, \ldots, x_n \alpha_n) = 0$$

for all $(x_1, \ldots, x_n) \in C^n$ satisfying

$$\prod_{i=1}^n x_i^{\lambda_i^{(j)}} = 1, \qquad 1 \leq j \leq r.$$

Then the function

$$G(z) = Q(\alpha_1 e^{z_1}, \ldots, \alpha_n e^{z_n})$$

vanishes on the orthogonal V_0 in $\mathbf{C}^n$ of $\mathbf{C}\lambda^{(1)} + \ldots + \mathbf{C}\lambda^{(r)}$. Therefore, for all $\zeta = (\zeta_1, \ldots, \zeta_n) \in V_0$, we have

$$\sum_{i=1}^{n} \zeta_i \frac{\partial}{\partial z_i} G(0) = 0,$$

which can be written

$$\sum_{i=1}^{n} \zeta_i X_i \frac{\partial}{\partial X_i} Q(\alpha) = 0.$$

For each $\zeta = (\zeta_1, \ldots, \zeta_n) \in \mathbf{C}^n$, we define a derivation Δ_ζ on $\mathbf{C}[X_0, \ldots, X_n]$ by

$$\Delta_\zeta = \sum_{i=1}^{n} \zeta_i X_i \frac{\partial}{\partial X_i}.$$

We denote by W_0 the orthogonal in $\mathbf{C}^n$ of $\mathbf{C}(\beta_1, \ldots, \beta_n)$. Notice that $\dim W_0 = n - 1$ and $\dim W_0/W_0 \cap V_0 = \sigma_0 - 1$. We choose a basis $f_{\sigma_0+1}, \ldots, f_n$ of $W_0 \cap V_0$, we complete it as a basis $f_2, \ldots, f_n$ of W_0, and we write

$$D'_1 = D_0, \quad D'_i = \Delta_{f_i} \quad 2 \le i \le n.$$

The system (*) is clearly equivalent to

$$D'^{\tau_1}_1 \circ \ldots \circ D'^{\tau_n}_n P(s, e^{s\ell_1}, \ldots, e^{s\ell_n}) = 0, \quad \tau_1 + \ldots + \tau_n < T, \ 0 \le s \le S.$$

We now consider the following linear system

(**) $$D'^{\tau_1}_1 \circ \ldots \circ D'^{\tau_{\sigma_0}}_{\sigma_0} P(s, x_1 e^{s\ell_1}, \ldots, x_n e^{s\ell_n}) = 0,$$

where $\tau_1 + \ldots + \tau_{\sigma_0} < T$, $0 \le s \le S$, $\prod_{i=1}^{n} x_i^{\lambda_i^{(j)}} = 1$ and P is as above. Since $D'_{\sigma_0+1}, \ldots, D'_n$ are associated with vectors orthogonal to $\mathbf{C}\lambda^{(1)} + \ldots + \mathbf{C}\lambda^{(r)}$, the rank of the system (**) is at least the rank of (*), and we only consider the derivatives in σ_0 directions. On the other hand we need to eliminate the x'_is. Rewriting each equation of (**) as a polynomial $Q_{s,\tau}$ in $x_1, \ldots, x_n$ modulo the equations $\prod_{i=1}^{n} x_i^{\lambda_i^{(j)}} = 1$, $1 \le j \le r$, each condition $Q_{s,\tau} \equiv 0$ gives, thanks to Lemma 3.2, at most

$$((4\sqrt{n})^{n-r}/(n-r)!) \, H(\mathcal{L}; L_1, \ldots, L_n)$$

equations. It is then clear that the rank of (**) is bounded by

$$((4\sqrt{n})^{n-r}/(n-r)!) \binom{T+\sigma_0}{\sigma_0} . S . H(\mathcal{L}; L_1, \ldots, L_n),$$

hence the Lemma 3.3.

Remark. If we choose for $\mathcal{L}$ the canonical basis of $\mathbf{C}^n$, we find the trivial upper bound $\binom{T+n}{n} . S$ for the rank of $\mathcal{A}$. Notice that a basis of W is $e_0, \ldots, e_{n-1}$, with $e_0 = (1, 0, \ldots, 0)$ and $e_i = (0, \delta_{i1}, \ldots, \delta_{i,n-1}, \beta_i)$, $1 \le i \le n-1$, hence $e_1, \ldots, e_{n-1}$ is a basis of W_0; for this special case, in the preceding proof if we choose $f_i = e_{i+1}$, then $D'_i = D_{i+1}$.

We need another upper bound for the rank of $\mathcal{A}$. Let r be an integer, $0 < r \le n$, and let $\lambda^{(1)}, \ldots, \lambda^{(r)}$ be linearly independent elements of $\mathbf{Z}^n$. Define $\mathcal{L} = (\lambda^{(1)}, \ldots, \lambda^{(r)})$,

$$\sigma_1(\mathcal{L}) = \sigma_1 = \begin{cases} r-1 & \text{if } \sigma_0 = r \text{ and } \beta_0 = 0, \\ r & \text{otherwise.} \end{cases}$$

Define a mapping

$$\varphi : (\mathbf{C}^\times)^n \longrightarrow (\mathbf{C}^\times)^r$$

by

$$\varphi(u_1, \ldots, u_n) = \left[\prod_{i=1}^n u_i^{\lambda_i^{(1)}}, \ldots, \prod_{i=1}^n u_i^{\lambda_i^{(r)}} \right],$$

and put

$$E(S) = \varphi(\Gamma(S)),$$

where

$$\Gamma(S) = \{(e^{\ell_1 s}, \ldots, e^{\ell_n s}), \ 0 \le s \le S\}.$$

Lemma 3.4. *For each $\mathcal{L}$ as above, the rank of the matrix $\mathcal{A}$ is at most*

$$\frac{(4\sqrt{n})^{n-r}}{(n-r)!} \cdot \binom{T + \sigma_1}{\sigma_1} . \operatorname{Card} E(S) . (L_0 + 1) . H(\mathcal{L}; L_1, \ldots, L_n).$$

Proof. Let V_1 denote the orthogonal in $\mathbf{C}^{n+1}$ of $\{0\} \times (\mathbf{C}\lambda^{(1)} + \ldots + \mathbf{C}\lambda^{(r)})$. Hence $\dim W/W \cap V_1 = \sigma_1$. For each $\zeta = (\zeta_0, \ldots, \zeta_n) \in \mathbf{C}^{n+1}$, we define a derivation Δ_ζ on $\mathbf{C}[X_0, \ldots, X_n]$ by

$$\Delta_\zeta = \zeta_0 \frac{\partial}{\partial X_0} + \sum_{i=1}^n \zeta_i X_i \frac{\partial}{\partial X_i}.$$

Let now $D'_{\sigma_1+1}, \ldots, D'_n$ be derivatives associated with a basis of $W \cap V_1$, and $D'_1, \ldots, D'_{\sigma_1}$ derivatives associated with vectors completing the previous as a basis of W. We consider the following linear system

$$(**) \qquad D_1'^{\tau_1} \circ \ldots \circ D'^{\tau_{\sigma_1}}_{\sigma_1} P(x_0, x_1 e^{s\ell_1}, \ldots, x_n e^{s\ell_n}) = 0,$$

where $\tau_1 + \ldots + \tau_{\sigma_1} < T$, $0 \le s \le S$, $\prod_{i=1}^n x_i^{\lambda_i^{(j)}} = 1$ and P is the general polynomial of degrees $L_0, \ldots, L_n$ in $X_0, \ldots, X_n$. First let's remark that it is equivalent to restrict s to card $E(S)$ values in the range $\{0, \ldots, S\}$. Since $D'_{\sigma_1 + 1}, \ldots, D'_n$ are associated with vectors orthogonal to $C\lambda^{(1)} + \ldots + C\lambda^{(r)}$, the rank of the system $(**)$ is larger than the rank of $(*)$. Rewriting each equation of $(**)$ as a polynomial $Q_{s,\tau}$ in $x_0, \ldots, x_n$ modulo the equations $\prod_{i=1}^n x_i^{\lambda_i^{(j)}} = 1$, $1 \le j \le r$, each condition $Q_{s,\tau} \equiv 0$ gives, thanks again to Lemma 3.2, at most

$$((4\sqrt{n})^{n-r}/(n-r)!) H(\mathcal{L}; L_1, \ldots, L_n) . (L_0 + 1)$$

equations. It is then clear that the rank of $(**)$ is bounded by

$$((4\sqrt{n})^{n-r}/(n-r)!) \binom{T + \sigma_1}{\sigma_1} . \operatorname{card} E(S) . (L_0 + 1) . H(\mathcal{L}; L_1, \ldots, L_n),$$

hence the Lemma 3.4.

Here is the main result of this section: the zero estimate.

Proposition 3.5. *Assume that the rank of $\mathcal{A}$ is $(L_0 + 1) \ldots (L_n + 1)$. Then there exists an integer r, $0 \le r \le n$, and there exist $\lambda^{(1)}, \ldots, \lambda^{(r)}$ linearly independent in $\mathbf{Z}^n$, such that if we set*

$$T_1 = [T/(n+1)] \quad and \quad S_1 = [S/(n+1)],$$

then either

(i) $\qquad \binom{T_1 + \sigma_0}{\sigma_0} . S_1 . H(\mathcal{L}; L_1, \ldots, L_n) \le (n+1)! . L_0 \ldots L_n$

or

(ii) *$r > 0$ and*

$$\binom{T_1 + \sigma_1}{\sigma_1} . \operatorname{Card} E(S_1) . H(\mathcal{L}; L_1, \ldots, L_n) \le \frac{(n+1)!}{n+1-r} . L_1 \ldots L_n.$$

Proof. See appendix.

We need some further simple properties of H. For the rest of this section, we assume $L_i \ge 1$, $1 \le i \le n$, and we denote by a and b two positive numbers satisfying

$$T \ge a \max_{1 \le j \le n} L_j.$$

and

$$TS \geq bL_0.$$

Lemma 3.6. *We have the following inequalities.*

(i) $$\frac{T^{r+1}}{(r+1)!} \cdot S \cdot H(\mathcal{L}, \mathbf{L}) \geq a^r b \frac{(n-r)!}{(r+1)!} L_0 \ldots L_n.$$

(ii) $$\frac{T^r}{r!} \cdot H(\mathcal{L}, \mathbf{L}) \geq a^r \frac{(n-r)!}{r!} L_1 \ldots L_n.$$

Proof. We have

$$H(\mathcal{L}, \mathbf{L}) \geq (n-r)! \, L_1 \ldots L_n / (\max_{1 \leq i \leq n} L_i)^r.$$

This completes the proof of Lemma 3.6.

Lemma 3.7. *Define* $A = \max\{|\ell_j|, \ 1 \leq j \leq n\}$, $L = \max\{|\lambda_i^{(j)}|,$ $1 \leq j \leq r, \ 1 \leq i \leq n\}$. *Assume* $\sigma_1 = r - 1$, $\operatorname{Card} E(S) < S$ *and that* $\beta_1, \ldots, \beta_n$ *are algebraic numbers in a field of degree* $\leq D$. *Then either the number*

$$\Lambda = \beta_1 \ell_1 + \ldots + \beta_n \ell_n$$

vanishes or

$$|\Lambda| \geq \exp\left\{-D\left[\sum_{j=1}^{n} h(\beta_j) + \log(n^{n+2} A S L^n)\right]\right\}.$$

Proof. Since $\sigma_1 = r - 1$ we have $\beta_0 = 0$ and

$$\beta_i = \sum_{j=1}^{r} c_j \lambda_i^{(j)}, \qquad 1 \leq i \leq n,$$

for some $c_j \in \overline{\mathbf{Q}}$. The assumption on $E(S)$ means that there exists $s \in \mathbf{Z}$, $1 \leq s \leq S$, such that

$$\prod_{i=1}^{n} \alpha_i^{\lambda_i^{(j)} s} = 1 \qquad \text{for } 1 \leq j \leq r.$$

Define $k_j \in \mathbf{Z}$, $1 \leq j \leq r$, by

$$\sum_{i=1}^{n} \lambda_i^{(j)} s \ell_i = 2\pi k_j \sqrt{-1}.$$

Hence

$$\Lambda = \sum_{i=1}^{n} \beta_i \ell_i = \sum_{j=1}^{r} c_j 2\pi k_j \sqrt{-1}/s$$

and

$$|\Lambda| = 2\pi \left| \sum_{j=1}^{r} c_j k_j / s \right|.$$

It is therefore sufficient to use Liouville inequality (e.g. Lemma 2.2 of [11]). We have

$$2\pi |k_j| \leq S. \sum_{i=1}^{n} |\lambda_i^{(j)}| \cdot |\ell_i| \leq ALSn.$$

Now we compute the c_j by solving the system

$$\beta_i = \sum_{j=1}^{r} c_j \lambda_i^{(j)}, \; 1 \leq i \leq n,$$

which gives

$$c_j = \Delta_j / \Delta,$$

where Δ and Δ_j are certain determinants; Δ is a non-zero integer of absolute value at most $(rL)^r$, and Δ_j are linear combinations of $\beta_1, \ldots, \beta_n$ with integral coefficients of absolute values at most $(rL)^{r-1}$. So

$$\sum_{j=1}^{r} c_j k_j = \left(\sum_{i=1}^{n} a_i \beta_i \right)/\Delta,$$

where a_i are integers of absolute values at most $n^{r+1} ASL^r$, and

$$|\Lambda| \geq \left| \sum_{i=1}^{n} a_i \beta_i \right|/\Delta S.$$

The desired estimate easily follows from Lemma 2.2 of [11].

4. Proof of Theorem 2.1

We use the notations of [11]. We first refine Proposition 3.1 of [11].

Let c_0, c_0', c_1, c_2, c_3, c_4 be positive real numbers satisfying the inequalities (3.1), (3.2) and (3.3) of [11]. We also assume $c_0 \leq 29$ and $c_3 \leq 2^{10}$.

We assume $V_1 \geq n/D$, and we define

$$E_1 = \min\{\exp(qDV_1/n), \min_{1 \leq j \leq n}\{2qDV_j|\log\alpha_j|^{-1}\}\},$$

$$E^* = \max\{2^{5n+4}q^{n+1}n^{2n}DE_1^n, E_1^{n^2}\},$$

$$W^* = \max\{W, n \cdot \log(2^{11}nq^2DV_n^+), \frac{q}{nD} \cdot \log E_1\},$$

$$U_1 = 2^{22}n^2q^{n+1}D^2 \max\{W^*, V_n^+, W^*V_n^+(\log E_1)^{-1}, \log E_1\}$$

and

$$U_2 = c_0'c_1c_2^nc_3c_4q^{3n}(q-1)n^{2n+1}(n!)^{-1}D^{n+2}V_1\ldots V_n$$
$$W^*(\log E^*)(\log E_1)^{-n-1},$$

and

$$U = \max\{U_1, U_2\}.$$

Proposition 4.1. *With the assumptions of Theorem 2.1, let* $K = \mathbb{Q}(\alpha_1,\ldots,\alpha_n,\beta_0,\ldots,\beta_{n-1})$*, and assume* q *is a prime number such that*

$$[K(\alpha_1^{1/q},\ldots,\alpha_n^{1/q}):K] = q^n.$$

Assume also $\beta_n = -1$*. Then*

$$|\Lambda| > e^{-U}.$$

We go back to the proof in §3 of [11]. We replace V_{n-1}^* by E^*. We replace (3.7) of [11] by

$$\log E^* \leq (19/2)qn^2D\log E_1.$$

We replace (3.10) of [11] by

$$q^3 2^{5n+8}n^{2n+1}E_1 \leq (E^*)^{1+1/n},$$

and (3.14) of [11] by

$$3q^{n+2} \cdot 2^{5n+10}n^{2n+1}E_1 S \leq (L_{-1}+1)(E^*)^{1+2/n},$$

because $C_3 \leq 2^{10}$. The inequality (3.17) of [11] is satisfied for $0 \leq t \leq T$ and $|z| \leq q^{n+2}2^{5n+7}n^{2n+1}E_1 S$.

Next we take Q in the interval $1 \leq Q \leq q^3 3^{5n+82} n^{2n+1}$, and in (3.18) of [11] we replace qL_n by $q^3 2^{5n+82} n^{2n+1}$.

This means that if we set

$$N = q^2 2^{5n+8} n^{2n+1},$$

then we replace L_n by N in (3.10), (3.14), (3.17) and (3.18) of [11]. In §3.4 of [11] we restrict J to the interval

$$0 \leq J \leq \left[\frac{\log N}{\log q} \right] + 1.$$

All the estimates in §3.3 of [11] will remain valid; however there are two important modifications which we now explain: the construction of the auxiliary function, and the contradiction.

a) *Preliminaries to the proof of Proposition 4.1*

In [11], we use the polynomials

$$\Delta(X, k) = (X + 1) \ldots (X + k)/k!.$$

Let us set $\tilde{L}_0 = (L_0 + 1)(L_{-1} + 1)$. Then the polynomials

$$\Delta(z + \lambda_{-1}, L_{-1} + 1)^{\lambda_0 + 1}, \qquad 0 \leq \lambda_{-1} \leq L_{-1}, \ 0 \leq \lambda_0 \leq L_0$$

give a basis of a space of polynomials of degree $\leq \tilde{L}_0$. This change of basis gives trivial changes in the matrix which we will consider. We will use the results of our §3 above, but now the space of polynomials we consider is of dimension $\tilde{L}_0$ (instead of $L_0 + 1$ in our §3 above).

When $\tilde{L}_1, \ldots, \tilde{L}_n$ are positive integers, we are interested in the rank of the matrix $\tilde{A}$ whose entries are

$$\sum_{\tau_0' + \tau_0'' = \tau_0} \frac{\tau_0!}{\tau_0'! \tau_0''!} \left(\frac{d^{\tau_0'}}{dz_0^{\tau_0'}} \Delta(s + \lambda_{-1}, L_{-1} + 1)^{\lambda_0 + 1} \right) \prod_{r=1}^{n-1} (\lambda_r + \lambda_n \beta_r)^{\tau_r} \prod_{i=1}^{n} \alpha_i^{\lambda_i s},$$

which can be written, with the notations of [11],

$$\Lambda_0(s, \tau) . \alpha_1^{\lambda_1 s} \ldots \alpha_n^{\lambda_n s}.$$

The index of a row is (s, τ), with

$$\tau = (\tau_0, \ldots, \tau_{n-1}), \ \tau_0 + \ldots + \tau_{n-1} < T, \quad \text{and} \quad 0 \leq s < S, \ (s, q) = 1,$$

while the index of a column is $\lambda = (\lambda_{-1}, \lambda_0, \ldots, \lambda_n)$ with

$$0 \le \lambda_{-1} \le L_{-1}, \ 0 \le \lambda_0 \le L_0, \quad \text{and} \quad 0 \le \lambda_i \le \tilde{L}_i, \qquad 1 \le i \le n.$$

We will choose $\tilde{L}_i = L_i$, $1 \le i \le n-1$, but we will take for $\tilde{L}_n$ the smallest integer such that we can construct the auxiliary function. More precisely, for each real number $\tilde{U} \le U$, we define

$$L_n^\sharp = \tilde{U}/c_1 c_2 n q^{n+1} D S V_n,$$

and

$$\tilde{L}_n = [L_n^\sharp].$$

Notice that in the case $\tilde{U} = U$, we find $\tilde{L}_n = L_n$, and also in this case

$$(L_{-1}+1)(L_0+1)\ldots(L_{n-1}+1)L_n^\sharp \ge c_0(1 - \frac{1}{q})\binom{T+n}{n} . S \qquad (4.2)$$

(compare with inequality (3.6) in [11]).

Let $i \in \{0,1\}$, $0 \le r \le n$, with $r > 0$ in case $i = 1$, and $\mathcal{L} = (\lambda^{(1)}, \ldots, \lambda^{(r)}) \in \mathbf{Z}^{nr}$, with $\lambda^{(1)}, \ldots, \lambda^{(r)}$ linearly independent. We set

$$L^\sharp = (L_1, \ldots, L_{n-1}, L_n^\sharp)$$

and

$$\tilde{L} = (L_1, \ldots, L_{n-1}, \tilde{L}_n).$$

The function $H(\mathcal{L}; L^\sharp)$ (see §3) is of the form

$$H(\mathcal{L}; L^\sharp) = A(\mathcal{L})\tilde{U} + B(\mathcal{L}).$$

Lemma 4.3. *Assume* $\sigma_i = r - i$. *Then* $B(\mathcal{L}) > 0$.

Proof. Assume $B(\mathcal{L}) = 0$. This means $\det \mathcal{L}_\theta = 0$ for all $\theta = (\theta_1, \ldots, \theta_{r-1}, n)$ with $0 < \theta_1 < \ldots < \theta_{r-1} < n$:

$$\mathcal{L}_\theta = \begin{vmatrix} \lambda_{\theta_1}^{(1)} & \cdots & \lambda_{\theta_1}^{(r)} \\ \vdots & \ddots & \vdots \\ \lambda_{\theta_{r-1}}^{(1)} & \cdots & \lambda_{\theta_{r-1}}^{(r)} \\ \lambda_n^{(1)} & \cdots & \lambda_n^{(r)} \end{vmatrix}.$$

As $\lambda^{(1)}, \ldots, \lambda^{(r)}$ are linearly independent, we have $\lambda_n^{(1)} = \ldots = \lambda_n^{(r)} = 0$. But the assumption $\beta_n = -1$ gives

$$(\beta_1, \ldots, \beta_n) \notin C\lambda^{(1)} + \ldots + C\lambda^{(r)},$$

hence $\sigma_0 = r + 1$ if $i = 0$ and $\sigma_1 = r$ if $i = 1$. This completes the proof of Lemma 4.3.

Now we are going to choose $\tilde{U}$ as the smallest positive number such that there exists $i_0 \in \{0, 1\}$, r_0, and $\mathcal{L}^0 = (\lambda_0^{(1)}, \ldots, \lambda_0^{(r_0)})$, with $\sigma_i^0 = \sigma_i(\mathcal{L}^0) = r_0 - i_0$, satisfying

$$c_0 \left(1 - \frac{1}{q}\right) \frac{(8\sqrt{n})^{n-r_0}}{(n - r_0)!} \binom{T + \sigma_i^0}{\sigma_i^0} S \cdot H(\mathcal{L}^0, L^\sharp)$$
$$\leq \tilde{L}_0^{1-i_0} \cdot (L_1 + 1) \ldots (L_{n-1} + 1) L_n^\sharp. \qquad (4.4)$$

We do this in the following way. We define $C(\mathcal{L})$ by

$$C(\mathcal{L}) \cdot c_0 \cdot \left(1 - \frac{1}{q}\right) \frac{(8\sqrt{n})^{n-r_0}}{(n - r_0)!} \binom{T + \sigma_i}{\sigma_i} \cdot c_1 c_2 n q^{n+1} D S^2 V_n$$
$$= \tilde{L}_0^{1-i_0} \cdot (L_1 + 1) \ldots (L_{n-1} + 1)$$

so that (4.4) can be written

$$A(\mathcal{L}^0) \cdot \tilde{U} + B(\mathcal{L}^0) \leq C(\mathcal{L}^0) \cdot \tilde{U}.$$

We define $\tilde{U}$ as the minimum of

$$B(\mathcal{L})/(C(\mathcal{L}) - A(\mathcal{L}))$$

for $(i, \mathcal{L})$ running over the set of $i \in \{0, 1\}$, and $\mathcal{L} = (\lambda^{(1)}, \ldots, \lambda^{(r)})$, for which $\sigma_i = r - i$ and $C(\mathcal{L}) > A(\mathcal{L})$. This set is not empty; thanks to (4.2) we can choose for $\mathcal{L}$ the canonical basis of $\mathbf{Z}^n$, with $i = 0$. Of course we choose a value $(i_0, \mathcal{L}^0)$ which gives the minimum, and we get (4.4) with equality. Notice that, if $r_0 = n$, then $\mathbf{Z}\lambda_0^{(1)} + \ldots + \mathbf{Z}\lambda_0^{(n)} = \mathbf{Z}^n$ and $i = 0$ (this follows from the definition of H: this is the only case where $H = 1$). If $r_0 < n$, then $\left(1 - \frac{1}{q}\right) \cdot 2^{n-r_0} \geq 1$.

Moreover, from this choice of $\tilde{U}$, we deduce, using Lemma 4.3,

$$A(\mathcal{L}) \cdot \tilde{U} + B(\mathcal{L}) \geq C(\mathcal{L}) \cdot \tilde{U} \qquad (4.5)$$

for all $(i, \mathcal{L})$ with $\sigma_i = r - i$. This means

$$c_0 \cdot \left(1 - \frac{1}{q}\right) \cdot \frac{(8\sqrt{n})^{n-r}}{(n-r)!} \cdot \binom{T + \sigma_i}{\sigma_i} \cdot S \cdot H(\mathcal{L}, L^\sharp)$$

$$\geq \tilde{L}_0^{1-i} \cdot (L_1 + 1) \dots (L_{n-1} + 1) \cdot L_n^\sharp. \qquad (4.6)$$

Let us show that $L_n^\sharp \geq 1$. Otherwise, the matrix $\tilde{A}$ (which corresponds to our choice of $\tilde{U}$) does not involve λ_n, and from the Remark 3.1 we deduce that its rank is $\tilde{L}_0 \cdot (L_1 + 1) \dots (L_{n-1} + 1)$. We now use Lemmas 3.3 and 3.4 with $n - 1$ variables instead of n (because α_n is not involved) to get a contradiction with (4.4).

Now we have $L_n^\sharp \geq 1$:

$$\tilde{U} \geq c_1 c_2 n q^{n+1} D S V_n,$$

hence $\tilde{L}_n \geq 1$, and we can use Lemmas 3.3 and 3.4 (with $L_0, \dots, L_n$ replaced by $\tilde{L}_0, \dots, \tilde{L}_n$, and $\tilde{L}_i = L_i$ for $1 \leq i \leq n - 1$) to deduce:

Lemma 4.7. *The rank of the matrix $\tilde{A}$ is at most*

$$\frac{1}{c_0} \tilde{L}_0 \cdot (L_1 + 1) \dots (L_{n-1} + 1) \cdot (\tilde{L}_n + 1)$$

b) *Construction of the auxiliary function.*

Lemma 4.8. *In Lemma 3.2 of [11] p. 266, one may restrict λ to run over the $(n+2)$-tuples $(\lambda_{-1}, \dots, \lambda_n)$ with $0 \leq \lambda_j \leq L_j, (-1 \leq j \leq n-1)$ and $0 \leq \lambda_n \leq \tilde{L}_n$.*

Proof. The proof is the same as in [11], apart from the fact that we use Lemma 4.7 in place of (3.6) of [11].

We now continue the proof as in §3.3 and §3.4 of [11]. We keep the estimates of [11] as they stand; we do not modify the parameters T, S, U, L_{-1}, ..., L_{n-1}, but the parameter L_n is replaced by $\tilde{L}_n$ which may be smaller, and therefore the upper bounds in [11] §3.3 and §3.4 are valid. Also it is important to notice that $L_n^{(J+1)} \leq L_n^{(J)}/q$, hence $L_n^{(J)} \leq q^{-J} \tilde{L}_n$.

c) *End of the proof of Proposition 4.1.*

We write the main inductive argument (p. 268 of [11]) for $J_0 = \left[\frac{\log N}{\log q}\right]$. Then we need to modify the argument in §3.5 of [11], because we cannot claim that $L_n^{(J_0)}$ vanishes.

Let us show that the numbers

$$\varphi_{J_0, \tau}(s) = \sum_{(\lambda)} \sum_{d=1}^{D} p_d^{(J_0)}(\lambda) \xi_d \Lambda_{J_0}(s, \tau) \alpha_1^{\lambda_1 s} \ldots \alpha_n^{\lambda_n s}$$

satisfy

$$\varphi_{J_0, \tau}(s) = 0 \quad \text{for } 0 \le s \le q^{J_0} S \text{ and } |\tau| \le q^{-J_0} T. \tag{4.9}$$

This is plain if $(s, q) = 1$, since this is the step before the last in the inductive argument. If q divides s, this has been proved at the last step of the induction (proof of Lemma 3.6 of [11]).

From Remark 3.1 above (or from the argument in §3.5 of [1]) we have $L_n^{(J_0)} \ge 1$, hence $L_n^{(J)} \ge q^J$ for $1 \le J \le J_0$. We define

$$J_1 = \left[J_0 - \frac{\log(5n)}{\log q}\right];$$

thus

$$q^{J_0 - J_1} \ge 5n \quad \text{and} \quad q^{J_1} \ge q^{J_0}/5nq \ge N/5nq^2.$$

We now use the zero estimate (Proposition 3.4 above). Define

$$T_1 = [[q^{-J_1}T]/(n+1)], \qquad S_1 = [q^{J_1}S/(n+1)].$$

From (4.9) we deduce that there exists an integer r, $0 \le r \le n$, and there exist $\lambda^{(1)}, \ldots, \lambda^{(r)}$ linearly independent in $\mathbf{Z}^n$, such that either

$$\binom{T_1 + \sigma_0}{\sigma_0} \cdot S_1 \cdot H(\mathcal{L}; L_1^{(J_1)}, \ldots, L_n^{(J_1)}) \le (n+1)! \prod_{j=-1}^{n} (L_j^{(J_1)} + 1), \tag{4.10}$$

or $r > 0$ and

$$\binom{T_1 + \sigma_1}{\sigma_1} \cdot \operatorname{Card} E(S_1) \cdot H(\mathcal{L}; L_1^{(J_1)}, \ldots, L_n^{(J_1)})$$

$$\le \frac{(n+1)!}{n-r+1} \cdot \prod_{j=1}^{n} (L_j^{(J_1)} + 1), \tag{4.11}$$

where $\mathcal{L}$ stands for $(\lambda^{(1)}, \ldots, \lambda^{(r)})$.

It is readily checked that

$$S_1 \cdot \binom{T_1 + \sigma_i}{\sigma_i} \geq \frac{4}{5} \cdot \frac{T^{\sigma_i} \cdot S}{q^{J_1(\sigma_i - 1)} \cdot (n+1)^{\sigma_i + 1} \cdot \sigma_i!}$$

and

$$\binom{T_1 + \sigma_i}{\sigma_i} \geq \frac{4}{5} \cdot \frac{T^{\sigma_i}}{q^{J_1 \sigma_i} \cdot (n+1)^{\sigma_i} \cdot \sigma_i!}$$

for $i = 0$ and $i = 1$. We write

$$\frac{H(\mathcal{L}; L^{(J_1)})}{\prod_{j=-1}^{n} L_j^{(J_1)}} \geq q^{J_1 r} \cdot \frac{H(\mathcal{L}; \tilde{L})}{\prod_{i=0}^{n} \tilde{L}_j},$$

where $\tilde{L}$ stands for $(\tilde{L}_1, \ldots, \tilde{L}_n)$. Since $L_j^{(J_1)} \geq 5n$, $1 \leq j \leq n$, we have $L_j^{(J_1)} + 1 \leq (1 + 1/5n) L_j^{(J_1)}$, hence

$$\prod_{j=1}^{n} (L_j^{(J_1)} + 1) \leq \frac{5}{4} \cdot \prod_{j=1}^{n} L_j^{(J_1)}.$$

Therefore (4.10) gives

$$q^{J_1(r - \sigma_0 + 1)} \cdot \frac{T^{\sigma_0}}{\sigma_0!} \cdot S \cdot H(\mathcal{L}; \tilde{L}) \leq \frac{7}{4} \cdot (n+1)^{\sigma_0 + 1}(n+1)! \cdot \prod_{j=0}^{n} \tilde{L}_j, \quad (4.12)$$

while (4.11) gives

$$q^{J_1(r - \sigma_1)} \cdot \frac{T^{\sigma_1}}{\sigma_1!} \cdot \operatorname{Card} E(S_1) \cdot H(\mathcal{L}; \tilde{L}) \leq \frac{7}{4} \cdot (n+1)^{\sigma_1} \frac{(n+1)!}{n - r + 1} \cdot \prod_{j=1}^{n} \tilde{L}_j.$$

$$(4.13)$$

Our assumption $\log E_1 \leq qDV_1/n$ gives

$$T/\tilde{L}_j \geq \tfrac{1}{2} \cdot c_2 n^2 q^2 DV_j / \log E_1 > 14 \cdot n^3,$$

while the assumption $E^* \geq E_1^{n^2}$ gives

$$TS/\tilde{L}_0 \geq \tfrac{1}{2} \cdot c_4 nqD \log E^* / \log E_1 > 14 \cdot n^3.$$

We consider different cases depending on whether $i = 0$ or $i = 1$, and $\sigma_i = r - i + 1$ or $\sigma_i = r - i$.

α) Assume (4.12) holds with $\sigma_0 = r + 1$. We get

$$\frac{T^{r+1}}{(r+1)!} \cdot S \cdot H(\mathcal{L}; \tilde{L}) \leq \frac{7}{4} \cdot (n+1)^{r+3} n! \cdot \prod_{j=0}^{n} \tilde{L}_j.$$

However, it is readily checked that

$$(n+1)^{r+3} n! (r+1)! \leq 2^{2r+3} n^{3r+3} (n-r)!, \qquad (4.14)$$

and we get a contradiction with lemma 3.6 (i) with $a > 14n^3$, $b > 14n^3$.

β) Assume (4.13) holds with $\sigma_1 = r$. We get

$$\frac{T^r}{r!} \cdot H(\mathcal{L}; \tilde{L}) \leq \frac{7}{4} \cdot \frac{(n+1)^{r+1}}{n-r+1} \cdot n! \cdot \prod_{j=1}^{n} \tilde{L}_j$$

$$< \frac{7}{2(n+1)} (4n^3)^r \cdot \frac{(n-r)!}{r!} \cdot \prod_{j=1}^{n} \tilde{L}_j,$$

(from (4.14) with r replaced by $r - 1$) which contradicts Lemma 3.6 (ii) with $a > 14n^3$.

γ) Assume (4.12) holds with $\sigma_0 = r$. We get

$$q^{2J_1} \cdot \frac{T^r}{r!} \cdot S \cdot H(\mathcal{L}; \tilde{L}) \leq \frac{7}{4} \cdot (n+1)^{r+2} n! \cdot \prod_{j=0}^{n} \tilde{L}_j.$$

We use our choice of N, with the bounds

$$(n+1)^{r+2} n! \cdot \frac{(8\sqrt{n})^{n-r}}{(n-r)!} \leq 2^{5n} n^{2n}$$

and $C_0 \leq 29$; we get a contradiction with (4.6).

δ) Assume (4.13) holds with $\sigma_1 = r - 1$. From Lemma 3.7 we deduce $\operatorname{Card} E(S_1) = S_1$, hence we get

$$q^{2J_1} \cdot \frac{T^{r-1}}{(r-1)!} \cdot S \cdot H(\mathcal{L}; \tilde{L}) \leq 2 \cdot (n+1)^{r+1} n! \cdot \prod_{j=1}^{n} \tilde{L}_j.$$

Once more, we get a contradiction with (4.6).

This completes the proof of Proposition 4.1.

d) *End of the proof of Theorem 2.1.*

We now deduce from Proposition 4.1 that in Proposition 3.8, p. 274 of [11] one may replace $\log(EDV_{n-1}^+)$ by $\log(ED)$, provided that $V_1 \geq n/D$.

This is clear if $n \leq 12$, because in this case we still have, with our value of E^*,

$$\log E^* \leq n \log(2^{13} q^2 n) \cdot \log(ED).$$

If $n \geq 13$, we have

$$\log E^* \leq n^2 (1 + \log q) \cdot \log(ED),$$

and

$$\log(2^{13} n q^2) < (3 + \log q)\sqrt{n},$$

while

$$(3 + \log q)(1 + \log q) < 2q^2/(q - 1),$$

and this is sufficient for our estimates of [11] p. 275.

Finally, it follows that in the theorem p. 258 of [11], one may replace $\log(EDV_{n-1}^+)$ by $\log(ED)$, provided that $V_1 \geq n/D$. This completes the proof of Theorem 2.1.

5. Proof of Theorem 2.2

We introduce parameters $c_0, c_1, \ldots, c_5$ which satisfy the following requirements: c_0 is a sufficiently large absolute constant, and

$$c_5 \geq c_0^2 2^n (n + 1)^{n+2} n!, \quad c_1 \geq c_0 \log c_5,$$

$$c_2 \geq c_0 c_5, \quad c_3 \geq c_0 c_5, \quad c_4 \geq c_0 c_5.$$

We could of course choose $c_2 = c_3 = c_4$, but the above notations will enable us to use the computations already made in [11]; also, if one wishes to provide good numerical values, it may be better to have more freedom.

We will denote by $f_1, \ldots, f_8$ positive numbers which can be explicitly computed in terms of n and $c_0, \ldots, c_5$, and which satisfy the property that $c_0 c_1 f_i$ are bounded by an absolute constant independent of c_0.

Next we define

$$S = \left[c_3 DW (\log E)^{-1} \right],$$

$$U = c_0 c_1 c_2^n c_3 c_4 \cdot \frac{n^n}{n!} \cdot D^{n+2} V_1 \ldots V_n W \left(\log(DE) \right) (\log E)^{-n-1},$$

and

$$T = [U/c_1 c_3 DW],$$

$$L_{-1} = [W],$$

$$L_{-2} = [U/c_1 c_4 D \log(DE)(L_{-1} + 1)],$$

$$L_0 = (L_{-1} + 1)(L_{-2} + 1).$$

Further, for each real number $\tilde{U} \geq c_0$, we define real numbers $L_1^\sharp, \ldots, L_n^\sharp$, and integers $\tilde{L}_1, \ldots, \tilde{L}_n$ by

$$L_j^\sharp = \tilde{U}/c_1 c_2 n D S V_j, \qquad 1 \leq j \leq n,$$

and

$$\tilde{L}_j = [L_j^\sharp], \qquad 1 \leq j \leq n.$$

We denote by $L_1, \ldots, L_n$ the values of $\tilde{L}_1, \ldots, \tilde{L}_n$ corresponding to $\tilde{U} = U$.

We assume $0 < |\Lambda| < e^{-U}$, and we shall eventually reach a contradiction.

Let us recall that

$$\Delta(z; k; \ell, m) = \frac{d^m}{dz^m} (\Delta(z; k))^\ell,$$

where

$$\Delta(z; k) = (z + 1)(z + 2) \ldots (z + k)/k!, \quad \Delta(z; 0) = 1.$$

We introduce the functions

$$\Lambda(z; \tau) = \Delta(z + \lambda_{-1}; L_{-1} + 1; \lambda_{-2} + 1; \tau_0) \cdot \prod_{r=1}^{n-1} \Delta(b_n \lambda_r - b_r \lambda_n; \tau_r)$$

for $0 \leq \lambda_j \leq \tilde{L}_j$, $(j = -2, -1, 1, \ldots, n)$, $\tau = (\tau_0, \ldots, \tau_{n-1})$, and $z \in \mathbb{C}$. Notice that the dependence in λ_{-2}, λ_{-1}, λ_1, $\ldots$, λ_n is hidden in the notation $\Lambda(z; \tau)$.

Our auxiliary functions will be of the form

$$f_\tau(z) = \sum_{(\lambda)} p(\lambda) \Lambda(z; \tau) \alpha_1^{\gamma_1 z} \ldots \alpha_{n-1}^{\gamma_{n-1} z},$$

where $\gamma_j = \lambda_j - \lambda_n b_j / b_n$, $(1 \leq j \leq n)$, and

$$\varphi_\tau(z) = \sum_{(\lambda)} p(\lambda) \Lambda(z; \tau) \alpha_1^{\lambda_1 z} \ldots \alpha_n^{\lambda_n z};$$

here, λ stands for $(\lambda_{-2}, \lambda_{-1}, \lambda_1 \ldots, \lambda_n)$.

We need analytical and arithmetical estimates on $\Lambda(z; \tau)$ (cf. [11], and [6]).

Lemma 5.1. *For* $|z| \leq c_5 ES$ *and* $|\tau| \leq T$,

$$\left|\Lambda(z; \tau)\right| \leq e^{f_1 U/D}.$$

Moreover, for $s \in \mathbf{Z}$, $0 \leq s \leq c_5 S$, *and* $|\tau| < T$, $\Lambda(s; \tau)$ *is a rational number with a denominator at most* $e^{f_2 U/D}$.

Now we consider the matrix $\mathcal{B}$ whose entries are

$$\Lambda(s; \tau)\alpha_1^{\lambda_1 s} \ldots \alpha_n^{\lambda_n s}$$

where the index of row is

$$(\tau_0, \ldots, \tau_{n-1}, s), \quad \text{with } |\tau| = \tau_0 + \ldots + \tau_{n-1} < T, 0 \leq s < S,$$

and the index of column is

$$(\lambda_{-2}, \lambda_{-1}, \lambda_1, \ldots, \lambda_n), \quad \text{with } 0 \leq \lambda_j \leq \tilde{L}_j.$$

We recall that the polynomials

$$\left(\Delta(z + r; k)\right)^\ell, \qquad 0 \leq r \leq R - 1, 1 \leq \ell \leq L$$

(with $k > R > 0$, $L > 0$) are linearly independent and of degrees $\leq kL$.

First step. We choose for $\tilde{U}$ the smallest positive real number with the following property:

There exist $i \in \{0, 1\}$, $r \in \mathbf{Z}$, $0 \leq r \leq n$, with $r > 0$ if $i = 1$, and there exist $\lambda^{(1)}, \ldots, \lambda^{(r)}$ linearly independent in $\mathbf{Z}^n$, such that

$$c_0(4\sqrt{n})^{n-r}\binom{T + \sigma_i}{\sigma_i} \cdot \operatorname{Card} E(S) \cdot L_0^i \cdot H(\mathcal{L}; L_1^\sharp, \ldots, L_n^\sharp)$$

$$\leq L_0^{1-i} L_1^\sharp \ldots L_n^\sharp,$$

where $\mathcal{L} = (\lambda^{(1)}, \ldots, \lambda^{(r)})$, and where we set $\operatorname{Card} E(S) = S$ in case $i = 0$. Note that for $\tilde{U}$ we have

$$c_0(4\sqrt{n})^{n-r}\binom{T + \sigma_i}{\sigma_i} \cdot \operatorname{Card} E(S) \cdot H(\mathcal{L}; L_1^\sharp, \ldots, L_n^\sharp)$$

$$\geq L_0^{1-i} L_1^\sharp \ldots L_n^\sharp,$$

with all $\mathcal{L}$ and $i = 0,\ 1$ and we have equality with at least one $\mathcal{L}$ and one $i \in \{0,1\}$. From Lemma 3.3 or Lemma 3.4, accordingly to $i = 0$ or $i = 1$, we deduce the existence of a non-zero polynomial of degree $\leq L_0 \cdot [L_1^\sharp] \ldots [L_n^\sharp]$ satisfying

$$D_0^{\tau_0} \ldots D_{n-1}^{\tau_{n-1}} P(s, \alpha_1^s, \ldots, \alpha_n^s) = 0.$$

And from Remark 3.1 we deduce $[L_n^\sharp] \neq 0$ hence $\tilde{L}_j \geq 1$ for $1 \leq j \leq n$.

Step 2. Among the numbers $\alpha_1^{i_1} \ldots \alpha_n^{i_n}$, $0 \leq i_j \leq [\mathbb{Q}(\alpha_j) : \mathbb{Q}]$, $1 \leq j \leq n$, $i_1 + \ldots + i_n \leq D$, we choose a basis $\xi_1, \ldots, \xi_D$ of the field $\mathbb{Q}(\alpha_1, \ldots, \alpha_n)$ over $\mathbb{Q}$. Then there exist rational integers $p_d(\lambda)$, $1 \leq d \leq D$, $0 \leq \lambda_j \leq \tilde{L}_j$, $j = -2, -1, 1, \ldots, n$, not all zero, bounded in absolute value by $\exp(f_3 U / D)$, such that if we set

$$p(\lambda) = \sum_{d=1}^{D} p_d(\lambda) \xi_d,$$

then for all $(n+1)$-tuple $(\tau_0, \ldots, \tau_{n-1}, s) \in \mathbb{N}^{n+1}$ satisfying $|\tau| < T$ and $0 \leq s < S$, the equation $\varphi_\tau(s) = 0$ holds.

Proof: We follow the proof of Lemma 3.2 of [11]. The equation $\varphi_\tau(s) = 0$ can be written

$$\sum_{(\lambda)} \sum_{d=1}^{D} p_d(\lambda) \xi_d \Lambda(s, \tau) \alpha_1^{\lambda_1 s} \ldots \alpha_n^{\lambda_n s} = 0.$$

The rank of the linear system obtained when $|\tau| < T$ and $0 \leq s < S$ vary is equal to the rank of $\mathcal{B}$ which is bounded above, thanks to Lemma 3.3 or Lemma 3.4, according to $i = 0$ or $i = 1$, by

$$(4\sqrt{n})^{n-r} \binom{T + \sigma_i}{\sigma_i} \cdot \operatorname{Card} E(S) \cdot (L_0 + 1)^i \cdot H(\mathcal{L}; L_1^\sharp, \ldots, L_n^\sharp)$$

for any $\mathcal{L}$ and $i = 0, 1$ and where we set $\operatorname{Card} E(S) = S$ in case $i = 0$. Hence by the choice of $\tilde{U}$ in step 1, rank $\mathcal{B}$ is less than $2L_0 L_1^\sharp \ldots L_n^\sharp / c_0$. We select rank $\mathcal{B}$ equations from the above linear system and we apply Lemma 2.1 of [11] to this sub-system, with

$$d = D,$$

$$n = D(L_0 + 1)([L_1^\sharp] + 1) \ldots ([L_n^\sharp] + 1),$$

$$m \leq 2L_0 L_1^\sharp \ldots L_n^\sharp / c_0,$$

and finally,

$$X \leq \exp(f_3 U/2D),$$

thanks to Lemma 5.1.

Step 3. For $|z| < c_5 S$ and $|\tau| < T$, we have

$$|f_\tau(z) - \varphi_\tau(z)| \leq |\Lambda| . e^{f_4 U}.$$

Proof: See Lemma 3.3 of [11] with $J = 0$.

Step 4. For $s \in \mathbf{Z}$, $1 \leq s < c_5 S$, and $|\tau| < T$, we have either $\varphi_\tau(s) = 0$ or

$$\log |\varphi_\tau(s)| > -f_5 U.$$

Proof. See part 1 of Lemma 3.4 in [11] with $J = 0$.

Step 5. For $|z| < c_5 S$ and $|\tau| < T/2$, we have

$$\log |f_\tau(z)| \leq -(\frac{1}{2c_1} - f_8)U.$$

Proof: This is essentially Lemma 3.5 of [11], again with $J = 0$. We use the extrapolation procedure of Baker together with Steps 2 and 3. In the estimate which is provided by Lemma 2.3 of [11], the main term comes from the quantity $(T/2)S \log(R/4r)$; we choose $r = c_5 S$, $R = Er$, and this gives the term $U/2c_1$.

We also need an upper bound for $|f_\tau|_R = \sup\{|f_\tau(z)|; |z| = R\}$; this estimate involves

$$\sum_{j=1}^{n} L_j R |\log \alpha_j| \leq ec_5 U/c_1 c_2$$

which is less than $f_8 U/2$. We also need an upper bound for $(T/2)S \log(18r/S)$, and we use our assumption $c_1 \geq c_0 \log c_5$.

Step 6. For $|\tau| < T/2$ and $0 \leq s < c_5 S$, $s \in \mathbf{Z}$, we have

$$\varphi_\tau(s) = 0.$$

This is an easy consequence of the three preceding steps; see [11] Lemma 3.6 with $J = 0$.

Step 7. We now reach the desired contradiction. We use Proposition 3.5. with $T_1 = [T/2(n+1)]$ and $S_1 = [c_5 S/(n+1)]$: there exists $i = 0$ or 1, and there exists $\mathcal{L}$ such that

$$\binom{T_1 + \sigma_i}{\sigma_i} . \operatorname{Card} E(S_1) . H(\mathcal{L}; \tilde{L}) \leq (n+1)! . L_0^{1-i} \tilde{L}_1 \ldots \tilde{L}_n. \quad (5.2)$$

We notice that, for $0 \leq \sigma \leq n$, we have

$$\binom{T_1 + \sigma}{\sigma} . S_1 \geq \binom{T + \sigma}{\sigma} . S . c_5/c_0 2^\sigma (n+1)^{\sigma+1}.$$

Notice also that $T/\tilde{L}_j \geq 8n^3$ $(1 \leq j \leq n)$ and $TS/L_0 \geq 8n^3$. Therefore Lemma 3.6 gives $\sigma_0 = r - 1$ if $i = 0$, and $\sigma_1 = r$ if $i = 1$. Next, Lemma 3.7 yields $\operatorname{Card} E(S_1) = S_1$ if $i = 1$. Now from (5.2) we deduce

$$c_5 . \binom{T + \sigma_i}{\sigma_i} . S . H(\mathcal{L}; \tilde{L}) \leq c_0 (n+1)! \, 2^{\sigma_i} (n+1)^{\sigma_i+1} . L_0^{1-i} \tilde{L}_1 \ldots \tilde{L}_n,$$

and for $c_5 \geq c_0^2 (n+1)^{n+2} n! \, 2^n$, this gives a contradiction with the first step (minimality of $\tilde{U}$).

Appendix: Algebraic subgroups of a torus

Let us consider an algebraic group G which is the product of the n-th power of the multiplicative group $\mathbf{G}_m$ with the additive group $\mathbf{G}_a$. We embed $\mathbf{G}_a$ and $\mathbf{G}_m$ in the projective line $\mathbf{P}^1$ in the natural way, that is we identify $\mathbf{G}_a$ with the affine line and $\mathbf{G}_m$ with this affine line but one point. Any algebraic subgroup G' of G is then a quasi-projective subvariety of the multiprojective space $\mathbf{P} = \prod_{i=0}^n \mathbf{P}_{(i)}^1$. There exist notions of multidegrees on $\mathbf{P}$ which we recall now.

Let V be a quasi-projective subvariety of $\mathbf{P}$ of dimension d and $\theta = (\theta_1, \ldots, \theta_d)$ an increasing sequence of $\{0, \ldots, n\}$. There exists a hypersurface of $\prod_{i=1}^d \mathbf{P}_{(\theta_i)}^1$ with the following property: for each point P of $\prod_{i=1}^d \mathbf{P}_{(\theta_i)}^1$ outside this hypersurface, the number of points in V whose projection is P is finite and independent of P. We denote this number by $\deg_\theta V$. The characteristic function of V is then the following homogeneous polynomial of degree d in the variables $X_0, \ldots, X_n$

$$\mathcal{H}(V; X_0, \ldots, X_n) = d! \sum_\theta \deg_\theta V . X_{\theta_1} \ldots X_{\theta_d},$$

where θ runs over the set $\varphi_{d,n+1}$ of all increasing sequences of d elements of $\{0, \ldots, n\}$.

We will see soon that this polynomial, which occurs in the zeros estimate of [9], is closely related to the polynomial H introduced in §3. Then we will establish Proposition 3.5.

First we recall that any algebraic subgroup of $G = \mathbf{G}_a \times \mathbf{G}_m^n$ splits in a product $G_0' \times G_1'$ where G_0' (resp. G_1') is a subgroup of $\mathbf{G}_a$ (resp. $\mathbf{G}_m^n$). So we consider two cases, either $G_0' = \{0\}$ and we will say that we are in case I, or $G_0' = \mathbf{G}_a$ and we will say that we are in case II. In case I it follows from Lemma 3.4 of [9]

$$\mathcal{H}(G'; X_0, \ldots, X_n) = \mathcal{H}(G_1'; X_1, \ldots, X_n),$$

while in case II we have, with $d + 1 = \dim G'$,

$$\mathcal{H}(G'; X_0, \ldots, X_n) = (d + 1) . X_0 . \mathcal{H}(G_1'; X_1, \ldots, X_n).$$

Lemma A.1. *For any connected algebraic subgroup G' of G there exists $\mathcal{L} = (\lambda^{(1)}, \ldots, \lambda^{(r)})$, with $r = n - \dim G'$ in case I and $r = n + 1 - \dim G'$ in case II, and $\lambda^{(i)} \in \mathbf{Z}^n$ such that*

$$\mathcal{H}(G'; X_0, \ldots, X_n) = (\dim G' . X_0)^i H(\mathcal{L}; X_1, \ldots, X_n),$$

where $i = 0$ in case I and $i = 1$ in case II.

Proof: From the remark above, it is enough to prove

$$\mathcal{H}(G_1'; X_1, \ldots, X_n) = H(\mathcal{L}; X_1, \ldots, X_n),$$

for some $\mathcal{L}$. Note that $n - r$ is always the dimension of G_1'. And according to the definitions of the functions $\mathcal{H}$ and H it all comes down to verify the equalities

$$\deg_\theta G_1' = |\det \mathcal{L}_{\theta'}|,$$

where $\theta \in \varphi_{n-r,n}$ and θ' stands for the complement of θ in $\varphi_{r,n}$. This is Proposition 4 of [3], we repeat the proof here for the convenience of the reader. By symmetry it is enough to deal with the index $\theta = (r + 1, \ldots, n)$. Let Λ be the subgroup of $\mathbf{Z}^n$ of rank r which is orthogonal to $T_{G_1'}$. By Theorem I (p.11) of [4] we can find, as in [7], p.434, generators $\lambda^{(1)}, \ldots, \lambda^{(r)}$ of Λ such that $\lambda_j^{(i)} = 0$ for $i > j$. If $\mathcal{L}$ is a basis of Λ the quantity $|\det \mathcal{L}_{\theta'}|$ is invariant by a change of basis, so we have $|\det \mathcal{L}_{\theta'}| = |\lambda_1^{(1)}| \ldots |\lambda_r^{(r)}|$. But G_1' is defined in G_1 by the equations

$$X_i^{\lambda_i^{(i)}} = \prod_{j>i} X_j^{-\lambda_j^{(i)}}, \quad i = 1, \ldots, r,$$

so, if we fix $X_{r+1}, \ldots, X_n$ each non zero, the number $\deg_\theta G_1'$ of points in G_1' over $X_{r+1}, \ldots, X_n$ is equal to $|\lambda_1^{(1)}| \ldots |\lambda_r^{(r)}|$, and we deduce the equality

$$\deg_\theta G_1' = |\det \mathcal{L}_{\theta'}|,$$

which establishes the lemma.

Proof of Proposition 3.5. If the rank of the matrix $\mathcal{A}$ is strictly less than $(L_0 + 1) \ldots (L_n + 1)$ there exists a polynomial P of degrees $L_0, \ldots, L_n$ which vanishes at each point $(s, e^{s\ell_1}, \ldots, e^{s\ell_n})$ for $s = 0, \ldots, S$ with order at least T with respect to the derivatives $D_0, \ldots, D_{n-1}$. The main zeros estimate of [9] (Theorem 2.1) applied to this situation exhibits a connected algebraic subgroup G' of G satisfying

$$\binom{T_1 + \dim W/W \cap G'}{\dim W/W \cap G'} \cdot \operatorname{card}(\Sigma + G')/G'.\mathcal{H}(G'; L_0, \ldots, L_n)$$

$$(\ast\ast\ast) \qquad\qquad\qquad\qquad\qquad \leq (n+1)! \, L_0 \ldots L_n,$$

where Σ stands for the set $\{(s, e^{s\ell_1}, \ldots, e^{s\ell_n}); 0 \leq s \leq S_1\}$ and W stands for the image of the analytic subgroup

$$(z_0, \ldots, z_{n-1}) \longrightarrow (z_0, e^{z_1}, \ldots, e^{z_{n-1}}, e^{\beta_0 z_0 + \cdots + \beta_{n-1} z_{n-1}}).$$

Let Λ be the subgroup of $\mathbf{Z}^n$ of rank r which is orthogonal to $T_{G_1'}$ and $\mathcal{L}$ a basis of Λ. In Case I, i.e. $G_0' = \{0\}$, we have $\dim G' = n - r$ and we compute $\operatorname{card}(\Sigma + G')/G' = S_1$ and $\dim W/W \cap G' = r$ if $G' \subseteq W$ or $= r + 1$ if $G' \not\subseteq W$. In Case II (i.e. $G_0' = \mathbf{G}_a$) we have $\dim G' = n + 1 - r$ so that $\operatorname{card}(\Sigma + G')/G' = \operatorname{card} E(S_1)$ and $\dim W/W \cap G' = r - 1$ if $G' \subseteq W$ or $= r$ if $G' \not\subseteq W$. Putting these calculations together with Lemma A.1, Proposition 3.5 follows at once from $(\ast\ast\ast)$.

References

[1] Baker A., The theory of linear forms in logarithms, in: *Transcendence theory; advances and applications*, A. Baker and D. W. Masser ed., Academic Press (1977), Chap. 1, p.1-27.

[2] Blass J., Glass A., Meronk D. and Steiner R., A lower bound for linear forms in logarithms, to appear.

[3] Bertrand D., and Philippon P., Sous-groupes algébriques de groupes algébriques commutatifs, *Illinois J. Math.*, to appear.

[4] Cassels J. W. S., *An introduction to the geometry of numbers*, Springer (1959).

[5] Gramain F., Sur le lemme de Siegel (d'après E. Bombieri et J. Vaaler), in *Problèmes diophantiens* 1983/84, *fasc. 1*, No 2, D. Bertrand and M. Waldschmidt ed, *Publications Math. Paris VI* No 64, (1984).

[6] Loxton J., Mignotte M., van der Poorten A., and Waldschmidt M., A lower bound for linear forms in the logarithms of algebraic numbers, *C. R. Math. Acad. Sci. Canada = Math. Report Acad. Sci.*, **11** (1987), 119–124.

[7] Masser D. W., and Wüstholz G., Fields of large transcendence degree, *Inv. Math.* **72**, (1983), 407–464.

[8] Nesterenko Y. V., Estimates for the characteristic function of a prime ideal, *Math. Sbornik* **123** (165) No 1, (1984) *Math. USSR Sbornik* **51** No 1, (1985), 9–32.

[9] Philippon P., Lemmes de zéros dans les groupes algébriques commutatifs, *Bulletin Soc. Math. France* **114**, tome III, (1986), 355–383.

[10] Philippon P., and Waldschmidt M., Formes linéaires de logarithmes sur les groupes algébriques commutatifs, *Illinois J. Math.*, to appear.

[11] Waldschmidt M., A lower bound for linear forms in logarithms, *Acta Arithmetica* **37**, (1980), 257–283.

19

REDUCIBILITY OF LACUNARY POLYNOMIALS, IX

A. Schinzel

The aim of this paper is to extend the results of the papers [1] and [4] concerning reducibility of trinomials and quadrinomials over $\mathbf{Q}$ to the case where their coefficients are arbitrary algebraic numbers. We shall use the following notation.

If $\mathbf{K}$ is a field, $f \in \mathbf{K}[x_1, \ldots, x_k]$ then

$$f \overset{\text{can}}{=\!=\!=} \text{const} \prod_{\sigma=1}^{s} f_\sigma^{e_\sigma}$$

means, besides the equality, that the polynomials $f_\sigma \in \mathbf{K}[x_1, \ldots, x_k]$ are irreducible over $\mathbf{K}$ and prime to each other. Constants are considered neither reducible nor irreducible.

If $\phi = f \prod_{i=1}^{k} x_i^{\alpha_i}$, where f is a polynomial prime to $x_1 x_2 \ldots x_k$ and α_i are integers, then we set

$$J\phi = f.$$

A polynomial g such that

$$Jg(x_1^{-1}, \ldots, x_k^{-1}) = \pm g(x_1, \ldots, x_k)$$

is called reciprocal. Let

$$J\phi \overset{\text{can}}{=\!=\!=} \text{const} \prod_{\sigma=1}^{s} f_\sigma^{e_\sigma}.$$

We set

$$K\phi = \text{const} \, \Pi_1 f_\sigma^{e_\sigma}$$

where Π_1 is extended over all f_σ that do not divide $J(x_1^{\delta_1} \ldots x_k^{\delta_k} - 1)$ for any integer vector $[\delta_1, \ldots, \delta_k] \neq 0$. Moreover if $\mathbf{K} = \mathbf{Q}$ we set

$$L\phi = \text{const} \, \Pi_2 f_\sigma^{e_\sigma}$$

where Π_2 is extended over all f_σ that are non-reciprocal. The leading coefficients of $K\phi$ and $L\phi$ are assumed equal to that of $J\phi$. In particular, for $k = 1$, $K\phi(x)$, called the kernel of ϕ, equals $J\phi(x)$ deprived of all its factors $x - \zeta$, where ζ is a root of unity. ζ_q is a primitive qth root of unity, $N_{K/Q}$ denotes the norm from K to Q or from $K(x_1, \ldots, x_k)$ to $Q(x_1, \ldots, x_k)$. For a matrix M, $h(M)$ denotes the maximum of the absolute values of the elements of M; vectors are treated as matrices with one row and denoted by boldface letters or underlined; $B_1(\underline{\alpha})$, $B_2(\underline{\alpha})$, ... denote real numbers dependent on the vector $\underline{\alpha}$ only.

If a, b are complex numbers different from 0 and roots of unity then the equation $a^m = b^n$ if solvable in integers $m, n \neq 0$ determines uniquely the rational numbers m/n. The height of this number is denoted in the sequel by $C(a, b)$. If the equation $a^m = b^n$ is insolvable in integers $m, n \neq 0$ or if one of a, b is a root of unity we set $C(a, b) = 0$.

The following theorems will be proved.

Theorem 1. Let α_0, α_1, α_2 be non-zero algebraic numbers such that $\alpha_0 \in Q(\alpha_1/\alpha_0, \alpha_2/\alpha_0) = K_0$. For every trinomial $t(x) = \alpha_0 + \alpha_1 x^{n_1} + \alpha_2 x^{n_2}$, $0 < n_1 < n_2$, we have the following two possibilities.

(i) $Kt(x)$ is irreducible over K_0.

(ii) There exist integers ν_1, ν_2, v such that

$$0 < \nu_1 < \nu_2 \leq B_1(\underline{\alpha}),$$

$$n_i = v\nu_i \quad (i = 1, 2)$$

and

$$K(\alpha_0 + \alpha_1 x^{\nu_1} + \alpha_2 x^{\nu_2}) \frac{\text{can}}{K_0} \prod_{\sigma=1}^{s} f_\sigma^{e_\sigma}$$

implies

$$Kt(x) \frac{\text{can}}{K_0} \prod_{\sigma=1}^{s} f_\sigma(x^v)^{e_\sigma}.$$

Theorem 2. Let α_0, α_1, α_2, α_3 be non-zero algebraic numbers such that $\alpha_0 \in Q(\alpha_1/\alpha_0, \alpha_2/\alpha_0, \alpha_3/\alpha_0) = K_0$. For every quadrinomial $q(x) = \alpha_0 + \sum_{j=1}^{3} \alpha_j x^{n_j}$, $0 < n_1 < n_2 < n_3$, we have the following five possibilities.

(iii) $Jq(x^{-1}) = kq^\sigma(x)$ for an automorphism σ of K_0 and a $k \in K_0$.

(iv) $Kq(x)$ is irreducible.

(v) $q(x)$ can be represented in one of the forms

$$k(U^2 + 2UV + V^2 - W^2) = k(U + V + W)(U + V - W),$$
$$k(U^3 + V^3 + W^3 - 3UVW) = k(U + V + W)$$
$$(U^2 + V^2 + W^2 - UV - UW - VW), \qquad (1)$$
$$k(T^2 - 4TUVW - U^2V^4 - 4U^2W^4) =$$
$$k(T - 2UVW - UV^2 - 2UW^2)(T - 2UVW + UV^2 + 2UW^2),$$

where $k \in \mathbf{K}_0$, T, U, V, W are monomials in $\mathbf{K}_0[x]$. The factors on the right-hand-side have kernels irreducible over $\mathbf{K}_0$ unless $\zeta_3 \in \mathbf{K}_0$, when

$$K(U^2 + V^2 + W^2 - UV - UW - VW) \overset{\text{can}}{\underset{\mathbf{K}_0}{=\!=}} K(U + \zeta_3 V + \zeta_3^2 W)$$
$$K(U + \zeta_3^2 V + \zeta_3 W).$$

(vi) $q(x)$ is the sum of two binomials with the highest common divisor $d(x)$ being a binomial. We have $Kd = d$ and either Kqd^{-1} is irreducible over $\mathbf{K}_0$ or it is a binomial.

(vii) There exist integers ν_1, ν_2, ν_3, v such that

$$0 < \nu_1 < \nu_2 < \nu_3 \leq B_2(\underline{\alpha}), \qquad (2)$$

$$n_i = v\nu_i, \quad i = 1, 2, 3, \qquad (3)$$

and

$$K\left(\alpha_0 + \sum_{j=1}^{3} \alpha_j x^{\nu_j}\right) \overset{\text{can}}{\underset{\mathbf{K}_0}{=\!=}} \text{const} \prod_{\sigma=1}^{s} f_\sigma(x)^{e_\sigma}$$

implies

$$Kq(x) \overset{\text{can}}{\underset{\mathbf{K}_0}{=\!=}} \text{const} \prod_{\sigma=1}^{s} f_\sigma(x^v)^{e_\sigma}.$$

Theorem 1 gives at once a necessary and sufficient condition for reducibility of $Kt(x)$ over $\mathbf{K}_0$; Theorem 2 gives such a condition only for $q(x)$ not fulfilling (iii). We formulate the latter in the following.

Corollary. Assume that $q(x)$ satisfies the conditions of Theorem 2, is not a product of two binomials and $Jq(x^{-1}) \neq kq^\sigma(x)$ for every automorphism σ of $\mathbf{K}_0$ and every $k \in \mathbf{K}_0$. Then $Kq(x)$ is reducible over $\mathbf{K}_0$ if and only if at least one of the following situations occurs.

(v′) $q(x)$ can be represented in one of the forms (1), where $k \in \mathbf{K}_0$, T, U, V, W are monomials in $\mathbf{K}_0[x]$ and the factors on the right-hand-side have non-constant kernels.

(vi′) $q(x)$ is the sum of two binomials with the highest common divisor $d(x)$ satisfying $Kd \notin \mathbf{K}_0$, $Kqd^{-1} \notin \mathbf{K}_0$.

(vii′) there exist integers v, ν_1, ν_2, ν_3 satisfying (2) and (3) such that $K(\alpha_0 + \sum_{j=1}^{3} \alpha_j x^{\nu_j})$ is reducible over $\mathbf{K}_0$.

In the case $\mathbf{K}_0 = \mathbf{Q}$ and under additional assumptions the corollary has been given in [7], but the formulation of the conditions corresponding to (v′) and (vi′) has been incomplete.

The proofs of both theorems lean heavily on earlier papers in this series, especially on [5]. The constants depending on $\underline{\alpha}$ taken from that paper are denoted, as there, by $C_i(\underline{\alpha})$ with suitable i.

Lemma 1. If $k > 1$, $\alpha_j \in \mathbf{C} \setminus \{0\}$, $0 \leq j \leq k$, and $K\left(\sum_{j=0}^{k} \alpha_j x^{n_j}\right) \in \mathbf{C}$, $0 = n_0 < n_1 < \ldots < n_k$, then there is a linear relation

$$\sum_{j=1}^{k} \gamma_j n_j = 0$$

where $\underline{\gamma} = [\gamma_1, \ldots, \gamma_k] \in \mathbf{Z}^k$ and $0 < h(\underline{\gamma}) \leq 4$.

Proof. Take in Lemma 7 of [5]

$$F(x_1 \ldots, x_k) = \left(\alpha_0 + \sum_{j=1}^{k} \alpha_j x_j\right)\left(\overline{\alpha}_0 + \sum_{j=1}^{k} \overline{\alpha}_j x_j\right),$$

where the bar denotes the complex conjugation. We have clearly

$$JF(x_1, \ldots, x_k) \neq \pm JF(x_1^{-1}, \ldots, x_k^{-1}).$$

On the other hand, if $K\left(\sum_{j=0}^{k} \alpha_k x^{n_j}\right) \in \mathbf{C}$ every zero of $\sum_{j=0}^{k} \alpha_j x^{n_j}$ is a root of unity from a finite set, say E. If

$$\sum_{j=0}^{k} \alpha_j x^{n_j} = \alpha_k \prod_{\zeta \in E} (x - \zeta)^{m(\zeta)}$$

we have

$$\sum_{j=0}^{k} \overline{\alpha}_j x^{n_j} = \overline{\alpha}_k \prod_{\zeta \in E} (x - \overline{\zeta})^{m(\zeta)}.$$

Hence

$$F(x^{n_1}, \ldots, x^{n_k}) = \alpha_k \overline{\alpha}_k \prod_{\zeta \in E} (x^2 - (\zeta + \overline{\zeta})x + 1)^{m(\zeta)}.$$

Thus

$$JF(x^{n_1}, \ldots, x^{n_k}) = JF(x^{-n_1}, \ldots, x^{-n_k})$$

and since the degree of F with respect to each variable is 2, the desired conclusion follows from Lemma 7 of [5].

Lemma 2. For every algebraic number field **K** and every polynomial $f \in \mathbf{K}[x]$ there exists a constant $c(f)$ with the following property. Every positive integer n has a divisor $\nu \leq c(f)$ such that

$$K f(x^{\nu}) \overset{\mathrm{can}}{=\!=\!=} \text{const} \prod_{\sigma=1}^{s} f_\sigma(x)^{e_\sigma}$$

implies

$$K f(x^n) \overset{\mathrm{can}}{=\!=\!=} \text{const} \prod_{\sigma=1}^{s} f_\sigma(x^{n/\nu})^{e_\sigma}.$$

Proof. If f is irreducible over **K** and $Kf = f$ the lemma is a special case of Theorem 27 in [6]. In the general case let

$$K f \overset{\mathrm{can}}{=\!=\!=} \text{const} \prod_{\rho=1}^{r} \phi_\rho^{e_\rho}. \tag{4}$$

We apply the above special case to every ϕ_ρ and find a $\nu_\rho \leq c(\phi_\rho)$, $\nu | n$ such that

$$K \phi_\rho(x^{\nu_\rho}) \overset{\mathrm{can}}{=\!=\!=} \text{const} \prod_{i=1}^{i_\rho} \phi_{\rho i}(x)$$

implies

$$K \phi_\rho(x^n) \overset{\mathrm{can}}{=\!=\!=} \text{const} \prod_{i=1}^{i_\rho} \phi_{\rho i}(x^{n/\nu_\rho}).$$

Let us take

$$c(f) = \left(\max_{\rho \leq r} c(\phi_\rho)\right)!, \quad \nu = \text{l.c.m.}_{\rho \leq r} \nu_\rho.$$

Clearly $\nu \leq c(f)$, $\nu | n$ and

$$K\phi_\rho(x^\nu) \stackrel{\text{can}}{\overline{\overline{K}}} \text{const} \prod_{i=1}^{i_\rho} \phi_{\rho i}(x^{\nu/\nu_\rho}). \tag{6}$$

The lemma follows from (4), (5) and (6) after a suitable change of notation.

Proof of Theorem 1. If $Kt(x) \in \mathbf{K}_0$ we have by Lemma 1

$$\gamma_1 n_1 + \gamma_2 n_2 = 0; \quad \underline{\gamma} = [\gamma_1, \gamma_2] \in \mathbf{Z}^2, \tag{7}$$

where $0 < h(\underline{\gamma}) \leq 4$.

If $Kt(x)$ is irreducible over $\mathbf{K}_0$ we have (i). Therefore, assume that $Kt(x)$ is irreducible over $\mathbf{K}_0$. Then

$$T(x) = K N_{\mathbf{K}_0/\mathbf{Q}}\left(\alpha_0 + \sum_{j=1}^{2} \alpha_j x^{n_j}\right)$$

is reducible over $\mathbf{Q}$. By Lemma 5 of [5], if $T(x)$ has in $\mathbf{Q}[x]$ at least two, counting multiplicity, irreducible non-reciprocal factors then there is a linear relation (7), where

$$0 < h(\underline{\gamma}) < C_9(\underline{\alpha}).$$

If, on the other hand, $T(x)$ has in $\mathbf{Q}[x]$ a reciprocal factor then there is a $\xi \in \mathbf{C}^*$, ξ different from roots of unity, such that for suitable conjugates α_j' and α_j'' of α_j, $0 \leq j \leq 2$,

$$\sum_{j=0}^{2} \alpha_j' \xi^{n_j} = \sum_{j=0}^{2} \alpha_j'' \xi^{-n_j} = 0, \quad n_0 = 0.$$

Let $\langle \eta_{i1}, \eta_{i2} \rangle$ $(i = 1, 2)$ be the two solutions of the system of equations

$$\alpha_0' + \alpha_1' y_1 + \alpha_2' y_2 = \alpha_0'' + \alpha_1'' y_1^{-1} + \alpha_2'' y_2^{-1} = 0. \tag{8}$$

It follows that for an $i \leq 2$

$$\xi^{n_1} = \eta_{i1}, \quad \xi^{n_2} = \eta_{i2},$$

hence

$$\eta_{i1}^{n_2} = \eta_{i2}^{n_1}.$$

Since ξ is not a root of unity, η_{ij} are not either, and it follows from the definition of $C(a, b)$ that

$$\frac{\max\{n_1, n_2\}}{(n_1, n_2)} = C(\eta_{i1}, \eta_{i2}).$$

This implies a relation of the type (7) with

$$h(\gamma) \leq B_0(\underline{\alpha}) = \max C(\eta_1, \eta_2),$$

where $\langle \eta_1, \eta_2 \rangle$ runs through the solutions of all systems of equations of the type (8).

Now, if (7) holds then assuming without loss of generality that $(\gamma_1, \gamma_2) = 1$, $\gamma_1 > 0$ we get $\gamma_2 < 0$

$$n_1 = -\gamma_2 n, \quad n_2 = \gamma_1 n,$$

where n is a positive integer.

Put in Lemma 2

$$f(x) = \alpha_0 + \alpha_1 x^{-\gamma_2} + \alpha_2 x^{\gamma_1}. \tag{9}$$

By virtue of that lemma there exists a constant $c(f)$ and a positive integer $\nu \leq c(f)$ such that $\nu | n$ and

$$K f(x^\nu) \frac{\text{can}}{K_0} \text{const} \prod_{\sigma=1}^{s} f_\sigma(x)^{e_\sigma}$$

implies

$$K f(x^n) \frac{\text{can}}{K_0} \text{const} \prod_{\sigma=1}^{s} f_\sigma(x^{n/\nu})^{e_\sigma}.$$

Since $t(x) = f(x^n)$ we get the case (ii) putting

$$\nu_1 = -\nu \gamma_2, \quad \nu_2 = \nu \gamma_1, \quad v = n/\nu,$$

$$B_1(\underline{\alpha}) = \max c(f),$$

where the maximum is taken over all polynomials f of the form (9) with $\underline{\gamma} = [\gamma_1, \gamma_2]$ satisfying

$$h(\underline{\gamma}) \leq \max\{4, C_9(\underline{\alpha}), B_0(\underline{\alpha})\}.$$

The proof of Theorem 2 is based on several lemmata.

Lemma 3. If $k = 3$ or 4, α_{ij}, $1 \le i \le k$, $1 \le j \le 2$, are integers, the matrix

$$\mathbf{A} = \begin{pmatrix} 1 & \alpha_{11} & \alpha_{12} \\ \vdots & \vdots & \vdots \\ 1 & \alpha_{k1} & \alpha_{k2} \end{pmatrix}$$

is of rank 3 and a permutation π of $\{1, \ldots, k\}$ has the property that for suitable β_j

$$\alpha_{ij} + \alpha_{\pi(i)j} = \beta_j, \qquad 1 \le i \le k, \quad 1 \le j \le 2, \tag{10_1}$$

then $k = 4$, π is the product of two cycles $(i_1, i_2)\,(i_3, i_4)$ and

$$\alpha_{i_1 j} + \alpha_{i_2 j} = \alpha_{i_3 j} + \alpha_{i_4 j}, \qquad j = 1, 2. \tag{10_2}$$

Proof. We have

$$\alpha_{ij} + \alpha_{\pi(i)j} = \beta_j = \alpha_{\pi(i)j} + \alpha_{\pi^2(i)j},$$

hence

$$\alpha_{ij} = \alpha_{\pi^2(i)j}. \tag{11}$$

The permutation π^2 cannot contain a cycle of length 4 nor just one cycle of length 2. If it contains a cycle of length 3 or two cycles of length 2 then by (11) rank$(\mathbf{A}) \le 2$, contrary to the assumption. Hence π^2 is the identity and either π is the identity or $k = 3$, $\pi = (i_1, i_2)(i_3)$ or $k = 4$, $\pi = (i_1, i_2)(i_3)(i_4)$ or $k = 4$, $\pi = (i_1, i_2)(i_3, i_4)$. In the first three cases we get from (10)

$$\alpha_{i_1 j} + \alpha_{i_2 j} = 2\alpha_{i_3 j} = 2\alpha_{i_k j}$$

hence rank$(\mathbf{A}) \le 2$. In the last case (10) implies (11).

Lemma 4. Assume that integer vectors $[\nu_{1i}, \nu_{2i}]$, $1 \le i \le 3$, are distinct and different from $[\nu_{10}, \nu_{20}] = [0, 0]$, the matrix $[\nu_{ij}]$ is of rank 2

$$Q(y_1, y_2) = J\left(\sum_{i=0}^{3} \alpha_i y_1^{\nu_{1i}} y_2^{\nu_{2i}}\right) = B_1(y_1, y_2) + B_2(y_1, y_2), \tag{12}$$

where $B_i(y_1, y_2)$, $i = 1, 2$ are binomials and

$$B_0 = (B_1, B_2) \ne 1. \tag{13}$$

If for a certain injection ρ of $\mathbf{K}_0$ into $\mathbf{C}$ and a $c \in \mathbf{C}$ we have

$$J \frac{Q(y_1^{-1}, y_2^{-1})}{B_0(y_1^{-1}, y_2^{-1})} = c \frac{Q^\rho(y_1, y_2)}{B_0^\rho(y_1, y_2)}, \tag{14}$$

then QB_0^{-1} is a binomial.

Proof. Changing if necessary y_2 to y_2^{-1} we may assume that

$$B_0(y_1, y_2) = y_1^{p_1} y_2^{p_2} - a, \quad p_1 \geq 0, \quad p_2 \geq 0, \quad \langle p_1, p_2 \rangle \neq \langle 0, 0 \rangle.$$

From (13) we infer that for suitable integers $q_i > 0$, $(q_1, q_2) = 1$

$$B_i(y_1, y_2) = b_i(y_1^{p_1 q_i} y_2^{p_2 q_i} - a^{q_i}) \prod_{j=1}^{2} y_j^{\beta_{ij}}, \tag{15}$$

while

$$\min\{\beta_{1j}, \beta_{2j}\} = 0, \qquad j = 1, 2. \tag{16}$$

Let us give y_1 the weight p_2, y_2 the weight $-p_1$. Then the weight of B_i equals

$$w(B_i) = \beta_{i1} p_2 - \beta_{i2} p_1.$$

From $w(B_1) = w(B_2)$ it would follow

$$(\beta_{11} - \beta_{21}) p_2 = (\beta_{12} - \beta_{22}) p_1,$$

hence in view of (12), (15) and (16) the rank of the matrix $[\nu_{ij}]$ would be 1, contrary to the assumption. Hence $w(B_1) \neq w(B_2)$ and we may assume without loss of generality that $w(B_1) < w(B_2)$. The sum of the terms with the highest weight on the left-hand-side of (14) equals up to a factor $\prod_{j=1}^{2} y_j^{\gamma_j}$

$$J \frac{B_1(y_1^{-1}, y_2^{-1})}{B_0(y_1^{-1}, y_2^{-1})}$$

and the relevant sum on the right-hand-side equals up to a similar factor

$$J \frac{B_2^\rho(y_1, y_2)}{B_0^\rho(y_1, y_2)}.$$

Comparing the degrees of both polynomials with respect to y_i we get from (15)

$$p_i q_1 = p_i q_2, \qquad i = 1, 2$$

hence $q_1 = q_2 = 1$ and

$$\frac{Q}{B_0} = \sum_{i=1}^{2} b_i \prod_{j=1}^{2} y_j^{\beta_{ij}}.$$

Lemma 5. Assume that integer vectors $[\nu_{1i}, \nu_{2i}]$, $1 \le i \le 3$ are distinct and different from $[\nu_{10}, \nu_{20}] = [0,0]$, the matrix $[\nu_{ij}]$ is of rank 2,

$$Q(y_1, y_2) = J\left(\sum_{i=0}^{3} \alpha_i y_1^{\nu_{1i}} y_2^{\nu_{2i}}\right), \tag{17}$$

$$D(y_1, y_2) = \left(N_{K_0/Q} Q(y_1, y_2), J N_{K_0/Q} Q(y_1^{-1}, y_2^{-1})\right).$$

If $KD \notin K_0$ then either there exists an automorphism σ of K_0 and a $c \in K_0$ such that

$$J Q(y_1^{-1}, y_2^{-1}) = c Q^{\sigma}(y_1, y_2) \tag{18}$$

or $Q(y_1, y_2)$ is the sum of two binomials with the highest common divisor B satisfying $KB \notin K_0$. In the latter case either $N_{K_0/Q} Q B^{-1}$ has no reciprocal factors or $Q B^{-1}$ is a binomial.

Proof. If Q is irreducible over C, so is $J Q(y_1^{-1}, y_2^{-1})$ and all their conjugates over $Q(y_1, y_2)$. Hence $KD \notin K_0$ implies the existence of two automorphisms ρ, τ of the normal closure of K_0 over Q such that for a suitable $c \in C$

$$c Q^{\tau}(y_1, y_2) = J Q^{\rho}(y_1^{-1}, y_2^{-1}).$$

Taking $\sigma = \tau \rho^{-1}$ we get (18) and hence

$$(a_i/a_0)^{\sigma} \in K_0, \quad 1 \le i \le 3.$$

Since $K_0 = Q(a_1/a_0, a_2/a_0, a_3/a_0)$ it follows that σ is an automorphism of K_0 and $c \in K_0$.

If Q is reducible over C we have by Theorem 1 of [1] the following two possibilities.

(viii) $Q(y_1, y_2)$ can be represented in one of the forms

$$c(U^2 + 2UV + V^2 - W^2) = c(U + V + W)(U + V - W), \tag{19a}$$

$$c(U^3 + V^3 + W^3 - 3UVW) = c(U + V + W)(U + \zeta_3 V + \zeta_3^2 W)$$
$$(U + \zeta_3^2 V + \zeta_3 W), \tag{19b}$$

$$c(T^2 - 4TUVW - U^2V^4 - 4U^2W^4) = c(T - 2UVW$$
$$+ UV^2 + 2UW^2)(T - 2UVW - UW^2 - 2UW^2), \tag{19c}$$

where T, U, V, W are monomials in $\mathbf{C}[y_1, y_2]$, $c \in \mathbf{C}$ and the factors in brackets on the right-hand-side are irreducible over $\mathbf{C}$.

(ix) $Q(y_1, y_2)$ is the sum of two binomials with the highest common divisor $B_0(y_1, y_2) \notin \mathbf{K}_0$. The quotient QB_0^{-1} either is a binomial or is irreducible over $\mathbf{C}$.

We shall consider the cases (viii) and (ix) successively. In the case (viii) $JQ(y_1^{-1}, y_2^{-1})$ is representable in the same form as $Q(y_1, y_2)$. If $KD \notin \mathbf{K}_0$, it follows from the irreducibility over $\mathbf{C}$ of the factors on the left-hand-side of (19) that for some automorphisms ρ, τ of the algebraic closure of $\mathbf{K}_0$, some roots of unity ε, ζ and a suitable $c_0 \in \mathbf{C}$ we have either

$$J(U + V + \varepsilon W)^\rho(y_1^{-1}, y_2^{-1}) = c_0(U + V + \zeta W)^\tau$$

or

$$J(U + \varepsilon V + \varepsilon^2 W)^\rho(y_1^{-1}, y_2^{-1}) = c_0(U + \zeta V + \zeta^2 W)^\tau, \quad (20)$$

or

$$J(T - 2UVW + \varepsilon UV^2 + 2\varepsilon UW^2)^\rho(y_1^{-1}, y_2^{-1}) =$$
$$c_0(T - 2UVW + \zeta UV^2 + 2\zeta UW^2)^\tau.$$

Let us put in the first two cases

$$k = 3, \quad U = uy_1^{\alpha_{11}}y_2^{\alpha_{12}}, \quad V = vy_1^{\alpha_{21}}y_2^{\alpha_{22}}, \quad W = wy_1^{\alpha_{31}}y_2^{\alpha_{32}}$$

and in the third case

$$k = 4, \quad T = tuy_1^{\alpha_{11}}y_2^{\alpha_{12}}, \quad UVW = uy_1^{\alpha_{21}}y_2^{\alpha_{22}},$$

$$UV^2 = vy_1^{\alpha_{31}}y_2^{\alpha_{32}}, \quad UW^2 = wy_1^{\alpha_{41}}y_2^{\alpha_{42}}$$

where $t, u, v, w \in \mathbf{C}$. The rank of the matrix

$$\begin{pmatrix} 1 & \alpha_{11} & \alpha_{12} \\ \vdots & \vdots & \vdots \\ 1 & \alpha_{k1} & \alpha_{k2} \end{pmatrix}$$

is the same as the rank of the matrix $\mathbf{B}$ consisting of exponents on the left-hand-side of (19) supplemented by the column $\begin{pmatrix} 1 \\ 1 \\ 1 \end{pmatrix}$: we verify this easily for each of the three cases separately. By (17)

$$\mathrm{rank}(\mathbf{B}) = 1 + \mathrm{rank}[\nu_{ij}] = 3$$

hence $\mathbf{A}$ satisfies the assumptions of Lemma 3. In particular for $k = 3$ the vectors $[\alpha_{1i}, \alpha_{2i}]$, $1 \leq i \leq k$, are all distinct. The same is true for $k = 4$ since otherwise two terms on the left-hand-side of (19) would be similar, contrary to the assumption that $[\nu_{1i}, \nu_{2i}]$, $0 \leq i \leq 3$, are all distinct. The formulae (20) imply the existence of a permutation π of $\{1, 2, \ldots, k\}$ and of a vector $[\beta_1, \beta_2]$ such that

$$\beta_j - \alpha_{ij} = \alpha_{\pi(i)j}, \quad j = 1, 2.$$

By Lemma 3 we have $k = 4$, $\pi = (i_1, i_2)(i_3, i_4)$ and

$$\alpha_{i_1 j} + \alpha_{i_2 j} = \alpha_{i_3 j} + \alpha_{i_4 j}, \quad j = 1, 2.$$

This gives the following three possibilities: $TUVW/U^2V^2W^2 \in \mathbf{C}$ or $TUV^2/U^2VW^3 \in \mathbf{C}$ or $TUW^2/U^2V^3W \in \mathbf{C}$, hence $T^2/UVW \in \mathbf{C}$ or $TUVW/U^2W^4 \in \mathbf{C}$ or $TUVW/U^2W^4 \in \mathbf{C}$, contrary to (viii) and the assumption that $[\nu_{1i}, \nu_{2i}]$, $0 \leq i \leq 3$, are all distinct.

In the case (ix) let

$$Q(y_1, y_2) = B_1(y_1, y_2) + B_2(y_1, y_2),$$

where B_1, B_2 are binomials, $B_0 = (B_1, B_2) \notin \mathbf{K}_0$. If QB_0^{-1} is a binomial B_0' then $KD \notin \mathbf{K}_0$ implies $KN_{\mathbf{K}_0/\mathbf{Q}} B_0 \notin \mathbf{K}_0$ or $KN_{\mathbf{K}_0/\mathbf{Q}} B_0' \notin \mathbf{K}_0$, hence either $KB_0 \notin \mathbf{K}_0$ or $KB_0' \notin \mathbf{K}_0$. In any case Q is the sum of two binomials with the highest common divisor B (equal either B_0 or B_0') satisfying $KB \notin \mathbf{K}_0$.

If QB_0^{-1} is not a binomial it is irreducible over $\mathbf{C}$; if $N_{\mathbf{K}_0/\mathbf{Q}} QB_0^{-1}$ had a reciprocal factor this factor would be common to it and to $N_{\mathbf{K}_0/\mathbf{Q}} JQ(y_1^{-1}, y_2^{-1}) B_0(y_1^{-1}, y_2^{-1})^{-1}$. Hence we would get for some automorphisms ρ, τ of the normal closure of $\mathbf{K}_0$ and for a suitable $c \in \mathbf{C}$

$$JQ^\rho(y_1^{-1}, y_2^{-1}) B_0^\rho(y_1^{-1}, y_2^{-1})^{-1} = cQ^\tau(y_1, y_2) B_0^\tau(y_1, y_2)^{-1}.$$

By virtue of Lemma 4 this implies that QB_0^{-1} is a binomial and the obtained contradiction proves that $N_{\mathbf{K}_0/\mathbf{Q}} QB_0^{-1}$ has no reciprocal factors.

Lemma 6. Let Q satisfy the assumption of Lemma 5. If $F \in \mathbf{K}_0[y_1, y_2]$ is a factor of Q irreducible over $\mathbf{K}_0$ then either $N_{\mathbf{K}_0/\mathbf{Q}} F$ is irreducible over $\mathbf{Q}$ or Q is the sum of two binomials divisible by F.

Proof. If Q is irreducible over $\mathbf{C}$ we have $F = \mathrm{const}\, Q$ and the irreducibility of $N_{\mathbf{K}_0/\mathbf{Q}} F$ follows from Lemma 10 of [3]. If Q is reducible over $\mathbf{C}$, but irreducible over $\mathbf{K}_0$ then again $F = \mathrm{const}\, Q$ and we have the case (viii) considered in the proof of Lemma 5, where T, U, V, W occurring in the formulae (19) are monomials over $\mathbf{C}$, but for no choice of $c \in \mathbf{C}$ monomials over $\mathbf{K}_0$. In the case (19a) or (19b) or (19c) the field $\mathbf{K}_1$ generated by the ratios of the coefficients of $U + V + W$ or $U + V + W$ or $T - 2UVW - UV^2 - 2UW^2$, respectively is an extension of degree $d = 2$ or 3 or 2 of the field $\mathbf{K}_0$ generated by the ratios of the coefficients of $U^2 + 2UV + V^2 - W^2$ or $U^3 + V^3 + W^3 - 3UVW$ or $T^2 - 2TUVW - U^2V^4 - 4U^2W^4$, respectively. By a suitable choice of $c_1 \in \mathbf{C}$ we can achieve that

$$P = \begin{cases} c_1(U + V + W) & \text{if } (19a) \text{ or } (19b) \\ c_1(T - 2UVW - UV^2 - 2UW^2) & \text{if } (19c) \end{cases}$$

satisfies $P \in \mathbf{K}_1[y_1, y_2]$ and then

$$Q = cc_1^{-d} N_{\mathbf{K}_1/\mathbf{K}_0} P,$$

$$N_{\mathbf{K}_0/\mathbf{Q}} F = \text{const } N_{\mathbf{K}_0/\mathbf{Q}} Q = \text{const } N_{\mathbf{K}_1/\mathbf{Q}} P.$$

However by Theorem 1 of [1] P is irreducible over $\mathbf{C}$ and since $\mathbf{K}_1$ is generated by the ratios of the coefficients of P $N_{\mathbf{K}_1/\mathbf{Q}} P$ is irreducible over $\mathbf{Q}$ by Lemma 10 of [3].

If Q is reducible over $\mathbf{K}_0$ we have by Theorem 1 of [1] the following three possibilities.

viii') Q can be represented in one of the forms (19) where $c \in \mathbf{K}_0$, T, U, V, W are monomials in $\mathbf{K}_0[y_1, y_2]$ and factors on the right-hand-side, irreducible over $\mathbf{C}$ have coefficients in $\mathbf{K}_0$.

viii'') Q can be represented in the form (19b), where $c \in \mathbf{K}_0$, U, V, W are monomials in $\mathbf{K}_0[y_1, y_2]$ and $\zeta_3 \notin \mathbf{K}_0$.

(ix) holds.

In the case (viii') F differs by a constant factor from one of the brackets on the right-hand-side of (19). Since the ratios of the coefficients of each factor in brackets generate $\mathbf{K}_0$, we infer by Lemma 10 of [3] that $N_{\mathbf{K}_0/\mathbf{Q}} F$ is irreducible over $\mathbf{Q}$. In the case (viii'') we have either

$$F = \text{const}(U + V + W),$$

or

$$F = \text{const}(U^2 + V^2 + W^2 - UV - UW - VW).$$

In the former case F is irreducible over $\mathbf{C}$, the ratios of its coefficients generate $\mathbf{K}_0$, hence the irreducibility of $N_{\mathbf{K}_0/\mathbf{Q}} F$ follows from Lemma

10 of [3]. In the latter case we have

$$F = \text{const } N_{K_0(\zeta_3)/K_0}(U + \zeta_3 V + \zeta_3^2 W)$$

hence

$$N_{K_0/Q}F = \text{const } N_{K_0(\zeta_3)/Q}(U + \zeta_3 V + \zeta_3^2 W)$$

and since $U + \zeta_3 V + \zeta_3^2 W$ is irreducible over $\mathbf{C}$ and the ratios of the coefficients generate $\mathbf{K}_0(\zeta_3)$ Lemma 10 of [3] applies again and gives the desired conclusion.

In the case (ix) let $Q = B_1 + B_2$, where B_1, B_2 are binomials, $B_0 = (B_1, B_2) \neq 1$. By Theorem 1 of [1] either QB_0^{-1} is a binomial B_0' or it is irreducible over $\mathbf{C}$. If $F|B_0$ or $Q = B_0 B_0'$ and $F|B_0'$ we have the desired conclusion. Otherwise

$$F = \text{const } QB_0^{-1}$$

and changing, if necessary, y_2 to y_2^{-1} we may use formula (15).

We get

$$QB_0^{-1} = \sum_{i=1}^{2} b_i \sum_{r=0}^{q_i-1} a^r y_1^{p_1(q_i-r-1)} y_2^{p_2(q_i-r-1)} \prod_{j=1}^{2} y_j^{\beta_{ij}}.$$

The ratios of the coefficients of QB_0^{-1} include b_2/b_1 and a (since $q_1 > 1$ or $q_2 > 1$), hence they generate $\mathbf{K}_0 = \mathbf{Q}(b_2/b_1, a^{q_1}, a^{q_2})$. Since QB_0^{-1} is irreducible over $\mathbf{Q}$ Lemma 10 of [3] gives the desired conclusion.

Lemma 7. If $KN_{K_0/Q}q(x)$ has in $\mathbf{Q}[x]$ an irreducible reciprocal factor then either (iii) holds or there is a permutation $\langle g, h, i, j \rangle$ of $\langle 0, 1, 2, 3 \rangle$

$$\frac{\max\{|n_i - n_g|, |n_j - n_h|\}}{(n_i - n_g, n_j - n_h)} \leq B_3(\underline{\alpha}). \qquad (21)$$

Proof. Let $n_0 = 0$, $(n_1, n_2, n_3) = d$, $n_i = dm_i$, $0 \leq i \leq 3$. Since $KN_{K/Q}(\sum_{j=0}^{3} \alpha_j x_j^{n_j})$ has an irreducible reciprocal factor, say $f(x)$, $KN_{K_0/Q}(\sum_{j=0}^{3} \alpha_j x^{m_j})$ also has one. Indeed, otherwise we should have

$$KN_{K_0/Q}\left(\sum_{j=0}^{3} \alpha_j x^{n_j}\right) \overset{\text{can}}{=\!=} \text{const} \prod_{\rho=1}^{r} F_\rho(x)^{e_\rho},$$

where
$$(F_\rho(x), JF_\rho(x^{-1})) = 1$$

and thus
$$KN_{K_0/Q}\left(\sum_{j=0}^{3} \alpha_j x^{n_j}\right) = \text{const} \prod_{\rho=1}^{r} F_\rho(x^d)^{e_\rho},$$

where
$$\left(F_\rho(x^d), JF_\rho(x^{-d})\right) = 1, \qquad 1 \le \rho \le r,$$

contrary to the fact that $f(x)|F_\rho(x^d)$ for at least one $\rho \le r$.

On the other hand
$$J\left(\sum_{j=0}^{3} \alpha_j x^{-m_j}\right) = k \sum_{j=0}^{3} \alpha_j^\sigma x^{m_j}$$

implies (iii) and
$$\frac{\max\{|m_i - m_g|, |m_j - m_h|\}}{(m_i - m_g, m_j - m_h)} \le B_3(\underline{\alpha})$$

implies (21). Therefore we may assume that
$$(n_1, n_2, n_3) = 1. \tag{22}$$

Let $\mathbf{n} = [n_1, n_2, n_3]$. By Lemma 1 of [5] there exist linearly independent vectors $\mathbf{p}, \mathbf{q} \in \mathbf{Z}^3$ and numbers $u, v \in \mathbf{Q}$ such that
$$\mathbf{n} = u\mathbf{p} + v\mathbf{q} \tag{23}$$

and
$$h(\mathbf{p})h(\mathbf{q}) \le c_0(3)n_3^{1/2}, \tag{24}$$

where $c_0(3)$ is a constant. Let
$$u = \frac{u_0}{w}, \quad v = \frac{v_0}{w}, \quad \text{where } u_0, v_0, w \in \mathbf{Z}, \ w > 0, \ (u_0, v_0, w) = 1. \tag{25}$$

By (22) we infer from $\mathbf{n} = u\mathbf{p} + v\mathbf{q}$ that
$$(u_0, v_0) = 1. \tag{26}$$

Put $\langle p_0, q_0 \rangle = \langle 0, 0 \rangle$,
$$G_0(y, z) = J\left(\sum_{j=0}^{3} \alpha_j y^{p_j} z^{q_j}\right), \qquad G = N_{K_0/Q} G_0;$$

$$H_0(y, z) = J\left(\sum_{j=0}^{3} \alpha_j y^{-p_j} z^{-q_j}\right), \qquad G = N_{K_0/Q} H_0; \tag{27}$$

$$D(y,z) = (G(y,z), H(y,z)).$$ (28)

By Lemma 5 either there exists an automorphism σ of $\mathbf{K}_0$ and a $c \in \mathbf{K}_0$ such that

$$H_0(y,z) = cG_0(y,z)$$ (29)

or

$$G_0(y,z) = B_1(y,z) + B_2(y,z), \quad B_0 = (B_1, B_2), \quad KB_0 \notin \mathbf{K}_0,$$ (30)

where B_1, B_2 are binomials, or finally

$$KD(y,z) \in \mathbf{K}_0.$$ (31)

We shall consider these cases successively. We have by (23), (25) and (26)

$$q(x^w) = JG_0(x^{u_0}, x^{v_0}),$$
$$Jq(x^{-w}) = JH_0(x^{u_0}, x^{v_0}),$$ (32)

hence (29) implies

$$Jq(x^{-w}) = cq(x^w),$$

which gives (iii). In the case (30) let

$$B_1(y,z) = \alpha_g y^{p_g} z^{q_g} + \alpha_i y^{p_i} z^{q_i},$$
$$B_2(y,z) = \alpha_h y^{p_h} z^{q_h} + \alpha_j y^{p_j} z^{q_j},$$

where $\langle g, h, i, j \rangle$ is a permutation of $\langle 0, 1, 2, 3 \rangle$. The condition $B_0 = (B_1, B_2) \notin \mathbf{K}_0$ implies that for a suitable $\varepsilon = \pm 1$

$$\frac{p_i - p_g}{(p_i - p_g, q_i - q_g)} = \varepsilon \frac{p_j - p_h}{(p_j - p_h, q_j - q_h)},$$
$$\frac{q_i - q_g}{(p_i - p_g, q_i - q_g)} = \varepsilon \frac{q_j - q_h}{(p_j - p_h, q_j - q_h)},$$
$$\left(\frac{-\alpha_i}{\alpha_g} \right)^{(p_j - p_h, q_j - q_h)} = \left(\frac{-\alpha_j}{\alpha_h} \right)^{\varepsilon(p_i - p_g, q_i - q_g)}$$ (33)

and the condition $KB \notin \mathbf{K}_0$ implies that α_i/α_g, α_j/α_h are not roots of unity. By the definition of $C(a,b)$ we have

$$\frac{\max\{(p_i - p_g, q_i - q_g), (p_j - p_h, q_j - q_h)\}}{(p_i - p_h, q_i - q_g, p_j - p_h, q_j - q_h)} = C\left(\frac{-\alpha_i}{\alpha_g}, \frac{-\alpha_j}{\alpha_h} \right).$$ (34)

On the other hand, by (23)

$$n_i - n_g = u(p_i - p_g) + v(q_i - q_g)$$
$$n_j - n_h = u(p_j - p_h) + v(q_j - q_h),$$

hence by (33) and (34)

$$\frac{\max\{(|n_i - n_g|, |n_j - n_h|)\}}{(n_i - n_g, n_j - n_h)} = \frac{\max\{(p_i - p_g, q_i - q_g), (p_j - p_h, q_j - q_h)\}}{(p_i - p_h, q_i - q_g, p_j - p_h, q_j - q_h)}$$

$$= C\left(\frac{-\alpha_i}{\alpha_g}, \frac{-\alpha_j}{\alpha_h}\right). \tag{35}$$

It remains to consider the case (31). Let f be an irreducible reciprocal factor of $KN_{K_0/Q}q(x)$. Then

$$f(x^w)\big| KN_{K_0/Q}q(x^w);$$

and in view of (32)

$$f(x^w)\big| (KG(x^{u_0}, x^{v_0}), KH(x^{u_0}, x^{v_0})).$$

Since $KD(x^{u_0}, x^{v_0}) \in K_0$ every zero ξ of $f(x^w)$ satisfies

$$\frac{G}{D}(\xi^{u_0}, \xi^{v_0}) = \frac{H}{D}(\xi^{u_0}, \xi^{v_0}) = 0.$$

By virtue of Lemma 4 of [2] the number of common zeros of G/D and H/D does not exceed the degree of their resultant with respect to y, thus it does not exceed

$$\deg_y G \deg_z H + \deg_y H \deg_z G \le 8[K_0 : Q]^2 h(p)h(q).$$

Hence by (24) the number of pairs $\langle \xi^{n_0}, \xi^{v_0}\rangle$, where ξ is a zero of $f(x^w)$ does not exceed $8[K_0 : Q]^2 c_0(3)n_3^{1/2}$. However since f is irreducible the zeros of $f(x^w)$ are all simple and by (27) to distinct zeros ξ correspond distinct pairs $\langle \xi^{u_0}, \xi^{v_0}\rangle$. Therefore

$$w \deg f \le 8c_0(3)[K_0 : Q]^2 n_3^{1/2}.$$

We may assume without loss of generality that f is primitive with the leading coefficient positive. If f is monic the assumptions of Lemma 3 of [5] are satisfied and thus we have

$$\max\{n_1, n_3 - n_2\} < C_1(\underline{\alpha})n_3^{1/2}(\log n_3)^3(\log \log en_3)^{-3} \tag{36}$$

and either

$$\max\{n_1, n_3 - n_2\} < C_2(\underline{\alpha})(n_1, n_3 - n_2) \tag{37}$$

or

$$\min\{n_2, n_3 - n_1\} < C_3(\underline{\alpha})n_3^{1/2}(\log n_3)^4(\log\log en_3)^{-3}. \tag{38}$$

However (36) and (38) together give

$$n_3 = \max\{n_1, n_3 - n_2\} + \min\{n_2, n_3 - n_1\}$$
$$\leq (C_1(\underline{\alpha}) + C_3(\underline{\alpha}))n_3^{1/2}(\log n_3)^4(\log\log en_3)^{-3}$$

and thus for a suitable $B_4(\underline{\alpha})$

$$n_3 < B_4(\underline{\alpha}). \tag{39}$$

If f_0 is not monic then by Lemma 4 of [5] we have either

$$n_3 < C_7(\underline{\alpha}) \tag{40}$$

or there exist four distinct nonnegative indices g, h, i, $j \leq 3$ such that

$$\frac{\max\{|n_i - n_g|, |n_j - n_h|\}}{(n_i - n_g, n_j - n_h)} \leq C_8(\underline{\alpha}). \tag{41}$$

The alternative (35), (39), (40) or (41) gives (21) with

$$B_3(\underline{\alpha}) = \max\left\{\max C\left(\frac{-\alpha_i}{\alpha_g}, \frac{-\alpha_j}{\alpha_h}\right), B_4(\underline{\alpha}), C_7(\underline{\alpha}), C_8(\underline{\alpha})\right\},$$

where the inner maximum is taken over all permutations $\langle g, h, i, j \rangle$ of $\langle 0, 1, 2, 3 \rangle$.

Remark. It is here that the theory of linear forms in logarithms intervenes via Lemma 3 of [5].

Lemma 8. If a quadrinomial is representable in one of the forms (1), where k, T, U, V, W are monomials in $K_0(x)$ then it is also representable in the same form, where $k \in K_0$, T, U, V, W are monomials in $K_0[x]$, T, U, V W are monomials in $K_0[x]$ and the factors on the right-hand-side of (1) differ from the original ones by monomial factors.

Proof is the same as the proof of the corresponding assertion for $K_0 = Q$ included in Lemma 18 of [3].

Proof of Theorem 2. The proof follows closely the proof of Theorem 4 in [3].

If $Kq(x) \in \mathbf{K}_0$ we have by Lemma 1

$$\gamma_1 n_1 + \gamma_2 n_2 + \gamma_3 n_3 = 0, \quad \gamma = [\gamma_1, \gamma_2, \gamma_3] \in \mathbf{Z}^3, \qquad (41)$$

where $h(\gamma) \leq 4$. If $Kq(x)$ is irreducible over $\mathbf{K}_0$ we have (iv).

Assume that $Kq(x)$ is reducible over $\mathbf{K}_0$. Then $KN_{\mathbf{K}_0/\mathbf{Q}}q(x)$ is reducible over $\mathbf{Q}$. By Lemma 5 of [5] if $KN_{\mathbf{K}_0/\mathbf{Q}}q(x)$ has at least two, counting multiplicity, irreducible non-reciprocal factors then there is a linear relation (41), where $h(\gamma) \leq C_9(\underline{\alpha})$. On the other hand, by Lemma 7 above, if $N_{\mathbf{K}_0/\mathbf{Q}}q(x)$ has at least one irreducible reciprocal factor we have (iii) or (21). Thus unless (iii) holds we have (41) with

$$h(\gamma) \leq \max\{C_9(\underline{\alpha}), B_3(\underline{\alpha})\} = B_5(\underline{\alpha}).$$

Integer vectors perpendicular to γ form a lattice, say Λ. We have $[\gamma_2, -\gamma_1, 0], [\gamma_3, 0, -\gamma_1], [0, \gamma_3, -\gamma_2] \in \Lambda$ and since $\gamma \neq 0$ two among these three vectors are linearly independent. By Lemma 6 of [2] Λ has a basis which written in the form of a matrix $\Delta = [\delta_{ij}]_{\substack{i \leq 2 \\ j \leq 3}}$ satisfies

$$h(\Delta) \leq 2h(\gamma) \leq 2B_5(\underline{\alpha}). \qquad (42)$$

Moreover

$$\text{rank}\,\Delta = 2 \qquad (43)$$

and since $\mathbf{n} \in \Lambda$

$$\mathbf{n} = [m_1, m_2]\Delta, \quad [m_1, m_2] \in \mathbf{Z}^2 \setminus \{0\}. \qquad (44)$$

Since $0 < n_1 < n_2 < n_3$ the vectors $[\delta_{1j}, \delta_{2j}]$, $j = 1, 2, 3$, are distinct and different from $\mathbf{0}$. Let us set

$$Q_0(y_1, y_2) = J\left(\alpha_0 + \sum_{j=1}^{3} \alpha_j y_1^{\delta_{1j}} y_2^{\delta_{2j}}\right). \qquad (45)$$

By (44) we have

$$q(x) = JQ_0(x^{m_1}, x^{m_2}). \qquad (46)$$

Since $Jq(x^{-1}) \neq cq^\sigma(x)$ for every automorphism σ of $\mathbf{K}_0$ and every $c \in \mathbf{K}_0$ we have also

$$JQ_0(y_1^{-1}, y_2^{-1}) \neq cQ^\sigma(y_1, y_2).$$

Hence by Lemma 5 we have either

$$\left(KN_{\mathbf{K}_0/\mathbf{Q}}Q_0(y_1, y_2), KN_{\mathbf{K}_0/\mathbf{Q}}Q_0(y_1^{-1}, y_2^{-1})\right) = 1 \qquad (47)$$

or

$$Q_0 = B_1 + B_2, B_0 = (B_1, B_2), KB_0 \notin \mathbf{K}_0 \qquad (48)$$

where B_1, B_2 are binomials and

$$\left(KN_{\mathbf{K}_0/\mathbf{Q}}Q_0 B_0^{-1}(y_1, y_2), KN_{\mathbf{K}_0/\mathbf{Q}}Q_0 B_0^{-1}(y_1^{-1}, y_2^{-1})\right) = 1 \qquad (49)$$

or finally

$$Q_0 B_0^{-1} \text{ is a binomial } B_0'. \qquad (50)$$

We shall deal first with the last case, which is the simplest. In this case we have by (46)

$$q(x) = JB_0(x^{m_1}, x^{m_2})JB_0'(x^{m_1}, x^{m_2}).$$

The factors on the right-hand-side are either binomials or constants and since the left-hand-side is a quadrinomial they are actually binomials. Moreover if $b(x) = JB_0(x^{m_1}, x^{m_2})$ we infer from $KB_0 \notin \mathbf{K}_0$ that $Kb(x) \notin \mathbf{K}_0$. Thus (vi) holds.

In the case (47) or (48)–(49) we apply Theorem 3 of [2] setting there

$$F(x_1, x_2) = mN_{\mathbf{K}_0/\mathbf{Q}}Q_0(x_1, x_2) \text{ or } mN_{\mathbf{K}_0/\mathbf{Q}}Q_0 B_0^{-1}(x_1, x_2), \qquad (51)$$

respectively, where m is the least positive integer such that $F \in \mathbf{Z}[x_1, x_2]$. The assumption of the said theorem is satisfied, since by (47) or (49)

$$LF(x_1, x_2) = KF(x_1, x_2)$$

and by virtue of it there exists an integral matrix $\mathbf{M} = [\mu_{ij}]_{\substack{i \leq r \\ j \leq 2}}$ of rank $r \leq 2$ and a vector $\mathbf{v} = [v_1, v_r] \in \mathbf{Z}^r$ such that

$$h(\mathbf{M}) \leq E(r, F),$$

$$[m_1, m_2] = \mathbf{vM} \qquad (53)$$

and

$$KF\left(\prod_{i=1}^{r} y_i^{\mu_{i1}}, \prod_{i=1}^{r} y_i^{\mu_{i2}}\right) \underset{\mathbf{Q}}{\overset{\text{can}}{=\!=\!=}} \text{const} \prod_{\sigma=1}^{s} F_\sigma(y_1, y_r)^{e_\sigma} \qquad (54)$$

implies

$$KF(x^{m_1}, x^{m_2}) \overset{can}{\underset{\mathbf{Q}}{=}} const \prod_{\sigma=1}^{s} KF_\sigma(x^{v_1}, x^{v_r})^{e_\sigma} \qquad (55)$$

In (52) $E(r, F)$ is a certain function of r, of the degree and the coefficients of F, explicitly given in [2]. Since by virtue of (42), (45), (48) and (51) F runs through a finite set depending on $\underline{\alpha}$ we get from (52)

$$h(M) \leq B_6(r, \underline{\alpha}). \qquad (56)$$

Let us set

$$\mathbf{N} = [\nu_{ij}]_{\substack{i \leq r \\ j \leq 3}} = \mathbf{M}\Delta. \qquad (57)$$

It follows from (43) that $\mathbf{N}$ is of rank r and from (44) and (53) that

$$\mathbf{n} = \mathbf{v}\mathbf{N}. \qquad (58)$$

Consider first the case $r = 2$ and put

$$Q(y_1, y_2) = JQ_0\left(\prod_{i=1}^{2} y_i^{\mu_{i1}}, \prod_{i=1}^{2} y_i^{\mu_{i2}}\right). \qquad (59)$$

By (45) and (58)

$$Q(y_1, y_2) = J\left(\alpha_0 + \sum_{j=1}^{3} \alpha_j y_1^{\nu_{1j}} y_2^{\nu_{2j}}\right). \qquad (60)$$

By (58) the vectors $[\nu_{1j}, \nu_{2j}]$ are distinct and different from $\mathbf{0}$; moreover

$$q(x) = JQ(x^{v_1}, x^{v_2}). \qquad (61)$$

Now by Theorem 1 of [1] we have the following possibilities:

Q is irreducible over $\mathbf{K}_0$. $\qquad (62)$

Q can be represented in one of the forms (1), where $k \in \mathbf{K}_0$ and $T, U, V\ W$ are monomials in $\mathbf{K}_0[y_1, y_2]$. The factors on the right-hand-side are irreducible over $\mathbf{K}_0$ unless $\zeta_3 \in \mathbf{K}_0$, when

$$\qquad (63)$$

$$U^2 + V^2 + W^2 - UV - UW - VW \overset{can}{\underset{\mathbf{K}_0}{=}} (U + \zeta_3 V + \zeta_3^2 W)(U + \zeta_3^2 V + \zeta_3 W).$$

Q is the sum of two binomials with the highest common divisor B being a binomial and QB^{-1}, irreducible over $\mathbf{K}_0$, not being one.

$$\qquad (64)$$

Q is the product of two binomials. $\qquad$ (65)

The case (65) is easily settled. From (61) we infer that

$$q(x) = \alpha_3 b_1(x) b_2(x),$$

where $b_i \in \mathbf{K}_0[x]$ are monic binomials.

If for at least one $i \leq 2$ we have $Kb_i(x) \notin \mathbf{K}_0$ the case (vi) holds.

If $Kb_i(x) \in \mathbf{K}_0$, $i = 1, 2$ we have

$$b_i(x) = x^{\beta_i} - \zeta_q^{a_i}.$$

Hence $\mathbf{K}_0 = \mathbf{Q}(\zeta_q^{a_1}, \zeta_q^{a_2}) = \mathbf{Q}(\zeta_q^{(a_1, a_2)})$ and taking the automorphism $\sigma : \zeta_q^{(a_1, a_2)} \to \zeta_q^{-(a_1, a_2)}$ of $\mathbf{K}_0$ we get

$$Jq(x^{-1}) = \zeta^{a_1 + a_2} q^\sigma(x),$$

contrary to the assumption.

It remains to consider the possibilities (62), (63) and (64). they are mutually exclusive since, in the cases (62) or (63), Q has no factor, irreducible or not, which would be a binomial. On the other hand, in view of (59), (48) implies (64), while (64) implies (48) if $KB \notin \mathbf{K}_0$ and (47) if $KB \in \mathbf{K}_0$ (in virtue of Lemma 5). Hence we have

$$m^{-1}F = \begin{cases} N_{\mathbf{K}_0/\mathbf{Q}} Q_0 B_0^{-1} & \text{if (64) holds with } KB \notin \mathbf{K}_0 \\ N_{\mathbf{K}/\mathbf{Q}} Q_0, & \text{otherwise.} \end{cases}$$

Let $b(x) = JB_0(x^{m_1}, x^{m_2}) = JB(x^{v_1}, x^{v_2})$. In view of (52) and (59) the implication (54) $\to$ (55) takes the form

$$KN_{\mathbf{K}_0/\mathbf{Q}} QB^{-1}(y_1, y_2) \overset{\text{can}}{\underset{\mathbf{Q}}{=\!=\!=}} \text{const} \prod_{\sigma=1}^{s} F_\sigma(y_1, y_2)^{e_\sigma} \qquad (66)$$

implies

$$KN_{\mathbf{K}_0/\mathbf{Q}} q(x) b(x)^{-1} \overset{\text{can}}{\underset{\mathbf{Q}}{=\!=\!=}} \prod_{\sigma=1}^{s} F_\sigma(x^{v_1}, x^{v_2})^{e_\sigma} \qquad (67)$$

if (64) holds with $KB \notin \mathbf{K}_0$ and

$$KN_{\mathbf{K}_0/\mathbf{Q}} Q(y_1, y_2) \overset{\text{can}}{\underset{\mathbf{Q}}{=\!=\!=}} \text{const} \prod_{\sigma=1}^{s} F_\sigma(y_1, y_2)^{e_\sigma} \qquad (68)$$

implies

$$KN_{\mathbf{K}_0/\mathbf{Q}}q(x)\overset{\mathrm{can}}{\underset{\mathbf{Q}}{=\!=\!=}}\prod_{\sigma=1}^{s}KF_\sigma(x^{v_1},x^{v_2})^{e_\sigma} \qquad (69)$$

otherwise. Now we apply Lemmata 5 and 6. By virtue of them if (64) holds $N_{\mathbf{K}_0/\mathbf{Q}}QB^{-1}$ is irreducible over $\mathbf{Q}$ and non-reciprocal. Hence on the right-hand-side of (66) we have exactly one irreducible factor with multiplicity 1 and if $KB \notin \mathbf{K}_0$ the same happens in (67). It follows that $KN_{\mathbf{K}_0/\mathbf{Q}}q(x)b(x)^{-1}$ is irreducible over $\mathbf{Q}$ and thus $Kq(x)b(x)^{-1}$ is irreducible over $\mathbf{K}_0$. If $Kb(x) \in \mathbf{K}_0$ we have the case (iv); if $Kb(x) \notin \mathbf{K}_0$ the case (vi). Similarly, if (62) holds, or $KB \in \mathbf{K}_0$ in (64), $KN_{\mathbf{K}_0/\mathbf{Q}}Q$ is irreducible over $\mathbf{Q}$ and non-reciprocal, hence on the right-hand-side of (68) we have exactly one irreducible factor with multiplicity 1 and the same happens in (69). It follows that $KN_{\mathbf{K}_0/\mathbf{Q}}q(x)$ is irreducible over $\mathbf{Q}$ and thus $Kq(x)$ is irreducible over $\mathbf{K}_0$, i.e. (iv) holds. Finally if (63) holds, by Lemmata 5 and 6 on the right-hand-side of (68) we have either two of three irreducible factors $N_{\mathbf{K}_0/\mathbf{Q}}\phi_i$ corresponding to the factors ϕ_i, $(i \leq i_0 = 2$ or 3) irreducible over $\mathbf{K}_0$ on the right-hand-side of (1). Whether $N_{\mathbf{K}_0/\mathbf{Q}}\phi_i$ are equal or distinct the implication (68) $\to$ (69) gives that $KN_{\mathbf{K}_0/\mathbf{Q}}\phi_i(x^{v_1},x^{v_2})$, $(i \leq i_0)$ is irreducible over $\mathbf{Q}$. Now by (61) to the representation of Q in any one of the forms (1) there corresponds a representation of $q(x)$ in the same form, where k, T, U, V, W are now monomials in $\mathbf{K}_0(x)$ and the factors on the right-hand-side of (1) are $\phi_i(x^{v_1},x^{v_2})$, $(i \leq i_0)$. By Lemma 8 there exists a representation of $q(x)$ in the form in question in which $k \in \mathbf{K}_0$ and T, U, V, W are monomials in $\mathbf{K}_0[x]$. Since by Lemma 8 the relevant factors $\psi_i(x)$ differ from $\phi_i(x^{v_1},x^{v_2})$ only by monomial factors we infer that $KN_{\mathbf{K}_0/\mathbf{Q}}\psi_i(x)$, $(i \leq i_0)$ is irreducible over $\mathbf{Q}$. It follows that $K\psi_i(x)$ is irreducible over $\mathbf{K}_0$ and (v) holds.

It remains to consider the case $r = 1$. Changing if necessary $\mathbf{v}$ to $-\mathbf{v}$ we can achieve that $v_1 > 0$ and by (42), (56) and (57)

$$0 < \nu_{11} < \nu_{12} < \nu_{13} \leq B_7(\underline{\alpha}). \qquad (70)$$

Let us put in Lemma 2

$$f(x) = \alpha_0 + \sum_{j=1}^{3}\alpha_j x^{\nu_{1j}}. \qquad (71)$$

By virtue of that lemma there exists a constant $c(f)$ and a positive integer $\nu \leq c(f)$ such that $\nu|v_1$ and

$$Kf(x^\nu)\overset{\mathrm{can}}{\underset{\mathbf{K}_0}{=\!=\!=}}\mathrm{const}\prod_{\sigma=1}^{s}f_\sigma(x)^{e_\sigma}$$

implies

$$Kf(x^{v_1})\overset{\text{can}}{\underset{\mathbf{K}_0}{=\!=\!=}} \text{const} \prod_{\sigma=1}^{s} f_\sigma(x^{v_1/\nu})^{e_\sigma}.$$

Since $q(x) = f(x^{v_1})$ we get the case (vii) putting

$$\nu_j = \nu\nu_{1j}, \qquad 1 \le j \le 3, \qquad v = v_1/\nu$$

and $B_2(\underline{\alpha}) = \max c(f)$, where the maximum is taken over all polynomials f of the form (71) with ν_{1j} satisfying (70).

References

[1] M. Fried and A. Schinzel, Reducibility of quadrinomials. *Acta. Arith.* **21** (1972), 153–171.

[2] A. Schinzel, Reducibility of lacunary polynomials I, *ibid* **16** (1969), 123–159.

[3] A. Schinzel, Reducibility of lacunary polynomials III, *ibid* **34** (1978), 227–266.

[4] A. Schinzel, Reducibility of lacunary polynomials VI, *ibid* **47** (1984), 89–105.

[5] A. Schinzel, Reducibility of lacunary polynomials VII, *Mhf Math.* **102** (1986), 309–337.

[6] A. Schinzel, *Selected topics on polynomials*, The University of Michigan Press, Ann Arbor 1982.

[7] A. Schinzel, Reducible lacunary polynomials, *Sém. Théorie des nombres*, Université de Bordeaux, 1983–1984, exposé no 29.

20

THE NUMBER OF SOLUTIONS OF THUE EQUATIONS
W. M. Schmidt

1. New bounds have recently been found for the number of solutions of Thue equations by Bombieri and Schmidt [3], by Schmidt [18], and by Mueller and Schmidt [16], [17]. We will give a survey of these results. By a Thue equation we shall understand a Diophantine equation

$$F(x,y) = h \qquad (1)$$

where $F(x,y) = a_0 x^r + a_1 x^{r-1}y + \ldots + a_r y^r$ is a form of degree $r \geq 3$ with coefficients in $\mathbf{Z}$, and irreducible over $\mathbf{Q}$, and where h is a non-zero integer. It was shown by Thue [22] in 1908 that such an equation has only finitely many solutions in rational integers x, y. Thue's method was ineffective in the sense that it provided no bounds for the size

$$|\underline{x}| = \max(|x|, |y|)$$

of possible solutions, but explicit bounds were first given by A. Baker [1] in 1967.

Here we shall be concerned with bounds for the *number* $N_F(h)$ of solutions, rather than with bounds for their size. Such bounds can be derived in principle from Thue's method. Explicit bounds were given e.g. by Lewis and Mahler [12]. These bounds depended on h, r and H, where H is the *height* of F, i.e. the maximum modulus of its coefficients. But Siegel [20] in the case $r = 3$, as well as in the binomial case $F = ax^r - by^r$, gave bounds independent of the coefficients of F, i.e. bounds depending on h, r only. This led to the question whether such bounds existed in general. This question was answered in the affirmative by Evertse [7] in 1983. Evertse's bounds were rather large; they grew more than exponentially in r.

Better bounds were discovered by Bombieri and Schmidt [3] in 1985. Let us initially discuss the case $h = 1$, i.e. equations

$$F(x,y) = 1. \qquad (2)$$

We have

Theorem A. *The number of solutions of (2) is $\ll r$.*

Here and throughout, the constants implicit in $\ll$ are absolute and are effectively computable. The form

$$F(x,y) = x^r + c(x - y)(2x - y)(3x - y)\ldots(rx - y)$$

is irreducible for suitable values of c, and with this form the equation (2) has solutions $(1,1)$, $(1,2)$,..., $(1,r)$, hence it has $\geq r$ solutions. Thus except for the value of the implicit constant, Theorem A is best possible. It had been suggested that a bound might exist which depends only on the number of *real* roots of $f(x) = F(x,1)$. But as was pointed out by Waldschmidt, the equation (2) with $F = x^{2n} + c(x-y)^2(2x-y)^2\ldots(nx-y)^2$ has the $r = 2n$ solutions $\pm(1,1)$, $\pm(1,2)$,..., $\pm(1,n)$, yet $f(x)$ has no real root when $c > 0$.

2. Write

$$f(x) = F(x,1) = a_0(x - \alpha_1)\ldots(x - \alpha_r) \tag{3}$$

with complex $\alpha_1,\ldots,\alpha_r$. The basic strategy for Theorem A consists of two steps.

Step 1. To show that when (2) holds with $y \neq 0$, then x/y is a "good" rational approximation to some root α_i.

Step 2. To show that the algebraic number α_i cannot have many good rational approximations.

In the proof of Theorem A we have to distinguish between *large* and *small* solutions, the large solutions being those with $|\underline{x}| \geq Y_L$, and the small ones the ones with $|\underline{x}| < Y_L$, where Y_L is a certain quantity defined in terms of F. The definition of Y_L is cumbersome, but essentially we have

$$Y_L \approx M^{1+\gamma_1 r^{-1/2}} e^{\gamma_1 r^{1/2}}, \tag{4}$$

where M is the Mahler height of F, and $\gamma_1 > 0$ is an absolute constant. (At times it is more convenient to work with the Mahler height rather than the height H introduced above.)

By a slight variation of an inequality of Lewis and Mahler [12] it is seen that when (2) holds with $y \neq 0$, there is an α_i with

$$\min\left(1, \left|\alpha_i - \frac{x}{y}\right|\right) \leq (2r^{1/2}M)^r|\underline{x}|^{-r}, \tag{5}$$

and this is taken as Step 1 for the large solutions. On the other hand, from the theory of Diophantine approximation, in particular from work of Thue [22], Siegel [19], Dyson [5], Bombieri [2], one may conclude that the number of reduced fractions x/y with (5) and with $|\underline{x}| \geq Y_L$ is $\ll 1$. Multiplying by the number r of roots α_i, we obtain the bound $\ll r$ for the number of large solutions.

In order to deal with small solutions we may suppose that F is *reduced* in the following sense. Write $F \sim F'$ if F' is obtained from F by a unimodular linear substitution of the variables. If (2) has a solution, then F is equivalent to a form F' with $F'(1,0) = 1$. Now F is reduced if $F(1,0) = 1$ and if F has minimum Mahler height among the forms $F' \sim F$ with $F'(1,0) = 1$. A further preparatory step allows us to suppose that

$$M \geq \gamma_2^r \tag{6}$$

with suitable $\gamma_2 > 1$.

With these suppositions, one can show that every solution of (2) with $y \neq 0$ has

$$\left| \alpha_i - \frac{x}{y} \right| \leq \frac{1}{2M^{1/(2r)}y^2} \tag{7}$$

for some α_i. Let us see how far we can get if we take this as Step 1 for small solutions. Let (x_0, y_0), $(x_1, y_1), \ldots, (x_\nu, y_\nu)$, be the small solutions with $y > 0$ and with (7) for some given α_i, ordered such that $y_0 \leq y_1 \leq \ldots \leq y_\nu \leq Y_L$. We have for $j = 1, \ldots, \nu$,

$$\frac{1}{y_{j-1}y_j} \leq \left| \alpha_i - \frac{x_{j-1}}{y_{j-1}} \right| + \left| \alpha_i - \frac{x_j}{y_j} \right| \leq \frac{1}{M^{1/(2r)}y_{j-1}^2},$$

from which we get the "gap principle"

$$y_j \geq M^{1/(2r)}y_{j-1}.$$

Therefore $M^{\nu/(2r)} \leq y_\nu \leq Y_L$, so that by (4) and (6), $\nu \ll r$. This could be taken as Step 2 for small solutions. Multiplying by the number r of roots α_i we get a bound $\ll r^2$ for the number of solutions of (2).

In order to improve this to $\ll r$, one has to modify Steps 1 and 2. It turns out that for each solution of (2), either (7) may be significantly sharpened for some α_i, or else (7) holds simultaneously for many roots α_i. Essentially we give an upper bound for

$$\prod_{i=1}^{r} \min\left(1, \left| \alpha_i - \frac{x}{y} \right| \right).$$

We cannot go into details here.

3. A solution of (1) is called *primitive* if x, y are coprime. Using p-adic arguments, Bombieri and Schmidt [3] generalize Theorem A to

Theorem B. *The number of primitive solutions of* (1) *is* $\ll r^{1+t}$, *where t is the number of distinct prime factors of* h.

In the cubic case this had essentially been established by Hooley [8]. In particular, it follows that for given F and $\epsilon > 0$, the number of solutions of (1) has

$$N_F(h) \le \gamma_3(F, \epsilon) h^\epsilon.$$

It has been conjectured that the right-hand-side here may be replaced by $\gamma_3(F) \log h$, or at least by $\gamma_3(F)(\log h)^{\gamma_4}$ with a suitable γ_4. As for lower bounds, Mahler [14], generalizing and much improving results of Chowla [4], showed that for cubic forms F we have $N_F(h) = \Omega_F\left((\log h)^{1/4}\right)$, and this was further strengthened by Silverman [21] to

$$N_F(h) = \Omega_F\left((\log h)^{1/3}\right). \tag{8}$$

If (as is generally believed) there are elliptic curves $F(x, y) = h$ whose Mordell-Weil group over $\mathbf{Q}$ has arbitrarily high rank, then given $\epsilon > 0$ there are cubic forms $F = F_\epsilon$ with

$$N_F(h) = \Omega_{F,\epsilon}\left((\log h)^{1-\epsilon}\right).$$

As long as we use only the archimedian absolute value, Diophantine approximation methods deal equally well with the "Thue inequality".

$$|F(x, y)| \le h \tag{9}$$

as with the Thue equation (1).

Theorem C. (Schmidt [18]). *The number* $Z_F(h)$ *of solutions of* (9) *satisfies*

$$Z_F(h) \ll r h^{2/r} (1 + \log h^{1/r}). \tag{10}$$

It is easily seen that the factor $h^{2/r}$ on the right-hand-side here is necessary, but the logarithmic term probably is not. The estimate (10) is deduced from the fact that for a suitably reduced form F with Mahler height $M \ge \gamma_5^r h$, where $\gamma_5 > 1$ is an absolute constant, the number $P_F(h)$ of primitive solutions of (9) has

$$P_F(h) \ll r(1 + \log h^{1/r}).$$

This last fact is deduced from a variation on the proof of Theorem A.

4. Siegel [20] in his fundamental work on Diophantine equations $f(x,y) = 0$ makes the conjecture that when the curve defined by this equation is irreducible and of positive genus, then a bound for the number of integer solutions may be found which depends only on the "number of coefficients". If interpreted in a naïve way this is false already for cubic Thue equations, e.g. by (8), and since such an equation has at most 5 non-zero coefficients. However, it is conceivable that the number of primitive solutions of a cubic Thue equation remains under some absolute bound. It is further conceivable that Siegel's conjecture is true for curves of genus > 1.

Siegel then shows that the number of solutions of a cubic Thue equation may be bounded in terms of h only. (This is now contained in Evertse's work [7], as well as in Theorem B or C.) More generally, let F be a form as above of degree $r \geq 3$, having not more than $s + 1$ non-zero coefficients, so that

$$F(x,y) = \sum_{i=0}^{s} b_i x^{r_i} y^{r-r_i}. \tag{11}$$

Theorem D.
$$Z_F(h) \ll (rs)^{1/2} h^{2/r} (1 + \log h^{1/r}).$$

The class of forms of the type (11) is not closed under unimodular substitutions. It is therefore convenient to deal with the class $C(t)$ of forms for which $u(\partial F/\partial x) + v(\partial F/\partial y)$ with any real $u, v \neq 0, 0$ is divisible by at most t non-proportional linear forms with real coefficients. For forms in $C(t)$, we get $Z_F(h) \ll (rt)^{1/2} h^{2/r} (1 + \log h^{1/r})$. On the other hand the forms (11) lie in $C(4s - 2)$.

Roots α_i of $f(x) = F(x,1)$ with large imaginary parts do not admit good rational approximations. Crucial for Theorem D is the following auxiliary result on the number of roots with small imaginary parts.

Suppose F is a polynomial of degree r with integer coefficients and Mahler height $M > e^{2r}$. Suppose that $f(z)f'(z)$ has not more than q real roots where $q \geq 1$. Then given ϕ in $1700 r^{-1} (\log r)^3 \leq \phi \leq 1$, the number of roots $\alpha = x + iy$ with imaginary parts y in $|y| \leq M^{-\phi}$ is

$$\ll (rq/\phi)^{1/2}.$$

5. The modified version of Siegel's conjecture for $s = 1$, i.e. for binomial forms $F = ax^r - by^r$, had been proved by Hyyrö [9], Evertse [6], Mueller [15], and the case $s = 2$ had been shown by Mueller and Schmidt [16]. In these cases, the number of solutions of (1) or of (9) may be bounded in terms of h. The general case was recently settled by the same authors [17].

Theorem E. *The number of solutions of (9) is*

$$\ll s^2 h^{2/r}(1 + \log h^{1/r}). \tag{12}$$

Theorem C suggests that the factor s^2 in (12) may be replaced by s. But such an estimate may still not be optimal, since e.g. the factor γ_7 in (13) below is $\ll 1$ when $r \geq s \log s$, and also the right-hand-side of (16) is $\ll 1$ in this case. The unpleasant logarithmic term in (12) may be removed when $r \geq 4s$:

Theorem F. *Suppose that $r \geq 4s$. Then the number of solutions of (9) is*

$$\leq \gamma_6(r, s) h^{2/r}.$$

More precisely, suppose that $r \geq 2(1 + \epsilon^{-1})s$ where $0 < \epsilon \leq 1$. Then the number of solutions of (9) is

$$\ll \gamma_7 h^{2/r} + (\gamma_8/\epsilon) h^{(1+\epsilon)/r} \tag{13}$$

with

$$\gamma_7 = \gamma_7(r, s) = (rs^2)^{2s/r},$$
$$\gamma_8 = \gamma_8(r, s) = s^2 e^{(3300 s \log^3 r)/r}. \tag{14}$$

Here $\gamma_7 \ll 1$, $\gamma_8 \ll s^2$ when $r \geq s \log^3 s$. Thus when $r \geq \max(4s, s \log^3 s)$, the number of solutions of (9) is

$$\ll s^2 h^{2/r}.$$

Theorem E is an immediate consequence of this fact and of Theorem C.

In the proof of Theorem F, unimodular substitutions and reductions of forms no longer come into play. We distinguish large, medium and small solutions. With $|\underline{x}| = \max(|x|, |y|)$, $\langle \underline{x} \rangle = \min(|x|, |y|)$, a solution is called large when $|\underline{x}| \geq Y_L$, it is called small if $\langle \underline{x} \rangle \leq Y_S$, and it is called medium otherwise, i.e. when $|\underline{x}| < Y_L$, $\langle \underline{x} \rangle > Y_S$. Here Y_S, Y_L are defined in terms of F and h. Again, Y_L has (4), and $Y_S = y_0^{1/(r-2s)}$ with

$$Y_0 = (e^6 s)^r e^{1600 \log^3 r} h.$$

6. Consider the domain $\mathcal{D}$ of $(\xi, \eta) \in \mathbf{R}^2$ with

$$|F(\xi, \eta)| \leq h. \tag{15}$$

The large and medium solutions lie in the long "spidery legs" of $\mathcal{D}$, and the basic idea to deal with them is as in the outline given in §2. But we must avoid having to multiply our estimates by the number r of roots α_i of $f(x) = F(x, 1)$. Therefore after Step 1 we have to insert the additional

Step 1a. To show that in Step 1 we may restrict ourselves to just a few of the roots α_i. More precisely, we show that there is a set S of cardinality $\ll s$, consisting of roots of f, such that when $\underline{x}$ is a solution of (9) with $y \neq 0$, then x/y is a good rational approximation to some root α_i lying in S.

This step is accomplished via the following lemma.

Suppose f is a polynomial of degree r with real coefficients and having $\leq s+1$ non-zero coefficients. Then there is a set S of roots α_i of f, of cardinality $|S| \leq 6s + 4$, such that for every real ξ,

$$\min_{\alpha_i \in S} |\xi - \alpha_i| \leq R \min_{1 \leq i \leq r} |\xi - \alpha_i|,$$

where

$$R = e^{800 \log^3 r}.$$

This introduces the factor R into our approximation estimates, which however turns out to be fairly harmless. Instead of the *ad hoc* methods we use in this context, one could also use a theorem of Khovansky [11] on the (fairly uniform) distribution of the arguments of the complex roots of polynomials with $\leq s+1$ non-zero coefficients. But reference to this work would result in rather larger constants.

The medium solutions are the most difficult to deal with. In a slight variation of Step 1 we show that for any such solution, either a root α_i of $f(x) = F(x, 1)$ has

$$\left| \alpha_i - \frac{x}{y} \right| \leq \gamma_9(r, s) \left(\frac{h}{H^{1/2} |y|^r} \right)^{1/s},$$

or a root α_i^{-1} of $\hat{f}(y) = F(1, y) = y^r f(1/y)$ has

$$\left| \alpha_i^{-1} - \frac{y}{x} \right| \leq \gamma_9(r, s) \left(\frac{h}{H^{1/2} |x|^r} \right)^{1/s}.$$

For this we need the fact, derived with the aid of the Newton polygon, that the roots of f are contained in not more than s rather narrow annuli centred at 0 in the complex plane.

The small solutions lie in the rather "fat" part of $\mathcal{D}$ with $\langle \underline{x} \rangle \leq Y_s$; here the number of solutions is roughly equal to the area.

7. In fact, let $A_F(h)$ be the area of the domain $\mathcal{D}$ given by (15). It had been shown by Mahler [13] that $A_F(h)$ is finite; clearly

$$A_F(h) = A_F \cdot h^{2/r}$$

where $A_F = A_F(1)$. Moreover, according to Mahler, for a fixed form F we have

$$Z_F(h) = A_F h^{2/r} + O(h^{1/(r-1)})$$

as $h \to \infty$. The constant implicit in O depends on F in a way not specified by Mahler. We have by [17]:

Theorem G. *Suppose that $r \geq 4s$. Then*

$$A_F \ll (rs^2)^{2s/r}, \tag{16}$$

so that in particular $A_F \ll 1$ when $r \geq s \log s$. Moreover,

$$|Z_F(h) - A_F h^{2/r}| \ll \gamma_{10}(r, s)(h^{1/(r-2s)} + h^{1/r} \log h^{1/r})$$

with

$$\gamma_{10}(r, s) = \min \left(e^{3400 \log^3 r}, \gamma_{11}(s) \right),$$

where $\gamma_{11}(s)$ depends on s only.

In particular, when $r \geq 2(1 + \epsilon^{-1})s$ with $0 < \epsilon \leq 1$, then

$$|Z_F(h) - A_F h^{2/r}| \ll \epsilon^{-1} \gamma_{11}(s) h^{(1+\epsilon)/r}.$$

The quantity $\gamma_{11}(s)$ depends on work of Khovansky [10] and is rather large.

Thus in the case when $r > 4s$, we have an asymptotic formula

$$Z_F(h) \sim A_F \cdot h^{2/r}$$

with an error term estimate independent of the height of F. It remains an interesting problem to obtain such a formula also when $r \leq 4s$.

References

[1] A. Baker, Contributions to the theory of Diophantine equations, I. On the representation of integers by binary forms, *Philos. Trans. Roy. Soc.*, Ser. A **263** (1967/68), 173–191.

[2] E. Bombieri, On the Thue-Siegel-Dyson theorem. *Acta. Math.*, **148** (1982), 255–296.

[3] E. Bombieri and W. M. Schmidt, On Thue's equation. *Inv. Math.*, **88** (1987), 69–81.

[4] S. Chowla, Contributions to the analytic theory of numbers II. *J. Indian Math. Soc.*, **20** (1933), 120–128.

[5] F. Dyson, The approximation to algebraic numbers by rationals. *Acta. Math.*, **79** (1947), 225–240.

[6] J. H. Evertse, On the equation $ax^n - by^n = c$, *Compositio Math.*, **47** (1982).

[7] _____. Upper bounds for the number of solutions of Diophantine equations, *Math. Centrum*, Amsterdam (1983), 1–127.

[8] C. Hooley, On the representation of numbers by binary cubic forms, *Glasgow Math. J.*, **27** (1985), 95–98.

[9] S. Hyyrö. Über die Gleichung $ax^n - by^n = c$ und das Catalansche Problem, *Ann. Ac. Scient. Fenn.* Ser. A1, No. **355** (1964), 50 pp.

[10] A. G. Khovansky, On a class of transcendental equations, *Soviet Math. Dokl.*, **22** (1980), 762–765.

[11] _____. Sur les racines complexes des systèmes d'équations algébriques comportant peu de termes, *C. R. Acad. Sc. Paris*, **292** (1981), 937–940.

[12] D. J. Lewis and K. Mahler, Representation of integers by binary forms, *Acta. Arith.*, **6** (1961), 333–363.

[13] K. Mahler, Zur Approximation algebraischer Zahlen III., *Acta. Math.*, **62** (1934), 91–166.

[14] _____. On the lattice points on curves of genus 1, *Proc. London Math. Soc.*, (2) **39** (1935), 431–466.

[15] J. Mueller, Counting solutions of $|ax^r - by^r| \leq h$. *Quarterly J. Oxford*, (to appear).

[16] J. Mueller and W. M. Schmidt. Trinomial Thue equations and inequalities, *J. f. Math.*, (to appear).

[17] _____. Thue equations and a conjecture of Siegel, *Acta. Math.*, (to appear).

[18] W. M. Schmidt, Thue equations with few coefficients, *Trans. A.M.S.*, (to appear).

[19] C. L. Siegel, Approximation algebraischer Zahlen, *Math. Zeitschr.*, **10** (1921), 173–213.

[20] _____. Über einige Anwendungen Diophantischer Approximationen, *Abh. Preuss. Akad. Wiss. Phys.-math. Kl.*, (1929), Nr. 1.

[21] J. Silverman, Integer points on curves of genus 1. *J. London Math. Soc..*, (2) **28** (1983), 1–7.

[22] A. Thue, Über Annäherungswerte algebraischer Zahlen, *J. reine ang. Math.*, **135** (1909), 284–305.

21

ON ARITHMETIC PROPERTIES
OF THE VALUES OF *E*-FUNCTIONS

A. B. Shidlovsky

In what follows, let Q, C and A be the fields of the rationals, complex numbers, and all algebraic numbers respectively, let K be an algebraic field of finite degree over Q, and let I be an imaginary quadratic field.

The entire function

$$f(z) = \sum_{n=0}^{\infty} \frac{c_n}{n!} z^n$$

is called an *E*-function if:

1) $c_n \in K$, $n = 0, 1, 2, \ldots$;
2) for any $\epsilon > 0$ the absolute values of the coefficients c_n and their conjugates grow as $O(n^{\epsilon n})$, $n \to \infty$;
3) there exists a sequence of natural numbers $\{q_n\}$ such that for each $n = 1, 2, \ldots$, all numbers $q_n c_k$, $k = 0, 1, \ldots n$ are integers of K and for any $\epsilon > 0$, $q_n = O(n^{\epsilon n})$, $n \to \infty$.

The concept of *E*-function was introduced by C. Siegel in 1929 in [1], in which he developed a new method of proof of transcendence and algebraic independence of the values of *E*-functions in algebraic points, generalizing the classical method of Hermite-Lindemann. Siegel applied his method to Bessel functions and some related functions. In the monograph [2] this method was enunciated as a general theorem.

In the middle of the 1950s (see [3], [4]), Siegel's method was substantially generalized. In the last 30 years, in the papers of various mathematicians, general theorems on algebraic independence and on estimates of the algebraic independence measure of the values of *E*-functions were proved, and these were applied to many concrete functions (see [3–8]).

We shall state some general results on the transcendence an algebraic independence of the values of *E*-functions and note some unsolved problems.

Consider the set of E-functions

$$f_1(z),\ldots,f_m(z), \tag{1}$$

comprising the solution of the system of linear differential equations

$$y'_k = Q_{k,0} + \sum_{i=1}^{m} Q_{k,i} y_i, \qquad k = 1,\ldots,m, \quad Q_{k,i} \in \mathsf{C}(z). \tag{2}$$

In particular, they may be the functions $f(z)$, $f'(z)$, $\ldots$, $f^{(m-1)}(z)$, where $f(z)$ is the E-function satisfying a linear differential equation of order m with coefficients in $\mathsf{C}[z]$.

Let $T(z) \in \mathsf{C}[z]$ be the common denominator of all functions $Q_{k,i}$ in the system (2). In what follows, K will denote a field to which all the coefficients of the power series in z of all E-functions (1) belong, as well as the number $\xi \in \mathsf{A}$ considered below. In a number of cases, the field K will coincide with I.

1. In [3] it is proved that if, for $m \geq 1$ the functions (1) are algebraically independent over $\mathsf{C}(z)$, and $\xi \in \mathsf{A}$, $\xi T(\xi) \neq 0$, then the numbers

$$f_1(\xi),\ldots,f_m(\xi) \tag{3}$$

are algebraically independent.

2. Let $m \geq 2$, the degree of transcendence of the set of functions (1) over $\mathsf{C}(z)$ be ℓ, $0 \leq \ell \leq m$, and let $\xi \in \mathsf{A}$, $\xi T(\xi) \neq 0$. In [4] it is shown that in this case the degree of transcendence of the numbers (3) is also equal to ℓ.

3. If ℓ satisfies the condition $1 \leq \ell < m$, and the functions

$$f_1(z),\ldots,f_\ell(z) \tag{4}$$

are algebraically independent over $\mathsf{C}(z)$, then for all but finitely many $\xi \in \mathsf{A}$, the numbers

$$f_1(z),\ldots,f_\ell(\xi) \tag{5}$$

are algebraically independent (see [4]).

4. Any transcendental E-function, satisfying a linear differential equation with polynomial coefficients takes a transcendental value at all except a finite number of algebraic points (see [4]).

The number of exceptional algebraic points in which the numbers (5) are algebraically dependent (or $f(\xi) \in \mathsf{A}$) includes zero and, perhaps, the zeros of the polynomial $T(z)$, besides additional points. The additional points depend on the structure of the algebraic equations which connect the functions (1) over $C[z]$, or on the properties of the algebraic field K.

For the given set of E-functions (1), the degree of transcendence of which over $C(z)$ equals ℓ, $1 \leq \ell < m$, let $\Lambda \subset A$ denote the set of all points $\xi \in A$ for which the numbers (5) are algebraically dependent, and let $\Lambda^* \subset A$ be a finite set such that $\Lambda \subset \Lambda^*$.

The problem naturally arises of determining the set Λ, or at least some set Λ^*, for a set of functions (1) from certain subclasses of E-functions. In a number of cases problems of this type have been solved. We give some of these results. Note that in Section 1 a case when Λ is determined is considered.

5. If the degree of transcendence of the set of functions (1) over $C(z)$ and over C are equal, then the set Λ consists only of zero and of the zeros of $T(z)$ (see [4]).

6. If the degree of transcendence over $C(z)$ of the E-functions (1) is ℓ, where $1 \leq \ell \leq m$, and the functions (4) are algebraically independent over $C(z)$, then they are connected with the functions $f_{\ell+1}(z), \ldots, f_m(z)$ by equations

$$P_\nu\big(f_1(z), \ldots, f_\ell(z), f_\nu(z)\big) = 0, \qquad \nu = \ell + 1, \ldots, m,$$

where $P_\nu = P_\nu(y_1, \ldots, y_\ell, y_\nu)$ is an irreducible primitive polynomial over $C[z][y_1, \ldots, y_\ell, y_\nu]$, the terms of which are ordered lexicographically by the powers of y_ν, $\ldots$, y_1, and where $\deg P_\nu = k_\nu$.

In [4] it was proved that if the highest terms of the polynomials P_ν are of the form $A_\nu(z)f_\nu^{k_\nu}(z)$, where $A_\nu(z) \in C[z]$, and $\xi \in A$, $\xi T(\xi) \prod_{\nu=\ell+1}^m A_\nu(\xi) \neq 0$, then the numbers (5) are algebraically independent. A set Λ^* is thus determined in this case.

In the case where $K = I$, a set Λ^* of another type will be given below, and the set Λ for $\ell = m - 1$ is determined.

In the general case of an arbitrary field K, there is so far no success in describing the set Λ, or even a Λ^*. There is only the following result.

7. For the functions (1) a finite set of algebraic equations has been determined connecting these functions over $C[z]$, which are called the minimal equations for these functions over $C(z)$. These equations represent a basis of the ideal in the ring of polynomials $C[z][y_1, \ldots, y_m]$, consisting of all polynomials $P(y_1 \ldots, y_m)$ such that $P\big(f_1(z), \ldots, f_m(z)\big) = 0$ (see [7]).

Let the left-hand parts of the minimal equations be arranged in lexicographic order according to the powers of $f_m(z), \ldots, f_1(z)$, and let M denote the set of zeros of the highest coefficients of all the minimal equations. If $[K : Q] = h$, then let $\xi_1, \ldots, \xi_h$ denote the conjugates of ξ over K, and $f_{1,i}(z), \ldots, f_{m,i}(z)$ be E-functions obtained from (1) when all the coefficients of their power series in powers of z are replaced by the numbers conjugate to them in the field K_i conjugate to K, $i = 1, \ldots, h$.

It has been proved that if the degree of transcendence of the functions (1) over $C(z)$ is ℓ, $1 \le \ell < m$, the functions (4) are algebraically independent over $C(z)$, and $\xi \in A$, $\xi \notin M$, $\xi T(\xi) \ne 0$, then there exists i, $1 \le i \le h$, such that the numbers $f_{1,i}(\xi_i), \ldots, f_{\ell,i}(\xi_i)$ are algebraically independent.

8. If in Section 7 we additionally suppose that the minimal equations have the property that no one of the functions (4) is contained in all their highest terms, then the numbers (5) are algebraically independent. A set Λ^* is thus determined in the case considered.

In the case where $K = I$, the properties of I allow one to obtain better results.

9. Let $K = I$, $1 \le \ell < m$, suppose that the functions (4) are algebraically independent over $C(z)$, let M be the set of zeros of all the highest terms of the minimal equations of the functions (1) over $C(z)$, and let $\xi \in I$, $\xi \notin M$, $\xi T(\xi) \ne 0$. Then the numbers (5) are algebraically independent. Consequently, a set Λ^* is determined in this case.

10. Let $K = I$, $\ell = m - 1$, and let the functions (4) be algebraically independent over over $C(z)$. Then the functions (1) are connected by only one equation $P = P\big(f_1(z), \ldots, f_m(z)\big) = 0$, where P is an irreducible primitive polynomial over $C[z][f_1(z), \ldots, f_m(z)]$ containing $f_m(z)$. Denote by H the set of the common zeros of all the coefficients of the polynomial P of terms containing $f_m(z)$.

It has been proved that for $\xi \in I$, the numbers (5) are algebraically independent if and only if $\xi \notin H$ and $\xi T(\xi) \ne 0$. This means that, in the case considered, Λ is determined. Analogous assertions for any ℓ, $1 \le \ell < m$, about the determination of Λ have as yet not been proved.

For $K = I$ the following assertion has been proved in the general case.

11. Let $K = I$, $1 \le \ell < m$, suppose that the functions (4) be algebraically independent over $C(z)$, and let $\xi \in I$, $\xi T(\xi) \ne 0$. Then if the numbers (5) are algebraically independent, there exists an irreducible primitive polynomial $P(y_1, \ldots, y_m) \in C[z][y_1, \ldots, y_m]$ in which all the

coefficients of terms containing at least one of the variables $y_{\ell+1}, \ldots, y_m$ have a zero at $z = \xi$ and such that $P(f_1(z), \ldots, f_m(z)) = 0$.

Quantitative analogues of the above results with bounds on the degrees of the algebraic independence of the values of the E-functions considered are contained in [6–8].

References

[1] C. L. Siegel, Über einige Anwendungen Diophantischer Approximationen, *Abh. Preuss. Acad. Wiss.*, 1929–1930, No. 1, 1–70.

[2] C. L. Siegel, *Transcendental Numbers*, Princeton University Press, 1949.

[3] A. B. Shidlovsky, On the criterion for algebraic independence of the values of entire functions of a certain class, *Dokl. Akad. Nauk SSSR*, 1955, **100**, No. 2, 221–224, *Izv. Akad. Nauk SSSR Ser. Mat.* 1959, **23**, 35–66.

[4] A. B. Shidlovsky, Towards a general theorem on the algebraic independence of the values of E-functions, *Dokl. Akad. Nauk SSSR*, 1966, **171**, No. 4, 810–813.

[5] A. B. Shidlovsky, On the arithmetic properties of the values of analytic functions, *Trudy Mat. Inst. V. A. Steklov Akad. Nauk SSSR*, 1973, **132**, 169–202.

[6] A. B. Shidlovsky, On bounds for the transcendence measures of the values of E-functions, *Mat. Zametki* 1967, **2**, No. 1, 33–44.

[7] A. B. Shidlovsky, On bounds for polynomials in the values of E-functions, *Mat. Sbornik*, 1981, **115** (157), Issue 1 (5), 3–39.

[8] A. B. Shidlovsky, On the arithmetic properties of polynomials from the values of E-functions connected with algebraic equations in the field of rational functions, *Acta. Arithmetica*, 1980, **37**, 405–426.

22

SOME EXPONENTIAL DIOPHANTINE EQUATIONS

T. N. Shorey

§1. We consider the equation

$$(m+1)\ldots(m+k) = y^l \quad \text{in integers } m \geq 1,\ y \geq 1,\ k \geq 2,\ l \geq 2. \quad (1)$$

Erdős [6] and Rigge [19], independently, proved in 1939 that equation (1) with $l = 2$ has no solution. For $l \geq 3$, Erdős [7] and Rigge (unpublished), independently, showed that equation (1) has no solution in integers $m \geq 1$, $y \geq 1$ and $k \geq k_0(l)$ where $k_0(l)$ is a number depending only on l. The proof depends on a theorem of Thue [29]. Erdős and Siegel (unpublished) showed that equation (1) has no solution in integers $m \geq 1$, $y \geq 1$, $l \geq 3$ and $k \geq k_0$ where k_0 is an absolute constant. Erdős [8] gave an elementary proof. Finally, Erdős and Selfridge [9] developed this elementary proof to confirm an old conjecture that equation (1) has no solution. The first contribution in the direction of this conjecture dates back to 1724 when Goldbach, in a letter to D. Bernoulli, showed that equation (1) with $k = 3$ and $l = 2$ has no solution. For subsequent contributions before 1939, we refer to Dickson [5] and Obláth [18].

For a positive integer $\nu > 1$, we define $P(\nu)$ to be the greatest prime factor of ν and we write $P(1) = 1$. Let $b, d, m, y, k \geq 2$ and $l \geq 2$ be positive integers such that $P(b) \leq k$ and $(m, d) = 1$. Let $d_1, \ldots, d_t$ be distinct positive integers not exceeding k. We put

$$\nu_l = \frac{1}{2}\left(1 + \frac{4l^2 - 8l + 7}{2(l-1)(2l^2 - 5l + 4)}\right), \quad l \geq 3.$$

Observe that

$$\nu_3 = \frac{47}{56}, \quad \nu_4 = \frac{45}{64} \quad \text{and} \quad \nu_l \leq \frac{299}{464} < \frac{2}{3}, l \geq 5. \quad (2)$$

We consider the equation

$$(m+d_1)\ldots(m+d_t) = by^l. \quad (3)$$

Putting $t = k$ and $b = 1$ in (3), we obtain (1). If $m \leq k^l$, we see from (3) that

$$P(m + d_i) \leq k, \quad 1 \leq i \leq t,$$

which implies that

$$t \leq (k \log k)(\log m)^{-1} + \pi(k). \tag{4}$$

See Erdős and Turk [10], Lemma 2.1. If $m > k^l$, the author [22], [25] proved the following result.

Theorem 1. (i). *Equation (3) with*

$$l \geq 3, \quad m > k^l, \quad t \geq \nu_l k$$

implies that k is bounded by an effectively computable absolute constant.
(ii). *Let $\epsilon > 0$. Equation (3) with*

$$l = 2, \quad m > k^2, \quad t \geq k - (1 - \epsilon)k \frac{\log \log k}{\log k}$$

implies that k is bounded by an effectively computable number depending only on ϵ.
(iii). *Equation (3) with*

$$m > k^l, t \geq kl^{-1/11} + \pi(k) + 2$$

implies that the minimum of k and l is bounded by an effectively computable absolute constant.

For a given k, we see from [27], p. 64 and [2] that equation (3) has only finitely many solutions in all the variables under necessary restrictions. For the proof of Theorem 1, we borrow several elementary arguments of Erdős. The proof of Theorem 1(ii) utilizes a theorem of Baker [2] on the integer solutions of a hyperelliptic equation. The proof of Theorem 1(iii) depends on the theory of linear forms in logarithms. Apart from Baker's estimate [4], better lower bounds for linear forms in logarithms with α_i's very close to one are required. See Lemma 2 of [22]. In view of Theorem 1 (iii), it remains to prove Theorem 1(i) for only finitely many l. For these finitely many l, we apply the method of Roth [20] as elaborated in Halberstam and Roth [12] on difference between consecutive ν-free integers and the effective irrationality measures of Baker [1] obtained by hypergeometric method. Further, we combine (2) and (4) to derive the following result from Theorem 1(i).

Corollary 1. *Let* $\epsilon > 0$. *Equation (3) with*

$$l \geq 3, \quad m > k^{(56/47)+\epsilon}, \quad t \geq (47/56)k$$

implies that k is bounded by an effectively computable number depending only on ϵ.

Next, we consider another generalization of equation (1), namely,

$$(m + d) \ldots (m + kd) = by^l, \qquad (m, d) = 1. \tag{5}$$

Erdős conjectured that equation (5) with $b = 1$ implies that k is bounded by an absolute constant. Marszalek [16] followed the elementary method of Erdős [8] to prove that equation (5) with $b = 1$ implies that $k \leq k_0(d)$ where $k_0(d)$ is an explicitly given number depending only on d. We shall prove the following result in §3.

Theorem 2. *Equation (5) with $l \geq 3$ implies that k is bounded by an effectively computable absolute constant* * *if at least one of the following conditions is satisfied:*

(i) $m + d > k$ *and* $\left(l, \prod_{p|d}(p - 1)\right) = 1$ *where the product is taken over all the prime divisors of d.*

(ii) $m + d > k$ *and* $P(d)$ *is fixed.*

(iii) $m + d \leq k$ *and* $b = 1$.

The proofs of Theorem 2(i) and Theorem 2(iii) are elementary. An empty product is considered to be equal to one and thus Theorem 2(i) includes the result of Erdős [8] stated above. Furthermore, Theorem 2(i) is applied in the proof of Theorem 2(ii). The proof of Theorem 2(ii) depends on the following estimate of Győry [11] on the magnitude of integer solutions on Thue-Mahler equation.

Theorem A. *Let $P \geq 2$ and denote by S the set of all positive integers composed of primes not exceeding P. Let $f(X, Z)$ be a binary form with integer coefficients such that $f(X, 1)$ has at least three distinct roots. For integers $x, z, u \neq 0$ and $s \in S$ with $(x, z) = 1$, the equation*

$$f(x, z) = us$$

implies that

$$\max(|x|, |z|, s) \leq (|u| + 1)^C$$

---

* In Theorem 2(ii), we understand that the absolute constant depends on $P(d)$.

where C is an effectively computable number depending only on f and P.

§2. It has been pointed out in §1 that the theory of linear forms in logarithms and the effective irrationality measures of Baker [1] are combined in the proof of Theorem 1(i). This feature has also been applied to the equation

$$y^q = \frac{x^n - 1}{x - 1} \quad \text{in integers } x > 1, y > 1, q > 1, n > 2. \tag{6}$$

Equation (6) asks for perfect powers with all the digits equal to one with respect to base x. There are such perfect powers

$$11^2 = 1 + 3 + 3^2 + 3^3 + 3^4, \quad 20^2 = 1 + 7 + 7^2 + 7^3, \quad 7^3 = 1 + 18 + 18^2.$$

Equation (6) has only finitely many solutions if at least one of the following conditions is satisfied:

(i) $q = 2$ (Nagell [17], Ljunggren [15]).

(ii) x is fixed (Shorey and Tijdeman [26]).

(iii) n has a fixed prime divisor (Shorey and Tijdeman [26]).

Furthermore, this assertion is effective. In fact, Ljunggren showed that the only solutions of (6) with $q = 2$ correspond to $y = 11$ and $y = 20$.

One would like to prove that (6) has only finitely many solutions. For this, it suffices to show that (6) has only finitely many solutions with n restricted to prime powers. This observation is a consequence of (ii) of the above assertion. The author [22], [23] proved

Theorem 3. *Equation (6) has only finitely many solutions in integers $x > 1$, $y > 1$, $q > 1$ and $n > 2$ satisfying $\omega(n) > q - 2$. Furthermore, the result is effective.*

Here $\omega(n)$ denotes the number of distinct prime factors on n. For proving Theorem 3, we combine an estimate of Baker [4] on linear forms in logarithms and a theorem of Baker [1] on the approximations of certain algebraic numbers by rationals to show that (6) has only finitely many solutions with $n \equiv 1 \pmod{q}$. In fact, this assertion is a consequence of the following result: If x exceeds a certain absolute constant, there is at most one positive integer n in a residue class mod q such that $(x^n - 1)/(x - 1)$ is a q-th perfect power. This feature of looking up the exponents in a residue class has been applied by the author [24] to prove the following result.

Theorem 4. *Let $a > 0$, $b > 0$, $k \neq 0$, $x > 1$ and $y > 1$ be integers. Suppose that (m_1, n_1) and (m_2, n_2) are distinct pairs in positive integers satisfying*

$$\max(ax^{m_i}, by^{n_i}) > 953k^6, \qquad i = 1, 2.$$

Assume that

$$ax^m - by^n = k \qquad (7)$$

is satisfied by $m = m_1$, $n = n_1$ and $m = m_2$, $n = n_2$. Then either $m_1 \not\equiv m_2 \pmod 3$ or $n_1 \not\equiv n_2 \pmod 3$.

The proof of Theorem 4 depends again on a result of Baker [1]. For integers A and X with $1 \leq A < X$, we denote by $S_X(A)$ the set of integers with all the digits equal to A with respect to base X. Then the author [24] applied Theorem 4 to derive

$$|S_X(1) \cap S_Y(1)| \leq 17, \qquad X \neq Y$$

and

$$|S_X(A) \cap S_Y(B)| \leq 24, \qquad A(Y - 1) \neq B(X - 1).$$

Combining Theorem 4 with a result of Schinzel and Tijdeman [21], we see that there exists an effectively computable number C_1 depending only on a, b and k such that for integers $x > 1$ and $y > 1$ with $\max(x, y) \geq C_1$, the number of distinct pairs (m, n) in integers $m > 1$ and $n > 1$ satisfying (7) is at most 4. For integers $a > 0$, $b > 0$, $x \geq 4$ and $y \geq 4$, the author [24] applied Theorem 4 to conclude that there are at most 9 distinct pairs (m, n) in integers $m \geq 3$ and $n \geq 3$ satisfying

$$ax^m - by^n = 1. \qquad (8)$$

If $a = b = 1$ in (8) i.e. the case of Catalan's equation, LeVeque [14] proved the above assertion with 9 replaced by 1. Finally, we recall that Tijdeman [30] confirmed, in principle, the conjecture of Catalan that 9 and 8 are the only perfect powers that differ by one. Here Baker's sharpenings [3] on linear forms in logarithms played a crucial role and this is also the case with respect to the results described in this article. Further, we remark that Baker's effective irrationality measures obtained by hypergeometric method play a similar crucial role in the applications mentioned in this paper. Finally, we refer to [27] and [28] for a survey account of exponential Diophantine equations.

§3. In this section, we shall prove Theorem 2. Suppose that equation (5) with $m + d > k$ and $l \geq 3$ is satisfied. Since $m + d > k$ and $(m, d) = 1$, it is well-known that the product of the left hand side of

(5) is divisible by a prime $p > k$. See Langevin [13]. Now we see from (5) that there is r with $1 \leq r \leq k$ such that $m + rd$ is divisible by p^l. Consequently,

$$m + kd \geq (k+1)^l \qquad (9)$$

which implies that

$$m + d > k^{l-1}. \qquad (10)$$

For $\mu = 1, \ldots, k$, we see from (5) that

$$m + \mu d = a_\mu x_\mu^l \qquad (11)$$

where a_μ and x_μ are positive integers satisfying

$$P(a_\mu) \leq k, \quad \left(x_\mu, \prod_{p \leq k} p\right) = 1 \quad \text{and} \quad (x_\mu, x_\nu) = 1, \quad \mu \neq \nu. \qquad (12)$$

The letter p always denotes a prime number. Put

$$S_1 = \{a_1 \ldots, a_k\}.$$

Proof of Theorem 2(i). Suppose that equation (5) with $m + d > k, l \geq 3$ and

$$\left(l, \prod_{p|d}(p-1)\right) = 1 \qquad (13)$$

is satisfied. Then there is no loss of generality in assuming that $l \geq 3$ is a prime number. Denote by $c_1, c_2, \ldots c_5$ effectively computable absolute positive constants. We may assume that $k \geq c_1$ with c_1 sufficiently large.

First, we show that the elements of S_1 are distinct. Let

$$a_\mu = a_\nu, \qquad 1 \leq \nu < \mu \leq k.$$

Then we see from (11) and (12) that $x_\mu > x_\nu$, $(x_\mu, x_\nu) = 1$ and

$$(\mu - \nu)d = a_\nu(x_\mu^l - x_\nu^l) \qquad (14)$$

which, by $(m, d) = 1$, implies that

$$d | (x_\mu^l - x_\nu^l). \qquad (15)$$

Now we apply (13) to derive

$$p | (x_\mu - x_\nu), \qquad p | d. \qquad (16)$$

By (15) and (16),

$$d \leq l(x_\mu - x_\nu). \tag{17}$$

We see from (14) that

$$kd > l(x_\mu - x_\nu)a_\nu x_\nu^{l-1} \geq l(x_\mu - x_\nu)(a_\nu x_\nu^l)^{(l-1)/l}$$

which, together with (17) and (10), implies that

$$k^l > (m + d)^{l-1} > k^{(l-1)^2}. \tag{18}$$

This is not possible, since $l \geq 3$. Hence $|S_1| = k$.

For every prime $p \leq k$, we choose an $f(p) \in S_1$ such that p does not appear to a higher power in the factorisation of any other element of S_1. Denote by S_2 the subset of S_1 obtained by deleting from S_1 all $f(p)$ with $p \leq k$. Then

$$\prod_{a_\mu \in S_2} a_\mu \leq \prod_{\substack{p \leq k \\ (p,d)=1}} p^{[k/p]+[k/p^2]+\cdots} \leq k!.$$

Therefore

$$\prod_{a_\mu \in S_2} a_\mu \leq k^k, \qquad |S_2| \geq k - \pi(k).$$

Consequently, there exists a subset S_3 of S_2 such that

$$a_\mu \leq c_2 k, \quad a_\mu \in S_3 \quad \text{and} \quad |S_3| > k/2. \tag{19}$$

Here we have used that the elements of S_2 are distinct and the product of ν distinct elements of S_2 is greater than or equal to $\nu!$. By (11), (19), (10) and (12), we see that

$$x_\mu > k, \quad a_\mu \in S_3. \tag{20}$$

Denote by S_3' the set of $a_\mu \in S_3$ such that either $\mu \leq k/8$ or $a_\mu \leq k/8$. Further, we write S_4 for the complement of S_3' in S_3. Observe that $|S_3'| \leq k/4$ and consequently,

$$|S_4| > k/4. \tag{21}$$

Further we see from (11) and (20) that

$$m + \mu d > k^{l+1}/8, \quad a_\mu \in S_4.$$

Consequently, we sharpen (9) to

$$m + kd > k^{l+1}/8. \tag{22}$$

Let a_i, a_j, a_μ and a_ν be elements of S_4 such that $i \neq \mu$, $i \neq \nu$ and

$$a_i a_j = a_\mu a_\nu. \tag{23}$$

Put

$$\Delta = (m + id)(m + jd) - (m + \mu d)(m + \nu d). \tag{24}$$

By (24), (11) and (23),

$$\Delta = a_\mu a_\nu \left((x_i x_j)^l - (x_\mu x_\nu)^l \right). \tag{25}$$

Then we see from (12) and (20) that $\Delta \neq 0$ and $(x_i x_j, x_\mu x_\nu) = 1$. For simplicity, we suppose that $\Delta > 0$. The proof for the case $\Delta < 0$ is similar. By (24), (25) and $(m, d) = 1$, we see that

$$d \big| \left((x_i x_j)^l - (x_\mu x_\nu)^l \right).$$

Now we argue as in the proof of (17). We derive from (13) that

$$d \leq l(x_i x_j - x_\mu x_\nu)$$

which, together with (25), implies that

$$\Delta \geq d \left((a_\mu x_\mu^l)(a_\nu x_\nu^l) \right)^{(l-1)/l}.$$

Now, since $a_\mu, a_\nu \in S_4$, we derive

$$\Delta \geq d \left(m + \frac{kd}{8} \right)^{2(l-1)/l} \geq \frac{d}{64}(m + kd)^{2(l-1)/l}. \tag{26}$$

On the other hand, we see from (24) that

$$\Delta \leq 2kd(m + kd). \tag{27}$$

Combining (26) and (27), we obtain

$$(m + kd)^{(l-2)/l} \leq 128k$$

which, together with (22) and $l \geq 3$, implies that $k \leq c_3$. This is not possible if $c_1 > c_3$.

Thus we have shown that there is no non-trivial relation (23) among the elements of S_4. Now we apply Lemma 4 of Erdős [8] and (19) to conclude that

$$|S_4| \leq c_4 k / \log k$$

which, together with (21), implies that $k \leq c_5$. This completes the proof of Theorem 1(i).

Proof of Theorem 2(ii). Suppose that equation (5) with $m + d > k$ and $l \geq 3$ is satisfied. Then there is no loss of generality in assuming that either $l = 4$ or $l \geq 3$ is a prime number. If $l \geq P(d)$ and $l \neq 4$, then (13) is satisfied and hence we conclude from Theorem 2(i) that k is bounded by an effectively computable absolute constant. Thus we may assume that either $l < P(d)$ or $l = 4$. In particular, there are only finitely many possibilities for l since $P(d)$ is bounded. We denote by $e_1, \ldots, e_7$ effectively computable positive numbers depending only on the greatest prime factor of d and l. We may assume that $k \geq e_1$ with e_1 sufficiently large. Let $0 < \eta < 1$. We denote by S_5 the set of all the elements $a_\mu \in S_1$ such that $a_\mu > k^{1-\eta}$ and we write T for the set of all μ with $1 \leq \mu \leq k$ such that $a_\mu \notin S_5$.

First, we show that the elements of S_5 are distinct. Let $1 \leq \nu < \mu \leq k$ such that $a_\mu, a_\nu \in S_5$ and $a_\mu = a_\nu$. Then we see from (14) and $(d, a_\nu) = 1$ that $a_\nu < k$. Now we derive from (11), (10) and (12) that $x_\mu > x_\nu > k$ and $(x_\mu, x_\nu) = 1$. Further, we observe from (14) that

$$k^{l-1} < x_\mu^{l-1} < dk$$

which, since $l \geq 3$, implies that

$$d > k. \tag{28}$$

We rewrite (14) as

$$x_\mu^l - x_\nu^l = b_{\mu,\nu} d, \quad b_{\mu,\nu} = (\mu - \nu) a_\nu^{-1}. \tag{29}$$

Observe that $b_{\mu,\nu}$ is a positive integer, since $(d, a_\nu) = 1$. Moreover, since $a_\nu \in S_5$, we have

$$b_{\mu,\nu} \leq k^\eta.$$

Now we apply Theorem A to (29) and we conclude that

$$d \leq k^{e_2 \eta}. \tag{30}$$

Let $\eta = (2e_2)^{-1}$. Then (28) and (30) are inconsistent. Hence the elements of S_5 are distinct. Therefore

$$|S_5| = k - |T|. \tag{31}$$

Now we show that

$$|T| \leq \pi(k). \tag{32}$$

Suppose that $|T| > \pi(k)$. Then there exists a subset T_1 of T such that

$$|T_1| \geq e_3 \pi(k) \tag{33}$$

and

$$x_\mu \equiv x_\nu \,(\mathrm{mod}\, 4Q^*(d)), \qquad \mu \in T_1, \, \nu \in T_1 \tag{34}$$

where $Q^*(d)$ denotes the product of all distinct prime factors > 2 of d. Let $\mu > \nu$ with $\mu, \nu \in T_1$, and $a_\mu = a_\nu$. Then we derive (17) from (15) and (34). Now we proceed, as in the proof of Theorem 2(i), to obtain (18) to arrive at a contradiction. Therefore the elements a_μ with $\mu \in T_1$ are distinct. Consequently

$$|T_1| \leq k^{1-\eta}$$

which contradicts (33) if e_1 is sufficiently large. This completes the proof of (32).

We combine (31) and (32) to obtain

$$|S_5| \geq k - \pi(k).$$

Now we argue, as in the proof of (19) and (20), to conclude that there exists a subset S_6 of S_5 satisfying

$$a_\mu \leq e_4 k, \quad a_\mu \in S_6 \quad \text{and} \quad |S_6| > k/2 \tag{35}$$

and

$$x_\mu > k, \quad a_\mu \in S_6. \tag{36}$$

In view of (35), we derive from Lemma 1 of Erdős [7] and (11) that there exist $a_\mu, a_\nu \in S_6$ with $\mu > \nu$ such that

$$(a_\mu, a_\nu) \geq e_5 k \tag{37}$$

and

$$A_1 x_\mu^l - A_2 x_\nu^l = A_3 d, \quad (x_\mu, x_\nu) = 1, \tag{38}$$

where

$$A_1 = \frac{a_\mu}{(a_\mu, a_\nu)}, \quad A_2 = \frac{a_\nu}{(a_\mu, a_\nu)}, \quad A_3 = \frac{\mu - \nu}{(a_\mu, a_\nu)}. \tag{39}$$

Observe that $A_3 > 0$ is an integer, since $(m, d) = 1$. Further we see from (39), (35) and (37) that

$$\max(A_1, A_2, A_3) \leq e_6.$$

Now we apply Theorem A and (36) to equation (38) to conclude that $k \leq e_7$. This completes the proof of Theorem 2(ii).

Proof of Theorem 2(iii). Suppose that equation (5) with $m + d \leq k$, $b = 1$ and $l \geq 3$ is satisfied. We may assume that $k \geq c_6$ where c_6 is a sufficiently large effectively computable absolute constant. Then it is easy to see that $d \geq 3$. See Marszalek [16], p. 217–18. Let r with $1 \leq r \leq k$ be the integer such that

$$m + rd \leq k, \quad m + (r + 1)d > k. \tag{40}$$

Then

$$r \leq k/3. \tag{41}$$

As observed in the beginning of §3, we see that the product $(m + (r + 1)d) \ldots (m + kd)$ is divisible by a prime $p > k - r$. Let μ with $r < \mu \leq k$ be the integer such that $m + \mu d$ is divisible by p. Then it follows from (5) with $b = 1$ and (40) that $m + \mu d \geq p^{l-1}$. Therefore

$$m + kd > (k - r)^{l-1}.$$

On the other hand, we observe from (40) that

$$m + kd \leq k + (k - r)d.$$

Combining these estimates, we see from (40) and (41) that

$$k/r > d > (k - r)^{l-2} - 2$$

which, together with (41) and $l \geq 3$, implies that $r = 1$ if c_6 is sufficiently large.

Thus we may assume that $r = 1$. Then the product $(m + 2d) \ldots (m + kd)$ is divisible by a prime $q \geq k$. First, suppose that $q > m + d$. Then we see from (5) with $b = 1$ that $m + kd \geq q^l \geq k^l$. Consequently, since $l \geq 3$,

$$k \geq m + d \geq k^{l-1}/2 \geq k^2/2.$$

Thus we may suppose that $q \leq m + d$. Then $k \leq q \leq m + d \leq k$. Therefore $m + d = q = k$. Now we derive from (5) with $b = 1$ and $l \geq 3$ that

$$m + kd \geq q^{l-1} = k^{l-1} \geq k(m + d)$$

which is not possible. This completes the proof of Theorem 2(iii).

References

[1] A. Baker, Rational approximations to $\sqrt[3]{2}$ and other algebraic numbers, *Quart. J. Math. Oxford* (2) **15** (1964), 375–383.

[2] A. Baker, Bounds for the solutions of the hyperelliptic equation, *Math. Proc. Cambridge Philos. Soc.* **65** (1969), 439–444.

[3] A. Baker, A sharpening of the bounds for linear forms in logarithms I, II, III, *Acta Arith.* **21** (1972), 117–129; **24** (1973), 33–36; **27** (1974), 247–252.

[4] A. Baker, The theory of linear forms in logarithms, In *Transcendence Theory: Advances and Applications*, edited by A. Baker and D. W. Masser, Academic Press, 1977, 1–27.

[5] L. E. Dickson, History of the theory of numbers, Vol. 2: *Diophantine analysis*, 679–680, reprint, Chelsea, New York, 1966.

[6] P. Erdős, Note on the product of consecutive integers, *J. London Math. Soc.* **14** (1939), 194–198.

[7] P. Erdős, Note on the product of consecutive integers (II), *J. London Math. Soc.* **14** (1939), 245–249.

[8] P. Erdős, On the product of consecutive integers (III), *Indag. Math.* **17** (1955), 85–90.

[9] P. Erdős and J. L. Selfridge, The product of consecutive integers is never a power, *Illinois J. Math.* **19** (1975), 292–301.

[10] P. Erdős and J. Turk, Products of integers in short intervals, *Acta Arith.* **44** (1984), 147–174.

[11] K. Győry, Explicit upper bounds for the solutions of some Diophantine equations, *Ann. Acad. Sci. Fenn. Ser. A.I.* **5** (1980), 3–12.

[12] H. Halberstam and K. F. Roth, On the gaps between consecutive k-free integers, *J. London Math. Soc.* **26** (1951), 268–273.

[13] M. Langevin, Plus grand facteur premier d'entiers en progression arithmétique, *Séminaire Delange-Pisot-Poitou.* 18e année. Fasc. 1, Exp. 3 (1977).

[14] W. J. LeVeque, On the equation $a^x - b^y = 1$, *Amer. J. Math.* **74** (1952), 325–331.

[15] W. Ljunggren, Some theorems on indeterminate equations of the forms $(x^n - 1)/(x - 1) = y^q$ (Norwegian), *Norsk. Mat. Tidsskr.* **25** (1943), 17–20.

[16] R. Marszalek, On the product of consecutive elements of an arithmetic progression, *Monatsh. für Math.* **100** (1985), 215–222.

[17] T. Nagell, Sur l'équation indéterminée $(x^n - 1)(x - 1) = y^2$, *Norsk. Mat. Forenings Skrifter* **1** (1921), no. 3, 17 pp.

[18] R. Obläth, Über Produkte aufeinanderfolgender Zahlen, *Tohoku Math. J.* **38** (1933), 73–92.

[19] O. Rigge, Über ein Diophantisches Problem, *9th Congress Math. Scand.*, Helsingfors, 1938, Mercator, Helsingfors, 1939, 155–160.

[20] K. F. Roth, On the gaps between square-free numbers, *J. London Math. Soc.* **26** (1951), 263–268.

[21] A. Schinzel and R. Tijdeman, On the equation $y^m = P(x)$, *Acta Arith.* **31** (1976), 199–204.

[22] T. N. Shorey, Perfect powers in values of certain polynomials at integer points, *Math. Proc. Camb. Phil. Soc.* **99** (1986), 195–207.

[23] T. N. Shorey, On the equation $z^q = (x^n - 1)/(x - 1)$, *Indag. Math.*, **89** (1986), 345–351.

[24] T. N. Shorey, On the equation $ax^m - by^n = k$, *Indag. Math.*, **89** (1986), 353–358.

[25] T. N. Shorey, Perfect powers in products of integers from a block of consecutive integers, *Acta Arith.*, **49** (1987), 71–79.

[26] T. N. Shorey and R. Tijdeman, New applications of Diophantine approximations to Diophantine equations, *Math. Scand.* **39** (1976), 5–18.

[27] T. N. Shorey, A. J. van der Poorten, R. Tijdeman and A. Schinzel, Applications of Gel'fond-Baker method to Diophantine equations, In *Transcendence Theory: Advances and Applications*, Academic press, 1977, 59–77.

[28] T. N. Shorey and R. Tijdeman, Exponential Diophantine equations, *Cambridge Tracts in Mathematics*, 1986.

[29] A. Thue, Über Annäherungswerte algebraischer Zahlen, *J. reine angew. Math.* **135** (1909), 284–305.

[30] R. Tijdeman, On the equation of Catalan, *Acta Arith.* **29** (1976), 197–209

23

ARITHMETIC SPECIALIZATIONS THEORY

V. G. Sprindžuk

To the set of numerous results from different branches of mathematics which were obtained on the grounds of the ideas and the methods of transcendental number theory, meaningful facts on the specializations in polynomials and in the fields of algebraic functions were added recently. Although the source of these results may be found in problems concerning the effectivization of Hilbert's irreducibility theorem and the explicit construction of universal Hilbert sets, further spontaneous development of the basic ideas and their interaction with new and old techniques have gone far from the initial aims.

In this short survey we discuss only a few of the main points of the theory and supply the reader with reference to further results.

Let $\mathbf{K}$ be a field of algebraic numbers of degree k over the field of rational numbers $\mathbf{Q}$, let $I_{\mathbf{K}}$ be its ring of integers, and let $F = F(x, y) \in I_{\mathbf{K}}[x, y]$ be irreducible. Suppose $0 \neq x_0 \in \mathbf{K}$ and let

$$F(x_0, y) = F_1(y) \ldots F_r(y) \tag{1}$$

be the decomposition of $F(x_0, y)$ into irreducible factors $F_j = F_j(y)$ in $\mathbf{K}[y]$. We intend to characterize in some way the arithmetic structure of $F(x_0, y)$ by means of the arithmetic structure of x_0, and first of all to describe the set $(\deg F_1, \ldots, \deg F_r)$. The following Theorem 1 shows that under certain conditions that can really be done.

Denote $S = S_{\mathbf{K}}$ the full system of non-equivalent valuations v of the field $\mathbf{K}$, $\mathbf{K}_v$ the completion of $\mathbf{K}$ in the metric v, $k_v = [\mathbf{K}_v : \mathbf{Q}_v]$. For $0 \neq \kappa \in \mathbf{K}$ we set

$$(\kappa)_v = \max(1, |\kappa|_v^{-k_v}), \qquad H_{\mathbf{K}}(\kappa) = \prod_{v \in S} (\kappa)_v.$$

Suppose further that the polynomial $F(x, y)$ satisfies

$$\deg_y F = n \geq 2, \qquad F(0, 0) = 0, \qquad \frac{\partial}{\partial y} F(0, 0) \neq 0 \tag{2}$$

and the power series

$$f = f(x) = \sum_{\nu=1}^{\infty} f_\nu x^\nu \qquad (3)$$

is defined from the equation $F(x,y) = 0$ with initial conditions $x = 0$, $y = 0$. Then $f_\nu \in K$ and in accordance with the well-known Eisenstein theorem there exists a natural number $E = E_f$ such that

$$E^\nu f_\nu \in I_K \qquad (\nu = 1, 2, \ldots). \qquad (4)$$

It is obvious that $f(x)$ is algebraic over $K[x]$ of degree n. The following fact having its origin in the generalization of the well-known theorem of Ostrowski deserves special attention (see [19], Lemma 3.4 and Lemma 3.5).

Lemma. *Let* q *be a prime ideal of* I_K *and* F_q *the field of residue classes* $I_K(\mathrm{mod}\, q)$. *If the norm of* q *exceeds a computable bound* C_1 *depending only on* F, *and* q *does not occur in* E_f *then the power series over* F_q

$$\overline{f} = \sum_{\nu=1}^{\infty} \overline{f}_\nu x^\nu$$

which is obtained from (3) by means of the mapping

$$f_\nu \to \overline{f}_\nu = f_\nu(\mathrm{mod}\, q)$$

is algebraic over $F_q[x]$ *of degree* n.

Denote by $R_v(f)$ the radius of convergence of the series (3) in the metric v. Then it follows from (4) that $R_v(f) > 0$ while the Lemma yields $R_v(f) = 1$ for all v except special ones which may be explicitly determined. Hence, if $x_0 \in K$ and has a big enough height $H_K(x_0)$ then there exists $v \in S$ with

$$|x_0|_v < R_v(f). \qquad (5)$$

That means that by changing the metrics v we can attribute a numerical value to the series (3) practically at any point $x_0 \in K$. Let $S(x_0)$ be the set of all v for which (5) holds and $\theta_v = f(x_0)_v$ the corresponding sum of the series (3) in the metric v.

Turning to the decomposition (1) we note that in view of the equality

$$F(x, f(x)) = 0 \qquad (6)$$

the numbers θ_v for $v \in S(x_0)$ are the roots of the polynomial $F(x_0, y)$ and hence they are distributed among the roots of the polynomials $F_j(y)$. We write $\theta_v \in F_j(y)$ if θ_v is the root of $F_j(y)$. The theorem that follows shows that there is an "ergodic law" of the distribution of the values θ_v among the roots of the polynomials $F_j(y)$ ([19], [20]).

Theorem 1. *Under the conditions specified above, the approximate formulae as follow hold:*

$$\frac{1}{n} \deg F_j(y) = \sum_{\theta_v \in F_j} \frac{\log(x_0)_v}{\log H_{\mathbf{K}}(x_0)} + O\left(\sqrt{\frac{\log\lceil F \rceil}{\log H_{\mathbf{K}}(x_0)}}\right), \qquad (7)$$

where the sum is taken over all $v \in S(x_0)$ for which θ_v is the root of $F_j(y)$ (with fixed $j = 1, 2, \ldots, \tau$), $\lceil F \rceil$ is the size of F and the symbol O involves the value depending only on n and k.

A number of consequences from this theorem concerning Diophantine equations, effective versions of Hilbert's irreducibility theorem and the inverse problem of Galois theory is discussed in [19] and [6]. For example, if x_0 is a primary number, i.e. there exists such a v_0 that $|x_0|_{v_0} < 1$ and $H_{\mathbf{K}}(x_0)$ is big enough, then $F(x_0, y)$ is irreducible in $\mathbf{K}[y]$. If x_0 is a power of a rational prime p and the metric v is a prolongation of the p-adic metric, then

$$[f(x_0)_v : \mathbf{K}] \geq \frac{n}{(n, k)}, \qquad (8)$$

where (n, k) is the greatest common divisor of n and k. The demonstration of Theorem 1 in the article [19] is a generalization of the initial results ([15], [16], [18]) concerning the case $\mathbf{K} = \mathbf{Q}$. It is easy to realize basic ideas of the arguments involved therein by turning to the theory of Siegel's G-functions [14].

Let $f_1, \ldots, f_{n-1}$ be algebraic power series of x with coefficients from $\mathbf{K}$; we may suppose that they are, together with 1, linearly independent over $\Re = \mathbf{Q}(x)$ (say, powers of the series (3)). Denote by $\mathcal{L}$ the ring generated by $f_1, \ldots, f_{n-1}$ over $\Re$,

$$\mathcal{L} = \Re[f_1, \ldots, f_{n-1}],$$

and let $\omega_1, \ldots, \omega_m$ be a basis of $\mathcal{L}$ over $\Re$ generated by 1, $f_1, \ldots, f_{n-1}$ and its powers. Then $\mathcal{L}$ is a differential ring and

$$\frac{d}{dx}\omega_k = g_{1k}\omega_1 + \ldots + g_{mk}\omega_m \qquad (k = 1, 2, \ldots, m),$$

where $g_{ik} \in \mathsf{K}(x)$ and there exists such a polynomial $d(x) \in \mathsf{K}[x]$ that $g_{ik}d(x) \in \mathsf{K}[x]$, $d(x)$ is defined by $f_1, \ldots, f_{n-1}$. Consequently,

$$\omega_k^{(s)} = g_{1ks}\omega_1 + \ldots + g_{mks}\omega_m \quad (s = 1, 2, \ldots),$$

$$g_{iks} \in \mathsf{K}(x), \quad g_{iks}d^s(x) \in \mathsf{K}[x].$$

We may accept that

$$g_{iks} = \frac{G_{iks}(x)}{d^s(x)},$$

where $G_{iks}(x)$ and $d(x)$ have integer coefficients in K. All the coefficients of the power series for $\omega^{(s)}d^s(x)$ are divisible by "almost s!" (taking into account the Eisenstein number), and the coefficients of the series

$$\overline{\omega}_k^{(s)}\overline{d}^s(x) = \omega_k^{(s)}d^s(x) \pmod{\mathbf{q}}$$

are zeros for the corresponding prime ideals $\mathbf{q}$. Hence, we have for such ideals

$$\overline{G}_{iks}\overline{\omega}_1 + \ldots + \overline{G}_{mks}\overline{\omega}_m = 0.$$

It may be deduced from the Lemma that for $\mathbf{q}$ with the norms exceeding C_1 and not occurring in the corresponding Eisenstein numbers all the polynomials $\overline{G}_{iks}$ must vanish. Iterated application of similar arguments shows that there exist such natural numbers d_s that

$$\frac{d_s}{s!}G_{iks}(x) \in I_{\mathsf{K}}[x], \qquad d_s \le C_2^s,$$

where C_2 may be determined explicitly through $f_1, \ldots, f_{n-1}$ (cf. [19], Lemma 4.5).

Thus we see the "factorials reduction" in differentiations of ω_k, the phenomenon which Siegel [14] and Galochkin [10] postulated in general conditions.

In view of (6) in our case we can take

$$\mathcal{L} = \Re[f, f^2, \ldots, f^{n-1}] = \Re[f]$$

and use the scheme of arguing by Siegel's method, supplementing it by considerations in all the metrics $v \in S$ (instead of traditional analysis in one metric only). Nevertheless none of the proofs of Theorem 1 which were obtained up to now follows exactly this way, leaving apparent perspective to further progress. In particular, the above mentioned proof of Theorem 1, although being on the lines of the indicated ideas, in fact utilizes the specific properties of algebraic power series in the construction and analysis of the auxiliary linear forms which allows one to overcome easily the crucial points of Siegel's method (multiplicity of zeros, etc.).

Being proved by this way, Theorem 1 rests on the ideas of Diophantine approximation theory, and in fact it is equivalent to the assertion on the Diophantine approximations to the values of algebraic power series (3) [20].

Theorem 2. *Set for* $v \in S(x_0)$

$$\lambda_v = \frac{nk}{k_v} \cdot \frac{\log(x_0)_v}{\log H_{\mathbf{K}}(x_0)} - e_v + \epsilon,$$

where $e_v = 1$ *in case of archimedean* v, *and* $e_v = 0$ *in case of nonarchimedean* v. *Then the system of inequalities*

$$|P(\theta_v)|_v < \lceil P \rceil^{-\lambda_v} \qquad (v \in S(x_0)) \qquad (9)$$

with a non-zero polynomial $P(y) \in I_{\mathbf{K}}[y]$ *of degree not exceeding* $n - 1$, *for* $\epsilon > 0$ *has only a finite number of solutions in polynomials* $P(y)$, *provided that*

$$H_{\mathbf{K}}(x_0) > (1 + \lceil F \rceil)^{C_3/\epsilon^2} \qquad (10)$$

where C_3 *is a computable value depending only on* n *and* k. *For* $\epsilon < 0$ *the system (9) has infinitely many solutions in* $P(y)$ *(with any* x_0 *for which* $S(x)$ *is not empty).*

It is possible to consider as well the non-integral polynomials

$$P(y) = \pi_0 + \pi_1 y + \ldots + \pi_{n-1} y^{n-1} \in \mathbf{K}[y]$$

defining the height of the polynomial by the product

$$H_{\mathbf{K}}(P) = \prod_{v \in S} H_v, \qquad H_v = \max\left(|\pi_0|_v, \ldots, |\pi_{n-1}|_v\right)^{k_v},$$

and setting

$$\mu_v = \frac{n}{k_v} \cdot \frac{\log(x_0)_v}{\log H_{\mathbf{K}}(x_0)}(1 + \epsilon).$$

Then for x_0 satisfying the condition like (10) for $\epsilon > 0$ the system

$$|P(\theta_v)|_v < H_{\mathbf{K}}(P)^{-\mu_v} H_v \qquad (v \in S(x_0))$$

has only a finite number of solutions in polynomials $P(y) \in \mathbf{K}[y]$, and for $\epsilon < 0$ it has infinitely many solutions.

It is interesting to compare Theorem 2 with the results of Schmidt-Schlickewei on linear forms with algebraic coefficients. Their interaction may lead to some new finiteness theorems on Diophantine equations.

Two basically different approaches to the demonstration of assertions like Theorems 1 and 2 were given by Bombieri [2]. The first approach is founded on the wide development of Siegel's G-function theory ([1]; see also [11]). Basic ideas of this approach are similar to those mentioned above where the "factorials reduction" principle is replaced by a recent general theorem due to Dwork and Robba. The second approach rests on Weil's "Théorème Décomposition" [22] and the theory of heights on abelian varieties ([2]; see also [7]). Yet another proof was given by Dèbes [6] who applied Gel'fond's method. Earlier Schneider ([12], [13]) and Bundschuh [3] used Gel'fond's method to analyze the arithmetic nature of the values of algebraic functions (in one metric).

It is interesting to note that in spite of the different approaches to the proof of assertions like Theorem 1, the remainder terms in the formulae (7) occur of the same order of magnitude. An important problem is to improve these terms essentially since then the totality of arithmetic consequences from Theorem 1 may be extended.

The second and third conditions in (2) may be weakened and replaced by the supposition that the polynomial $F(0, y)$ has a simple root of degree not exceeding $n - 1$. The conclusions which may be obtained are then less precise although still important (e.g. in just this way the universal Hilbert sets were constructed ([17], [9]). If the polynomial $F(0, y)$ is irreducible, the direct conclusions which follow are trivial, and in fact just the case of irreducibility of $F(0, y)$ arises in deep problems of Diophantine equations theory and in the arithmetic theory of integral polynomials. As an example, we consider the problem of representation of primary numbers by polynomials [21].

Let $P(y)$ be a polynomial of degree $n \geq 2$ with rational integer coefficients, **P** the set of all rational primary numbers, i.e. such rational numbers $x_0 \neq 0$ that there is only one metric v in the field **Q** with $|x_0|_v < 1$; hence, if $x_0 = a/b$, $(a, b) = 1$, then in the case of the archimedean metric v we have $a = \pm 1$, $|b| \geq 2$ (primary numbers of the first kind) and if v is non-archimedean then a is a power of a rational prime, $|a| \geq |b|$ (primary numbers of the second kind). Denote $h(x_0) = \max(|a|, |b|)$.

Setting $F(x, y) = x - P(y)$ we find from Theorem 1 that if $P(y)$ has a simple rational root then for $x_0 \in$ **P** with big enough $h(x_0)$ the polynomial $P(y) - x_0$ is irreducible in **Q**$[y]$. If $P(y)$ has a simple root of degree k then under the same restrictions on x_0 all the irreducible factors of $P(y) - x_0$ are of degree no less than n/k. The last assertion is trivial in the case $k = n$, i.e. when $P(y)$ is irreducible.

Suppose now that $P(y)$ is irreducible and consider the Diophantine equation

$$P(y_0) = x_0; \qquad y_0 \in \mathbf{Q}, \quad x_0 \in \mathbf{P}, \qquad (11)$$

in unknown numbers y_0, x_0. If x_0 is of the first kind, the equation (11) is reduced to the equation of Thue, and in the case of $n \geq 3$ an explicit bound for its solution may be obtained by the Gel'fond-Baker method. If x_0 is of the second kind, then we have no grounds to make any conclusion on the finiteness of the number of its solutions, since x_0 may range through all prime numbers and we then have to determine whether $P(y)$ represents infinitely many primes (Bounijakowski problem). Thus, an analysis of the equation (11) becomes quite meaningful. To approach the phenomena hidden in the equation (11) it is reasonable to investigate the specialization $\mathcal{K} \to \mathcal{K}_0$ of the splitting field of $P(y) - x$ in the algebraic closure of $\mathbf{Q}(x)$ when $x \to x_0$. Using inequalities like (8) one can obtain in this way an assertion as follows [21].

Theorem 3. *Let $P(y)$ be an integral irreducible polynomial of degree $n \geq 2$, κ one of its roots, $\mathbf{K} = \mathbf{Q}(\kappa)$, G its splitting field, $\mathcal{K}$ the splitting field of $P(y) - x$ over $\mathbf{Q}(x)$,*

$$[\mathcal{K} : \mathbf{Q}(x)] = [\mathcal{K} : \overline{\mathbf{Q}}(x)] = N$$

(i.e. $\mathcal{K}$ is regular over $\mathbf{Q}$). Then there exists a power series φ with rational coefficients such that the degree of φ over $\mathbf{Q}(x)$ is N and

$$G(x, \varphi) = G\mathcal{K}.$$

If the equation (11) has a solution x_0 with big enough height $h(x_0)$ then under the specialization $x \to x_0$ we have $\varphi \to \varphi_0$, $\mathcal{K} \to \mathcal{K}_0$ where

$$[\varphi_0 : \mathbf{Q}] = N, \qquad [\varphi_0 : \mathbf{K}] = \frac{N}{n},$$

$$[\mathcal{K}_0 : \mathbf{Q}][G : \mathbf{Q}] \geq N$$

and the equality holds only in the case

$$\mathbf{Q}(\varphi_0) = G \times \mathcal{K}_0.$$

In the course of proving Theorem 3 it is ascertained that the series φ satisfies over $\mathbf{Q}(x)$ an absolutely irreducible equation

$$\Phi(x, \varphi) = 0,$$

while
$$\Phi\big(P(y), z\big) = \mathrm{Nm}_{\mathbf{K}/\mathbf{Q}}(\Psi)$$

where Ψ lies exactly in $\mathbf{K}[y, z]$ and is absolutely irreducible. These facts do not yield directly the finiteness of the solutions of the equation (11) but they may serve as a basis for further consideration.

In general, we see the theory of arithmetic specializations as a vast territory for further exploration which will undoubtedly lead to new important results.

REFERENCES

[1] E. Bombieri, On G-functions in *Recent progress in Analytic Number Theory*, H. Halberstam and C. Hooley ed., Academic Press (1981), vol. 2, 1–67.

[2] E. Bombieri, On Weil's "Théorème de Décomposition", *Amer. J. Math.* **105** (1983), 295–308.

[3] P. Bundschuh, Une nouvelle application de la méthode de Gel'fond. Se. Delange-Pisot-Poitou. *Théorie des Nombres*, 19ème année (1977–78), N 42.

[4] P. Dèbes, Une version effective du théorème d'irreducibilité de Hilbert, *Sém. Anal. Ultramétrique, Amice-Christol-Robba*, 10ème année (1982–83), N 42.

[5] P. Dèbes, Spécialisations de polynómes, *Math. rep. Acad. Sc.*, Royal Soc. Canada, vol. *V*, N 6 (Dec. 1983).

[6] P. Dèbes, Valeurs algebriques de fonctions algebriques et théorème d'irreducibilité de Hilbert, These 3ème cycle, Univ. P. et M. Curie (Paris VI), 1984.

[7] P. Dèbes, Quelques remarques sur un article de Bombieri concernant le théorème de décomposition de Weil, *Amer. J. Math.*, **107** (1985), 39–44.

[8] P. Dèbes, Parties hilbertiennes et progressiones géometriques, *C. R. Acad. Sc. Paris*, t. **302**, Série I, N 3 (1986), 87–90.

[9] M. Fried, On the Sprindžuk-Weissauer approach to universal Hilbert subsets, *Israel J. Math.*, vol. 51, N 4 (1985), 347–363.

[10] A. I Galochkin, Lower bounds of polynomials in the values of a certain class of analytic functions, *Math. Sb.* **95** (1974), 396–417.

[11] E. M. Matveev, Linear forms in the values of G-functions and Diophantine equations, *Math. Sb.* **117** (1982), 379–396.

[12] T. Schneider, Rationale Punkte über einer algebraischen Kurve, Sem. Delange-Pisot-Poitou, Théorie des Nombres, $15^{\text{ème}}$ année (1973/74), N 20.

[13] T. Schneider, Eine bemerkung zu einem Satz von C. L. Siegel. *Comm. pure and applied Math.*, **29** (1976), 775–782.

[14] C. L. Siegel, Über einige Anwendungen diophantischer Approxmationen, *Abh. Preuss. Akad. Wiss., Phys.-Math. Kl.* **1** (1929), 14–67.

[15] V. G. Sprindžuk, Hilbert's irreducibility theorem and rational points on algebraic curves, *Doklady Acad. Nauk SSSR*, **247** (1979), 285–289.

[16] V. G. Sprindžuk, Reducibility of polynomials and rational points on algebraic curves, *Doklady Acad. Nauk SSSR*, **250** (1980), 1327–1330.

[17] V. G. Sprindžuk, Diophantine equations involving unknown primes, *Trudy M.I.A.N.* SSSR, **148** (1981), 180–196.

[18] V. G. Sprindžuk, *Classical Diophantine equations in two unknowns*, Nauka, Moscow, 1982.

[19] V. G. Sprindžuk, Arithmetic specializations in polynomials, *J. reine und angew. Math.*, **340** (1983), 26–52.

[20] V. G. Sprindžuk, Diophantine approximations to the values of algebraic functions, *Doklady Akad. Nauk Byelorussian SSR*, **29** (1985), 101–103.

[21] V. G. Sprindžuk, Arithmetic specializations in the polynomial inversion fields, *Doklady Akad. Nauk Byelorussian SSR*, **30** (1986), 581–584.

[22] A. Weil, Arithmetic on algebraic varieties, *Annals of Math.*, **53** (1951), 412–444.

24

ON THE TRANSCENDENCE METHODS OF GELFOND
AND SCHNEIDER IN SEVERAL VARIABLES

M. Waldschmidt

1. Introduction

The methods we consider here were introduced by Gelfond and Schneider in their solutions of Hilbert's seventh problem on the transcendence of α^β (for algebraic α and β). Gelfond's proof [5] involved the two functions e^z and $e^{\beta z}$, with their derivatives, at the multiples of $\log \alpha$, while Schneider's proof [12] involved the two functions z and α^z, evaluated at the points $\mathbf{Z} + \mathbf{Z}.\beta$ (without derivatives).

Both methods have been extensively developed later. In his Bourbaki lecture [2], D. Bertrand pointed out a similarity between two of the most recent results which have been obtained, one by the method of Gelfond - Baker [16], and the other by Schneider's method [15].

The purpose of this paper is to prove a theorem which contains the two above-mentioned results, by combining the methods of Gelfond and Schneider.

Here is a corollary of our main result. Let G be a commutative algebraic group of dimension $d \geq 1$ which is defined over the field of algebraic numbers. We denote by $T_G(\mathbf{C})$ the tangent space of G at the origin, and by $\exp_G : T_G(\mathbf{C}) \longrightarrow G(\mathbf{C})$ the exponential map of the Lie group $G(\mathbf{C})$. Let d_0 (resp. d_1) be the dimension of the maximal unipotent (resp. multiplicative) factor of G, so that $G = \mathbf{G}_a^{d_0} \times \mathbf{G}_m^{d_1} \times G_2$, where G_2 is of dimension $d_2 = d - d_0 - d_1$.

Theorem 1.1 *Let V be a hyperplane of $T_G(\mathbf{C})$, W a subspace of V of dimension $t \geq 0$ over $\mathbf{C}$, and $Y = \mathbf{Z}y_1 + \ldots + \mathbf{Z}y_m$ a finitely generated subgroup of V of rank m over $\mathbf{Z}$. Assume that W is defined over $\overline{\mathbf{Q}}$ in $T_G(\mathbf{C})$, and that $\Gamma = \exp_G Y$ is contained in $G(\overline{\mathbf{Q}})$. Assume further*

$$m > (d_1 + 2d_2).(d - 1 - t). \tag{1.2}$$

Then V contains a non-zero algebraic Lie sub-algebra of $T_G(\mathbf{C})$ which is defined over $\overline{\mathbf{Q}}$.

The arrangement of this paper is as follows. In §2 we give a refinement of the six exponentials theorem. In §3 we derive further corollaries from Theorem 1.1. In §4 we state our main theorem, and in §5 we show that it contains Theorem 1.1. The proof of the main theorem is given in §7, using an auxiliary function which is constructed in §6.

The main part of this work was done at the Institute for Advanced Study of Princeton, in the fall 1985. The author is grateful to E. Bombieri. He wishes also to thank K. and R. Murty who gave him the opportunity to lecture on this subject at Montreal early 1986.

§2. A refinement of the six exponentials theorem.

a) *A strong version of the six exponentials theorem*

A well-known open problem is to prove that if t is a real number such that 2^t and 3^t are both rational integers, then t is rational. More generally, the *four exponentials conjecture* [1], [6], [11], [13] states that if x_1, x_2 are Q-linearly independent complex numbers, and y_1, y_2 are Q-linearly independent complex numbers, then one at least of the four numbers

$$e^{x_i y_j}, \qquad i = 1, 2; \ j = 1, 2$$

is transcendental.

The best known result in this direction is the so-called *six exponentials theorem* [1], [6], [11]: if x_1, x_2 (resp. y_1, y_2, y_3) are Q-linearly independent complex numbers, then one at least of the six numbers

$$e^{x_i y_j}, \qquad i = 1, 2; \ j = 1, 2, 3$$

is transcendental.

We refine this result in the following way.

Corollary 2.1 *Let x_1, x_2 be two complex numbers which are Q-linearly independent, and let y_1, y_2, y_3 be three complex numbers which are Q-linearly independent. Further let α_{ij}, $i = 1, 2$; $j = 1, 2, 3$, be six algebraic numbers. Assume that the six numbers*

$$\exp(x_i y_j - \alpha_{ij}), \qquad i = 1, 2; \ j = 1, 2, 3,$$

are algebraic. Then

$$x_i y_j = \alpha_{ij} \qquad \text{for } i = 1, 2 \text{ and } j = 1, 2, 3.$$

If one takes for granted that $\mathbf{Q}$-linearly independent logarithms of algebraic numbers are algebraically independent (a weak form of Schanuel's conjecture), then it is sufficient to consider in corollary 2.1 two numbers y_1, y_2 instead of three (a strong form of the four exponentials conjecture).

We first deduce Corollary 2.1 from Theorem 1.1, then we give some consequences.

b) *Proof of Corollary 2.1*

We choose $G = \mathbf{G}_a^2 \times \mathbf{G}_m^2$, which means $d = 4$, $d_0 = 2$, $d_1 = 2$, $d_2 = 0$. We identify $T_G(\mathbf{C})$ with $\mathbf{C}^4$ by

$$\exp_G(u_1, u_2, u_3, u_4) = (u_1, u_2, e^{u_3}, e^{u_4}) \in \mathbf{C}^2 \times (\mathbf{C}^\times)^2,$$

and we consider the hyperplane V of $\mathbf{C}^4$ of equation

$$x_2(u_3 + u_1) = x_1(u_4 + u_2).$$

This hyperplane is the image of the linear map of $\mathbf{C}^3$ into $\mathbf{C}^4$:

$$(z_1, z_2, z_3) \longrightarrow (z_1, z_2, x_1 z_3 - z_1, x_2 z_3 - z_2).$$

It contains the points

$$\eta_j = (\alpha_{1j}, \alpha_{2j}, x_1 y_j - \alpha_{1j}, x_2 y_j - \alpha_{2j}), \qquad j = 1, 2, 3.$$

We take
$$Y = \mathbf{Z}\eta_1 + \mathbf{Z}\eta_2 + \mathbf{Z}\eta_3.$$

We have $m = rk_{\mathbf{Z}}\, Y = 3$, because a relation

$$h_1\eta_1 + h_2\eta_2 + h_3\eta_3 = 0$$

with rational integers h_1, h_2, h_3 implies

$$h_1\alpha_{i1} + h_2\alpha_{i2} + h_3\alpha_{i3} = 0, \qquad i = 1, 2,$$

and
$$h_1 y_1 + h_2 y_2 + h_3 y_3 = 0,$$

which gives $h_1 = h_2 = h_3 = 0$.

If the six numbers

$$\delta_{ij} = \exp(x_i y_j - \alpha_{ij}), \qquad i = 1, 2; \; j = 1, 2, 3,$$

are all algebraic, then

$$\exp_G \eta_j \in G(\overline{\mathbf{Q}}), \qquad j = 1, 2, 3.$$

Finally, we put

$$W = \mathbf{C}(1, 0, -1, 0) + \mathbf{C}(0, 1, 0, -1).$$

This is a vector space of dimension $t = 2$, which is defined over $\mathbf{Q}$ in $\mathbf{C}^4$, and which is contained in V.

We use Theorem 1.1: the inequality

$$m > (d_1 + 2d_2)(d - 1 - t)$$

is satisfied; therefore V contains a non-zero $\overline{\mathbf{Q}}$-Lie sub-algebra $T_H(\mathbf{C})$ of $T_G(\mathbf{C})$. Since G is linear, V contains such a $T_H(\mathbf{C})$ of dimension 1.

The assumption that x_1, x_2 are $\mathbf{Q}$-linearly independent means that V does not contain a non-zero element of the form $(0, 0, a_1, a_2)$ with rational a_1, a_2. Hence V contains a non-zero element $(\gamma_1, \gamma_2, 0, 0)$ with algebraic γ_1, γ_2. Therefore $\gamma_1 x_2 = \gamma_2 x_1$, and the number $\gamma = x_2/x_1$ is algebraic and irrational.

Define

$$\log \delta_{ij} = x_i y_j - \alpha_{ij}, \qquad i = 1, 2; \; j = 1, 2, 3.$$

Then

$$\gamma \log \delta_{1j} - \log \delta_{2j} = \alpha_{2j} - \gamma \alpha_{1j}, \; j = 1, 2, 3.$$

Since γ is irrational, we deduce from Baker's theorem (see Corollary 3.3 below):

$$\log \delta_{1j} = 0, \; \log \delta_{2j} = 0, \; \text{and} \; \alpha_{2j} = \gamma \alpha_{1j}$$

for $j = 1, 2, 3$, which is the desired conclusion

c) *Some consequences of the strong six exponentials theorem.*

The next result can be referred to as the *five exponentials theorem.*

Corollary 2.2 *Let x_1, x_2 be two $\mathbf{Q}$-linearly independent complex numbers, and y_1, y_2 be also two $\mathbf{Q}$-linearly independent complex numbers. Further let η be a non-zero algebraic number. Then one at least of the five numbers*

$$e^{x_1 y_1}, e^{x_1 y_2}, e^{x_2 y_1}, e^{x_2 y_2}, e^{\eta x_2/x_1}$$

is transcendental.

Remark. Here is the *strong five exponentials conjecture*: under the hypotheses of Corollary 2.2, if $\alpha_{11}, \alpha_{12}, \alpha_{21}, \alpha_{22}, \beta$ are algebraic numbers, and if the five numbers $\exp(x_i y_j - \alpha_{ij})$, $i = 1, 2$; $j = 1, 2$, and $\exp(\eta \frac{x_2}{x_1} - \beta)$ are all algebraic then

$$x_i y_j = \alpha_{ij} \quad i = 1, 2; \ j = 1, 2, \quad \text{and} \quad \eta x_2 = \beta x_1.$$

This is clearly a weaker statement than the strong four exponentials conjecture, but still it contains non trivial open problems; for instance, if $\log a$, $\log b$, $\log c$ are non-zero logarithms of algebraic numbers, is it true that

$$(\log a).(\log b) \neq \log c?$$

(Choose $x_1 = 1$, $x_2 = \log a$, $y_1 = 1 + \log b$, $y_2 = \log b$, $\alpha_{11} = \eta = 1$, $\alpha_{12} = \alpha_{21} = \alpha_{22} = \beta = 0$).

Proof of Corollary 2.2 Apply Corollary 2.1 with

$$y_3 = \eta/x_1, \ \alpha_{11} = \alpha_{12} = \alpha_{21} = \alpha_{22} = \alpha_{23} = 0, \ \alpha_{13} = \eta.$$

If the two numbers

$$\gamma_1 = e^{x_1 y_1} \text{ and } \gamma_2 = e^{x_1 y_2}$$

are algebraic, then the theorem of Hermite-Lindemann (which is the case $n = 1$ of Corollary 3.3) implies that η, $\log \gamma_1$ and $\log \gamma_2$ are $\mathbf{Q}$-linearly independent.

Let us give a few special cases of Corollary 2.2.

(2.2.3) Let α_1, α_2, β be non-zero algebraic numbers, with $\log \alpha_1$, $\log \alpha_2$ $\mathbf{Q}$-linearly independent. Let $t \in \mathbf{C}$, $t \neq 0$. Then one at least of the numbers

$$\alpha_1^t, \alpha_2^t, e^{\beta t}$$

is transcendental, and also one at least of the numbers

$$\alpha_1^t, \alpha_2^t, e^{\beta/t}.$$

is transcendental.

(2.2.4) Let α and β be non-zero algebraic numbers with $\log \alpha \neq 0$, and let $t \in \mathbf{C}$ be irrational. Then one at least of

$$\alpha^t, \alpha^{t^2}, e^{\beta t},$$

and one at least of

$$\alpha^t, \alpha^{t^2}, e^{\beta/t}$$

is transcendental. If, further, $\beta t / \log \alpha$ is irrational, then one at least of

$$\alpha^t, \alpha^{t^2}, e^{\beta t^2}$$

is transcendental.

(2.2.5). Let $\alpha_1, \alpha_2, \gamma, \eta$ be non-zero algebraic numbers with $\log \alpha_1$, $\log \alpha_2$ $\mathbf{Q}$-linearly independent and $\log \gamma \neq 0$. Then one at least of

$$\alpha_1^{\eta \log \gamma}, \alpha_2^{\eta \log \gamma},$$

and at least one of

$$\alpha_1^{\eta / \log \gamma}, \alpha_2^{\eta / \log \gamma}$$

is transcendental.

For instance, if α and β are non-zero algebraic numbers with $\log \alpha \neq 0$ and $\log \beta \neq 0$, then the numbers

$$\alpha^{\log \beta} \text{ and } \alpha^{(\log \beta)^2}$$

are not both algebraic. In this result, only the case $\alpha = \beta$ was known, as a consequence of some results on algebraic independence [4].

§3. Further corollaries to Theorem 1.1

We first consider the case $t = d - 1$ (Gelfond's method), next the case $t = 0$ (Schneider's method), and finally we give an example with $t = 1$.

a) *Gelfond's method*

If, in Theorem 1.1, the hyperplane V itself is defined over $\overline{\mathbf{Q}}$, then one can choose $W = V$, $t = d - 1$, and the assumption on m reduces to $m > 0$, which means $Y \neq 0$. One deduces the following corollary, which is Wüstholz's result announced in [16] (see [2] Th. 4).

Corollary 3.1 *Let G be a commutative algebraic group defined over $\overline{\mathbf{Q}}$, and let $u \in T_G(\mathbf{C})$ be such that $\exp_G u \in G(\overline{\mathbf{Q}})$. Then the smallest subspace of $T_G(\mathbf{C})$ defined over $\overline{\mathbf{Q}}$ which contains u is an algebraic Lie sub-algebra of $T_G(\mathbf{C})$, defined over $\overline{\mathbf{Q}}$.*

Proof. (See [2] p. 36–37). Let W_0 be the smallest subspace of $T_G(\mathbf{C})$ defined over $\overline{\mathbf{Q}}$ which contains u. We want an algebraic subgroup H_0 of G, defined over $\overline{\mathbf{Q}}$, such that $W_0 = T_{H_0}(\mathbf{C})$. If $W_0 = T_G(\mathbf{C})$ (resp. $W_0 = 0$), take $H_0 = G$ (resp. $H_0 = 0$). Otherwise choose any hyperplane W of $T_G(\mathbf{C})$ defined over $\overline{\mathbf{Q}}$, which contains W_0. By Theorem 1.1, there exists an algebraic subgroup H of G, of positive dimension, such that $T_H(\mathbf{C}) \subset W$. Let H_W be the largest connected algebraic subgroup of G, defined over $\overline{\mathbf{Q}}$, for which

$$T_{H_W} \subset W.$$

By Theorem 1.1 on G/H_W, we deduce that u belongs to $T_{H_W}(\mathbf{C})$. Finally, we define H_0 as the intersection of H_W, when W runs over the hyperplanes of $T_G(\mathbf{C})$, defined over $\overline{\mathbf{Q}}$, which contain W_0. We get $T_{H_0}(\mathbf{C}) \subset W_0$, hence $T_{H_0} = W_0$. This proves Corollary 3.1.

Let us remark that our proof of Corollary 3.1 does not use Baker's method: we do not perform an extrapolation involving the Schwarz lemma on the one-dimensional complex line $\mathbf{C}.u$ in $T_G(\mathbf{C})$; also we do not need to introduce in our proof suitable division points of u, even if $\exp_G u$ is of finite order in $G(\overline{\mathbf{Q}})$ (compare with [2] p. 37). However, if one looks for effective estimates, one gets sharper results in the situation of Corollary 3.1 than in the general case of Theorem 1.1 if one combines the present approach with Baker's extrapolation procedure (see [10]).

b) *Schneider's method*

If we have no arithmetic assumption on V, we can always choose $W = 0$, which means $t = 0$, and the hypothesis on $m = rk_{\mathbf{Z}}Y$ is

$$m > (d_1 + 2d_2)(d - 1).$$

The corresponding statement for the multiplicative case $(d = d_1)$ is given in [2]. Here is an example involving a power of an elliptic curve $(d = d_2)$.

Let $\wp$ be a Weierstrass elliptic function with algebraic invariants g_2, g_3:

$$\wp'^2 = 4\wp^3 - g_2\wp - g_3.$$

A complex number u is an algebraic point of $\wp$ if either u is a pole of $\wp$ or else $\wp(u)$ is an algebraic number. Let k be the field of endomorphisms of the corresponding elliptic curve.

Corollary 3.2. *Let* u_{ij}, $1 \le i \le n$, $1 \le j \le \ell$, *be algebraic points of* $\wp$, *with* $n \ge 1$ *and* $\ell > \frac{2}{[k:\mathbf{Q}]} \cdot n(n + 1)$. *Further, let* $t_1, \ldots, t_n$ *be complex* *numbers. Assume that for $1 \leq j \leq \ell$, the point*

$$\sum_{i=1}^{n} t_i u_{ij}$$

is an algebraic point of $\wp$.

a) *If the ℓ points*

$$(u_{1j}, \ldots, u_{nj}), \qquad 1 \leq j \leq \ell,$$

in $\mathbf{C}^n$ are k-linearly independent, then the numbers $1, t_1, \ldots, t_n$ are k-linearly independent.

b) *If the ℓn numbers*

$$u_{ij}, \qquad 1 \leq i \leq n, \ 1 \leq j \leq \ell,$$

are k-linearly independent, then $t_1, \ldots, t_n$ are all in k.

Proof.

(a) Let E be the elliptic curve in $\mathbf{P}_2$ whose exponential map is given by

$$\exp_E(z) = \bigl(1, \wp(z), \wp'(z)\bigr).$$

Consider the algebraic group $G = E^{n+1}$ of dimension $d = d_2 = n + 1$. We identify $T_G(\mathbf{C})$ with $\mathbf{C}^{n+1}$ by

$$\exp_G(z_1 \ldots z_{n+1}) = (\exp_E z_1, \ldots, \exp_E z_{n+1}).$$

Let V be the hyperplane $z_{n+1} = t_1 z_1 + \ldots + t_n z_n$. Let

$$u_{n+1,j} = \sum_{i=1}^{n} t_i u_{ij}, \qquad 1 \leq j \leq \ell,$$

and

$$y_j = (u_{1j}, \ldots, u_{n+1,j}) \in \mathbf{C}^{n+1}, \qquad 1 \leq j \leq \ell.$$

We denote by σ the ring of endomorphisms of E, and by Y the σ-module generated by $y_1, \ldots, y_\ell$. Plainly we have

$$Y \subset V \quad \text{and} \quad rk_{\mathbf{Z}} Y = \ell \,.\, [k : \mathbf{Q}].$$

From Theorem 1.1 we deduce that V contains a non-zero $\overline{\mathbf{Q}}$ algebraic Lie sub-algebra of $T_G(\mathbf{C})$. Since $G = E^{n+1}$, V contains such a $\overline{\mathbf{Q}}$-Lie sub-algebra of dimension 1, hence there exists $(b_1, \ldots, b_{n+1}) \in k^{n+1}$ such that $0 \neq (b_1, \ldots, b_{n+1}) \in V$. This proves (a).

(b) There is no loss of generality in assuming that the k-vector space $k + kt_1 + \ldots + kt_n$ is generated by $1, t_1, \ldots, t_r$. Assume $r \geq 1$. Write

$$t_i = b_{i0} + \sum_{\rho=1}^{r} b_{i\rho} t\rho, \qquad r < i \leq n$$

where $b_{i\rho}$, $0 \leq \rho \leq r$, are in k. Define

$$u'_{\rho j} = u_{\rho j} + \sum_{i=r+1}^{n} b_{i\rho} u_{ij}, \qquad 1 \leq \rho \leq r, \ 1 \leq j \leq \ell,$$

and apply (a) with n replaced by r to get a contradiction. Hence $r = 0$ and $t_1 \in k$ for $1 \leq i \leq n$.

c) *Baker's theorem.*

We will deduce from Theorem 1.1 the following result of Baker [1] Chap. 2.

Corollary 3.3. *Let $\alpha_1, \ldots, \alpha_n$ be non-zero algebraic numbers such that $\log \alpha_1, \ldots, \log \alpha_n$ are linearly independent over $\mathbf{Q}$. Then the numbers $1, \log \alpha_1, \ldots, \log \alpha_n$ are linearly independent over $\overline{\mathbf{Q}}$.*

Of course Corollary 3.3 is a special case of Corollary 3.1 (see [16], [2]): we take $G = \mathbf{G}_a \times \mathbf{G}_m^n$, and

$$u = (1, \log \alpha_1, \ldots, \log \alpha_n) \in \mathbf{C} \times \mathbf{C}^n;$$

if there is a non-trivial relation

$$\beta_0 + \beta_1 \log \alpha_1 + \ldots + \beta_n \log \alpha_n = 0,$$

then Corollary 3.1 shows that the hyperplane

$$\beta_0 z_0 + \beta_1 z_1 + \ldots + \beta_n z_n = 0$$

contains the smallest $\overline{\mathbf{Q}}$ algebraic Lie sub-algebra of $T_G(\mathbf{C})$ which contains u, hence the point $(\log \alpha_1, \ldots, \log \alpha_n)$ in $\mathbf{C}^n$ belongs to a hyperplane which is defined over $\mathbf{Q}$.

We give another proof of Corollary 3.3, which is more close to Schneider's method (see [14] §8.3.b, [7], [17]).

a) We first use Schneider's method to prove that $\log \alpha_1, \ldots, \log \alpha_n$ are $\overline{\mathbf{Q}}$-linearly independent. Assume

$$\log \alpha_n = \beta_1 \log \alpha_1 + \ldots + \beta_{n-1} \log \alpha_{n-1},$$

where $\beta_1, \ldots, \beta_{n-1}$ are algebraic (and not all rational). Consider the algebraic group $G = \mathbf{G}_a^{n-1} \times \mathbf{G}_m$, of dimension $d = d_1 = n$, and the hyperplane V of equation

$$z_n = z_1 \log \alpha_1 + \ldots + z_{n-1} \log \alpha_{n-1}$$

in $\mathbf{C}^{n-1} \times \mathbf{C}$. Further, let

$$Y = \{(h_1 + h_n \beta_1, \ldots, h_{n-1} + h_n \beta_{n-1}, h_1 \log \alpha_1 + \ldots + h_n \log \alpha_n);$$
$$(h_1, \ldots, h_n) \in \mathbf{Z}^n\}.$$

Therefore Y is of rank n and is contained in V. From Theorem 1.1 with $t = 0$ we deduce that V contains a non-zero element $(\gamma_1, \ldots, \gamma_{n-1}, 0)$ where $\gamma_1, \ldots, \gamma_{n-1}$ are algebraic. This contradicts the assumption that $\log \alpha_1, \ldots, \log \alpha_{n-1}$ are linearly independent over $\overline{\mathbf{Q}}$.

b) We now start from a relation

$$\log \alpha_n = \beta_0 + \beta_1 \log \alpha_1 + \ldots + \beta_{n-1} \log \alpha_{n-1},$$

where $1, \log \alpha_1, \ldots, \log \alpha_{n-1}$ are $\overline{\mathbf{Q}}$-linearly independent. We take $G = \mathbf{G}_a^n \times \mathbf{G}_m$, and V is the hyperplane

$$z_n = \beta_0 z_0 + z_1 \log \alpha_1 + \ldots + z_{n-1} \log \alpha_{n-1},$$

while

$$Y = \{(h_n, h_1 + h_n \beta_1 \ldots, h_{n-1} + h_n \beta_{n-1}, h_1 \log \alpha_1 + \ldots + h_n \log \alpha_n);$$
$$(h_1, \ldots, h_n) \in \mathbf{Z}^n\}.$$

We now take
$$W = \mathbf{C}(1, 0, \ldots, 0, 1) \subset \mathbf{C}^n \times \mathbf{C},$$

and we use Theorem 1.1 with $d = n + 1$, $d_0 = n$, $d_1 = 1$, $t = 1$, $m = n$. We find in V a non-zero element $(\gamma_0, \gamma_1, \ldots, \gamma_{n-1}, 0)$, where the γ's are algebraic. Therefore

$$\beta_0 \gamma_0 + \gamma_1 \log \alpha_1 + \ldots + \gamma_{n-1} \log \alpha_{n-1} = 0.$$

From the initial assumption we deduce $\gamma_1 = \ldots = \gamma_{n-1} = 0$, hence $\beta_0 = 0$, which is what we wanted.

Remark. This new proof of Baker's theorem can be refined into an effective lower bound for linear forms in logarithms (see [15] §6.d; compare with [1] Chap. 2). The estimate we get by this method is the same as can be achieved by Gelfond's method alone. As mentioned above, it means that it is weaker than the estimates which arise by combining Gelfond's and Baker's method. It would be interesting to deduce Baker's estimates from Schneider's method.

§4. The main result.

a) *The statement.*

In Theorem 1.1, we assumed that V was a hyperplane of $T_G(\mathbf{C})$. In some cases (e.g. [15]), it is interesting to deal with a subspace of $T_G(\mathbf{C})$ of any dimension $n < d$, and instead of the assumption

$$m > (d_1 + 2d_2)(d - 1 - t),$$

we require only

$$m > (d_1 + 2d_2) \cdot \frac{n - t}{d - n}.$$

Let us consider again an algebraic group $G = \mathbf{G}_a^{d_0} \times \mathbf{G}_m^{d_1} \times G_2$, defined over $\overline{\mathbf{Q}}$, of dimension $d = d_0 + d_1 + d_2 \geq 1$. Let

$$\pi_0 : G \longrightarrow \mathbf{G}_a^{d_0} \text{ and } \pi_1 : G \longrightarrow \mathbf{G}_m^{d_1}$$

be the corresponding projections. Here we do not assume that d_0 and d_1 are maximal.

Theorem 4.1 *Let V be a subspace of $T_G(\mathbf{C})$ of dimension $n < d$, W a subspace of V, and Y a finitely generated subgroup of V. Assume that W is defined over $\overline{\mathbf{Q}}$, and that $\Gamma = \exp_G Y$ is contained in $G(\overline{\mathbf{Q}})$. Finally, define*

$$\kappa = rk_{\mathbf{Z}}(Y \cap \mathrm{Ker}\exp_G).$$

Then there exists a connected algebraic subgroup G' of G, defined over $\overline{\mathbf{Q}}$, with $G' \neq G$, satisfying the following properties. Define

$$\delta = \dim G/G', \qquad \delta_0 = \dim \mathbf{G}_a^{d_0}/\pi_0(G'),$$

$$\delta_1 = \dim G_m^{d_1}/\pi_1(G'), \qquad \delta_2 = \delta - \delta_0 - \delta_1,$$
$$\lambda = rk_{\mathbf{Z}}\Gamma/\Gamma \cap G', \qquad \tau = \dim W/W \cap T_{G'}(\mathbf{C}).$$

Then $\delta > \tau$ and

$$(\lambda + \delta_1 + 2\delta_2)(d - n) \leq (\delta - \tau)(d_1 + 2d_2 - \kappa).$$

The conclusion holds trivially with $G' = 0$ if the inequality

$$m > (d_1 + 2d_2 - \kappa) \cdot \frac{n - t}{d - n} \qquad (4.2)$$

is not satisfied. On the other hand, if this inequality (4.2) holds, then $\dim G' > 0$.

The special case $t = d - 1$ of Theorem 1.1 (see Corollary 3.1) readily follows from Theorem 4.1: when W is a hyperplane of $T_G(\mathbf{C})$, the condition $\delta > \tau$ means $W \supset T_{G'}(\mathbf{C})$, and the assumption $m > 0$ gives $\dim G' > 0$.

The proof of Theorem 4.1 is given in §7 below. We will deduce Theorem 1.1 from Theorem 4.1 in §5. Now we give some further corollaries to Theorem 4.1.

b) *Schneider's method*

Here, we use only the case $t = 0$ of Theorem 4.1. Let us recall (cf. [14]) that

$$\mu(Y, V) = \min_{E \subset V} \frac{rk_{\mathbf{Z}} Y/Y \cap E}{\dim_{\mathbf{C}} V/E},$$

where E runs over the set of vector subspaces of V with $E \neq V$.

Corollary 4.3 *With the assumptions of Theorem 4.1, if $\exp_G V$ is Zariski dense in $G(\mathbf{C})$, then*

$$\mu(Y, V) \leq (d_1 + 2d_2 - \kappa)/(d - n).$$

In the case $\dim V = 1$, the conclusion is simply

$$m \leq (d_1 + 2d_2 - \kappa)/(d - n),$$

which is equivalent to the results of [14] Chap. 4.

We will deduce Corollary 4.3 from Theorem 4.1 in section e below. We first deduce some consequence from Corollary 4.3.

c) *Algebraic points on the graph of an analytic homomorphism.*

Let us denote by G' a commutative algebraic group defined over $\overline{\mathbf{Q}}$, by $\Psi : \mathbf{C}^n \longrightarrow G'(\mathbf{C})$ an analytic homomorphism, and by Y a subgroup of $\overline{\mathbf{Q}}^n$ such that $\Psi(Y) \subset G'(\overline{\mathbf{Q}})$. We write

$$\rho = \rho(G') = \begin{cases} 1 & \text{if } G' \text{ is linear,} \\ 2 & \text{otherwise.} \end{cases}$$

Corollary 4.4 *Assume that* $\dim G' \geq 1$, *and that* G' *does not contain a non-zero unipotent linear subgroup. If* Ψ *is not constant, then*

$$\mu(Y, \overline{\mathbf{Q}}^n) \leq \rho.$$

We deduce Corollary 4.4 by applying Corollary 4.3 to $G = \mathbf{G}_a^n \times G''$, where G'' is the Zariski closure of $\Psi(\mathbf{C}^n)$ in $G'(\mathbf{C})$, with $d \geq n+1$, $d = d_\rho$, $d_0 = n$.

A special case of Corollary 4.4 was already given in [14] Prop. 8.1.2. Here is a consequence of Corollary 4.4.

Corollary 4.5 *Let* L *be the maximal (connected) unipotent linear algebraic subgroup of* G. *Then the image of* $\Psi(\overline{\mathbf{Q}}^n) \cap G'(\overline{\mathbf{Q}})$ *in* G'/L *has a finite rank* $\leq \rho n$.

Proof. We proceed by induction on n. Assume $y_1, \ldots, y_{\rho n+1}$ are in $\overline{\mathbf{Q}}^n$, with $\Psi(y_j) \in G'(\overline{\mathbf{Q}})$, $1 \leq j \leq \rho n + 1$, and that their images in G'/L are $\mathbf{Q}$-linearly independent. Let $Y = \mathbf{Z}y_1 + \ldots + \mathbf{Z}y_{\rho n+1}$. From Corollary 4.4 we deduce $\mu(Y, \overline{\mathbf{Q}}^n) \leq \rho$. Let W be a subspace of $\mathbf{C}^n$, defined over $\overline{\mathbf{Q}}$, such that

$$rk_{\mathbf{Z}} Y \cap W \geq 1 + \rho \dim_{\mathbf{C}} W,$$

and $W \neq \mathbf{C}^n$. The restriction of Ψ to W gives a contradiction with the induction hypothesis.

Thanks to Corollary 4.4, we can prove the following result, which was announced in [15] (6.7).

Corollary 4.6 *If* $rk_{\mathbf{Z}} Y \geq \rho n + 1$, *then there exists* $y \in Y$, $y \neq 0$, *such that the homomorphism* $t \longrightarrow \Psi(yt)$ *of* $\mathbf{C}$ *into* $G'(\mathbf{C})$ *is rational.*

This result was proved already in [14] Th. 8.1.1 under the assumption $Y \subset \mathbf{R}^n$, and in [14] Th. 6.3.2 under the assumption that G' is an extension by a linear group of an abelian variety which is isogeneous to a product of simple abelian varieties of C.M. type. We could also deduce Corollary 4.6 from Corollary 3.1 following [14] Chap. 6.

Proof of Corollary 4.6. Assume first $\mu(Y, \mathbf{C}^n) > \rho$. Then Corollary 4.4 shows that $\Psi(\mathbf{C}^n)$ is contained in the maximal unipotent linear subgroup of G', hence Ψ is rational. The general case follows from the arguments of [14] p. 150.

d) *The coefficient* $\mu^{\sharp}(\Gamma, G)$.

Let us introduce the following Dirichlet exponent: let K be a subfield of $\mathbf{C}$, G be a commutative algebraic group of dimension $d \geq 1$, Γ a finitely generated subgroup of $G(K)$, and

$$\pi_0 : G \longrightarrow \mathbf{G}_a^{d_0}, \quad \pi_1 : G \longrightarrow \mathbf{G}_m^{d_1}$$

two surjective morphisms, with $d_0 \geq 0$, $d_1 \geq 0$. Further, we set $d_2 = d - d_0 - d_1$. Therefore $G = \mathbf{G}_a^{d_0} \times \mathbf{G}_m^{d_1} \times G_2$, where $\dim G_2 = d_2$. We define

$$\mu^{\sharp}(\Gamma, G) = \min_{G' \subsetneq G} (\lambda + \delta_1 + 2\delta_2)/\delta,$$

where G' runs over the set of algebraic subgroups of G, defined over K, with $G' \neq G$, and

$$\delta = \dim G/G', \quad \delta_0 = \dim \mathbf{G}_a^{d_0}/\pi_0(G'),$$

$$\delta_1 = \dim \mathbf{G}_m^{d_1}/\pi_1(G'), \quad \delta_2 = \delta - \delta_0 - \delta_1,$$

$$\lambda = rk_{\mathbf{Z}}\Gamma/\Gamma \cap G'.$$

It should be noted that $\mu^{\sharp}(\Gamma, G)$ depends not only on Γ and G, but also on K, π_0 and π_1. If G' is any algebraic subgroup of G, $G' \neq G$, *we define* $\mu^{\sharp}(\Gamma/\Gamma \cap G', G/G')$ by choosing

$$\pi_0' : G' \longrightarrow \mathbf{G}_a^{d_0'}, \quad \pi_1' : G' \longrightarrow \mathbf{G}_m^{d_1'},$$

with $\delta_0 = \dim \mathbf{G}_a^{d_0}/\pi_0(G')$ and $\delta_1 = \dim \mathbf{G}_m^{d_1}/\pi_1(G')$ so that we get commutative diagrams:

$$
\begin{array}{ccc}
G & \xrightarrow{\pi_0} & \mathbf{G}_a^{d_0} \\
\downarrow & & \downarrow \\
G/G' & \xrightarrow{\pi_0'} & \mathbf{G}_a^{d_0}/\pi_0(G')
\end{array}
$$

and

$$
\begin{array}{ccc}
G & \xrightarrow{\pi_1} & G_m^{d_1} \\
\downarrow & & \downarrow \\
G/G' & \xrightarrow{\pi_1'} & G_m^{d_1}/\pi_1(G')
\end{array}
$$

Also we define $\mu^\sharp(\Gamma \cap G', G')$ by choosing the restrictions $G' \longrightarrow \pi_0(G')$ and $G' \longrightarrow \pi_1(G')$ with $\pi_0(G') \cong G_a^{d_0 - \delta_0}$ and $\pi_1(G') \cong G_m^{d_1 - \delta_1}$, where $\cong$ means *isogeneous to*.

By taking $G' = 0$, we see that

$$
\mu^\sharp(\Gamma, G) \le \frac{\ell + d_1 + 2d_2}{d},
$$

where $\ell = \mathrm{rank}_\mathbf{Z} \Gamma$.

e) *Proof of Corollary 4.3*

The conclusion of Theorem 4.1 is

$$
\mu^\sharp(\Gamma, G) \le \frac{d_1 + 2d_2 - \kappa}{d - n}. \tag{4.7}
$$

If $\mu^\sharp(\Gamma, G) = (\ell + d_1 + 2d_2)/d$, then (4.7) gives

$$
\frac{\ell + \kappa}{n} \le \frac{d_1 + 2d_2 - \kappa}{d - n}.
$$

and Corollary 4.3 follows. Otherwise, we write

$$
\mu^\sharp(\Gamma, G) = (\lambda + \delta_1 + 2\delta_2)/\delta
$$

for some algebraic subgroup G' of G, $G' \ne G$, of dimension $d - \delta > 0$. Clearly we have

$$
\mu^\sharp(\Gamma/\Gamma \cap G', G/G') = (\lambda + \delta_1 + 2\delta_2)/\delta.
$$

We define

$$
E = V \cap T_{G'}, \qquad V' = V/E, \qquad Y' = Y/Y \cap E,
$$
$$
n' = \dim V', \qquad m' = rk_\mathbf{Z} Y', \qquad \kappa' = rk_\mathbf{Z}(Y' \cap \ker \exp_{G/G'});
$$

therefore

$$
\mu(Y, V) \le m'/n'.
$$

Further, let

$$
\Gamma' = \exp_{G/G'} Y' = \Gamma/\Gamma \cap G'.
$$

We notice that $m' = \lambda + \kappa'$. We apply (4.7) to Γ':

$$(\delta - n')\mu^\sharp(\Gamma', G/G') \leq \delta_1 + 2\delta_2 - \kappa'.$$

Hence

$$m'\delta \leq (\lambda + \kappa')\delta \leq n'(\lambda + \delta_1 + 2\delta_2).$$

We conclude that

$$\mu(\Gamma, V) \leq \frac{m'}{n'} \leq \frac{\lambda + \delta_1 + 2\delta_2}{\delta} \leq \frac{d_1 + 2d_2 - \kappa}{d - n}.$$

§5. Proof of Theorem 1.1

In this section we deduce Theorem 1.1 (with a slight refinement) from Theorem 4.1. We first introduce a generalization of the coefficient $\mu^\sharp$ of §4. Next we prove an auxiliary lemma concerning some problem which arises with the periods of the exponential map, and finally we complete the proof of Theorem 1.1

a) *The coefficient* $\mu^\sharp(\Gamma, G, W)$.

Let K be a subfield of $\mathbf{C}$, G be commutative connected algebraic group of dimension d, $\pi_0 : G \longrightarrow \mathbf{G}_a^{d_0}$ and $\pi_1 : G \longrightarrow \mathbf{G}_m^{d_1}$ two surjective morphisms of algebraic groups, $d_2 = d - d_0 - d_1$, Γ a finitely generated subgroup of $G(K)$, and W a subspace of $T_G(\mathbf{C})$, distinct from $T_G(\mathbf{C})$. We define

$$\mu^\sharp(\Gamma, G, W) = \min_{G'} \frac{\lambda + \delta_1 + 2\delta_2}{\delta - \tau},$$

where G' runs over the set of connected algebraic subgroups of G which are defined over K, with $G' \neq G$ and $\delta > \tau$, and where

$$\delta = \dim G/G', \qquad\qquad \delta_0 = \dim \mathbf{G}_a^{d_0}/\pi_0(G'),$$

$$\delta_1 = \dim \mathbf{G}_m^{d_1}/\pi_1(G'), \qquad \delta_2 = \delta - \delta_0 - \delta_1,$$

$$\lambda = rk_{\mathbf{Z}}\Gamma/\Gamma \cap G', \qquad \tau = \dim_{\mathbf{C}} W/W \cap T_{G'}.$$

Remarks

(1) Since $\tau = \dim(T_{G'} + W)/T_{G'}$, we have $\delta - \tau = \dim T_G/(T_{G'} + W)$, and therefore the condition $\tau = \delta$ is equivalent to $T_{G'} + W = T_G$. In any case $\mu^\sharp$ satisfies:

$$\mu^\sharp(\Gamma, G, W) \leq \frac{\ell + d_1 + 2d_2}{d - n},$$

where $\ell = rk_\mathbf{Z}\Gamma$ and $t = \dim_\mathbf{C} W$.

(2) The coefficient $\mu^\sharp$ depends not only on G, Γ and W, but also on the choice of π_0 and π_1. For G' connected algebraic subgroup of G, we define

$$\mu^\sharp(\Gamma/\Gamma \cap G', G/G', W/W \cap T_{G'}), \qquad \text{if } G' \neq G,$$

and

$$\mu^\sharp(\Gamma \cap G', G', W \cap T_{G'}), \qquad \text{if } G' \neq 0,$$

with the same conventions as in §4.c.

We need the following generalization of Lemma 1.3.1 of [14] and Lemma 3.2 of [15].

Lemma 5.1 *Assume*

$$\mu^\sharp(\Gamma, G, W) < \frac{\ell + d_1 + 2d_2}{d - t}.$$

Then there exists an algebraic subgroup G' of G, which is defined over K, of dimension $d' \geq 1$, such that either $W \supset T_{G'}$, or

$$\mu^\sharp(\Gamma \cap G', G', W \cap T_{G'}) = \frac{\ell' + d_1' + 2d_2'}{d' - t} > \frac{\ell + d_1 + 2d_2}{d - t},$$

where

$$t' = \dim W \cap T_{G'}, \qquad d_1' = \dim \pi_1(G'),$$
$$d_0' = \dim \pi_0(G'), \qquad \ell' = rk_\mathbf{Z}\Gamma \cap G',$$

and

$$d_2' = d' - d_0' - d_1'.$$

Proof. Assume that W does not contain a non-zero K-algebraic Lie subalgebra of $T_G(\mathbf{C})$. We will prove the desired conclusion by induction on d. If $d = 1$ then $t = 0$ and

$$\mu^\sharp(\Gamma, G, 0) = \ell + d_1 + 2d_2.$$

Assume Lemma 5.1 holds for all proper algebraic subgroups of G. By the definition of $\mu^\sharp$, there exists an algebraic subgroup G° of G such that

$$\mu^\sharp(\Gamma, G, W) = (\lambda^\circ + \delta_1^\circ + 2\delta_2^\circ)/(\delta^\circ - \tau^\circ),$$

where

$$\delta^\circ = \dim G/G^\circ, \qquad \delta_0^\circ = \dim G_a^{d_0}/\pi_0(G^\circ),$$
$$\delta_1^\circ = \dim G_m^{d_1}/\pi_1(G^\circ), \qquad \delta_2^\circ = \delta^\circ - \delta_0^\circ - \delta_1^\circ,$$
$$\lambda^\circ = rk_\mathbf{Z}\Gamma/\Gamma \cap G^\circ, \qquad \tau^\circ = \dim_\mathbf{C} W/W \cap T_{G^\circ}.$$

We define

$$d^\circ = \dim G^\circ, \qquad d_0^\circ = \dim \pi_0(G^\circ),$$
$$d_1^\circ = \dim \pi_1(G^\circ), \qquad d_2^\circ = d^\circ - d_0^\circ - d_1^\circ,$$
$$t^\circ = \dim W \cap T_{G^\circ}, \qquad \ell^\circ = rk_\mathbf{Z}\Gamma \cap G^\circ.$$

Hence

$$\delta^\circ + d^\circ = d, \qquad \delta_i^\circ + d_i^\circ = d_i, \qquad i = 0,1,2,$$
$$\lambda^\circ + \ell^\circ = \ell, \qquad t^\circ + \tau^\circ = t.$$

The assumption that W does not contain T_{G° gives $d^\circ > t^\circ$, and the hypothesis

$$\mu^\sharp(\Gamma, G, W) < (\ell + d_1 + 2d_2)/(d - t)$$

is equivalent to

$$\frac{\ell^\circ + d_1^\circ + 2d_2^\circ}{d^\circ - t^\circ} > \frac{\ell + d_1 + 2d_2}{d - t}.$$

If

$$\mu^\sharp(\Gamma \cap G^\circ, G^\circ, W \cap T_{G^\circ}) = (\ell^\circ + d_1^\circ + 2d_2^\circ)/(d^\circ - t^\circ),$$

then the lemma is proved with $G' = G^\circ$. Otherwise we can use the induction hypothesis, since $d^\circ < d$. We deduce that there exists an algebraic subgroup G' of G° such that

$$\mu^\sharp(\Gamma \cap G', G', W \cap T_{G'}) = \frac{\ell' + d_1' + 2d_2'}{d' - t'} > \frac{\ell^\circ + d_1^\circ + 2d_2^\circ}{d^\circ - t^\circ},$$

with

$$d_i' = \dim \pi_i(G'), \qquad (i = 0,1)$$
$$d' = \dim G', \qquad d_2' = d' - d_0' - d_1'.$$
$$t' = \dim W \cap T_{G'}, \qquad \ell' = rk_\mathbf{Z}\Gamma \cap G'.$$

This completes the proof of Lemma 5.1.

b) *Another auxiliary lemma.*

Let G be a commutative algebraic group over $\mathbf{C}$ of dimension $d = d_0 + d_1 + d_2 \geq 1$, as before. Further let $Y = \mathbf{Z}y_1 + \ldots + \mathbf{Z}y_m$ be a finitely generated subgroup of $T_G(\mathbf{C})$, and G' an algebraic subgroup of G. Define

$$\Gamma = \exp_G Y, \qquad Y' = Y \cap T_{G'}, \qquad \Gamma' = \exp_G Y'.$$

Of course we have $\Gamma' \subset \Gamma \cap G'$, but the rank of $\Gamma \cap G'$ may be larger than the rank of Γ', because of the periods of $\exp_G$.

Let us define $\Omega = \operatorname{Ker} \exp_G$, and

$$\kappa = rk_{\mathbf{Z}} Y \cap \Omega, \qquad \kappa' = rk_{\mathbf{Z}} Y' \cap \Omega.$$

Lemma 5.2. *We have*

$$rk_{\mathbf{Z}} \Gamma' \geq rk_{\mathbf{Z}} \Gamma \cap G' - (d_1 + 2d_2 - \kappa) + d_1' + 2d_2' - \kappa',$$

where, as before,

$$d' = \dim G', \qquad d_0' = \dim \pi_0(G'),$$
$$d_1' = \dim \pi_1(G'), \qquad d_2' = d' - d_0' - d_1'.$$

Proof Let Ω' be the kernel of the exponential map of G/G' in $T_{G/G'} \cong T_G/T_{G'}$. We first remark that G/G' is a product of $\mathbf{G}_a^{d_0 - d_0'} \times \mathbf{G}_m^{d_1 - d_1'}$ by an algebraic group of dimension $d_2 - d_2'$, hence

$$rk_{\mathbf{Z}} \Omega' \leq (d_1 + 2d_2) - (d_1' + 2d_2').$$

By considering the surjective map

$$Y/Y' \longrightarrow \Gamma/\Gamma \cap G'$$

given by $\exp_{G/G'}$, we find

$$rk_{\mathbf{Z}} Y - rk_{\mathbf{Z}} Y' \leq rk_{\mathbf{Z}} \Gamma - rk_{\mathbf{Z}} \Gamma \cap G' + rk_{\mathbf{Z}} \Omega'.$$

From

$$rk_{\mathbf{Z}} Y = rk_{\mathbf{Z}} \Gamma + \kappa$$

and

$$rk_{\mathbf{Z}} Y' = rk_{\mathbf{Z}} \Gamma' + \kappa'$$

we easily deduce Lemma 5.2.

c) *An upper bound for* $\mu^{\sharp}$.

We now come back to the arithmetic case where G is defined over $\overline{\mathbf{Q}}$. We can state Theorem 4.1 in the following way.

Corollary 5.3 *With the assumptions of Theorem 4.1,*

$$\mu^{\sharp}(\Gamma, G, W) \leq (d_1 + 2d_2 - \kappa)/(d - n).$$

d) *Proof of Theorem 1.1*

We proceed by induction on d, the case $d = 1$ being trivial. We assume that the hypotheses of Theorem 1.1 are satisfied, apart from (1.2) which we replace by the weaker assumption

$$m > (d_1 + 2d_2 - \kappa)(d - 1 - t), \tag{5.4}$$

with $\kappa = rk_{\mathbf{Z}}(Y \cap \mathrm{Ker}\,\exp_G) = \ell - m$.

From (5.4) we have

$$\ell + d_1 + 2d_2 \geq (d - t)(d_1 + 2d_2 - \kappa). \tag{5.5}$$

We assume that the conclusion of Theorem 1.1 does not hold, and we will deduce a contradiction.

By Corollary 5.3 (with $n = d - 1$) and assumption (5.5) we have

$$\mu^{\sharp}(\Gamma, G, W) \leq d_1 + 2d_2 - \kappa < \frac{\ell + d_1 + 2d_2}{d - t}.$$

Using Lemma 5.1 with the assumption that V (hence W) does not contain a non-zero $\overline{\mathbf{Q}}$-Lie sub-algebra of $T_G(\mathbf{C})$, we find an algebraic subgroup G' of G, of dimension $d' \geq 1$, such that

$$\mu^{\sharp}(\Gamma \cap G', G', W \cap T_{G'}) = \frac{\ell + d_1' + 2d_2'}{d' - t'} > \frac{\ell + d_1 + 2d_2}{d - t}, \tag{5.6}$$

From Lemma 5.2 we deduce that $\Gamma' = \exp_G Y'$, with $Y' = Y \cap T_{G'}$, satisfies

$$\mu^{\sharp}(\Gamma', G', W \cap T_{G'}) \geq \mu^{\sharp}(\Gamma \cap G', G', W \cap T_{G'}) - (d_1 + 2d_2 - \kappa) + d_1' + 2d_2' - \kappa'.$$

From (5.5) and (5.6) we get

$$\mu^{\sharp}(\Gamma', G', W \cap T_{G'}) > d_1' + 2d_2' - \kappa'.$$

Corollary 5.3 shows that $V \cap T_{G'}$ is not a hyperplane of $T_{G'}$, hence $V \supset T_{G'}$, which is the desired contradiction.

§6. The auxiliary function

The proof of Theorem 4.1 involves a refinement of Proposition 2.4 of [15], which we now give. We consider as before an algebraic group $G = G_a^{d_0} \times G_m^{d_1} \times G_2$ over $\overline{\mathbf{Q}}$, of dimension $d = d_0 + d_1 + d_2$, a vector subspace V of $T_G(\mathbf{C})$ of dimension $n < d$, a subspace W of V, of dimension $t \geq 0$, which is defined over $\overline{\mathbf{Q}}$ in $T_G(\mathbf{C})$, and a finitely generated subgroup $Y = \mathbf{Z}y_1 + \ldots + \mathbf{Z}y_m$ of V of rank m such that $\Gamma = \exp_G Y$ is contained in $G(\overline{\mathbf{Q}})$. For each integer $S \geq 1$ we write

$$Y(S) = \{ h_1 y_1 + \ldots + h_m y_m : (h_1, \ldots, h_m) \in \mathbf{Z}^m,$$

$$0 \leq h_j \leq S, \ 1 \leq j \leq m \},$$

and

$$\Gamma(S) = \exp_G Y(S).$$

Next, let κ satisfy

$$0 \leq \kappa \leq rk_{\mathbf{Z}} V \cap \operatorname{Ker} \exp_G.$$

Finally, we choose a basis $e_1, \ldots, e_t$ of W, defined over $\overline{\mathbf{Q}}$, and we denote by $\Psi : \mathbf{C}^t \longrightarrow G(\mathbf{C})$ the t-parameters subgroup defined by

$$\mathbf{C}^t \cong W \subset T_G(\mathbf{C}) \overset{\exp_G}{\longrightarrow} G(\mathbf{C}).$$

Given an embedding of G_2 into a projective space $\mathbf{P}_N$, and a polynomial P in $d_0 + d_1 + N + 1$ unknowns, which is homogeneous in the last $N + 1$ unknowns, we say that P vanishes at a point γ in $G(\mathbf{C})$ with multiplicity T along W if the function $z \longrightarrow P(\Psi(z) + \gamma)$ has a zero of order T at the point $z = 0$ in $\mathbf{C}^t$ (see [9] and [10]).

We choose two real numbers $a \geq 1$ and $b \geq 1$.

Proposition 6.1 *There exist an embedding of G_2 in a projective space $\mathbf{P}_N$ over $\overline{\mathbf{Q}}$, and a constant $C > 0$, satisfying the following properties.*

For each integer $S \geq 2$, define T, D_0, D_1, D_2, Δ as functions of S by

$$\Delta^{d-n} = C \cdot S^{d_1 + 2d_2 - \kappa} \cdot (\log S)^{bd_0}$$

and

$$T(\log S)^a = D_0(\log S)^b = D_1 S = D_2 S^2 = \Delta.$$

There exists a sequence $(P_S)_{S \geq S_0}$ of polynomials in the ring,

$$\overline{\mathbf{Q}}[X_1^0, \ldots, X_{d_0}^0, \ X_1^1, \ldots, X_{d_1}^1, \ X_0^2, \ldots, X_N^2],$$

where P_S

  –is of degree $\leq D_0$ in the variables $X_1^0, \ldots, X_{d_0}^0$,

  –is of degree $\leq D_1$ in the variables $X_1^1, \ldots, X_{d_1}^1$,

  –is homogeneous of degree $\leq D_2$ in the variables $X_0^2, \ldots, X_N^2$,

  –vanishes at all the points of $\Gamma(S)$ with multiplicity $\geq T$ along W,

  –but does not vanish everywhere on $G(\mathbf{C})$.

This result is proved in [15] Proposition 2.4 in the case $W = 0$ and $b = 1$. The estimates for the derivatives are provided by Lemma 7 of D. Bertrand in Appendix 1 of [14] (compare with [3] and [10]).

§7. Philippon's zero estimate.

We quote here a special case of the main result of [8] (see also [9]) which will enable us to complete the proof of Theorem 4.1.

Let K be a subfield of $\mathbf{C}$, $G = \mathbf{G}_a^{d_0} \times \mathbf{G}_m^{d_1} \times G_2$ a commutative connected algebraic group over K, $\Gamma = \mathbf{Z}\gamma_1 + \ldots + \mathbf{Z}\gamma_m$ a finitely generated subgroup of $G(K)$, and W a subspace of $T_G(\mathbf{C})$ defined over K. We fix an embedding of G_2 into a projective space $\mathbf{P}_N$, defined over K.

Proposition 7.1 *There exists a positive constant c with the following property. Let T, D_0, D_1, D_2, S be positive numbers, with $D_2 \leq D_0$ and $D_2 \leq D_1$. Assume that there exists a hypersurface of $\mathbf{A}_{d_0 + d_1} \times \mathbf{P}_N$, of degrees $\leq D_0, D_1, D_2$, which does not contain G, but vanishes along W with order $\geq T + 1$ at each point of $\Gamma(S)$.*

Then there exists a connected algebraic subgroup G' of G, defined over K, such that if we set

$$\delta = \dim G/G', \qquad\qquad \delta_0 = \dim \mathbf{G}_a^{d_0}/\pi_0(G'),$$
$$\delta_1 = \dim \mathbf{G}_m^{d_1}/\pi_1(G'), \qquad \delta_2 = \delta - \delta_0 - \delta_1,$$

$$\lambda = rk_{\mathbf{Z}}\Gamma/\Gamma \cap G', \qquad \tau = \dim_{\mathbf{C}} W/W \cap T_{G'},$$

then $\delta \geq 1$ *and*

$$T^\tau S^\lambda \leq c D_0^{\delta_0} D_1^{\delta_1} D_2^{\delta_2}. \tag{7.2}$$

Given our choice of parameters in Section 6, the inequality (7.2) yields

$$S^{\lambda+\delta_1+2\delta_2}(\log S)^{b\delta_0 - a\delta} \leq c\Delta^{\delta-\tau}. \tag{7.3}$$

Therefore

$$(\lambda + \delta_1 + 2\delta_2)(d - n) \leq (\delta - \tau)(d_1 + 2d_2 - \kappa).$$

It remains to check that $\delta > \tau$. But if $\delta = \tau$, then $\lambda + \delta_1 + 2\delta_2 = 0$, hence $\lambda = \delta_1 = \delta_2 = 0$ and $\delta_0 = \delta \geq 1$; then (7.3) gives a contradiction if we choose, say, $a = 1$ and $b > t$.

This completes the proof of Theorem 4.1.

References

[1] A. Baker, *Transcendental number theory*; Cambridge Univ. Press, 2nd Ed., 1979.

[2] D. Bertrand, Lemmes de zéros et nombres transcendants; Séminaire Bourbaki, 38ème année, 1985–86, no 652; *Astérisque* **145–146** (1987), 21–44.

[3] D. Bertrand, La théorie de Baker revisitée; *Publ. Math. Univ. P. et M. Curie*, no 73, Groupe d'Etude sur les Problèmes Diophantiens 1984–85, no 2, 25p.

[4] W. D. Brownawell, The algebraic independence of certain numbers related by the exponential function; *J. Number Theory*, **6** (1974), 22–31.

[5] A. O. Gelfond, Sur le septième probléme de Hilbert; *Dokl. Akad. Nauk S.S.S.R.*, **2** (1934), 1–6.

[6] S. Lang, *Introduction to transcendental number theory*; Addison Wesley, 1966.

[7] J. C. Moreau, Lemmes de Schwarz en plusieurs variables et applications arithmétiques; Sém. P. Lelong - H. Skoda (Analyse), 1978–79, *Springer Lecture Notes* **822** (1980), 174–190.

[8] P. Philippon, Lemmes de zéros dans les groupes algébriques commutatifs; *Bull. Soc. Math. France*, **114** (1986), 355–383.

[9] P. Philippon, Un lemme de zéros pour les groupes produits; *Publ. Math. Univ. P. et M. Curie*, no 73, Groupe d'Etude sur les Problèmes Diophantiens 1984–85, no 6, 4p.

[10] P. Philippon et M. Waldschmidt, Formes linéaires de logarithmes sur les groupes algébriques commutatifs, *Illinois J. Math.* to appear.

[11] K. Ramachandra, Contributions to the theory of transcendental numbers, *Acta Arith.*, **14** (1968), 65–88.

[12] Th. Schneider, Transzendenzuntersuchungen periodischer Funktionen, *J. reine angew. Math.*, **172** (1934), 65–69.

[13] Th. Schneider, *Introduction aux nombres transcendants*, Springer-Verlag 1957, Gauthier-Villars 1959.

[14] M. Waldschmidt, Nombres transcendants et groupes algébriques, Soc. Math. France, *Astérisque* 69–70, 1979.

[15] M. Waldschmidt, Sous-groupes analytiques de groupes algébriques, *Annals of Math.*, **117** (1983), 627–657.

[16] G. Wüstholz, Some remarks on a conjecture of Waldschmidt, *Approximations Diophantiennes et nombres transcendants*, Coll. Luminy 1982, Birkhäuser 1983, 329–336.

[17] Yu Kunrui, Linear forms in elliptic logarithms, *J. Number Theory* **20** (1985), 1–69.

25

A NEW APPROACH TO BAKER'S THEOREM
ON LINEAR FORMS IN LOGARITHMS III

G. Wüstholz

1. Introduction

1.1 We fix nonzero algebraic numbers $\alpha_1, \ldots, \alpha_n$ and algebraic numbers $\beta_1, \ldots, \beta_n$ not all zero and consider the linear form

$$L(z_1, \ldots, z_n) = \beta_1 z_1 + \ldots + \beta_n z_n.$$

Let the canonical heights of $\alpha_1, \ldots, \alpha_n$ be bounded by $A_1, \ldots, A_n \geq 4$ and the heights of the $\beta_1, \ldots, \beta_n$ by $B \geq 4$; then Baker in a famous series of papers obtained the remarkable result that if $\Lambda = L(\log \alpha_1, \ldots, \log \alpha_n) \neq 0$, $A_1 \leq \ldots \leq A_n$ and

$$\Omega = \log A_1 \ldots \log A_n = \Omega' \log A_n$$

we have

$$\log |\Lambda| > -(16nd)^{200n} (\log(B\Omega)) \Omega \log \Omega', \qquad (1.1.1)$$

where d denotes the degree of the field generated by $\alpha_1, \ldots, \alpha_n$ and $\beta_1, \ldots, \beta_n$ over the rationals. Furthermore Baker obtained

$$\log |\Lambda| > -(16nd)^{200n} (\log B) \Omega \log \Omega', \qquad (1.1.2)$$

if all the β's are rational integers. This substantial improvement of (1.1.1) has a lot of important consequences. For a detailed account see [1].

1.2 No substantial improvement of (1.1.1) or (1.1.2) has been made up to now. Looking at Baker's proof of (1.1.1) and (1.1.2), one can divide it into two parts: the constructive and the deconstructive part. Baker's method for the deconstructive part is the so-called Kummer theory, a very ingenious and sophisticated tool. If one studies the constructive part one easily notices that in order to make all arguments work there one only needs inequalities of the type

$$\log|\Lambda| > -(16nd)^{200n}(\log(B\Omega))\Omega \tag{1.2.1}$$

in the general case, and

$$\log|\Lambda| > -(16nd)^{200n}(\log B)\Omega \tag{1.2.2}$$

in the rational integers case. It turns out that the extra factor $\log \Omega'$ is introduced in the deconstructive part by Kummer theory.

1.3 Kummer theory also creates difficulties in another context as recently came to light through the work of K. Yu. After Baker's proof of (1.1.1) and (1.1.2), the p-adic analogues seemed to have been obtained by van der Poorten [2] and his results were the basis of a large quantity of papers including a recent book of T. N. Shorey and R. Tijdeman. However, recently Yu found a series of errors in van der Poorten's paper that break down the arguments given there. The most serious part is the Kummer theory which does not work in the p-adic case without additional restrictions, as was pointed out by Yu.

1.4 In the last couple of years a new method was developed in transcendence theory. This method, the so-called multiplicity estimates on group varieties, can be used to replace the Kummer theory efficiently. In [4] we described the multiplicity estimates in the special case of Baker's theorem and in [5] we showed how this can be applied by means of some geometry of numbers to the present situation. There we showed that we have the following improvement of Baker's result (1.1.1), namely

$$\log|\Lambda| > -c(\log(B\Omega))\Omega. \tag{1.4.1}$$

This was obtained by simple induction on the number of logarithms together with the just-mentioned geometry of numbers. However in the rational case Baker's estimate still was better in a certain range for B, a range that is crucial for application. Hence (1.4.1) is more of theoretical interest.

1.5 Before we state our new results it is useful to replace the classical height by the more comfortable Weil height as follows. Let K be an algebraic number field and v a place of K. Then we denote by K_v the completion of K at v and if $d = [\overline{K} : \mathbf{Q}]$ we set $d_v = [K_v : \mathbf{Q}_v]$; we write

$v|p$ if v is a finite place of K lying over the prime p and $v|\infty$ if v is an infinite place. We normalize for every place v of K the absolute value $| \ |_v$ as follows:

(i) $\qquad\qquad |p|_v = p^{-d_v/v} \qquad\qquad$ if $\qquad\qquad v|p,$

(ii) $\qquad\qquad |x|_v = |x|^{-d_v/v} \qquad$ if $v|\infty,\ x \in K_v.$

It follows that for $0 \neq x \in K$

$$\prod_v |x|_v = 1$$

which is the so-called *product formula*. Let $x = (x_1, \ldots, x_N)$ be in K^N. Then we put

$$H(x) := \prod_v \max_n (|x_n|_v) \qquad (1.5.1)$$

and the logarithmic height

$$h(x) := \sum_v \max_n \log(|x_n|_v). \qquad (1.5.2)$$

Both heights depend only on the projective coordinates of x because of the product formula. If $\alpha \in K^*$ is any algebraic number then we put

$$H(\alpha) := H\big((1, \alpha)\big) \qquad (1.5.3)$$

and

$$h(\alpha) := h\big((1, \alpha)\big) \qquad (1.5.4)$$

If $L = \beta_1 z_1 + \ldots + \beta_n z_n$ is a linear form we put $h(L) = h\big((\beta_1, \ldots, \beta_n)\big)$.

1.6 Theorem. *If $\beta_1, \ldots, \beta_n$ are rational integers, $\Lambda = L(\log \alpha_1, \ldots, \log \alpha_n) \neq 0$ then*

$$\log |\Lambda| > -c(n, d) h(L) h(\alpha_1) \ldots h(\alpha_n) \qquad (1.6.1)$$

for an effectively computable positive constant $c = c(n, d)$ depending only on n, d.

1.7 In a subsequent note we shall determine an explicit value for $c(n, d)$ which is much better than Baker's constant, $(16nd)^{200n}$

2. Preliminaries to the proof of Theorem 1.6

2.1 We assume that (1.6.1) does not hold for a sufficiently large constant $c > 0$. Then with the same parameters as in [5] we construct an auxiliary polynomial P. After the usual extrapolation procedure it is shown that P satisfies (1.5) in [4]. In the same way as in [5], section 5, we deduce two cases: In the first case there exist linear forms $L_1(z_1, \ldots, z_n)$, $\ldots, L_r(z_1, \ldots, z_n)$ with $1 \leq r < n$ such that

$$L_i(\log \alpha_1, \ldots, \log \alpha_n) = 0, \qquad 1 \leq i \leq r,$$

and $L_1, \ldots, L_r$ are linearly independent. We may assume now without loss of generality that

$$h(\alpha_1) \leq \ldots \leq h(\alpha_n). \tag{2.1.1}$$

Furthermore we may assume that for some indices $1 \leq i_1 < \ldots < i_{n-r} < n$, the rank of the system

$$L_1, \ldots, L_r, z_{i_1}, \ldots, z_{i_{n-r}}$$

is equal to n. If not then there exists a relation

$$0 = x_1 L_1 + \ldots + x_r L_r + x_{r+1} z_{i_1} + \ldots + x_n z_{i_{n-r}}$$

with integer coefficients and not all of $x_{r+1}, \ldots, x_n$ zero. Without loss of generality we may assume that $x_n \neq 0$. Since

$$L_i(\log \alpha_1, \ldots, \log \alpha_n) = 0, \qquad 1 \leq i \leq r,$$

this leads to

$$x_{r+1} \log_{\alpha_{i_1}} + \ldots + x_n \log \alpha_{i_{n-r}} = 0. \tag{2.1.2}$$

In other words there exists a linear form

$$M(z_{i_1}, \ldots, z_{i_{n-r}})$$

with integer coefficients such that

$$M(\log \alpha_{i_1}, \ldots, \log \alpha_{i_{n-r}}) = 0.$$

But then by Lemma 6 of [1] we may assume that

$$h(M) \leq c_1 \log h(\alpha_{i_{n-r}}).$$

It follows that we may eliminate one of the $\log \alpha_i$, $(1 \leq i < n)$, using (2.1.2). Without loss of generality let this be $\log \alpha_{n-1}$. From $L(\log \alpha_1, \ldots, \log \alpha_n)$ we get a new linear form $L'(\log \alpha_1, \ldots, \log \alpha_{n-2}, \log \alpha_n)$ such that for

$$\Lambda' = L'(\log \alpha_1, \ldots, \log \alpha_{n-2}, \log \alpha_n)$$

we get $0 \neq \Lambda'$ and, for suitable $c_2 > 0$,

$$\log |\Lambda'| \leq -c_2 h(L) h(\alpha_1) \ldots h(\alpha_n),$$

and by (2.1.1) again, for suitable $c_3, c_4 > 0$,

$$h(L') \leq c_3 h(L) + h(M) \leq c_4 h(L) h(\alpha_{n-1}).$$

By induction we obtain

$$\log |\Lambda'| > -c_5 h(L') h(\alpha_1) \ldots h(\alpha_{n-2}) h(\alpha_n)$$
$$\geq -c_6 h(L) h(\alpha_1) \ldots h(\alpha_n)$$

for suitable $c_5, c_6 > 0$. From this there follows easily a contradiction if c is sufficiently large.

2.2 Let therefore $1 \leq i_1 < \ldots < i_{n-r} < n$ be integers such that the rank of

$$L_1, \ldots, L_r, z_{i_1}, \ldots, z_{i_{n-r}}$$

is equal to n. Then we can write the linear form L in terms of this basis. Easy calculations lead to a new linear form L' in $L_1, \ldots, L_r, z_{i_1}, \ldots, z_{i_{n-r}}$:

$$L' = \beta_1' L_1 + \ldots + \beta_r' L_r + \beta_{i_1}'' z_{i_1} + \ldots + \beta_{i_{n-r}}'' z_{i_{n-r}},$$

such that for some $c_7 > 0$,

$$h(L') \leq c_7 h(L) + \log h(\alpha_n) \tag{2.2.1}$$

and, for a suitable c_8 and with

$$\Lambda' = L'(\log \alpha_1, \ldots, \log \alpha_n),$$

$$\log |\Lambda'| \leq -c_8 h(L) h(\alpha_1) \ldots h(\alpha_n). \tag{2.2.2}$$

But

$$\Lambda' = \beta_{i_1}'' \log(\alpha_{i_1}) + \ldots + \beta_{i_{n-r}}'' \log(\alpha_{i_{n-r}})$$

and therefore induction leads to

$$\log |\Lambda'| \geq -c_9 h(L') h(\alpha_{i_1}) \ldots h(\alpha_{i_{n-r}}) \qquad (2.2.3)$$
$$\geq -c_{10} h(L) h(\alpha_1) \ldots h(\alpha_n),$$

for suitable $c_9, c_{10} > 0$. Here we used (2.2.1) which leads to

$$h(L') \leq c_{11} h(L) \log h(\alpha_n) \leq c_{12} h(L) h(\alpha_n).$$

But for sufficiently large c the inequality in (2.2.3) contradicts (2.2.2).

2.3 If the case discussed in **2.1** does not hold then we are in the second case. This means that there exist linear forms with integer coefficients $L_1(z_1, \ldots, z_n)$, ..., $L_r(z_1, \ldots, z_n)$ with $1 \leq r < n$ and positive real numbers $\mu_1, \ldots, \mu_r$ such that

$$L \in (L_1, \ldots, L_r) \qquad (2.3.1)$$

and

$$h(e^{L_i(\log \alpha_1, \ldots, \log \alpha_n)}) \leq \mu_i \qquad (2.3.2)$$

with

$$\mu_1 \ldots \mu_r \leq c_{12} h(\alpha_1) \ldots h(\alpha_n) \qquad (2.3.3)$$

and

$$|\ell_{ij}| \leq |\ell_{ij}| h(\alpha_j) \leq r \mu_j, \quad 1 \leq i \leq n, \ 1 \leq j \leq r. \qquad (2.3.4)$$

2.4 We define $b_i = -\frac{\beta_i}{\beta_n}$, $(1 \leq i \leq n-1)$, and put

$$\partial_i = \frac{\partial}{\partial z_i} + b_i \frac{\partial}{\partial z_n}, \quad 1 \leq i \leq n-1. \qquad (2.4.1)$$

From (2.3.1) we deduce that the Jacobian matrix

$$J = \begin{pmatrix} \partial_1 L_1 & \ldots & \partial_{n-1} L_1 \\ \vdots & \ddots & \vdots \\ \partial_1 L_r & \ldots & \partial_{n-1} L_r \end{pmatrix}$$

has rank $r - 1$ and hence, without loss of generality,

$$\partial_{r-1+i}(L_j) = \left(\sum_{k=1}^{r-1} \xi_{i,k} \partial_k \right)(L_j) \qquad (2.4.2)$$

for $1 \leq i \leq n-r$, $1 \leq j \leq r$ and $\xi_{i,k} \in \mathbf{Q}$.

2.5 Next we put

$$L_i(z_1,\ldots,z_n) = l_{i1}z_1 + \ldots + l_{in}z_n, \qquad 1 \leq i \leq r$$

and

$$M_i(z_1,\ldots,z_{n-1}) = L_i - \frac{l_{in}}{\beta_n}, \qquad 1 \leq i \leq r.$$

Then we get explicitly

$$M_i(z_1,\ldots,z_{n-1}) = (l_{i1} + l_{in}b_1)z_1 + \ldots + (l_{i,n-1} + l_{in}b_{n-1})z_{n-1}$$

for $1 \leq i \leq r$.

3. Proof of Theorem 1.6

3.1 We define

$$\alpha_i' := \alpha_1^{l_{i1}} \ldots \alpha_n^{l_{in}}, \qquad 1 \leq i \leq r$$

and

$$\gamma_i = \alpha_1^{l_{i1}+l_{in}b_1} \ldots \alpha_{n-1}^{l_{i,n-1}+l_{in}b_{n-1}}$$

for $1 \leq i \leq r$. Then we have the following estimates:

$$\begin{aligned}
|\gamma_i - \alpha_i'| &= |\alpha_i'||\exp(\log\gamma_i - \log\alpha_i') - 1| \\
&\leq 3|\alpha_i'||\log\gamma_i - \log\alpha_i'| \\
&= 3|\alpha_i'||(M_i - L_i)(\log\alpha_1,\ldots,\log\alpha_n)| \\
&\leq 3H(\alpha_i') \cdot H(L_i)|\Lambda|.
\end{aligned}$$

by **2.5**. Also

$$|\gamma_i| \leq H(\alpha_i')\{1 + 3H(L_i) \cdot |\Lambda|\} \qquad (3.1.2)$$

and by (2.3.2)

$$h(\alpha_i') \leq \mu_i, \qquad 1 \leq i \leq r. \qquad (3.1.3)$$

We may further assume that $\mu_i \geq 1$ for $1 \leq i \leq r$ and $h(\alpha_1) \geq 1$, hence $h(\alpha_i) \geq 1$, $1 \leq i \leq n$.

3.2 We choose a sufficiently large integer k and define $\epsilon = \frac{1}{3r}$ and

$$S = [h(L)] \cdot k^{1/2}, \qquad S' = h(L) \cdot k^{(\frac{1}{2})\epsilon},$$

$$T' = [kh(\alpha_1)\ldots h(\alpha_n)], \qquad T = T'(1-\epsilon)^{3r},$$

$$S_i = \left[\frac{1}{r+1}S\right], \qquad 1 \le i \le r+1,$$

$$T_i = \left[\frac{1}{r+1}T\right] - 1, \qquad 1 \le i \le r+1,$$

$$D_{-1} = [h(L)] - 1,$$

$$D_0 = [k^{-\epsilon}T],$$

$$D_i = [k^{-\epsilon}T\mu_i^{-1}], \qquad 1 \le i \le r.$$

Finally, for non-negative integers $t_0, \ldots, t_{r-1}$, we put

$$\Phi(z_0,\ldots,z_{n-1}) := \sum_{\lambda_{-1}=0}^{D_{-1}} \ldots \sum_{\lambda_r=0}^{D_r} p(\lambda)\Lambda(z_0)e^{\lambda_1 M_1 + \ldots + \lambda_r M_r}$$

where $p(\lambda) = p(\lambda_{-1}, \lambda_0, \ldots, \lambda_r)$ are integers to be determined later and

$$\Lambda(z) = \Delta(z + \lambda_{-1}; [h(L)], \lambda_0 + 1, t_0) \prod_{i=1}^{r-1} \Delta\left(\sum_{j=1}^{r}(\beta_n l_{ij} - \beta_i l_{in})\lambda_j; t_i\right).$$

The function Φ is closely related to the function

$$\Psi(z_0,\ldots,z_n) := \sum_{\lambda_{-1}=0}^{D_{-1}} \ldots \sum_{\lambda_r=0}^{D_r} p(\lambda)\Lambda(z_0)e^{\lambda_1 L_1 + \ldots + \lambda_r L_r}.$$

We put

$$f(z) = \Phi(z, z\log\alpha_1,\ldots,z\log\alpha_{n-1})$$

and

$$g(z) = \Psi(z, z\log\alpha_1,\ldots,z\log\alpha_n).$$

3.3 From now on $c_1, c_2, \ldots$ will denote positive constants each of which can be determined explicitly from the context. Given then the parameters above, we determine rational integers $p(\lambda)$ such that

$$\max|p(\lambda)| \le e^{c_1 Th(L)} \tag{3.3.1}$$

and

$$g(s) = 0 \qquad (3.3.2)$$

for $0 \le s \le S'$, $0 \le t_0 + \ldots + t_{r-1} \le T'$. This is proved exactly in the same way as Lemma 7 in [1]. The only modification appears in the estimate of the Δ-functions. Here the estimates go as follows:

$$\left| \Delta \left(\sum_{j=1}^{r} (l_{ji}\beta_n - l_{jn}\beta_i)\lambda_j; t_i \right) \right| \le (2H(L))^{t_i} \exp \left(\sum_{j=1}^{r} (|l_{ji}| + |l_{jn}|)D_j \right).$$

Now we use the estimates in (2.3.4) and obtain

$$\sum_{j=1}^{r} (|l_{ji}| + |l_{jn}|)D_j \le 2 \sum_{j=1}^{r} H(L_j) \cdot D_j$$

$$\le 2r \sum_{j=1}^{r} \mu_j D_j$$

$$\le T'$$

for k sufficiently large. Hence altogether we obtain an upper bound of the form

$$e^{c_2 T \cdot h(L)}$$

and this has the desired form.

3.4 Now we make the extrapolation in the same way as in Lemma 9 of [1] and obtain

$$g(s) = 0 \qquad (3.4.1)$$

for $0 \le s \le S$, $0 \le t_0 + \ldots + t_{r-1} \le T$. Here the proofs need some modifications. In order to estimate the differences

$$|f(l) - g(l)|$$

as in Lemma 8 of [1], as well as the growth of the function $f(z)$, we use the estimates (3.1.1)–(3.1.3). The second modification is more sophisticated. It concerns the Hermite interpolation formula used in the proof of Lemma 9 in [1]. There we have to estimate terms of the type

$$\left(\frac{d}{dz} \right)^r f(s)$$

for certain ranges of τ and s. This reduces to estimates of the partial derivatives

$$\left(\frac{\partial}{\partial z_0}\right)^{\tau_0} \cdots \left(\frac{\partial}{\partial z_{n-1}}\right)^{\tau_{n-1}} \Phi(s, s\log\alpha_1, \ldots, s\log\alpha_{n-1})$$

for certain ranges of $s, \tau_0, \ldots, \tau_{n-1}$. This again reduces to estimates of

$$\left(\frac{\partial}{\partial z_0}\right)^{\tau_0} (\beta_n\partial_1)^{\tau_1} \ldots (\beta_n\partial_{n-1})^{\tau_{n-1}} \Psi(s, s\log\alpha_1, \ldots, s\log\alpha_n).$$

But by (2.4.2), this expression is a linear combination of expressions of the type

$$\left(\frac{\partial}{\partial z_0}\right)^{\tau_0} (\beta_n\partial_1)^{\sigma_1} \ldots (\beta_n\partial_{r-1})^{\sigma_{r-1}} \Psi(s, s\log\alpha_1, \ldots, s\log\alpha_n)$$

for $\sigma_1 + \ldots + \sigma_{r-1} = \tau_1 + \ldots + \tau_{n-1}$. These expressions finally are linear forms in functions Ψ with $t_0, \ldots, t_{r-1}$ replaced by $t_0' = t_0$, $t_1' \geq t_1$, $\ldots, t_{r-1}' \geq t_{r-1}$. But these functions vanish at s. It follows that

$$0 = \left(\frac{\partial}{\partial z_0}\right)^{\tau_0} (\beta_n\partial_1)^{\sigma_1} \ldots (\beta_n\partial_{r-1})^{\sigma_{r-1}} \Psi(s, s\log\alpha_1, \ldots, s\log\alpha_n).$$

So we can estimate in the same way the terms $\left(\frac{d}{dz}\right)^{\tau} f(s)$ as in Lemma 9 in [1]. A slight modification is necessary in Stewart's trick. Namely we have to estimate terms like

$$\frac{1}{\tau_i!} \left| \sum_{j=1}^{r} (\beta_n l_{ji} - \beta_i l_{jn}) \lambda_j \log\alpha_i \right|^{\tau_i}.$$

Now by (2.3.2), (2.3.4) we have the following estimates:

$$|l_{ji} \log\alpha_i| \leq |l_{ji}|(\pi + \log|\alpha_i|)$$
$$\leq (\pi + 1) r \mu_j$$

and, since $|\alpha_i| \leq H(\alpha_i) \leq H(\alpha_n)$,

$$|l_{jn} \log\alpha_i| \leq |l_{jn}|(\pi + \log|\alpha_i|)$$
$$\leq |l_{jn}|(\pi + h(\alpha_n))$$
$$\leq (\pi + 1) r \mu_j.$$

Hence the sum can be estimated by

$$2(\pi + 1)rH(L) \sum_{j=1}^{r} \mu_j \lambda_j \le H(L)T$$

for k sufficiently large. Therefore we get

$$\frac{1}{\tau_i!} \left| \sum_{j=1}^{r} (\beta_n l_{ji} - \beta_i l_{jn}) \lambda_j \log \alpha_i \right|^{\tau_i} \le e^T H(L)^{\tau_i}$$

and this is bounded by

$$e^{2Th(L)}$$

which is the order of magnitude needed.

3.5 We finish the proof of Theorem 1.6 as follows. For $t_0 = \ldots = t_{r-1} = 0$ the function Ψ is a polynomial

$$P(z_0, e^{L_1}, \ldots, e^{L_r})$$

of degrees bounded by $D_{-1}D_0$ in z_0 and $D_1, \ldots, D_r$ in $L_1, \ldots, L_r$. It vanishes at the points

$$(s, {\alpha'_1}^s, \ldots, {\alpha'_r}^s), \qquad 0 \le s < S,$$

to order at least T along the analytic subgroup defined by the relation

$$L = \beta'_1 L_1 + \ldots + \beta'_r L_r = 0$$

given by (2.3.1). One easily checks that, for k sufficiently large, the conditions of Theorem 1.1 in [4] are satisfied and we deduce the existence of ρ with $1 \le \rho \le r - 1$ and linear forms $L'_1, \ldots, L'_\rho$ in $L_1, \ldots, L_r$ with integer coefficients such that (i) or (ii) hold. Now we can apply induction and this proves the theorem.

References

[1] A. Baker, The theory of linear forms in logarithms, in *Transcendence theory: advances and applications*, edited by A. Baker and D. W. Masser, Academic Press, London (1977), 1–27.

[2] A. van der Poorten, Linear forms in logarithms in the p-adic case, *Transcendence theory: advances and applications*, Academic Press, London (1977), 29–57.

[3] T. N. Shorey, R. Tijdeman, *Exponential Diophantine Equations*, Cambridge University Press, Cambridge, 1986.

[4] G. Wüstholz, A new approach to Baker's Theorem on linear forms in logarithms I, *Springer Lecture Notes* **1290** (1987).

[5] G. Wüstholz, A new approach to Baker's Theorem on linear forms in logarithms II, *Springer Lecture Notes* **1290** (1987).

26

LINEAR FORMS IN LOGARITHMS
IN THE p-ADIC CASE

Kunrui Yu*

1. Introduction and results

The p-adic theory of transcendental numbers was initiated by Mahler in the 1930s. Mahler [15], [16] obtained in 1932 and 1935 the p-adic analogues of both the Hermite-Lindemann and the Gelfond-Schneider theorems, and during the course of the work he founded the p-adic theory of analytic functions.

In 1939, Gelfond [11] proved a quantitative result on linear forms in two p-adic logarithms; in 1967, Schinzel [20] improved Gelfond's result and computed explicitly all the constants. In 1975, Baker and Coates [7] established in the case $n = 2$ a p-adic analogue of a sharpened inequality of Baker [5].

Since Baker published in the 1960s his first series of papers [3], [4] on linear forms in $n \geq 2$ logarithms of algebraic numbers, his method has been employed in the investigation on linear forms in $n \geq 2$ p-adic logarithms of algebraic numbers. In 1967, Brumer [8] proved that if α_1, ..., α_n are multiplicatively independent $\wp$-adic units then any nontrivial form in $\wp$-adic logarithms:

$$\beta_1 \log \alpha_1 + \ldots + \beta_n \log \alpha_n$$

does not vanish. Subsequently, Coates [9] proved a p-adic analogue of Baker's result [4], Sprindžuk [22], [23] proved p-adic analogues of Baker's results [3], [4], and Kaufman [13] proved a $\wp$-adic analogue of Feldman's result [10]. In 1977, van der Poorten [19] published a paper containing four theorems on linear forms in p-adic logarithms, with much more generality than the previous work and with essentially the same degree of precision as Baker's result [6].

In order to state van der Poorten's results, we introduce some notation. Denote by α_1, ..., α_n, $n \geq 2$, non-zero algebraic numbers in

* Supported by an Alexander von Humboldt Fellowship

an algebraic number field K of degree D over $\mathbf{Q}$, and of heights respectively not exceeding $A_1, \ldots, A_n$ (with $A_j \geq e^e$, $1 \leq j \leq n$). Write $\Omega' = \log A_1 \ldots \log A_{n-1}$, $\Omega = \Omega' \log A_n$. Denote by $b_1, \ldots, b_n$, $b_n \neq 0$, rational integers with absolute values not exceeding B. Denote by $\wp$ a prime ideal of K lying above the rational prime p; write $e_\wp$ for the ramification index of $\wp$ and $f_\wp$ for its residue class degree, so $N\wp = N_{K/\mathbf{Q}}\wp = p^{f_\wp}$. Let $g_\wp = [\frac{1}{2} + e_\wp/(p-1)]$, $G_\wp = N\wp^{g_\wp} \cdot (N\wp - 1)$. For $\alpha \in K$, $\alpha \neq 0$, denote by $\mathrm{ord}_\wp \alpha$ the order to which $\wp$ divides the fractional ideal (α) and put $\mathrm{ord}_\wp 0 = \infty$. Then van der Poorten's [19] Theorem 1 (the main theorem) and Theorem 2 are as follows.

Theorem 1 (Van der Poorten). *The inequalities*

$$\infty > \mathrm{ord}_\wp(\alpha_1^{b_1} \ldots \alpha_n^{b_n} - 1)$$
$$> (16(n+1)D)^{12(n+1)} G_\wp \Omega \log \Omega' \log B$$

have no solutions in rational integers $b_1, \ldots, b_n$; $b_n \not\equiv 0 \pmod{p}$, with absolute values at most B.

Theorem 2 (Van der Poorten). *The inequalities*

$$\infty > \mathrm{ord}_\wp(\alpha_1^{b_1} \ldots \alpha_n^{b_n} - 1)$$
$$> (16(n+1)D)^{12(n+1)} (G_\wp/\log p)\Omega(\log B)^2$$

have no solutions in rational integers $b_1, \ldots, b_n$ with absolute values at most B.

Unfortunately, the proof in van der Poorten [19] involves several errors and inaccuracies, which we should like to remark on at the end of Section 3 of the present paper, so that it seems to be necessary to restudy thoroughly the whole p-adic theory of linear forms in logarithms of algebraic numbers. In this paper, we report on the results we have obtained at this stage, record the lemmas we require (in Section 2), and give a sketch of the proofs (in Sections 3 and 4). The full details will be published later [27]. Take now

$$K = \mathbf{Q}(\alpha_1, \ldots, \alpha_n)$$

and keep the notations D, $\wp$, p, $e_\wp$, $f_\wp$, $N\wp = N_{K/\mathbf{Q}}\wp$ and $\mathrm{ord}_\wp$ introduced above. Denote by $K_\wp$ the completion of K with respect to the (additive) valuation $\mathrm{ord}_\wp$; and the completion of $\mathrm{ord}_\wp$ will be denoted by $\mathrm{ord}_\wp$. Now let Σ be an algebraic closure of $\mathbf{Q}_p$. Write $\mathbf{C}_p$ for the completion of Σ with respect to the valuation of Σ, which is the unique extension of the valuation $|\ |_p$ of $\mathbf{Q}_p$. Denote by ord_p the additive form of the valuation of $\mathbf{C}_p$. According to Hasse [12], pp. 298–302, we can embed $K_\wp$ into $\mathbf{C}_p$: there exists a Q-isomorphism σ from K into Σ such that $K_\wp$ is value-isomorphic to $\mathbf{Q}_p(\sigma(K))$, whence we can identify $K_\wp$ with $\mathbf{Q}_p(\sigma(K))$. Obviously

$$\mathrm{ord}_\wp \beta = e_\wp \, \mathrm{ord}_p \beta \qquad \text{for all } \beta \in K_\wp.$$

Further, for an algebraic number α, write $h(\alpha)$ for its logarithmic absolute height (see, for example, Yu [28], pp. 6–7). Let $b_1, \ldots, b_n$ be rational integers and q be a rational prime such that

$$q \nmid p(p^{f_\wp} - 1). \tag{1}$$

Let $V_1, \ldots, V_n, V_{n-1}^+, V_{n-1}^*, B_0, B_n, B, B^*, W, W^*$ be real numbers satisfying the following conditions

$$V_j \geq \max\left(h(\alpha_j), \frac{f_\wp \log p}{D}\right), \qquad 1 \leq j \leq n,$$

$$V_1 \leq \ldots \leq V_{n-1}, \qquad V_{n-1}^+ = \max(1, V_{n-1}),$$

$$V_{n-1}^* = \max\left\{p^{f_\wp}, (2^{11} nq^{\frac{n+1}{n-1}} D^{\frac{n}{n-1}} V_{n-1}^+)^n\right\};$$

$$B_0 \geq \min_{1 \leq j \leq n, b_j \neq 0} |b_j|, \qquad B_n \geq |b_n|,$$

$$B \geq \max_{1 \leq j \leq n-1} |b_j|, \qquad B^* \geq \max\{|b_1|, \ldots, |b_n|, 2\};$$

$$W \geq \max\left\{\log\left(1 + \frac{3}{8n}\frac{f_\wp \log p}{D}\left(\frac{B_n}{V_1} + \frac{B}{V_n}\right)\right), \log B_0, \frac{f_\wp \log p}{D}\right\} \tag{2}$$

$$W^* = \max\{W, n\log(2^{11} nqD)\}.$$

Then we have

Theorem 1. *Suppose that*

$$\mathrm{ord}_\wp \alpha_j = 0, \qquad 1 \leq j \leq n, \tag{3}$$

$$[K(\alpha_1^{1/q}, \ldots, \alpha_n^{1/q}) : K] = q^n, \tag{4}$$

$$\mathrm{ord}_p b_n \leq \mathrm{ord}_p b_j, \qquad 1 \leq j \leq n-1 \tag{5}$$

and

$$\alpha_1^{b_1} \ldots \alpha_n^{b_n} \neq 1. \tag{6}$$

Then

$$\text{ord}_{\mathfrak{p}}(\alpha_1^{b_1}\ldots\alpha_n^{b_n}-1) < 2770008\left(\tfrac{8}{3}e\right)^n n^{n+\frac{5}{2}}.$$

$$\cdot q^{2n}(q-1)\log^2(nq)\frac{(p^{f_{\mathfrak{p}}}-1)(2+1/(p-1))^n}{(f_{\mathfrak{p}}\log p)^{n+2}}\cdot D^{n+2}.$$

$$\cdot V_1\ldots V_n\left(\frac{W}{6n}+\log(4D)\right)\left(\log(4DV_{n-1}^+)+\frac{f_{\mathfrak{p}}\log p}{8n}\right);$$

further, for $p > 2$, we have

$$\text{ord}_{\mathfrak{p}}(\alpha_1^{b_1}\ldots\alpha_n^{b_n}-1) < 311077\left(\tfrac{8}{3}e\right)^n n^{n+\frac{5}{2}}q^{2n}(q-1)\cdot$$

$$\cdot\log^2(nq).(p^{f_{\mathfrak{p}}}-1)\left(\frac{2+1/(p-1)}{f_{\mathfrak{p}}\log p}\right)^{n+2}\cdot D^{n+2}V_1\ldots V_n\cdot$$

$$\cdot\left(\frac{W}{6n}+\log(4D)\right)\left(\log(4DV_{n-1}^+)+\frac{f_{\mathfrak{p}}\log p}{8n}\right).$$

Finally, if

$$\text{ord}_p b_n = 0,$$

then (2) may be replaced by

$$W \geq \max\left\{\log\left(1+\frac{3}{8n}\frac{f_{\mathfrak{p}}\log p}{D}\left(\frac{B_n}{V_1}+\frac{B}{V_n}\right)\right),\frac{f_{\mathfrak{p}}\log p}{D}\right\}. \qquad (2')$$

In the following Theorem 2, we assume, instead of (2),

$$W \geq \max\left\{\log\left(1+\frac{2}{5n}\frac{f_{\mathfrak{p}}\log p}{D}\left(\frac{B_n}{V_1}+\frac{B}{V_n}\right)\right),\log B_0,\frac{f_{\mathfrak{p}}\log p}{D}\right\}. \qquad (7)$$

Let a_n, d_n be positive integers satisfying

$$a_2 = 338071, \quad a_3 = 244589, \quad a_4 = 202601, \qquad a_5 = 178202,$$

$$a_6 = 161998, \quad a_7 = 150321, \quad a_n = 141430, \quad n \geq 8;$$

$$d_2 = 14491, \quad d_3 = 10484, \quad d_4 = 8685, \qquad d_5 = 7639,$$

$$d_6 = 6944, \quad d_7 = 6444, \quad d_n = 6063, \qquad n \geq 8.$$

Theorem 2. *Suppose that* (3), (4), (5), (6) *hold. Then*

$$\operatorname{ord}_\mathfrak{p}(\alpha_1^{b_1}\dots\alpha_n^{b_n}-1)<$$

$$<\min\left(601719\left(\frac{5}{2}e\right)^n,a_n\left(\frac{8}{3}e\right)^n\right)n^{n+\frac{7}{2}}q^{2n}(q-1)\log^2(nq)\cdot$$

$$\cdot e_\mathfrak{p}(p^{f_\mathfrak{p}}-1)\frac{(2+1/(p-1))^n}{(f_\mathfrak{p}\log p)^{n+2}}D^{n+2}V_1\dots V_n\left(\frac{W}{6n}+\log(4D)\right)^2;$$

further, for $p>2$, *we have*

$$\operatorname{ord}_\mathfrak{p}(\alpha_1^{b_1}\dots\alpha_n^{b_n}-1)<$$

$$<\min\left(23273\left(\frac{5}{2}e\right)^n,d_n\left(\frac{8}{3}e\right)^n\right)n^{n+\frac{7}{2}}q^{2n}(q-1)\log^2(nq)\cdot$$

$$\cdot e_\mathfrak{p}(p^{f_\mathfrak{p}}-1)\frac{(2+1/(p-1))^{n+3}}{(f_\mathfrak{p}\log p)^{n+2}}D^{n+2}V_1\dots V_n\left(\frac{W}{6n}+\log(4D)\right)^2.$$

Finally, if

$$\operatorname{ord}_p b_n = 0, \tag{8}$$

then (7) *may be replaced by*

$$W\geq\max\left\{\log\left(1+\frac{2}{5n}\frac{f_\mathfrak{p}\log p}{D}\left(\frac{B_n}{V_1}+\frac{B}{V_n}\right)\right),\frac{f_\mathfrak{p}\log p}{D}\right\}. \tag{7'}$$

Corollary of Theorem 2. *One may remove in Theorem 2 the hypothesis* (5) *and replace* (8) *by*

$$\min_{1\leq j\leq n}\operatorname{ord}_p b_j = 0,$$

provided one replaces (7) *and* (7') *by*

$$W\geq\max\left\{\log\left(1+\frac{4B^*}{5n}\right),\log B_0,\frac{f_\mathfrak{p}\log p}{D}\right\}$$

and

$$W\geq\max\left\{\log\left(1+\frac{4B^*}{5n}\right),\frac{f_\mathfrak{p}\log p}{D}\right\},$$

respectively.

Remark. It is easy to verify that in Theorem 2

$$\min\left(601719\left(\frac{5}{2}e\right)^n, a_n\left(\frac{8}{3}e\right)^n\right) = \begin{cases} a_n\left(\frac{8}{3}e\right)^n, & 2 \leq n \leq 22; \\ 601719\left(\frac{5}{2}e\right)^n, & n \geq 23, \end{cases}$$

and

$$\min\left(23273\left(\frac{5}{2}e\right)^n, d_n\left(\frac{8}{3}e\right)^n\right) = \begin{cases} d_n\left(\frac{8}{3}e\right)^n, & 2 \leq n \leq 21; \\ 23273\left(\frac{5}{2}e\right)^n, & n \geq 22. \end{cases}$$

In a joint paper with G. Wüstholz, which is in preparation, we shall remove the Kummer condition (4) and the appearance of V_{n-1}^+ in the bounds of Theorem 1. This is achieved by recent work of Wüstholz concerning multiplicity estimates in connection with Baker's theory of linear forms in logarithms of algebraic numbers. (See Wüstholz [26]). Furthermore, in that joint paper, we shall show how a combination of Kummer theory with multiplicity estimates will yield very sharp effective bounds.

2. Preliminaries

We require Mahler's concepts on normal series and normal functions. A p-adic power series

$$\sum_{k=0}^{\infty} f_k z^k, \qquad f_k \in \mathbf{C}_p$$

is called a *normal series* if $\mathrm{ord}_p f_k \geq 0$, $k = 0, 1, \ldots$, and $\mathrm{ord}_p f_k \to \infty$ as $k \to \infty$. A p-adic function $f(z)$ representable by a normal series is called a *normal function*. The following lemma is fundamental.

Lemma 1. (Mahler [16])

(i) *A normal series converges to a normal function for all $z \in \mathbf{C}_p$ with $\mathrm{ord}_p z \geq 0$.*

(ii) *If $f(z) = \sum_{k=0}^{\infty} f_k z^k$ is a normal series then for each $z_0 \in \mathbf{C}_p$ with $\mathrm{ord}_p z_0 \geq 0$, we have*

$$f(z) = \sum_{k=0}^{\infty} f_k(z_0)(z - z_0)^k,$$

where $f_k(z_0) = \frac{f^{(k)}(z_0)}{k!} = \sum_{h=k}^{\infty} \binom{h}{k} f_h z_0^{h-k}$, $k = 0, 1, \ldots$, *and* $\mathrm{ord}_p f_k(z_0) \geq 0$, $k = 0, 1, \ldots$, $\mathrm{ord}_p f_k(z_0) \to \infty$ *as* $k \to \infty$.

(iii) *If f is a normal function with zeros $\beta_1, \ldots, \beta_h \in C_p$ of multiplicities at least $m_1, \ldots, m_h$ respectively, where $\operatorname{ord}_p \beta_j \geq 0$, $1 \leq j \leq h$, then*

$$f(z) = g(z) \prod_{j=1}^{h} (z - \beta_j)^{m_j},$$

where g is a normal function.

Remark. If $\delta \in C_p$ satisfies $\operatorname{ord}_p \delta > 1/(p-1)$, then the p-adic series

$$e^{\delta z} = \sum_{k=0}^{\infty} \frac{\delta^k}{k!} z^k$$

is a normal series.

Suppose $\theta = b/a$, where a, b are positive integers and $(a, b) = 1$. Let η be a fixed root of $x^a - p = 0$ in C_p and define

$$p^\theta = \eta^b.$$

Thus

$$\operatorname{ord}_p p^\theta = \theta.$$

If $\delta \in C_p$ satisfies $\operatorname{ord}_p \delta > \theta + 1/(p-1)$, then $e^{\delta z}$ has "supernormality" in the sense that

$$e^{\delta \cdot p^{-\theta} z} = \sum_{k=0}^{\infty} \frac{(\delta p^{-\theta})^k}{k!} z^k$$

is a normal function. It is easy to see that if $\beta \in K_\mathfrak{p}$ and $\operatorname{ord}_\mathfrak{p}(\beta - 1) = e_\mathfrak{p} \operatorname{ord}_p(\beta - 1) \geq 1$ then for a sufficiently large integer $\ell > 0$, the function

$$(\beta^{p^\ell})^z = \exp(\log(\beta^{p^\ell}).z)$$

has supernormality, where both the logarithmic and exponential functions are p-adic functions. The following lemma gives an upper bound for ℓ in terms of p and $e_\mathfrak{p}$ such that the function $(\beta^{p^\ell})^z$ has supernormality required for our p-adic analysis.

Lemma 2. *Let κ be the integer defined by*

$$p^{\kappa-1}(p-1) \leq (1 + (p-1)/p)e_\mathfrak{p} < p^\kappa(p-1) \tag{9}$$

and set

$$\theta = \begin{cases} 1, & \text{if } \kappa \geq 1 \text{ and } p^{\kappa-1}(p-1) > e_\wp; \\ \frac{p^\kappa}{(2+1/(p-1))e_\wp}, & \text{otherwise.} \end{cases} \tag{10}$$

If $\beta \in C_p$ satisfies

$$\mathrm{ord}_p(\beta - 1) \geq 1/e_\wp, \tag{11}$$

then

$$\mathrm{ord}_p(\beta^{p^\kappa} - 1) > \theta + \frac{1}{(p-1)}. \tag{12}$$

Proof. It is easy to verify that for $\gamma \in C_p$ and every positive integer h, the condition $\mathrm{ord}_p(\gamma - 1) \geq h/e_\wp$ implies

$$\mathrm{ord}_p(\gamma^p - 1) \geq \min\left(\frac{h}{e_\wp}p, \frac{h}{e_\wp} + 1\right). \tag{13}$$

We now show that if $\kappa \geq 1$, then

$$\mathrm{ord}_p(\beta^{p^j} - 1) \geq p^j/e_\wp \tag{14}$$

for $j = 0, 1, \ldots, \kappa - 1$. We may assume $\kappa \geq 2$, since (14) is obviously true when $\kappa = 1$. Now (14) is valid for $j = 0$. Assuming (14) holds for some j with $0 \leq j \leq \kappa - 2$, we see by (14), (13) and (9) that

$$\mathrm{ord}_p(\beta^{p^{j+1}} - 1) = \mathrm{ord}_p((\beta^{p^j})^p - 1)$$
$$\geq \min(p^{j+1}/e_\wp, 1 + p^j/e_\wp) \quad = p^{j+1}/e_\wp.$$

This proves (14) for $\kappa \geq 1$ and $j = 0, 1, \ldots, \kappa-1$. The lemma is evidently true when $\kappa = 0$. If $\kappa = 1$, by (13) and (14) with $j = \kappa - 1$, we obtain

$$\mathrm{ord}_p(\beta^{p^\kappa} - 1) = \mathrm{ord}_p((\beta^{p^{\kappa-1}})^p - 1)$$
$$\geq \min(p^\kappa/e_\wp, 1 + p^{\kappa-1}/e_\wp)$$
$$> \theta + 1/(p-1),$$

where the last inequality follows from (9) and (10). This completes the proof of the lemma.

For the p-adic analytic parts of the proofs of our theorems, instead of using the Schnirelman integral [21] (see also Adams [1]), which yields a p-adic analogue of the Cauchy integral formula, we introduce a Hermite interpolation formula; then we give, based on Lemma 1 (Mahler), and similarly to the work of Schinzel [20] and van der Poorten [19], a lemma for the extrapolation procedure (the following Lemma 4).

Lemma 3. (See Yu. [27]) *Suppose that* $\tau_1 > 0, \ldots, \tau_n > 0$ *are rational integers,* $T = \tau_1 + \ldots + \tau_n$, $\beta_1, \ldots, \beta_n$, $\beta_i \neq \beta_j$, $1 \leq i < j \leq n$, *and* $q_{i,t}$, $1 \leq i \leq n$, $0 \leq t < \tau_i$, *are given numbers in* C *(or* $\mathbf{C}_p$*). Then the unique polynomial* $Q(z)$ *of degree at most* $T - 1$ *satisfying*

$$Q^{(t-1)}(\beta_i) = q_{i,t-1}, \qquad 1 \leq i \leq n, \quad 1 \leq t \leq \tau_i,$$

is given by the formula

$$Q(z) = \sum_{i=1}^{n} \sum_{t=1}^{\tau_i} q_{i,t-1}(-1)^{\tau_i-t}\frac{(z-\beta_i)^{t-1}}{(t-1)!}\left\{\prod_{\substack{j=1\\j\neq i}}^{n}\left(\frac{z-\beta_j}{\beta_i-\beta_j}\right)^{\tau_j}\right\}.$$

$$\cdot \sum_{s=1}^{\tau_i-t}(-1)^{s-1}\sum_{\substack{\lambda_1+\ldots+\lambda_{\tau_i-t}=\tau_i-t\\\lambda_j=0(j<s),\lambda_j\geq1(j\geq s)}}\prod_{r=1}^{\tau_i-t}\frac{\left(\frac{\partial}{\partial\beta_i}\right)^{\lambda_r}\left\{(z-\beta_i)\prod_{\substack{k=1\\k\neq i}}^{n}(\beta_i-\beta_k)^{\tau_k}\right\}}{\lambda_r!\prod_{\substack{k=1\\k\neq i}}^{n}(\beta_i-\beta_k)^{\tau_k}}.$$

Remark. van der Poorten [18] gives a similar formula, but it is false; a simple counterexample can be obtained in the case $n = 2$, $\tau(1) = \tau(2) = 2$. Consequently, the interpolation formula in Lemma 1 of van der Poorten [19] is also false.

Lemma 4. *Suppose that* $\theta > 0$ *is a rational number,* $q > 0$ *is a rational prime with* $q \neq p$, *and* $M > 0$, $R > 0$ *are rational integers with* $q|R$. *Suppose further that* $F(z)$ *is a p-adic normal function and*

$$\min_{\substack{s=1,\ldots,R,(s,q)=1\\t=0,\ldots,M-1}}\left(\mathrm{ord}_p\frac{F^{(t)}(sp^\theta)}{t!}+t\theta\right)$$

$$\geq \left(1-\frac{1}{q}\right)RM\theta + M\,\mathrm{ord}_p(R!) + (M-1)\frac{\log R}{\log p}. \quad (15)$$

Then for all rational integers ℓ, *we have*

$$\mathrm{ord}_p\, F\left(\frac{\ell}{q}p^\theta\right) \geq \left(1-\frac{1}{q}\right)RM\theta.$$

We have given explicitly the condition (15), which is important for evaluating the constants in the bounds for $\mathrm{ord}_p(\alpha_1^{b_1} \ldots \alpha_n^{b_n} - 1)$.

We also require the following p-adic analogue of the Liouville inequality. Let E be an algebraic number field, $\wp'$ be a prime ideal of E with ramification index $e_{\wp'}$ and residue class degree $f_{\wp'}$, and p' be the unique rational prime contained in $\wp'$. Write $\mathrm{ord}_{p'} \beta = \frac{1}{e_{\wp'}} \mathrm{ord}_{\wp'} \beta$ for $\beta \in E$.

Lemma 5. *Let $P(x_1, \ldots, x_m) \in \mathbf{Z}[x_1, \ldots, x_m]$ with*

$$\deg_{x_k} P \leq N_k \ (\geq 1), \qquad 1 \leq k \leq m.$$

For $\beta_1, \ldots, \beta_m \in E$, the condition $P(\beta_1, \ldots, \beta_m) \neq 0$ implies

$$\left| \mathrm{ord}_{p'} P(\beta_1, \ldots, \beta_m) \right| \leq \frac{[E : \mathbf{Q}]}{e_{\wp'} f_{\wp'} \log p'} \left\{ \log L(P) + \sum_{k=1}^{m} N_k h(\beta_k) \right\},$$

where $L(P)$ denotes the length of P.

Using the Lemma in the Appendix of Anderson and Masser [2], we obtain the following version of Siegel's Lemma.

Lemma 6. *Let $\beta_1, \ldots, \beta_r$ be algebraic numbers in a number field of degree D. Let*

$$P_{i,j} \in \mathbf{Z}[x_1, \ldots, x_r], \qquad 1 \leq i \leq n, 1 \leq j \leq m, \text{ not all zero,}$$

satisfying

$$\deg_{x_k} P_{i,j} \leq N_{j,k}, \qquad 1 \leq i \leq n, 1 \leq j \leq m, 1 \leq k \leq r.$$

Define

$$X = \max_{1 \leq j \leq m} \left\{ \left(\sum_{i=1}^{n} L(P_{i,j}) \right) \exp \left(\sum_{k=1}^{r} N_{j,k} h(\beta_k) \right) \right\}$$

and

$$\gamma_{i,j} = P_{i,j}(\beta_1, \ldots, \beta_r), \qquad 1 \leq i \leq n, 1 \leq j \leq m.$$

If $n > mD$, then there exist rational integers $y_1, \ldots, y_n$ with

$$0 < \max_{1 \leq i \leq n} |y_i| \leq X^{mD/(n-mD)}$$

such that

$$\sum_{i=1}^{n} \gamma_{i,j} y_i = 0, \qquad 1 \le j \le m.$$

Remark. This is a slight refinement of Lemma 3 of Mignotte and Waldschmidt [17].

Set

$$\Delta(z;0) = 1, \quad \Delta(z;k) = (z+1)\dots(z+k)/k!, \quad \text{if } k \in \mathbf{Z}, \, k \ge 1, \quad (16)$$

and for ℓ, m non-negative integers

$$\Delta(z;k,l,m) = \frac{1}{m!} \left[\frac{d^m}{dx^m} \left(\Delta(x;k) \right)^\ell \right]_{x=z}. \qquad (17)$$

Let B, B_n, L_1, ..., L_n $(n \ge 2)$, T be positive integers. Put $L = \max_{1 \le j \le n-1} L_j$.

Lemma 7. *Suppose that* $b_1, \dots, b_n, \lambda_1, \dots, \lambda_n, \tau_1, \dots, \tau_{n-1}$ *are rational integers satisfying*

$$|b_j| \le B, \quad 1 \le j \le n-1, \quad |b_n| \le B_n,$$

$$0 \le \lambda_j \le L_j, \quad 1 \le j \le n,$$

$$\tau_j \ge 0, \quad 1 \le j \le n-1, \quad \tau_1 + \dots + \tau_{n-1} \le T.$$

Then

$$\prod_{j=1}^{n-1} |\Delta(b_n\lambda_j - b_j\lambda_n; \tau_j)| \le e^T \cdot \left(1 + \frac{(n-1)(B_nL + BL_n)}{T} \right)^T.$$

This is a slight refinement of an estimate of Loxton, Mignotte, van der Poorten and Waldschmidt [14]. For the estimates concerning $\Delta(z;k,l,m)$, we refer to Lemma T1 of Tijdeman [24] and Lemma 2.4 of Waldschmidt [25].

3. Sketch of the proof of Theorem 1

The proof of Theorem 1 follows Baker [6] and Waldschmidt [25]. In this section we use the notations introduced for Theorem 1 and let κ and θ be defined in Lemma 2. Put

$$G = N\wp - 1 = p^{f_\wp} - 1.$$

It is well-known (see Hasse [12], p. 220) that if m is a positive integer with $p \nmid m$, then $K_{\mathfrak{p}}$ contains the m-th roots of unity if and only if $m|G$. In particular, $K_{\mathfrak{p}}$ contains the G-th roots of unity. Let ζ be a fixed G-th primitive root of unity in $K_{\mathfrak{p}}$. By Hasse [12], p. 153, 155, 220 and the fact that $\operatorname{ord}_{\mathfrak{p}} \alpha_j = 0$, $1 \leq j \leq n$, (see (3)), there exist integers r'_j with $0 \leq r'_j < G$ such that

$$e_{\mathfrak{p}} \operatorname{ord}_p(\alpha_j \zeta^{r'_j} - 1) = \operatorname{ord}_{\mathfrak{p}}(\alpha_j \zeta^{r'_j} - 1) \geq 1, \quad 1 \leq j \leq n.$$

Let $r_1, \ldots, r_n$ be the integers such that

$$r_j \equiv p^\kappa r'_j \pmod{G}, \quad 0 \leq r_j < G, \quad 1 \leq j \leq n.$$

Thus we see, by Lemma 2, that

$$\operatorname{ord}_p(\alpha_j^{p^\kappa} \zeta^{r_j} - 1) > \theta + 1/(p-1), \quad 1 \leq j \leq n. \tag{18}$$

For later references, we give an expression (the following formula (20)) for

$$(\alpha_j^{p^\kappa} \zeta^{r_j})^{1/q} = \exp\left(\frac{1}{q} \log(\alpha_j^{p^\kappa} \zeta^{r_j})\right),$$

where the logarithmic and exponential functions are again p-adic functions, which are well-defined by (18) and the fact that $(q,p) = 1$ (see (1)). By (1), $(q,G) = 1$. Hence we can choose rational integers a, b such that

$$aG + bq = 1.$$

Let $\zeta_q \in \mathbf{C}_p$ be a fixed q-th primitive root of unity and put

$$\xi = \zeta_q^a \zeta^b.$$

It is easy to see that there exists a q-th root α'_j of α_j in $\mathbf{C}_p$, $1 \leq j \leq n$, such that

$$(\alpha_j^{p^\kappa} \zeta^{r_j})^{1/q} = {\alpha'_j}^{p^\kappa} \xi^{r_j} = {\alpha'_j}^{p^\kappa} \zeta_q^{ar_j} \zeta^{br_j}, \quad 1 \leq j \leq n. \tag{19}$$

By (1), for each j with $1 \leq j \leq n$ there exists a unique rational integer a_j such that

$$p^\kappa a_j \equiv ar_j \pmod{q}, \quad 0 \leq a_j < q.$$

Write $\alpha_j^{1/q}$ for $\alpha'_j \zeta_q^{a_j}$, which is obviously a q-th root of α_j in $\mathbf{C}_p$. Now by (19), we obtain

$$(\alpha_j^{p^\kappa} \zeta^{r_j})^{1/q} = (\alpha_j^{1/q})^{p^\kappa} \zeta^{br_j}, \quad 1 \leq j \leq n. \tag{20}$$

Let

$$\epsilon = \epsilon(n) = \begin{cases} 4 \cdot 10^{-20}, & n \geq 8, \\ 4.366 \cdot 10^{-5}, & 2 \leq n \leq 7, \end{cases}$$

and $c_0, \ldots, c_4$ be positive numbers given by the following two tables

Cases		c_0	c_1	c_2	c_3	c_4
$p = 2$	$2 \leq n \leq 7$	8	3.2119513	56/15	47.766502	79.102681
	$n \geq 8$	16	3.0703894	8/3	116.51153	192.64207

Cases		c_0	$c_1/\left(2 + \frac{1}{p-1}\right)$	c_2	$c_3/\left(2 + \frac{1}{p-1}\right)$	c_4
$p = 3$	$2 \leq n \leq 7$	8	1.0355914	56/15	12.8	32
	$n \geq 8$	16	1.0117649	8/3	12.8	35.671814
$p \geq 5$	$2 \leq n \leq 6$	8	1.0723192	56/15	16.457689	77.89776
	$n = 7$	8	1.0192191	56/15	16	69.994513
	$n \geq 8$	16	1.0234756	8/3	39.253842	192.63692

Put

$$U = (1 + \epsilon)c_0 c_1 c_2^n c_3 c_4 \frac{n^{2n+1}}{n!} q^{2n}(q - 1)\frac{G(2 + 1/(p - 1))^n}{e_p(f_p \log p)^{n+2}}.$$

$$\cdot D^{n+2} V_1 \ldots V_n W^* \log V_{n-1}^*.$$

It is readily verified that Theorem 1 follows from the following

Proposition 1. *Suppose that* (3), (4), (5) *and* (6) *hold. Then*

$$\text{ord}_p(\alpha_1^{b_1} \ldots \alpha_n^{b_n} - 1) < U.$$

If $\text{ord}_p b_n = 0$, *then* (2) *may be replaced by* (2').

Now we give a sketch of the proof of Proposition 1. Put

$$S = q\left[\frac{c_3 n D W^*}{f_p \log p}\right],$$

$$T = \left[\frac{U f_p \log p}{q^n D} \cdot \frac{1}{c_1 c_3 W^* \theta}\right],$$

$$L_{-1} = [W^*],$$

$$L_0 = \left[\frac{Ue_{\mathfrak{p}} f_{\mathfrak{p}} \log p}{q^n D} \cdot \frac{1}{c_1 c_4 (L_{-1}+1) \log V_{n-1}^*} \right],$$

$$L_j = \left[\frac{Ue_{\mathfrak{p}} f_{\mathfrak{p}} \log p}{q^n D} \cdot \frac{1}{c_1 c_2 n p^\kappa S V_j} \right], \quad 1 \le j \le n, \quad L = \max_{1 \le j \le n-1} L_j,$$

$$X_0 = \left\{ D \prod_{j=-1}^{n} (L_j+1) \right\} 3^{T(L_{-1}+1)} \left(2e \left(2 + \frac{S}{L_{-1}+1} \right) \right)^{(L_{-1}+1)(L_0+1)} \cdot$$

$$\cdot \left(1 + \frac{(n-1)(B_n L + B L_n)}{T} \right)^T \cdot \exp \left(p^\kappa S \sum_{j=1}^{n} L_j V_j + nD \max_{1 \le j \le n} V_j \right).$$

For $(J, \lambda_{-1}, \ldots, \lambda_n, \tau_0, \ldots, \tau_{n-1}) \in \mathbf{N}^{2n+3}$ set

$$\Lambda_J(z, \tau) = \Delta \big(q^{-J} z + \lambda_{-1}; L_{-1}+1, \lambda_0+1, \tau_0 \big) \prod_{j=1}^{n-1} \Delta(b_n \lambda_j - b_j \lambda_n; \tau_j),$$

where $\Delta(z; k)$, $\Delta(z; k, l, m)$ are defined by (16), (17). We abbreviate $(\lambda_{-1}, \ldots, \lambda_n)$ as λ, $(\tau_0, \ldots, \tau_{n-1})$ as τ and write $|\tau| = \tau_0 + \ldots + \tau_{n-1}$. Using a remark from Mignotte and Waldschmidt [17], §4.2, we can fix a basis $\xi_1, \ldots, \xi_D$ of $K = \mathbf{Q}(\alpha_1, \ldots, \alpha_n)$ over $\mathbf{Q}$ of the shape

$$\xi_d = \alpha_1^{k_{1,d}} \ldots \alpha_n^{k_{n,d}}, \quad (k_{1,d}, \ldots, k_{n,d}) \in \mathbf{N}^n,$$

$$k_{1,d} + \ldots + k_{n,d} \le D-1, \quad 1 \le d \le D.$$

For rational integers $r^{(J)}$, $L_j^{(J)} \ge 0$, $-1 \le j \le n$, and $p_d^{(J)}(\lambda) = p_d^{(J)}(\lambda_{-1}, \ldots, \lambda_n)$, which will be constructed in the main inductive argument, set

$$\varphi_J(z, \tau) = \sideset{}{'}\sum_{\lambda} \sum_{d=1}^{D} p_d^{(J)}(\lambda) \xi_d \Lambda_J(z, \tau) \prod_{j=1}^{n} (\alpha_j^{p^\kappa} \zeta^{r_j})^{\lambda_j z}, \tag{21}$$

where $\sum'_\lambda$ denotes the sum taken over the range of $\lambda = (\lambda_{-1}, \ldots, \lambda_n)$:

$$0 \le \lambda_j \le L_j^{(J)} \quad (-1 \le j \le n),$$
$$r_1 \lambda_1 + \ldots + r_n \lambda_n \equiv r^{(J)} \pmod{G}. \tag{22}$$

Note that, by (18), the p-adic functions

$$(\alpha_j^{p^\kappa} \zeta^{r_j})^{\lambda_j z} = \exp \big(\lambda_j z \log(\alpha_j^{p^\kappa} \zeta^{r_j}) \big), \quad 1 \le j \le n,$$

are normal. In the rest of Sect. 3, s always denotes a rational integer, and τ a point $(\tau_0, \ldots, \tau_{n-1})$ in $\mathbf{N}^n$. By an argument similar to that in Waldschmidt [25], Sect. 3.5, Proposition 1 follows from the following:

Main inductive argument. *Suppose that there are* $\alpha_1, \ldots, \alpha_n, b_1,$ $\ldots, b_n$ *satisfying* (3), (4), (5), (6) *such that*

$$\operatorname{ord}_p(\alpha_1^{b_1} \ldots \alpha_n^{b_n} - 1) \geq U. \tag{23}$$

Then for every rational integer J *with*

$$0 \leq J \leq \left\lceil \frac{\log L_n}{\log q} \right\rceil + 1,$$

there exist rational integers $r^{(J)}, L_j^{(J)}$ $(-1 \leq j \leq n)$ *with*

$$0 \leq r^{(J)} < G, \qquad \gcd(r_1, \ldots, r_n, G) | r^{(J)},$$

$$L_{-1}^{(J)} = L_{-1}, \quad L_0^{(J)} = L_0, \qquad 0 \leq L_j^{(J)} \leq q^{-J} L_j, \quad 1 \leq j \leq n,$$

and rational integers $p_d^{(J)}(\lambda)$ *for* $d = 1, \ldots, D$ *and* λ *in the range* (22), *not all zero, with absolute values not exceeding* $X_0^{\frac{1}{c_0 - 1}}$, *such that*

$$\varphi_J(s, \tau) = 0, \qquad 1 \leq s \leq q^J S, \qquad (s, q) = 1,$$

$$|\tau| \leq q^{-J} T.$$

In the rest of this section, we always keep the hypothesis (23). The main inductive argument is proved by an induction on J. The case $J = 0$ can be verified by means of Lemma 6, on taking $r^{(0)} = 0$, $L_j^{(0)} = L_j$, $-1 \leq j \leq n$. Suppose the main inductive argument is true for some J with $0 \leq J \leq \left\lceil \frac{\log L_n}{\log q} \right\rceil$; we have to prove it for $J + 1$. We proceed in three steps.

Step 1. Show that for $k = 0, \ldots, n - 1$, we have

$$\varphi_J(s, \tau) = 0, \qquad 1 \leq s \leq q^{J+k} S, \qquad (s, q) = 1,$$

$$|\tau| \leq \left(1 - \left(1 - \frac{1}{q}\right) \frac{k}{n}\right) q^{-J} T. \tag{24}$$

Step 2. Prove that

$$\varphi_J\left(\frac{s}{q}, \tau\right) = 0, \qquad 1 \leq s \leq q^{J+1} S, \qquad (s, q) = 1,$$

$$|\tau| \leq q^{-(J+1)} T. \tag{25}$$

Step 3. Deduce from (25) the main inductive argument for $J + 1$.

Let
$$\gamma_j = \lambda_j - \frac{b_j}{b_n}\lambda_n, \qquad 1 \le j \le n-1,$$

and put

$$f_J(z,\tau) = \sum_\lambda{}' \sum_{d=1}^{D} p_d^{(J)}(\lambda)\xi_d \Lambda_J(z,\tau) \prod_{j=1}^{n-1} (\alpha_j^{p^\kappa} \zeta^{r_j})^{\gamma_j z}.$$

Note that, by (18) and (5), the p-adic functions

$$(\alpha_j^{p^\kappa} \zeta^{r_j})^{\gamma_j p^{-\theta} z} = \exp\left(\gamma_j p^{-\theta} z \log(\alpha_j^{p^\kappa} \zeta^{r_j})\right), \quad 1 \le j \le n-1,$$

are normal. In Steps 1 and 2, we require

Lemma 8. *For any τ with $|\tau| \le T$ and any rational number $y > 0$ with* $\operatorname{ord}_p y \ge 0$, *we have*
$$\operatorname{ord}_p(\varphi_J(y,\tau) - f_J(y,\tau)) \ge U - T\log(L_{-1}+1)/\log p - \operatorname{ord}_p b_n.$$

To carry out Step 1, we argue by a further induction on k. By virtue of the main inductive hypothesis, (24) holds for $k = 0$. Now we assume (24) is valid for some k with $0 \le k \le n-2$, and we prove it for $k+1$. By Lemma 8, we see that

$$\operatorname{ord}_p f_J(s,\tau) \ge U - T\log(L_{-1}+1)/\log p - \operatorname{ord}_p b_n$$

$$1 \le s \le q^{J+k}S, \quad (s,q)=1, \quad |\tau| \le \left(1 - \left(1-\frac{1}{q}\right)\frac{k}{n}\right)q^{-J}T. \quad (26)$$

We apply Lemma 4 to each of the normal functions

$$F(z) = F_J(z,\tau) = p^{(L_{-1}+1)(L_0+1)(\theta + \frac{1}{p-1})}f_J(p^{-\theta}z,\tau) \quad (27)$$

with

$$|\tau| \le \left(1 - \left(1-\frac{1}{q}\right)\frac{k+1}{n}\right)q^{-J}T,$$

taking $R = q^{J+k}S$, $M = \left[\left(1-\frac{1}{q}\right)\frac{1}{n}q^{-J}T\right] + 1$. By (26) we get

$$\operatorname{ord}_p\left(\frac{1}{m!}\frac{d^m}{dz^m}F_J(sp^\theta,\tau)\right) \ge$$

$$\ge U - T\log(L_{-1}+1)/\log p - \left(\left(1-\frac{1}{q}\right)\frac{1}{n}q^{-J}T+1\right)\operatorname{ord}_p b_n$$

$$+(L_{-1}+1)(L_0+1)\left(\theta + \frac{1}{p-1}\right) - m\theta,$$

for

$$0 \le m \le \left(1 - \frac{1}{q}\right) \frac{1}{n} q^{-J}T, \quad 1 \le s \le q^{J+k}S, \quad (s,q) = 1,$$

$$|\tau| \le \left(1 - \left(1 - \frac{1}{q}\right) \frac{k+1}{n}\right) q^{-J}T.$$

Our setting of the parameters U, S, T, L_j $(-1 \le j \le n)$ and the choices of $c_0, \ldots, c_4, \epsilon$ make the condition (15) hold. Thus by Lemma 4 and (27) we obtain

$$\operatorname{ord}_p f_J(s,\tau) > \frac{1}{n} \left(1 - \frac{1}{q}\right)^2 q^k ST\theta - (L_{-1} + 1)(L_0 + 1)\left(\theta + \frac{1}{p-1}\right),$$

$$s \in \mathbf{Z}, \quad |\tau| \le \left(1 - \left(1 - \frac{1}{q}\right) \frac{k+1}{n}\right) q^{-J}T.$$

This together with Lemma 8 yields

$$\operatorname{ord}_p \varphi_J(s,\tau) > \frac{1}{n} \left(1 - \frac{1}{q}\right)^2 q^k ST\theta - (L_{-1}+1)(L_0+1)\left(\theta + \frac{1}{p-1}\right), \quad (28)$$

$$s \ge 1, \quad |\tau| \le \left(1 - \left(1 - \frac{1}{q}\right) \frac{k+1}{n}\right) q^{-J}T.$$

On the other hand, Lemma 5 gives an upper bound for $\operatorname{ord}_p \varphi_J(s,\tau)$ with $1 \le s \le q^{J+k+1}S$, $(s,q) = 1$, $|\tau| \le \left(1 - \left(1 - \frac{1}{q}\right) \frac{k+1}{n}\right) q^{-J}T$, provided $\varphi_J(s,\tau) \ne 0$. However, our setting of the parameters U, S, T, L_j, $-1 \le j \le n$, and the choices of $c_0, \ldots, c_4, \epsilon$ force this upper bound to be less than or equal to the right-hand side of (28). Hence we conclude

$$\varphi_J(s,\tau) = 0, \quad 1 \le s \le q^{J+k+1}S, \quad (s,q) = 1,$$

$$|\tau| \le \left(1 - \left(1 - \frac{1}{q}\right) \frac{k+1}{n}\right) q^{-J}T.$$

This completes Step 1.

To carry out Step 2, we apply Lemma 4 to the normal functions defined by (27) with

$$|\tau| \le q^{-(J+1)}T,$$

taking $R = q^{J+n-1}S$, $M = \left[\left(1 - \frac{1}{q}\right) \frac{1}{n} q^{-J}T\right] + 1$. By (24) with $k = n-1$, Lemma 8 and by our setting of the parameters and the choices of c_0,

$\ldots$, c_4, ϵ, which make (15) hold, Lemma 4 together with (27) yields a lower bound for $\operatorname{ord}_p f_J\left(\frac{s}{q},\tau\right)$, $s \in \mathbf{Z}$, $|\tau| \leq q^{-(J+1)}T$. Again by Lemma 8, we get

$$\operatorname{ord}_p \varphi_J\left(\frac{s}{q},\tau\right) > \frac{1}{n}\left(1-\frac{1}{q}\right)^2 q^{n-1}ST\theta-(L_{-1}+1)(L_0+1)\left(\theta+\frac{1}{p-1}\right),$$

$$s \geq 1, \qquad |\tau| \leq q^{-(J+1)}T. \tag{29}$$

On the other hand, Lemma 5 gives for s with $1 \leq s \leq q^{J+1}S$, $(s,q)=1$ and τ with $|\tau| \leq q^{-(J+1)}T$

$$\operatorname{ord}_p \varphi_J\left(\frac{s}{q},\tau\right) \leq$$

$$\leq \frac{q^n D}{e_p f_p \log p}\left\{\log\left(D\prod_{j=-1}^{n}(L_j+1)\right) + \frac{1}{c_0-1}\log X_0 + p^\kappa S\sum_{j=1}^{n}L_j V_j\right.$$

$$+nD\max_{1\leq j\leq n} V_j+(\log 3)\frac{1}{q}T(L_{-1}+1)+\frac{1}{q}T\log\left(1+\frac{(n-1)q(B_n L + BL_n)}{T}\right)$$

$$+(L_{-1}+1)(L_0+1)\log\left(2e\left(2+\frac{S}{L_{-1}+1}\right)\right)$$

$$\left.+2(L_{-1}+1)(L_0+1)\log(qL_n)\right\}, \tag{30}$$

provided $\varphi_J\left(\frac{s}{q},\tau\right) \neq 0$. As in Step 1, our setting of the parameters and the choices of $c_0, \ldots, c_4, \epsilon$ force the right-hand side of (30) to be less than or equal to that of (29), i.e. force (25) to hold. This finishes Step 2.

Here we should remark that, in our setting of the parameters, the inequality in unknowns $c_0, \ldots, c_4$, which resulted from

The right-hand side of (30) $\leq$ The right-hand side of (29), $\tag{31}$

is more severe than the corresponding inequalities in Step 1. The choices of $c_0, \ldots, c_4$ are determined by solving optimally a system of two inequalities in $c_0, \ldots, c_4$ resulting from (15) and (31).

Now we detail Step 3. By (20) and Hasse [12], p. 273, for rational integers $\mu_1, \ldots, \mu_n$ satisfying

$$r_1\mu_1 + \ldots + r_n\mu_n \equiv r^{(J)} \pmod{G},$$

we have

$$\prod_{j=1}^{n}(\alpha_j^{p^\kappa}\zeta^{r_j})^{\mu_j\frac{s}{q}} = \prod_{j=1}^{n}\left((\alpha_j^{p^\kappa}\zeta^{r_j})^{\frac{1}{q}}\right)^{\mu_j s}$$

$$= \prod_{j=1}^{n}(\alpha_j^{\frac{1}{q}})^{p^\kappa\mu_j s}\cdot\zeta^{bs(r_1\mu_1+\dots+r_n\mu_n)}$$

$$= \zeta^{bsr^{(J)}}\prod_{j=1}^{n}(\alpha_j^{\frac{1}{q}})^{p^\kappa\mu_j s}. \tag{32}$$

Writing

$$\mu_j = \lambda_j^0 + q\lambda_j, \qquad 0 \le \lambda_j^0 < q, \quad 1 \le j \le n,$$

we see that

$$(\alpha_j^{\frac{1}{q}})^{p^\kappa\mu_j s} = (\alpha_j^{\frac{1}{q}})^{p^\kappa\lambda_j^0 s}\cdot\alpha_j^{p^\kappa\lambda_j s}, \qquad 1 \le j \le n. \tag{33}$$

By (25), (32), (33), we have

$$\sum_{\lambda_1^0=0}^{q-1}\cdots\sum_{\lambda_n^0=0}^{q-1}\prod_{j=1}^{n}(\alpha_j^{\frac{1}{q}})^{p^\kappa\lambda_j^0 s}\sum_{\lambda_{-1}=0}^{L_{-1}^{(J)}}\sum_{\lambda_0=0}^{L_0^{(J)}}\sum_{\lambda_1,\dots,\lambda_n}\sum_{d=1}^{D}$$

$$p_d^{(J)}(\lambda_{-1},\lambda_0,\lambda_1^0+q\lambda_1,\dots,\lambda_n^0+q\lambda_n)\xi_d\Delta(q^{-(J+1)}s+\lambda_{-1};L_{-1}+1,\lambda_0+1,\tau_0)\cdot$$

$$\cdot\prod_{j=1}^{n-1}\Delta(q(b_n\lambda_j-b_j\lambda_n)+(b_n\lambda_j^0-b_j\lambda_n^0);\tau_j)\prod_{j=1}^{n}\alpha_j^{p^\kappa\lambda_j s} = 0, \tag{34}$$

$$1 \le s \le q^{J+1}S, \qquad (s,q)=1, \qquad |\tau| \le q^{-(J+1)}T,$$

where $\sum_{\lambda_1,\dots,\lambda_n}$ is taken over the rational integers $\lambda_1,\dots,\lambda_n$ satisfying

$$0 \le \lambda_j \le L_j^{(J+1)}(\lambda_1^0,\dots,\lambda_n^0) = \left[\frac{L_j^{(J)}-\lambda_j^0}{q}\right], \qquad 1 \le j \le n, \tag{35}$$

and

$$\sum_{j=1}^{n}r_j(\lambda_j^0+q\lambda_j) \equiv r^{(J)} \pmod{G}. \tag{36}$$

We emphasize that, by (1)

$$(q,G) = 1,$$

hence (36) is equivalent to

$$r_1\lambda_1 + \ldots + r_n\lambda_n \equiv r^{(J+1)}(\lambda_1^0,\ldots,\lambda_n^0) \quad (\text{mod } G), \qquad (36')$$

where $r^{(J+1)}(\lambda_1^0,\ldots,\lambda_n^0)$ is the unique solution of the congruence

$$qx \equiv r^{(J)} - (r_1\lambda_1^0 + \ldots + r_n\lambda_n^0) \quad (\text{mod } G)$$

in the range $0 \le x < G$. Now by the main inductive hypothesis for J, there exists a set of $\lambda_1^0, \ldots, \lambda_n^0$ with $0 \le \lambda_j^0 < q$, $1 \le j \le n$, such that the rational integers

$$p_d^{(J)}(\lambda_{-1},\lambda_0,\lambda_1^0 + q\lambda_1,\ldots,\lambda_n^0 + q\lambda_n)$$

with

$$1 \le d \le D, \qquad 0 \le \lambda_j \le L_j^{(J)}, \qquad j = -1, 0,$$

$$\lambda_1, \ldots, \lambda_n \text{ in the range defined by (35), (36'),}$$

are not all zero. Fix this set of $\lambda_1^0, \ldots, \lambda_n^0$, take

$$r^{(J+1)} = r^{(J+1)}(\lambda_1^0,\ldots,\lambda_n^0),$$

which is obviously divisible by g.c.d.$(r_1,\ldots,r_n,G)$, and set

$$L_j^{(J+1)} = L_j^{(J)} = L_j \qquad (j = -1,0),$$

$$L_j^{(J+1)} = L_j^{(J+1)}(\lambda_1^0,\ldots,\lambda_n^0), \qquad 1 \le j \le n,$$

$$p_d^{(J+1)}(\lambda_{-1},\lambda_0,\lambda_1,\ldots,\lambda_n) = p_d^{(J)}(\lambda_{-1},\lambda_0,\lambda_1^0 + q\lambda_1,\ldots,\lambda_n^0 + q\lambda_n),$$

$$1 \le d \le D, \qquad 0 \le \lambda_j \le L_j^{(J+1)}, \qquad -1 \le j \le n,$$

$$r_1\lambda_1 + \ldots + r_n\lambda_n \equiv r^{(J+1)} \quad (\text{mod } G). \qquad (37)$$

On noting the condition (4) and the fact that

$$(p^\kappa s, q) = 1,$$

we deduce from (34) that

$$\sum_\lambda{}' \sum_{d=1}^{D} p_d^{(J+1)}(\lambda)\xi_d \Delta\big(q^{-(J+1)}s + \lambda_{-1}; L_{-1} + 1, \lambda_0 + 1, \tau_0\big)\cdot$$

$$\cdot \prod_{j=1}^{n-1} \Delta\big(q(b_n\lambda_j - b_j\lambda_n) + (b_n\lambda_j^0 - b_j\lambda_n^0); \tau_j\big) \cdot \prod_{j=1}^{n} \alpha_j^{p^\kappa \lambda_j s} = 0, \qquad (38)$$

$$1 \le s \le q^{J+1}S, \qquad (s,q) = 1, \qquad |\tau| \le q^{-(J+1)}T,$$

where $\sum'_\lambda$ denotes the sum taken over λ's in (37). By virtue of the fact that, for each j with $1 \leq j \leq n - 1$, $\Delta(b_n\lambda_j - b_j\lambda_n; \tau_j)$ is a linear combination of the $\tau_j + 1$ numbers

$$\Delta\big(q(b_n\lambda_j - b_j\lambda_n) + (b_n\lambda_j^0 - b_j\lambda_n^0); k\big), \qquad k = 0, 1, \ldots, \tau_j,$$

with coefficients independent of $\lambda_1, \ldots, \lambda_n$, (38) implies

$$\zeta^{-sr^{(J+1)}} \varphi_{J+1}(s, \tau) = 0,$$

$$1 \leq s \leq q^{J+1}S, \qquad (s, q) = 1, \qquad |\tau| \leq q^{-(J+1)}T.$$

This completes Step 3, thereby establishing the main inductive argument.

In Section 2, below Lemma 3, we have mentioned an error in van der Poorten [18], [19]. We should like to make a further remark about [19]. Recall

$$g_\mathfrak{p} = \left[\frac{1}{2} + e_\mathfrak{p}/(p - 1)\right], \qquad G_\mathfrak{p} = N\wp^{g_\mathfrak{p}} \cdot (N\wp - 1)$$

and let ζ' be a $G_\mathfrak{p}$-th primitive root of unity in C_p. In [19], p. 35 van der Poorten asserts that for $\alpha \in K$ with $\text{ord}_\mathfrak{p}\, \alpha = 0$ there is an integer $r, 0 \leq r < G_\mathfrak{p}$ such that

$$\text{ord}_\mathfrak{p}(\alpha\zeta'^r - 1) \geq g_\mathfrak{p} + 1. \tag{39}$$

Note that this is false. A simple counterexample is the following. Take $K = \mathbf{Q}$, $\wp = 3\mathbf{Z}$, then $e_\mathfrak{p} = g_\mathfrak{p} = 1$, $G_\mathfrak{p} = 6$. Take $\alpha = \frac{2}{5}$ and let ζ' be a 6-th primitive root of unity. It is readily verified that

$$\text{ord}_\mathfrak{p}(\alpha\zeta'^r - 1) \leq 1 < g_\mathfrak{p} + 1, \qquad 0 \leq r < G_\mathfrak{p}.$$

We should indicate that the assertion (39) does hold for the special case where $g_\mathfrak{p} = 0$, by the discussion at the beginning of this section. But even in this special case, there are still some inaccuracies in [19]. For instance, in the proof of his Lemma 7 ([19], p. 46, 47), which corresponds to our Step 3, van der Poorten has not put an additional restriction on q that

$$(q, G_\mathfrak{p}) = 1, \tag{40}$$

which seems to be essential to make his proof work. On the other hand, if one does assume (40), then by Hasse [12], p. 220, $K_\mathfrak{p}$, whence K, does not contain the q-th roots of unity, and we cannot understand the arguments related to Kummer theory in Section 5 of [19], pp. 49–51. The same remark extends to the proofs of Theorems 2, 3, 4 of [19].

4. A few remarks on the proof of Theorem 2

The proof of Theorem 2 is similar to but simpler than that of Theorem 1. Here we indicate only the main points, which differ from the proof of Theorem 1. Let

$$U = (1 + \epsilon')c_0 c_1 c_2^n c_3^2 \frac{n^{2n+2}}{n!} q^{2n}(q - 1)$$

$$\frac{G(2 + 1/(p - 1))^n}{(f_\wp \log p)^{n+2}} D^{n+2} V_1 \dots V_n W^{*2}$$

and set the parameters $S, T, L_1, \dots, L_n$ by the formulae below the statement of Proposition 1 in Section 3. The choices of c_0, c_1, c_2, c_3, which are different from those in Section 3, are determined by solving optimally a certain system of two inequalities. For $(\lambda_1, \dots, \lambda_n, \tau_1, \dots, \tau_{n-1}) \in \mathbb{N}^{2n-1}$, we write λ for $(\lambda_1, \dots, \lambda_n)$, τ for $(\tau_1, \dots, \tau_{n-1})$ and

$$\Lambda(\tau) = \prod_{j=1}^{n-1} \Delta(b_n \lambda_j - b_j \lambda_n; \tau_j).$$

We set the auxiliary functions as follows. Put

$$\varphi_J(z, \tau) = \sum_\lambda{}' \sum_{d=1}^{D} p_d^{(J)}(\lambda) \xi_d \Lambda(\tau) \prod_{j=1}^{n} (\alpha_j^{p^\kappa} \zeta^{r_j})^{\lambda_j z}$$

and

$$f_J(z, \tau) = \sum_\lambda{}' \sum_{d=1}^{D} p_d^{(J)}(\lambda) \xi_d \Lambda(\tau) \prod_{j=1}^{n-1} (\alpha_j^{p^\kappa} \zeta^{r_j})^{\gamma_j z},$$

where

$$\gamma_j = \lambda_j - \frac{b_j}{b_n} \lambda_n, \qquad 1 \le j \le n - 1,$$

and $\sum_\lambda'$ denotes the sum taken over the range of $\lambda = (\lambda_1, \dots, \lambda_n)$:

$$0 \le \lambda_j \le L_j^{(J)}, \quad 1 \le j \le n, \qquad r_1 \lambda_1 + \dots + r_n \lambda_n \equiv r^{(J)} \pmod{G}.$$

Acknowledgements. The author is grateful to A. Baker for his encouragement during the course of the work. The work was done when the author was enjoying the hospitality of the Max-Planck-Institut für Mathematik, Bonn. He would like to express his gratitude to G. Wüstholz for suggesting the problem and helpful advice. He is also grateful to D. W. Masser, R. Tijdeman and M. Waldschmidt for their encouragement in the course of the Symposium.

References

[1] W. W. Adams, Transcendental numbers in the p-adic domain, *Amer. J. Math.* **88** (1966), 279–308.

[2] M. Anderson and D. W. Masser, Lower bounds for heights on elliptic curves, *Math. Zeitschrift*, **174** (1980), 23–34.

[3] A. Baker, Linear forms in the logarithms of algebraic numbers I, II, III, *Mathematika*, **13** (1966), 204–216; **14** (1967), 102–107, 220–228.

[4] A. Baker, Linear forms in the logarithms of algebraic numbers IV, *Mathematika*, **15** (1968), 204–216.

[5] A. Baker, A sharpening of the bounds for linear forms in logarithms II, *Acta Arith.* **24** (1973), 33–36.

[6] A. Baker, The theory of linear forms in logarithms, *Transcendence theory: advances and applications* edited by A. Baker and D. W. Masser, Academic Press, London, 1977, pp. 1–27.

[7] A. Baker and J. Coates, Fractional parts of powers of rationals, *Math. Proc. Camb. Phil. Soc.* **77** (1975), 269–279.

[8] A. Brumer, On the units of algebraic number fields, *Mathematika*, **14** (1967), 121–124.

[9] J. Coates, An effective p-adic analogue of a theorem of Thue I, II: The greatest prime factor of a binary form, *Acta Arith.* **15** (1969), 279–305; **16** (1970), 399–412.

[10] N. I. Feldman, An improvement of the estimate of a linear form in the logarithms of algebraic numbers, *Mat. Sbornik*, **77** (1968), 423–436; = *Math. USSR Sbornik*, **6** (1968), 393–406.

[11] A. O. Gelfond, Sur la divisibilité de la différence des puissances de deux nombres entières par une puissance d'un idéal premier, *Mat. Sbornik*, **7** (1940), 7–26.

[12] H. Hasse, *Number theory*, Springer-Verlag, Berlin, Heidelberg, New York 1980.

[13] R. M. Kaufman, Bounds for linear forms in the logarithms of algebraic numbers with p-adic metric, *Vestnik Moskov. Univ. Ser. I*, 26 (1971), 3–10.

[14] J. H. Loxton, M. Mignotte, A. J. van der Poorten and M. Waldschmidt, A lower bound for linear forms in the logarithms of algebraic numbers, *C. R. Math. Acad. Sci. Canada*, to appear.

[15] K. Mahler, Ein Beweis der Transzendenz der P-adischen Exponentialfunktion, *J. reine angew. Math.* 169 (1932), 61–66.

[16] K. Mahler, Über transzendente P-adische Zahlen, *Compositio Math.* 2 (1935), 259–275.

[17] M. Mignotte and M. Waldschmidt, Linear forms in two logarithms and Schneider's method, *Math. Annalen*, 231 (1978), 241–267.

[18] A. J. van der Poorten, Hermite interpolation and p-adic exponential polynomials, *J. Australian Math. Soc.* 22 (Series A) (1976), 12–26.

[19] A. J. van der Poorten, Linear forms in logarithms in the p-adic case, *Transcendence theory: advances and applications* edited by A. Baker and D. W. Masser, Academic Press, London, 1977, pp. 29–57.

[20] A. Schinzel, On two theorems of Gelfond and some of their applications, *Acta Arith.* 13 (1967), 177–236.

[21] L. G. Schnirelman, On functions in normed algebraically closed fields, *Izv. Akad. Nauk SSSR, Ser. Mat. 5/6*, 23 (1938), 487–496.

[22] V. G. Sprindžuk, Concerning Baker's theorem on linear forms in logarithms, *Dokl. Akad. Nauk BSSR*, 11 (1967), 767–769.

[23] V. G. Sprindžuk, Estimates of linear forms with p-adic logarithms of algebraic numbers, *Vescī Akad. Navuk BSSR, Ser. Fīz-Mat.* (1968), No. 4, 5–14.

[24] R. Tijdeman, On the equation of Catalan, *Acta Arith.* 29 (1976), 197–209.

[25] M. Waldschmidt, A lower bound for linear forms in logarithms, *Acta Arith.* 37 (1980), 257–283.

[26] G. Wüstholz, A new approach to Baker's theorem on linear forms in logarithms I, II, to appear; III, this volume.

[27] K. R. Yu, Linear forms in the p-adic logarithms, *Max-Planck-Institut für Mathematik*, Bonn, MPI/87-20.

[28] K. R. Yu, Linear forms in elliptic logarithms, *Journal of Number Theory*, 20 (1985), 1–69.